Lab Manual

Stoker's General, Organic, and Biological Chemistry

SEVENTH EDITION

H. Stephen Stoker
Weber State University

Prepared by

G. Lynn Carlson

CENGAGE
Learning

Australia • Brazil • Japan • Korea • Mexico • Singapore • Spain • United Kingdom • United States

For product information and technology assistance, contact us at **Cengage Learning Customer & Sales Support, 1-800-354-9706**

For permission to use material from this text or product, submit all requests online at **www.cengage.com/permissions** Further permissions questions can be emailed to **permissionrequest@cengage.com**

ISBN-13: 978-1-305-08109-3

Cengage Learning
20 Channel Center Street
Boston, MA 02210
USA

Cengage Learning is a leading provider of customized learning solutions with office locations around the globe, including Singapore, the United Kingdom, Australia, Mexico, Brazil, and Japan. Locate your local office at **www.cengage.com/global**

Cengage Learning products are represented in Canada by Nelson Education, Ltd.

To learn more about Cengage Learning Solutions, visit **www.cengage.com**

Purchase any of our products at your local college store or at our preferred online store **www.cengagebrain.com**

Printed in the United States of America
· 2 3 4 5 6   22 21 20 19 18

TABLE OF CONTENTS

EXPERIMENT	TITLE	PAGES

PREFACE

AN INTRODUCTION TO THE STUDY OF CHEMISTRY

Chemistry is the branch of science that examines the properties of substances, and how different forms of matter interact with each other to make new substances. It is so central to scientific inquiry, especially in these days when all the sciences seem no longer to be separate fields, that every student of biology or medicine needs some understanding of chemical concepts. Your lecture course will introduce you to the chemical concepts needed to be successful in the formal study of pathology and physiology, but the laboratory experience has several additional goals as well.

The laboratory for your course will illustrate some of the concepts you will cover in lecture, but there won't be time to have a lab experience for every concept you need to know. You are entering a phase of study where the amount of material you need to know to progress is so large that you won't have time to re-discover all the ideas that have been accepted for many years. But you can take the time to observe some of these concepts in action. The most important goals of the lab are these:

- To demonstrate some of the chemical concepts underlying the study of the human body
- To allow the student some experience manipulating scientific tools and instruments
- To develop the practice of critical thinking when confronted with an array of data
- To practice the art of reaching a reasoned conclusion from available data

Most of these goals could be reached by studying any of the sciences or humanities, but we have chosen to frame the goals in terms of the chemistry of the human body.

How much of what you do in lab will you **really** need as a professional? Who can tell? But in addition to the skills you practice, you should be able to develop an understanding of how health professionals get the laboratory results you will need to work with the patients under your care. And it's possible that some of the things you learn may help you spot a mistake on the part of another worker that could mean the life of a patient—a concentration made up wrong or an inappropriate chemical administered, for example. The more you learn now, the easier will be your later courses and the tasks assigned to you in your health career.

HOW TO GET THE MOST FROM YOUR CHEMISTRY CLASSES

How to prepare for lecture

Your instructor will undoubtedly have some insights on the best way to prepare for the material in the lecture syllabus. Usually, however, it is more than a good idea—it is essential—to read the assigned material **before** you come to class. Perhaps it doesn't all make sense yet. Perhaps it raises questions that you don't know how to find an answer for. Perhaps it puts you to sleep (although we hope not!). But read before lecture. And after. Things that make little sense the first time will be clarified in class, and you will have the opportunity to ask questions about other items. And once may not be enough—read the material **after** lecture as well, and even more will sink in that time. Does it sound like a lot of work? Yes, it is, but your future depends on it, and the lives of patients may depend one day on how well you understand and retain the material in your science classes, chemistry included.

How to prepare for lab

Preparing for a successful laboratory experience follows those general rules as well, but there are a few more things to consider. You will not be just a passive or interactive listener in lab. Instead, you will be doing most of the work yourself. Being prepared will shorten your time in the laboratory and increase your chances of producing a successful experiment.

The manual is set up so that the experiments can be done by individuals, partners, or teams. Your instructor will choose the format, but you need to be prepared to take on the burden of performing the experiment alone if necessary.

Being prepared for lab means more than being sure you have the manual with you. Yes, the directions are there and can often (not always!) be followed one step at a time. Being prepared means reading the experiment **carefully** before you arrive. Do you understand what the experiment is trying to demonstrate? Is the pre-lab exercise assigned? Read the experiment from beginning to end and **then** try to do the pre-lab. You will defeat your own goals if you try to skimp on this step. Look over the report sheet. Is there anything you can do in advance? Do you understand the layout of the report form so you can work efficiently as you collect data? Your instructor may give you a very detailed introduction to the experiment or a very brief one, and you need to be ready for either situation!

When your report is returned to you, don't just add up the points you received and then file the paper away. Check to see if there are any comments. What did you do right? What was incorrect? Especially if you failed to understand something, be sure you review it and do your best to understand where you went wrong. In the end, you yourself are responsible for what you learn from lab.

HOW THIS BOOK IS ORGANIZED

Each experiment has an introduction, a procedure, a page of pre-lab exercises about the concepts the lab illustrates, and a report form. Some have a scenario that places the experiment in a real-world context. In addition, each experiment has a link to a set of references and on-line resources that might help you succeed with the experiment.

Where new or unusual techniques are introduced, pictures within the text should help you envision the subject, and some of the web resources will help there as well.

Be especially aware of ① and ♥ boxes. These boxes alert you to special information about steps where students might be expected to have difficulty (①) or where you are working with a substance that might require special safety precautions (♥).

In fact, safety is so important that a separate safety introduction comes next.

G. Lynn Carlson

SAFETY IN THE CHEMISTRY LAB

Many students walk into the chemistry laboratory afraid to touch anything, but why? Do you walk into the kitchen certain that things will go wrong? That the materials you are working with (all chemicals and mixtures of chemicals) are going to damage you? And yet a kitchen is a relatively dangerous place, not even surrounded with the most basic safety precautions that are routine in a chemistry lab. Just as in your home, you cannot eliminate 100% of the potential hazards in a chemistry laboratory, but you can follow sensible rules of behavior and observe reasonable safety precautions to minimize the chance of a serious problem developing.

WHAT TO WEAR

Perhaps you haven't thought about appropriate laboratory attire, but it's a good idea to do so. Unsuitable clothing or hairstyles can get you into trouble. For instance, long hair (shoulder length or longer) can trail into beakers of chemicals or even into a burner flame if not tied back. Open-toed shoes can allow spills to dribble onto your toes. Bare feet can tread on broken glass. And unprotected eyes can be damaged, possibly permanently, by almost anything that gets into them.

Your laboratory instructor will have a specific set of clothing guidelines, but in general you can expect that you must wear approved eye protection (usually chemical splash goggles) at all times when you are in the laboratory, even if you personally are not working with chemicals. You should generally plan to wear something that protects your body from incidental spills—no exposed navels and no sandals on lab day, and tie your hair back if it is long. For actually handling chemicals, many people prefer to wear plastic gloves routinely, and for some of the experiments gloves may be required. A lab coat or an apron is often also required; follow your instructor's requirements for these.

Plan to wash your hands before leaving lab. It's a good habit to cultivate, no matter how innocuous the chemicals you have been working with.

Common sense rules. Take sensible precautions and then be careful what you do.

WHAT TO DO

Read the experiment! Ahead of time! Within each experiment in this manual, there are icon-marked boxes (♥) to help you be aware of potentially dangerous steps. Pay attention to those, and know what to do to minimize your risk.

Be serious in lab. Most accidents in beginning science labs happen because someone wasn't paying attention to the experiment and mixed the wrong chemicals, knocked something over, dropped a piece of glassware, or otherwise was not concentrating on the job at hand. Pay attention to what you are doing, and you will not have any problems with silly safety mistakes.

Report any accidents to your instructor, who will advise you on how to proceed.

KNOW YOUR SAFETY TOOLS

For some experiments, you will be expected to set up your work in a fume hood. It serves much the same function as the vent over your stove, but this piece of equipment is there to protect the worker (that means you!) from potentially harmful vapors. You have to use it correctly though. Set up your experiment at least 15-20 cm **inside** the hood, not right out at the edge. That way the vapors are carried to the back of the hood and out, not into the lab. When appropriate, close the clear door of the hood to the appropriate distance as well, so that the draft becomes even more efficient in the right direction. And never duck your head into the hood to see what's going on!

Most laboratories have four additional pieces of safety equipment available for serious emergencies. An eyewash shower is there for splashes of chemicals to the eyes, but if you are wearing your chemical splash goggles 100% of the time, you won't need it, will you? Regardless, you need to know where it is and how to work it, just in case. A full-body shower is also available in case someone has a major spill that can't be washed off in the sink. Fire extinguishers and fire blankets are for the worst-case scenario of an experiment that got out of hand, so that the chemicals caught fire or a student's clothing caught fire, and the student needed more than the standard "stop, drop, and roll" assistance. Your instructor will point out the locations of these tools and how to use them.

For minor spills, your best friend is water. Even corrosive chemicals usually become less harmful if diluted with lots of water, so your first precaution if you spill something on yourself is to wash it off. Acid spills on the work benches or floors can be treated first with baking soda, to neutralize them and make them safer to handle, and small base spills can be treated with vinegar. Your instructor can help you with those types of problems.

If you feel the need to find out more about specific precautions, there are many sources of information. Chemicals used in the laboratory all have a "materials safety data sheet" available. There should be copies in your classroom's stockroom or available online from several URLs. The site "Where to find MSDS on the Internet" *http://www.ilpi.com/msds/* (accessed June 2, 2011) is a good entry port. Also look at the American Chemical Society web site *www.acs.org* (accessed June 2, 2011) for many leads.

GLASSWARE HANDLING

Laboratory glassware generally requires no more special handling than common sense: try not to drop things, and be sure to sweep up broken glass if a break happens. Sometimes you may need to rinse the broken test tube or beaker (if it contains a highly corrosive substance, for instance) but usually the main precaution is to sweep up all the bits and deposit them in the designated container. Usually the container is a special one for broken glass rather than the general wastebasket.

Some of the experiments in this manual may call for "Pasteur" or "disposable" pipets. These are fragile glass pipets that can be used in a more controllable fashion than eyedroppers, but their disposal can present a problem if they are just dumped in the waste basket, because they can poke through the trash bags and may stick whoever is picking up the trash. For this reason, schools often designate disposable glass pipets as "sharps," and place them in plastic containers for disposal as if they were needles. Your instructor will let you know the procedure for handling disposable glass at your institution.

CHEMICAL DISPOSAL

Most of the chemicals you will work with can be safely discarded in the sink or trash, but there are still a few precautions to take. Small amounts of innocuous chemicals usually can be diluted as they are flushed down the sink and cause no harm, but larger amounts usually should be collected and treated before they are discarded. Solids may be either discarded in the wastebasket or collected by your instructor for appropriate disposal. Your institution will have an established set of rules for chemical disposal, and the rules will be guided by federal, state, and local regulations. Follow these rules and you should have no trouble doing the right thing.

FINAL NOTE

There is no way to eliminate all potential hazards in the chemistry laboratory—all human activity involves risk. The best we can do is be aware of the possible difficulties, minimize them by using prudent practice and proper protective equipment, and carry on with our activities. The reward is a good laboratory class and the satisfaction of learning some new and interesting things. Follow the recommended safety precautions, the recommended disposal procedures, and use common sense, and everything should go fine.

Have a good experience!

G. Lynn Carlson
carlson@uwp.edu

EXPERIMENT 1

INVESTIGATION OF MIXTURES

GOALS

1. To examine the properties of several types of mixtures
2. To separate a heterogeneous mixture into its components
3. To develop a flow chart for an experimental procedure
4. To practice the following laboratory techniques
    *   Mixing and pouring of solutions
    *   Gravity filtration
    *   Suction filtration
    *   Use of ring stand and Bunsen burner

INTRODUCTION

A lot of the study of chemistry and biochemistry has to do with mixtures. Usually the best way to study a mixture is to separate it into its individual components first, and then study them one by one. After that, it is possible to examine a sample of the mixture itself to determine how the individual components might interact. Many of the experiments in this manual, for example, will revolve around separation and study of mixtures of biological importance.

Before you begin to work with those, however, we will ask you to practice separating a few simple mixtures so you can apply the appropriate techniques when you need them. In general, there are two types of mixtures to consider, homogeneous mixtures and heterogeneous mixtures.

The separation of a homogeneous mixture, such as a liquid dissolved in a liquid or a solid dissolved in a liquid, usually takes advantage of the differences in physical properties of the substances under consideration. For two liquids, the difference in boiling temperatures may be sufficient, although it usually requires special apparatus to work well on a laboratory scale. The separation of solid-solid homogeneous mixtures follows similar principles although the apparatus required is somewhat different. For a homogeneous mixture of a solid dissolved in a liquid, it is often possible to simply evaporate the liquid away from the solid, either allowing the liquid to escape into the atmosphere or collecting it in a special container.

Techniques for the separation of solid-solid and liquid-solid heterogeneous mixures are usually simple and take advantage of the different physical properties of the components of the mixture. For instance, iron-fortified, sweetened cereal, a solid-solid heterogeneous mixture, can contain the iron as finely divided solid iron; the easiest way to take the iron away from the flakes is to run a magnet through the cereal. On the other hand, the easiest way to separate the sugar in the cereal from the flakes might be to stir the flakes with water, causing the sugar to dissolve and leave the (now very soggy) flakes behind. Evaporation of the water from the sugar solution should give you solid sugar particles.

But how would you separate the flakes from the sugar solution in order to carry out your final evaporation? The usual method is some type of filtration, either using gravity as the driving force to separate the liquid from the solid, or using a vacuum. A gravity filtration is usually done when the particles are large, not easily decomposed, and not easily redissolved in water. A vacuum filtration is faster, allowing sensitive substances to be filtered quickly, and usually requires less solvent to wash the solid obtained, so there is less likelihood that you will wash away the solid you are trying to save.

There are special techniques for examination and separation of gaseous mixtures and the odd homogeneous mixtures known as colloids, which you might study at a later time.

The first part of this experiment deals with a mixture of sand and colored water that falls under the classification of heterogeneous solid-liquid mixtures. You will use the mixture to practice both gravity and vacuum filtration; exact techniques are given in the experimental procedure section. Your object is to obtain a sample of sand contaminated **only** with deionized water and not with any coloring matter.

Solid samples containing relatively coarse particles, such as sand or large crystals, can usually be separated from liquids by gravity filtration, a process very similar to the separation of liquid coffee from its grounds by allowing the liquid to drip through a piece of porous paper. Solid samples that consist of very small particles are often filtered using the vacuum or suction filtration technique.

The second part of the experiment requires you to think a little harder and produce your own procedure for the separation of a more complex solid-solid mixture. Before you begin, you will be asked to provide a flow chart of your proposed sequence of operations. A flow chart can be a sequence of sketches showing the different stages of the procedure, or a series of phrases that highlight the individual steps in a sequence. For most people, the sketch or diagram works best. It represents the major steps in a sequence of events to reach the goal, either as words in a box or perhaps as a drawing of the apparatus to be used to perform the step.

REFERENCES

Stoker, H. S. *General, Organic, and Biological Chemistry,* Cengage Learning; Chapter 1.
   Review topics:
   - physical states of matter
   - chemical and physical changes
   - pure substances and mixtures

WEB RESOURCES

Separation Process-Wikipedia, the free encyclopedia.
http://en.wikipedia.org/wiki/separation_of_mixtures
Mixtures. *http://www.tutorvista.com/content/chemistry/chemistry-iii/chemistry-concepts/mixtures.php*

SAFETY NOTES

The substances used in this experiment are household substances and are generally considered safe to work with. However, follow your instructor's laboratory regulations about the use of chemical splash goggles and gloves; their use may be governed by state law in your area.

Materials from Parts A and B of this experiment may be discarded in the waste basket, or as your instructor directs.

EXPERIMENTAL PROCEDURE

PART A. SEPARATION OF A SOLID-LIQUID MIXTURE

Gravity filtration

Step 1. Measure about 100 mL of a mixture of sand and colored water into a 250-mL beaker. Be sure to shake the stock bottle well before you measure your sample so that you actually get a significant amount of sand as well as water.

Step 2. Divide this sample into two equal portions, and plan to use one portion in the gravity filtration and one portion in the suction filtration.

Step 3. Prepare a gravity filtration apparatus, using a ring stand, a ring or a funnel support, a narrow-stemmed glass or plastic funnel, and a piece of filter paper of the appropriate size. The completed setup should look like the drawing in Figure 1-1. To prepare the filter paper, fold the paper in quarters, and wet it with deionized water.

Step 4. Filter the sand from the water solution. For a gravity filtration, the best technique usually consists of pouring most of the liquid through the filter first, then adding as much of the solid as possible, and finally rinsing the original container with several small rinses of deionized water.

Step 5. Examine your sand sample and the water solution, and describe their appearance on the report form.

Figure 1-1. The apparatus for gravity filtration.

Suction filtration
    Step 1. Obtain a side-arm Erlenmeyer flask, a Büchner (suction) funnel, a vacuum seal, and a piece of filter paper of the appropriate size, and set up a suction filtration apparatus as described by your instructor or as shown in Figure 1-2.

> ⓘ Be sure the flask is clamped to a firm support. This setup can be tipped over very easily!

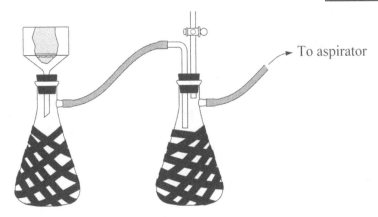

→ To aspirator

Figure 1-2. The apparatus for suction filtration.

    Step 2. Dampen the filter paper, turn on the vacuum source, and pour the solution into the funnel.
    Step 3. When the liquid has passed into the receiving flask, rinse the original container with small portions of deionized water, and rinse the solid cake of sand with this solution. Remember that the point of the rinses is to free the solid from the original solution, so that the solid is not contaminated by the colored material. Usually, three small, careful rinses are sufficient.

> ⓘ Check to be sure that the vacuum is really working before pouring!

    Step 4. Examine the sand, and record your observations and conclusions on the report form.

PART B. SEPARATION OF A SOLID-SOLID MIXTURE

    Step 1. Obtain about 150 mL of a heterogeneous mixture consisting of sand, table salt, iron filings, and wood shavings. Find the exact mass of the material you obtained, and record the mass.
    Step 2. Develop a procedure that will allow you to separate this mixture into its individual components. After you have decided on a procedure, draw a flow chart that describes what you plan to do, and record that flow chart in the space provided on the report form. Your instructor may wish to approve your plan before you go on to Step 3.
    Step 3. Carry out the separation using the techniques you have practiced in this laboratory exercise and any other techniques you might need. Some extra tools you might use are available from your instructor.

> ⓘ Careful! Your instructor probably has tools available that you will **not** need, too.

    Step 4. Find and record the mass of each substance isolated from the mixture. When all have been isolated and weighed, calculate the per cent recovery of the total solids you started with.

PART C. EXERCISES

Complete the exercises in the report section.

EXPERIMENT 1 PRELAB EXERCISES:

INVESTIGATION OF MIXTURES

Name_____ **Partner** _____**Date** _____

1. A student tried to separate a mixture of sand and red-colored water using a suction filtration apparatus as described in the directions for Experiment 1. From a mixture of 10.0 g of sand and 90.0 g of water, the student recovered 9.5 g of sand. What was the percent recovery of sand from the mixture?

2. The sand, originally white, was a light pink after drying. How could the student improve the color of the sand further?

3. A camper obtained a sample of water from a mountain stream for cooking, but the water appeared to be cloudy with silt. What would be the best method of separating the silt from the water?

4. Draw a flow chart for the separation you have described in Question 3, using the pictures in the instructions for Experiment 1 as a guide. Remember that the person is a camper, not a laboratory worker, and probably does not have laboratory equipment available.

5. The suction filtration flask shown in Figure 1-2 has been taped with strong black tape. Why was this safety precaution a good idea?

EXPERIMENT 1 REPORT: INVESTIGATION OF MIXTURES

Name_____ Partner _____Date _____

PART A. SEPARATION OF A SOLID-LIQUID MIXTURE

Gravity filtration

1. Describe the appearance of your recovered sand and the separate water solution.

2. Evaluate the success of your technique, using these questions as guides for your evaluation. Does the sand retain color? Is the solution cloudy? Did you get complete separation of the mixture? What evidence do you have for your conclusion?

3. What might you do better in subsequent gravity filtrations?

Suction filtration

4. Describe the appearance of your recovered sand and the water solution.

5. Evaluate the success of your technique as you did for Question 2 above.

6. What might you do better in subsequent suction filtrations, to improve your technique?

7. Compare the two techniques considering their ease of use, efficiency, and ability to separate solid-liquid mixtures.

PART B. SEPARATION OF A SOLID-SOLID MIXTURE

Draw below a flow chart for the separation of the mixture of solids.

What is the mass of the solid mixture you are working with?   _____

Record here the mass of any containers/filter papers
you used to isolate the individual solids.   _____

    Iron filings container   _____

    Salt container   _____

    Sand container   _____

    Wood shavings container   _____

Record here the mass of each solid isolated.

    Iron filings   _____

    Salt   _____

    Sand   _____

    Wood shavings   _____

What is the total mass of solids isolated?   _____

What percent of the initial mixture were you able to recover?   _____

Show how you calculated this.

PART C. EXERCISES

1. Would you be able to separate a mixture of table salt (sodium chloride) and table sugar (sucrose) by the methods you have used in this experiment? Explain your answer.

2. In a similar experiment, a student separated a 1.50 g sample of sand mixed with salt and obtained 0.70 g of salt and 0.65 g of sand. What percent recovery of the initial mixture was obtained?

EXPERIMENT 2

CHEMICAL CHANGES

GOALS

1. To identify several events that might be observed when a chemical change takes place
2. To distinguish between chemical and physical changes in a series of operations
3. To review several laboratory techniques for use in this class
4. To observe some chemical changes possible when copper reacts with other substances
5. To practice writing chemical names and formulas

INTRODUCTION

You have undoubtedly observed chemical reactions in your daily life. Have you observed a cake rise and bake in the oven? Wood reduced to ashes in a campfire or fireplace? A rust spot appear on your car and get steadily larger? All of these are examples of typical, everyday chemical changes.

When a chemical reaction takes place, a substance (reactant) changes to make a new substance (product). In most reactions, more than one reactant and more than one product are involved. These chemical changes are usually accompanied by some observable behaviors like those in the list below. Few reactions show all of these together; any one may be an indicator that a reaction has occurred.

Hallmarks of a Chemical Change
- Change in color
- Formation of a precipitate
- Dissolving of a solid
- Bubbling or fizzing due to the release of a gas
- Change in the temperature (i.e., the container gets warmer or colder)

Physical changes are usually much less exciting. You can see a change in form, or perhaps a separation of two or more objects, or the melting or boiling of a substance, but rarely are there flashy changes in color or heat to observe.

Observations That Identify a Physical Change
- Change in shape or size
- Melting
- Freezing
- Boiling
- Evaporation
- Condensation

In this experiment, you'll work with the element copper (symbol Cu). You will dissolve a small piece of copper wire in concentrated nitric acid (yes, you will want to be careful with that!) to form copper nitrate, which you will then convert to several other copper compounds before you recover the original copper at the end. The chemical changes you will carry out can be schematically represented as shown in Figure 2-1. Notice that these are not complete chemical equations. Other reactants also are needed to make these reactions happen, and sometimes additional products are formed. As you proceed through the experiment, you will see several events that you will need to classify as evidence of a physical or a chemical change. Use the lists on the previous page to help you decide what type of change you are seeing.

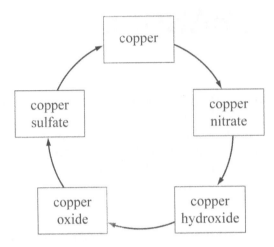

Figure 2-1. The cycle of reactions in the transformation of copper.

The notation used to describe a chemical change is the balanced chemical equation. Later in the semester you will be writing equations, but for now you will simply look at the equations related to this experiment with the goal of learning to read the chemical formulas and the equations. In these examples, you will see "coefficients" and two kinds of "subscripts." Coefficients precede a chemical formula in an equation, and show how many formula units take part in the reaction. Numerical subscripts are always within a chemical formula itself, and tell how many of a particular kind of atom the individual formula contains. Alphabetical subscripts in parentheses are sometimes also used, to show the physical state of a substance: solid, liquid, gas, or aqueous (*aq*) for "dissolved in water."

These are the balanced chemical reactions you will be doing during this experiment. And note that in Reactions 5 and 6, two chemical reactions are happening simultaneously!

Reaction 1. $Cu_{(s)} + 4HNO_{3(l)} \longrightarrow Cu(NO_3)_{2(aq)} + 2NO_{2(g)} + 2H_2O_{(l)}$ (Step 2, p 12)

Reaction 2. $Cu(NO_3)_{2(aq)} + 2NaOH_{(aq)} \longrightarrow Cu(OH)_{2(s)} + 2NaNO_{3(aq)}$ Step 4, p 12)

Reaction 3. $Cu(OH)_{2(s)} \longrightarrow CuO_{(s)} + H_2O_{(l)}$ (Step 5, p 12)_

Reaction 4. $CuO_{(s)} + H_2SO_{4(aq)} \longrightarrow CuSO_{4(aq)} + H_2O_{(l)}$ (Step 9, p 12)

Reaction 5. $CuSO_{4(aq)} + Mg_{(s)} \longrightarrow MgSO_{4(aq)} + Cu_{(s)}$ (Step 11, p 12)

Reaction 6. $Mg_{(s)} + H_2SO_{4(aq)} \longrightarrow MgSO_{4(aq)} + H_{2(g)}$ (Step 11, p 12)

REFERENCES

Stoker, H. S. *General, Organic, and Biological Chemistry*, Cengage Learning; Chapters 1 and 9.
   Review topics:
- Physical states of matter
- Chemical and physical changes
- Elements and compounds
- Chemical symbols and formulas

WEB RESOURCES

Watch a movie of a chemical reaction that has everything. Needs Quicktime 4 or later.
Ammonium Dichromate Volcano.
http://jchemed.chem.wisc.edu/jcesoft/cca/cca3/main/volcano/page1.htm

For a historical perspective on the inspiration for this experiment, go to Nitric Acid Acts Upon
Copper. *http://www.bluffton.edu/~bergerd/nitric_acid.html*

Or see also: *http://www.youtube.com/watch?v=ZlOfF92Z3GA*.

SAFETY NOTES

Wear disposable gloves and chemical splash goggles throughout the experiment. The first step of the
reaction sequence must be carried out in an efficient fume hood!
   Disposal procedures: Liquids from this experiment can usually be flushed down the sink with lots
of additional water. The solid copper collected may be collected for later disposal or discarded in the
trash, but follow your instructor's directions, which will be guided by local regulations.

EXPERIMENTAL PROCEDURE

PART A. CHEMICAL CHANGES OF COPPER

For each step of the experiment
    1. Watch the reaction closely, and record your observations on the report form.
    2. Note specifically those stages at which you see physical and/or chemical changes occurring.
    3. After completing the experiment, answer the questions at the end of the report.
    Reminder: Wear disposable gloves and chemical splash goggles throughout the experiment.

Step 1. Determine the mass of a piece of copper wire to .01 g, and record it on the report form. The mass should be approximately 0.5 g. Coil the copper wire and place it in a 150- mL beaker.

> ♥ CAUTION: Nitric acid is very corrosive. HANDLE WITH CARE!

Step 2. In a fume hood, add 5 mL of concentrated nitric acid to the beaker, and quickly cover the beaker with a watch glass. Observe what happens as the copper reacts with the nitric acid—there should be a lot to see.

Step 3. When the copper wire is completely dissolved, and the solution looks blue, not green, remove the watch glass and add 25 mL of deionized water to the beaker. The beaker may now be removed from the hood until the last step.

Step 4. To the solution of copper(II) nitrate formed in Step 2, add 20 mL of 5 M sodium hydroxide (NaOH), and stir to form a precipitate of copper(II) hydroxide. Again observe the changes and record what you see. "M" is a concentration unit you will learn more about later in the course.

Step 5. Place the beaker on a wire screen supported on a ring stand. While stirring the mixture constantly (see Figure 2-2), heat the beaker gently with a Bunsen burner until all of the blue copper(II) hydroxide is converted to black copper(II) oxide, then stop heating. Allow the solution to cool for a few minutes before continuing with Step 6. If you still see some blue powder on the sides, stir it down into the hot liquid. Water is another product of this reaction.

> ♥ This is another potential trouble spot. Heat <u>very</u> gently and stir constantly or the mixture will splash. This mixture contains hot lye. You don't want to get it on your skin!

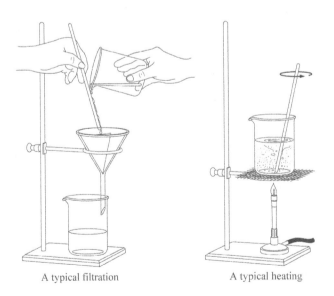

A typical filtration   A typical heating

Figure 2-2. Typical filtration and heating techniques.

Step 6. Fold a piece of filter paper into quarters, then open it out to act as a cup, and insert it into a funnel supported on a ring stand (See Fig. 2-2). Wet the paper with deionized water, and set an empty beaker under the funnel to catch the filtrate.

Step 7. Pour most of the liquid through the filter first and then stir the precipitate and pour it into the funnel. Use your wash bottle to rinse the sides of the beaker and the stirring rod with no more than about 5 mL of deionized water, and pour the rinse water into the funnel. It is <u>not</u> necessary to clean all the copper oxide from the sides of the beaker, since you will be returning the solid to the beaker for the next step.

> ♥ The filtration will go faster if you pour most of the liquid through first. Try pouring the liquid down a stirring rod to decrease dribbling.

Step 8. After filtration is complete, discard the liquid. Using your spatula, carefully remove the filter paper from the funnel and place it into the 150-mL beaker.

Step 9. Add 10 mL of 4 M sulfuric acid, and stir until all the copper oxide has dissolved, producing a solution of copper sulfate. Don't forget to record your observations and conclusions.

Step 10. Lift the filter paper from the solution with a forceps, hold it over the beaker, and use a wash bottle to rinse it thoroughly. If paper fibers can be seen in the solution, a second filtration should be performed in order to remove them.

> ⓘ 4 M (and a little later, 8 M, are concentration units for different solutions of sulfuric acid in water. Be sure you use the correct solution for each step!

Step 11. IN THE HOOD, add about 0.5 g of magnesium to the solution of copper(II) sulfate. The magnesium should be added slowly, a few pieces at a time, with stirring. In this step, two different reactions are occurring at the same time. Elemental copper is produced from the reaction of magnesium with copper(II) sulfate. However, some of the magnesium also reacts with the sulfuric acid to produce hydrogen gas.

Step 12 (if needed). If the solution is still noticeably blue after all the magnesium has reacted, add more magnesium until the blue color is gone. When all of the copper sulfate has been converted to copper (after the blue solution has become colorless), slowly add, with stirring, 10 mL of 8 M sulfuric acid to dissolve any unreacted magnesium.

> ⓘ This takes a while. Be patient for best results!

Step 13. Determine the mass of a piece of filter paper to 0.01 g, and record it on the report form. Place this filter into your funnel, and pour the solution and copper into it. Use a wash bottle to rinse all of the copper from the beaker. Thoroughly rinse the copper with deionized water.

Step 14. Remove the filter paper from the funnel, open it up and place it on a clean watch glass. Put the watch glass and contents into a drying oven for at least 15 minutes.

Step 15. When the sample and the paper are dry, determine and record the combined mass of the paper and copper. Calculate the percent recovery of the copper.

PART B. EXERCISES

Answer the questions on the report sheet concerning this sequence of reactions.

EXPERIMENT 2 PRELAB EXERCISES:

CHEMICAL CHANGES

Name_____ Partner _____ Date _____

1. Define a "chemical change" and give three specific examples of situations in which a chemical change is occurring.

2. Why do the directions recommend that the student wear plastic gloves throughout the experiment?

3. In a similar experiment to this one, a student began with a cube of solid copper with a mass of 2.85 g. After the cycle of reactions was finished, the student obtained 2.65 g of powdered copper. What was the percent recovery of copper in this experiment?

4. A different student also used a piece of copper with a mass of 2.85 g, and obtained 2.95 g of copper at the end of the experiment, an apparent recovery of 104%. How could the student account for this result, since the best to expect is 100%?

5. Describe the action needed to clean up a spill of concentrated nitric acid on the workbench. You may need to research this question using print or Internet sources before you answer it.

EXPERIMENT 2 REPORT: CHEMICAL CHANGES

Name_____ **Partner** _____ **Date** _____

PART A. CHEMICAL CHANGES OF COPPER

Mass of copper wire you started with (from Step 1) _____

Observations on the reactions

Step	Observations	Write the equation for what is happening in this step. See Introduction for the correct form.
1		
2		
3		
4		
5		

Mass of filter paper + copper (from Step 15) _____

Mass of filter paper (from Step 13) _____

Mass of recovered copper _____

Percent recovery of the copper you began with _____

PART B. EXERCISES

1. Show your calculation of the percent recovery for this experiment.

2. In what steps of the exercise did you see physical changes take place?

3. Consider the blue solution you obtained at the end of Step 2. Should it be classified as a pure substance? A homogeneous or heterogeneous mixture? Justify your answer.

4. Consider the liquid you obtained at the end of Step 13. If the final products at the end of the sequence of reactions were magnesium sulfate, copper, and hydrogen gas, what substances were in the liquid you poured down the sink? You may assume that you used exactly the right amount of sulfuric acid for all the reactions to go to completion, and that no sulfuric acid was left.

In the spaces below, list the names and formulas of all the substances that were used or produced in this experiment. Identify each as an element or a compound.

Substances Used in This Exercise

Name	Formula	Element or compound?

EXPERIMENT 3

MEASUREMENT IN THE LABORATORY

GOALS

1. To introduce the students to the laboratory measuring tools for length, mass, volume, and temperature
2. To review the concepts of significant figures in measurements and calculations
3. To allow the students to devise their own experiments to determine some physical quantities
4. To practice mathematical conversions from one measuring system to another

INTRODUCTION

When you study natural events, sometimes you simply observe the events and report on them. But to really understand and explain an event you have seen, you often have to take measurements and perform calculations to deduce how your observations relate to each other. There is a case to be made for the statement "If it can't be measured, it isn't science, it's opinion" (Heinlein, 2004).

For your observations and measurements in this class to have meaning, you will need to know how to use your tools to best advantage and learn how to tell whether a measurement or a calculation gives information or nonsense. This is why you will spend some time in this exercise examining some common laboratory tools for measuring substances. Then you will practice using these tools and the data obtained with them to reach a few conclusions.

The other main goal of this exercise is to practice using "significant figures" in measurements and calculations. There are several fairly rigid rules for the "correct" way to round off a measurement, and it's not always clear to students why these rules are applied the way they are. This exercise should help you with that difficulty. Your textbook has several pages of information about the significant figure rules and their application (Chapter 2); you should refer to them as this exercise progresses. We will not give a complete review here, just a few examples.

You also need to be familiar with the concepts of "precision" and "accuracy" before you begin this experiment. A "precise" measurement is one that is extremely reproducible. For example, if you keep a mileage record for your car for a year, and find that the car gets 32.3 miles to the gallon of gas every month for a year, that is a very precise measurement. On the other hand, if you find that some months you get 35.0 miles/gal, and some months you get 30.0 or 27.0 miles/gal, this is not a very reproducible figure, and so is not "precise."

Now suppose your method of tallying your mileage has a mistake in it somewhere, so that the average you calculate is 20.2 mi/gal every month for 12 months, but your actual usage is 32.3 mi/gal. Your calculated figure is "precise," because it is very reproducible, but it is not very accurate because it is only about 60% of the "true" value.

The number of significant figures you record in an experiment depends on the type of measuring tool you use, how precisely it is calibrated, and how accurately it measures something. A beaker that has markings for 50-mL increments is obviously not going to be as precise or accurate as a graduated cylinder that has markings for every milliliter. Whenever you use a measuring tool in the laboratory, you need to decide how well it does the job, and record the measurement to the limits of precision of the tool. You can always round off a measurement later if all the significant figures are not needed, but you may lose useful information if you round off too soon.

Typical measurements in this laboratory will be measurements of length, volume, mass, and temperature. As you look at the tools involved in these measurements, you'll see that the rulers for length measurements, for example, are not all equal. A plastic ruler may be calibrated only to the nearest millimeter, while a meter stick may be somewhat better than that. The situation is even more obvious when you look at different sizes and brands of graduated cylinders. Large ones may be calibrated only to the nearest 5 mL, while small ones may have markings to the nearest 0.1 mL.

On the other hand, thermometers generally are marked to the nearest degree, and the temperature should be estimated to the nearest 0.2 °C to make maximum use of the precision of the thermometer.

> ⓘ Remember, when you are paying attention to significant figures, a temperature of 32.0° is 32.0°, not 32°!

For mass measurements, a digital scale takes the certainty to even greater heights, always giving you two or three decimal places that are certainly precise, although occasionally a scale is not accurate.

The general rule of thumb is this: Read and record as accurately and precisely as you can, to the limits of the markings of your tool; then estimate one more place. In the case of a 500-mL graduated cylinder for measuring volume, for instance, with markings every 5 mL, you can tell the difference between (say) 375 mL and 380 mL easily, but the height of the liquid level between these two markings allows you to estimate if the volume is 376 mL or 378 mL. You probably cannot tell the difference between 376 mL and 377 mL, though, so we recommend estimation to the nearest 2 mL only. Mass readings taken on a digital readout balance, however, should be recorded exactly as shown, because there is no way to estimate another place.

What can you do with these precise and accurate measurements once you have them? Sometimes, such as when you check your own height or weight, the measurement is sufficient by itself. In the lab, you more often need to perform a calculation to determine the fact you are looking for. These calculations fall into roughly two categories, conversion from one measuring system to another, such as from metric to English, or calculations using the measurements themselves to find a new quantity.

An example of a conversion is the question "If I am 1 m 72.3 cm tall, how tall am I in feet? This one goes like this.

$$1.723 \; \cancel{m} \times \frac{39.4 \; \cancel{in}}{\cancel{m}} \times \frac{1 \; ft}{12 \; \cancel{in}} = 5.6572 \; ft$$

The 12 is a counted number, but the 39.4 is approximate, with only three significant figures. So the final answer is 5.66 ft. Would you prefer that in feet and inches? How many inches are in 0.6572 feet? The height works out to be 5 ft 7 in.

> ⓘ Notice that the unlike units did not cancel, so they are carried over as part of the answer.

More commonly, you will stay within the same measurement system for any given operation, and use several types of measurements in the same calculation. In this experiment, you will determine the density of a liquid, a measure of how tightly matter is packed into it, by measuring its volume and mass and reporting its mass per unit volume. In mathematical terms,

$$D = m/V$$

D is the density, m is the mass, usually measured in grams, and V is the volume in any convenient size, but usually in milliliters. If a liquid sample has a volume of 34.6 mL and a mass of 52.06 g, the density is 1.50 g/mL **to three significant figures**.

Use these ideas and the detailed discussion in your textbook to help you through this experiment.

REFERENCES

Stoker, H. S. *General, Organic, and Biological Chemistry;* Cengage Learning; Chapter 2.
   Review topics:
   - The metric system of measurement
   - Conversion factors
   - Significant figures in measurement

Other references:
Heinlein, R. A. *The Notebooks of Lazarus Long*; Baen: Riverdale, NY; 2004.

WEB RESOURCES

The Harvard Bridge. *http://www.mit.edu/people/stevenj/mitmap/harvard-bridge.html*

Math.com Formulas & Tables. *http://www.math.com/tables*

Nationmaster.com-Encyclopedia: Harvard Bridge.
http://www.nationmaster.com/encyclopedia/Harvard-Bridge

SAFETY NOTES

The substances used in this experiment are household substances and are generally considered safe when used as directed. However, follow your instructor's laboratory regulations about the use of chemical splash goggles and gloves; their use may be governed by state law in your area.

   Materials from this experiment may be discarded in the waste basket, in the sink, or as your instructor directs.

EXPERIMENTAL PROCEDURE

PART A. LABORATORY MEASURING TOOLS

Step 1. Examine the various measuring tools available in your lab; they will be similar to the ones in Figure 3-1. At each work station, there should be some combination of the following items:

- For length measurements: meter stick and plastic ruler
- For volume measurements: two beakers, two graduated cylinders, and two kinds of pipets
- For mass measurements: a digital scale
- For temperature measurements: a centigrade thermometer

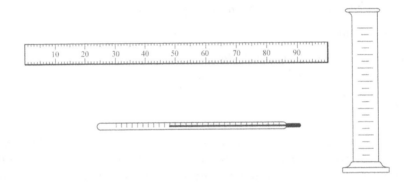

Figure 3-1. Several common laboratory measuring tools.

For each tool, observe the calibration marks, and decide how many significant figures can be measured with it. Record your conclusions on the table in the report form. Then go on to Step 2, where you will measure a number of objects using the tools you have already examined. Be sure as you record these measurements that you always use the correct number of significant figures, to the limits of precision available for whatever measuring device you use.

Step 2. Using the most accurate tool available to you, make the following measurements, and record them on the report form to the correct number of significant figures.

- Length and width and height of a textbook
- Mass of a paper clip and a 250-mL beaker
- Volume of milk in a bottle, and
- The mass of the same milk
- The temperature of the milk in the bottle in degrees C

ⓘ Use the beaker you already know the mass of, to simplify your measurements.

Step 3. Using the appropriate measurements from Step 2, calculate

- The volume of the textbook, in cubic mm, cubic cm, and cubic in
- The density of the milk in g/mL and kg/L
- The temperature of the milk in degrees Fahrenheit

Be sure to report your answers using the correct number of significant figures.

PART B. APPLICATIONS OF MEASUREMENTS AND CALCULATIONS

In this section, three applications of measurement are described. Your instructor will let you know which one or ones you should do. In each case you will be expected to devise your own procedure and perform the necessary calculations to report your answer using the correct number of significant figures.

Project 1. Devise an experiment to measure and calculate the volume of a basketball or soccer ball. You may need to look up the formula for volume in books or on the Internet. Perform the measurement three times, so you can take an average and examine the repeatability of your measurements.

Project 2. Devise an experiment to determine the average mass of a new pencil or pen, given a box of 12 to work with.

Project 3. Devise a way to measure the length of the hallway outside your lab in the unusual unit of your choice. Examples: student, ring stands, textbooks, paces, stepladders, crocodiles... Perform the experiment.

> ⓘ You are limited to tools available to you in your classroom!

PART C. EXERCISES

See the report form for exercises to practice what you have learned.

EXPERIMENT 3 PRELAB EXERCISES:

MEASUREMENT IN THE LABORATORY

Name_____ Partner _____ Date _____

1. Define the difference between precision and accuracy. Give an example of each concept; the examples should be <u>different</u> from the examples given in the Introduction.

2. A student used a 500-mL graduated cylinder to measure the volume of water in a 1-cup measure. Three trials of the measurement gave volumes of 240 mL, 242 mL, and 235 mL. What is the average of the three measurements? Were the measurements precise? Were they accurate? \You may wish to refer to conversion factor tables to decide this. Justify your answers.

3. A student used a standard laboratory thermometer to measure the temperature of boiling water. On the lab report sheet, the student recorded the boiling temperature as 94. There are several problems with this record. Describe two, and explain what the student should have written instead.

4, The mass of ten paper clips was found to be 10.135 g. What is the average mass of a paper clip in this sample?

5. In the laboratory, a student performed an experiment in which the final length of a piece of wire was a significant measurement. In three separate experiments, the student found the length of wire after the experiment to be 23.50 mm, 25.0 mm, and 35 mm. Find the average of the three trials, to the correct number of significant figures, and decide if these measurements gave an accurate result.

EXPERIMENT 3 REPORT: MEASUREMENT IN THE LABORATORY

Name_____   Partner _____   Date _____

PART A. LABORATORY MEASURING TOOLS

Observations on measuring tools

Type of measurement	Tool	Precision Possible
Length	_____	_____
Volume	_____	_____
	_____	_____
	_____	_____
	_____	_____
	_____	_____
Mass	_____	_____
Temperature	_____	_____

Measurements of a textbook

Length_____   Width _____   Height _____

Textbook volume calculations (V = l x w x h)

In mm^3_____   In cm^3_____   _____In in^3

Mass measurements

Paper clip _____   Beaker _____   Milk _____

Density of the milk (D = m/V)

In g/mL _____   In kg/L _____

Temperature measurements

Temperature of milk _____°C   _____°F

PART B. APPLICATIONS OF MEASUREMENTS AND CALCULATIONS

Which project did you select?   _____
Flow chart for your experiment design

Measurements (Include any formulas needed, such as volume formulas.)

Calculations

Final answer   _____

PART C EXERCISES

1. Shown in the table are the volume and mass measurements for three copper cubes. Find the density of copper from the data, using the correct significant figures.

Measurement of Copper

Cube #	Volume (mL)	Mass (g)
1	2.5	22.3
2	3.0	26.7
3	2.25	20.0

   Average density   _____

   Is this figure precise?   _____

2. Look up the density of copper in a reference source. Is the average density you obtained accurate?

                                                              _____

3. A student was found to be 5 ft 7 in tall. This student was used as the standard of length to measure the length of Harvard Bridge in Cambridge, MA in 1958. (See Web Resources section.) The length of the bridge was found to be 364 students and 1 ear (arbitrarily equated to exactly 4 in). What is the length of Harvard Bridge in meters?

                                           Length = _____

4. If the density of water is 1.00 g/mL, what is the weight in pounds of 1.00 gallon of water? You may assume that mass and weight are equivalent at the Earth's surface.

                                           Mass = _____

EXPERIMENT 4

DENSITY MEASUREMENTS ON SOLIDS AND LIQUIDS

GOALS

1. To investigate the concept of density as it applies to liquids and solids
2. To determine the identity of an unknown solid or liquid after its density is determined
3. To review the use of significant figures in laboratory measurements and calculations

INTRODUCTION

The density of an object (usually represented by D) depends on the substance it is made of and reflects how tightly matter is packed in that substance. The mathematical representation of density is

$$\text{Density} = \frac{\text{mass}}{\text{volume}} \qquad \text{or in shorthand} \qquad D = \frac{m}{V}$$

For example, a cubic centimeter (cm^3) of lead (D = 11.9 g/cm^3) has substantially more matter in terms of protons and neutrons, and thus more mass, than a cubic centimeter of magnesium (D = 1.74 g/cm^3). In terms of everyday items, a gallon of milk is slightly denser than a gallon of water. You can also distinguish between some gases because of their relative densities. For example, you may have seen carbon dioxide "smoke" at a theater or concert performance; it hugs the floor because it is more dense than air.

In fact, if you have a relatively limited number of options, you can use the density of an unknown substance as a clue for identifying it. For example, an unknown metal might be identifiable this way, since the densities of common metallic elements and alloys are well known. Usually other clues are needed as well, however, since there are many metallic mixtures and pure substances, and densities are generally all in the range of 0.6 g/cm^3 to 20 g/cm^3; many different substances have similar densities.

The concept of density finds an application in the home-brewing industry, when the alcohol and sugar content is judged by the density of the brew, and in medicine, when the density of a patient's urine is used as one gauge of health. Normal urine density is 1.03 g/cm^3 to 1.13 g/cm^3; any deviation from that range warrants attention.

In this exercise, you will learn two methods for determining density, one for a solid and one for a liquid. Both of them rely on determining the mass and the volume of the substance under investigation. Once you understand the technique and the calculation, you will have the opportunity to identify an unknown substance by density alone. There will be a limited list to choose from, so density will be a sufficient clue for you if you work carefully.

Finding the density of a liquid is easy. You really need only two measurements, the volume of the liquid and the mass of the sample. If you have previously recorded the mass of the measuring container, then a single operation is enough. Usually the mass measurement is not the limiting factor in the accuracy of the exercise; balances are accurate and precise. Volume is usually the least precise measurement in this type of exercise, so use a small graduated cylinder, with frequent graduations.

> ⓘ Be sure to take advantage of **all** the precision the cylinder can give.

You will work with a water solution whose density you don't know. So how do you know if your result is accurate or not? Try the experiment several times; if you get several results that agree well, your result is probably accurate. But which one do you report as the "correct" result? The average of several trials is usually a better estimate of the "true" value than any individual answer.

The determination of the density of a solid also requires a mass and volume measurement. You can measure the mass in the classical way, the same way Archimedes did, by weighing the object in water and correcting for buoyancy due to the water, but as long as the object is small enough and light enough to fit onto the balance pan, you can actually weigh it directly and save a calculation. The volume measurement can be done by calculation, if the object is a regular solid, such as a cube, sphere, or cylinder. If it is irregular, the process is a bit more complicated; you look at how much space it takes up when immersed in a known volume of water in a graduated container. Fortunately, the objects available in this exercise are small, and you can use the easier of the two methods.

Although you will not be working with a gas sample, you should be aware that the density of a gas can also be determined. Solid and liquid densities are measured in g/cm^3 or g/mL, while gas densities are usually reported in g/L, not g/cm^3.

REFERENCES

Stoker, H. S. *General, Organic, and Biological Chemistry;* Cengage Learning; Chapter 2.
  Review topics:
  • Density
  • Significant figures
  • Calculator use

WEB RESOURCES

BIO|ANALOGICS TECHNOLOGY-Body Composition Techniques.
http://www.bioanalogics.com/bctech.htm

Hot Air Balloon. *http://www.sciencebyjones.com/hot_air_balloon.htm*

NOVA Online|Balloon Race Around The World | Hot Science: Floating and Sinking.
http://www.pbs.org/wgbh/nova/balloon/science/density/

SAFETY NOTES

The substances used in this experiment are household substances and are generally considered safe to work with when used as directed. Some of the liquid unknowns are flammable liquids. Assume any unidentified liquid sample may be flammable and keep it away from open flames.

Follow your instructor's laboratory regulations about the use of chemical splash goggles and gloves; these regulations may be governed by state law in your area.

Materials from Parts A and B of this experiment may be discarded in the sink or waste basket, or as your instructor directs. Reusable solids may be returned to the instructor for storage.

EXPERIMENTAL PROCEDURE

In all of the measurements you make, record your data to the greatest number of significant figures that it is possible to read, given the measurement tool you are using.

PART A. THE DENSITY OF A LIQUID

Step 1. Find the mass of a graduated cylinder and record it.

Step 2. Measure into that cylinder approximately 10 mL of the liquid you are investigating, and make a note of the identification on the label. Record the exact volume you are using. Dry the outside of the cylinder if it is wet, because you only want to weigh the liquid **inside** the cylinder.

> ⓘ You should use the same cylinder each time!

Step 3. Find the combined mass of the cylinder and liquid, and calculate the experimentally determined density of the liquid.

Step 4. Repeat the measurements twice more, using a dry cylinder (dry inside and outside) and a new sample of the liquid each time. Complete the calculations.

PART B. THE DENSITY OF A SOLID

Step 1. Find the mass of the solid sample and record it. If the sample has any identifying marks on it, such as a name or sample number or color, record those as well.

Step 2. Select a graduated cylinder that will hold the solid object you are investigating. Fill the cylinder with enough water so that the water level will completely cover the object when it is immersed in the water. Measure and record the volume.

Step 3. Immerse the object in the water, and measure the new water level, as shown in Figure 4-1. Try to avoid allowing any air bubbles to cling to the object because they will affect the accuracy of your results. Calculate the density of the object.

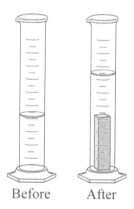

Before   After

Figure 4-1. How to determine the volume of a solid.

Step 4. Dry the object you are investigating. Repeat the measurements **twice** more, using the same solid sample, and a dry cylinder (dry inside and outside), but use a new sample of water each time. Complete the calculations.

PART C. IDENTIFICATION OF AN UNKNOWN SUBSTANCE

Your instructor will assign to you a sample of a substance for identification. Use the appropriate method you have already learned to determine its density, and identify the substance from the following list. Report all your data and calculations on the report sheet.

Density in g/mL (g/cm³) of Some Common Substances at Room Temperature

water	1.00
concentrated salt water	1.2
ethyl alcohol	0.79
ethylene glycol	1.11
isopropyl alcohol	0.785
aluminum	2.70
magnesium	1.74
iron	7.86
zinc	7.14
nickel	8.9

PART D. EXERCISES

Complete the exercises on the report form.

EXPERIMENT 4 PRELAB EXERCISES:

DENSITY MEASUREMENTS ON SOLIDS AND LIQUIDS

Name_____   Partner _____   Date _____

1. A student measured the density of ethanol four times, and obtained densities of 0.789, 0.810, 0.775, and 0.790.

  a. What were the units for these measurements?

  b. What was the average density obtained in the four trials?

  c. Were the results accurate and/or precise? Why or why not?

2. Chloroform is a liquid that is not soluble in water. It has a density of 1.498 g/mL. A student mixed a sample of chloroform and a sample of water and observed the result. Sketch what the student saw, and identify all components of the drawing. For a container, draw a test tube.

3. A 50.0-mL sample of ethylene glycol had a mass of 55.505 g. Find the density of ethylene glycol.

4. A lump of aluminum had a mass of 27.00 g. Since it was irregular in shape, the volume was obtained by displacement of water. A cylinder of water was loaded with 50.00 mL of water, and the aluminum piece placed inside the cylinder. The water level rose to 60.07 mL. What was the density of the aluminum block?

EXPERIMENT 4 REPORT: DENSITY MEASUREMENTS ON SOLIDS AND LIQUIDS

Name_____ Partner _____ Date _____

PART A. DENSITY OF A LIQUID

What liquid was used? _____

	Trial 1	Trial 2	Trial 3
Mass of cylinder	_____	_____	_____
Mass of cylinder + liquid	_____	_____	_____
Mass of liquid	_____	_____	_____
Volume of liquid	_____	_____	_____
Density of liquid	_____	_____	_____

Average density over three trials _____

PART B. DENSITY OF A SOLID

Identifying marks _____

	Trial 1	Trial 2	Trial 3
Mass of solid	_____	_____	_____
Volume of water in cylinder	_____	_____	_____
Volume of water + solid	_____	_____	_____
Volume of solid	_____	_____	_____
Density of solid	_____	_____	_____

Average density over three trials _____

PART C. IDENTIFICATION OF AN UNKNOWN SUBSTANCE

Is it a solid or a liquid? _____   Identification number _____

Describe any observable physical properties, such as color, odor, and viscosity.

Record your data for the density determinations here, following the pattern set in Part A or B. Do three trials of your substance, and report the average density. Don't forget the appropriate units!

The average density from the three trials is   _____

The most likely identity of the unknown substance is   _____

PART D. EXERCISES

1. How precise is your data for the determination of the density of a liquid? Justify your answer.

2. How precise is your data for the determination of the density of a solid? Justify your answer.

3. Use the Internet to determine which is the densest of the naturally occurring elements. Which is it, and what is its density?

4. Devise a flow chart for the determination of the density of a solid ball that floats on water. If your experiment works, and you find that the volume of the ball is 15.2 mL and the mass of the ball is 12.1 mL, what is the density?

5. Given that the density of fat is about 0.9 g/cm^3 and the density of muscle is about 1.10 g/cm^3, why do obese people sometimes find it easier to float in water than lean people?

EXPERIMENT 5

THE SPECIFIC HEAT OF A SOLID

GOALS

1. To determine how heat content of a substance can be measured
2. To measure the specific heat of two metals
3. To apply this information to practical problems

INTRODUCTION

Everyone knows that if you are cold, you can warm up by crowding close to a fireplace or other heat source. And if you are too warm, moving to a cooler room will help you cool off. This well-known practical example of heat transfer illustrates one way of describing the second law of thermodynamics: heat transfer occurs only from a warmer to a cooler object. In the first instance, heat is transferred from the fire and the room air to you. In the second instance, heat is transferred from you to the room air. Cold is never "transferred."

The most convenient way to measure the quantity of heat contained in an object or system is to measure its temperature. In laboratory exercises, heat content is generally measured in the Celsius or centigrade scale, in °C, although there are some situations that require the Kelvin scale (abbreviated K; note that the degree sign is not used with the Kelvin scale). Since these two scales have degrees that are the same size and measure the same amount of heat, interconversion is simple. Increasingly, in medical situations, the centigrade scale is also used, and if a nurse in your doctor's office takes your temperature, the temperature is as likely to be recorded in °C as in °F.

Under carefully controlled conditions, it is also possible to measure the amount of heat released by an object. This type of measurement usually involves a device known as a calorimeter. It is a closed container of some type, insulated from its surroundings so that all heat that is inside it remains inside it, and that heat from the surroundings does not enter it. In that way, the heat exchange taking place in the calorimeter can be assumed to occur due entirely to whatever process is going on inside. Think of it as a specialized picnic cooler holding warm food, and inside it the only heat present is due to the warm food, not to heat coming in from the atmosphere. Calorimeters can be quite sophisticated, or very simple.

For simple calorimetery, an object is heated to a designated temperature, and then placed into cooler water, which then warms up because of the heat transfer taking place. A fairly simple calculation then can reveal the heat given off by the object.

Heat released (q) = mass of substance x specific heat x change in temperature. Or

$$q_{released} = m_{object}\, c_{object}\, \Delta T_{object}$$

In this equation, c represents the "specific heat" of the substance under consideration. This is a constant for a given substance, and represents the quantity of heat needed to raise the temperature of 1.00 g of that substance 1.00 degree C. This is a well-known quantity for many common substances, but it also can be determined experimentally. ΔT_{object} indicates the temperature change of the object, $T_{final} - T_{initial}$.

Of course, the water heats up in the course of the experiment, and the amount of heat absorbed by the water is equal to the amount of heat given off by the experimental object. This can also be calculated in the same way.

$$q_{absorbed} = m_{H_2O}\, c_{H_2O}\, \Delta T_{H_2O}$$

A little mathematical rearrangement shows that

$$m_{object}\, c_{object}\, \Delta T_{object} = m_{H_2O}\, c_{H_2O}\, \Delta T_{H_2O}$$

If the mass or specific heat of the object is not known, you can use this relationship to figure it out. The specific heat of water is a well known quantity, equal to 1 cal/g·°C or 4.186 joules/g·°C.

In this experiment, you will be trying to find the specific heat of two metals, aluminum and copper, in two different experiments. You'll use one of the simple types of calorimeter, a pair of nested styrofoam cups. Using two cups instead of just one helps to insulate the process you are observing from the outside heat, and a lid improves that insulation. Once you have mixed the hot metal with the weighed water in the calorimeter, stirring the mixture gently (not with a thermometer, though—thermometers are fragile!) will help the mixture reach a steady temperature faster.

Five measurements are necessary: the temperature of the hot metal, the beginning temperature of the water in the calorimeter, the final temperature of the mixture of metal and water, and the masses of the water and the metal. You will measure the temperatures in degrees centigrade, the masses in grams, and calculate the the heat in joules.

Few experiments give perfect data; there are always ways to improve the situation. In this instance, you will make no allowances for how much heat is absorbed by the styrofoam. This is usually a very small number compared to the heat absorbed by the water and does not usually affect your result significantly. More sophisticated measurements always take that heat into account. Some heat is also lost to the atmosphere as you move the metal from the heating bath to pour it into the calorimeter. Again, if you move quickly the size of the error introduced is small, so we will neglect it in this experiment.

Another easy way to minimize errors in reporting is to perform an experiment several times and find the average of these several trials. Usually the average is a better representation of the "true" value of the quantity than any individual measurement is likely to be. Rather than repeat the same experiment several times yourself, you will share your results with a few other class members and report an average specific heat from the pooled data. You can also look up the well-known heat capacities for aluminum and copper, and compare your results with the "literature" value.

REFERENCES

Stoker, H. S. *General, Organic, and Biological Chemistry;* Cengage Learning; Chapter 2.
  Review topics:
  - Heat transfer between materials
  - Specific heat of water
  - Specific heat calculations
  - Significant figure rules

WEB RESOURCES

Specific Heat Capacities Table. *http://en.wikipedia.org/wiki/Heat_capacity*

Universal Industrial Gases, Inc...Nitrogen N2 Properties, Uses, Applications. *http://www.uigi.com/nitrogen.html*

SAFETY NOTES

The substances used in this experiment are household substances and are generally considered safe when used as directed. However, follow your instructor's laboratory regulations about the use of safety chemical splash goggles and gloves; their use may be governed by state law in your area.

Your main concern for safety in this experiment is to avoid spilling boiling water on yourself or touching the hot metal.

Materials from Parts A and B of this experiment may be returned to the stockroom or discarded in the waste basket, or as your instructor directs.

EXPERIMENTAL PROCEDURE

PART A. THE SPECIFIC HEAT OF ALUMINUM

Step 1. Your instructor will provide you with a sample of aluminum; it may be in the form of a single block of metal or a number of small pieces. You will need about 40-50 g of aluminum. Find and record the exact mass of the metal.

Step 2. Place the weighed aluminum inside a large, dry test tube to which you have attached a clamp. The clamp does not have to be attached to a ring stand because you will be using it as your handle to control the hot sample during the experiment.

Step 3. Prepare the apparatus for heating the metal. First set up a ring stand, and arrange it for heating as shown in Figure 5-1, either with a Bunsen burner or a hot plate, as your instructor directs. Next, fill a 600-mL beaker about 3/4 full of water, add several boiling chips, and heat the water to boiling.

> ♥ If the beaker is too full, it will boil over!

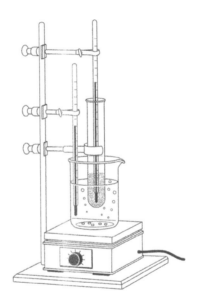

Figure 5-1. Heating the metal.

Step 4. Place the test tube full of metal in the heating water along with a thermometer. Heat the metal sample for at least 30 minutes to be sure it is at a final temperature equal to the temperature of the boiling water. You may have to add water to the bath to maintain the water level during this time.

While you are waiting for your aluminum to heat up, go on to Part B. In this experiment you will be testing a different metal, but the procedure and apparatus are the same, so you can start both Part A and Part B at the same time.

Step 5. While the metal samples are heating, prepare the calorimeter. This consists of two styrofoam cups nested inside each other, with a lid. The lid should have two holes, one for a thermometer and one for a stirring rod. Record the mass of the cups, and place 50 mL of water in the inner cup. Reweigh, and determine the mass of water. Place a thermometer and a stirring rod in the calorimeter through the holes in the lid (See Figure 5-2).

> ⓘ Note that the directions in Part B recommend that you set up the copper sample to heat simultaneously!

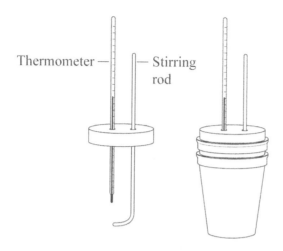

Thermometer — | | — Stirring rod

Figure 5-2. The parts of a styrofoam calorimeter.

Step 6. Once you place the thermometer in the calorimeter, the entire apparatus is rather top-heavy, so brace it by placing it inside a 400-mL beaker for support.

Step 7. During this waiting period you should also take the time to look up the accepted specific heats of copper and aluminum. There are many sources for this information.

Step 8. At the end of the heating time, record the temperature of the water in the calorimeter. At this point, it should be approximately room temperature. **Also** record the temperature of the boiling water; your metal samples should be at the same temperature as the water by now.

Step 9. Moving as quickly as is consistent with safety, remove the test tube containing the aluminum sample from the heating bath, open the calorimeter, and pour the metal sample into the calorimeter. Try not to drip any hot water from the outside of the tube into the calorimeter because this could contribute significantly to experimental error.

Step 10. Replace the lid, and use the stirring rod to mix the water in the calorimeter. Watch the temperature. Record the highest temperature reached by the water containing the aluminum sample.

Step 11. Complete the calculations to discover your experimental value for the specific heat of aluminum. Also record the results obtained by three other groups, and report the average of the trials.

PART B. THE SPECIFIC HEAT OF COPPER

Step 1. Place the aluminum sample from the previous experiment in a storage container designated by your instructor. Dry the calorimeter cups.

Step 2. Repeat the entire process described in Part A, except this time use copper metal instead of aluminum. For maximum use of time, you should heat the two samples, aluminum and copper, at the same time. There is plenty of room in the beaker of boiling water, and the copper sample can wait in the boiling water until you are ready for it.

PART C. EXERCISES

Refer to the report form for several exercises dealing with specific heat.

EXPERIMENT 5 PRELAB EXERCISES:

THE SPECIFIC HEAT OF A SOLID

Name_____ **Partner** _____ **Date** _____

1. A sample of silver (106.00 g) was heated inside a dry test tube in a water bath until the silver was the same temperature as the water. The temperature of the water was found to be 98.6 °C. The silver metal was quickly transferred to 50.0 g of room-temperature water (24.5 °C) and allowed to transfer its excess heat to this water. At the highest temperature reached, the water and metal sample reached a temperature of 32.4 °C. Find the specific heat of silver. The specific heat of water is 4.186 J/g·deg.

heat absorbed by the water $q = m_{H_2O}c_{H_2O}\Delta T_{H_2O}$

heat released by silver = heat absorbed by water = $m_{Ag}c_{Ag}\Delta T_{Ag}$

2. In the experiment described here, what would be the effect of hot water dripping into the calorimeter from the outside of the heated test tube? Would the error introduced into the calculation of a metal's specific heat cause the reported value to be higher or lower than if the experiment were done correctly?

3. In an experiment, 3500 kJ of heat were absorbed by a 10.0-g cube of copper. In a similar experiment, 3500 kJ of heat were absorbed by a 10.0-g sample of silver. Which metal reached a higher temperature? Prove your point with calculations.

4. On the diagram to the right, point out and label the air control and the gas control.

EXPERIMENT 5 REPORT: SPECIFIC HEAT OF A SOLID

Name_____ Partner _____ Date _____

PART A. THE SPECIFIC HEAT OF ALUMINUM

Mass of Al metal _____

Mass of calorimeter _____

Mass of calorimeter + water _____

Mass of water used _____

Temperature of hot Al metal (equal to the temperature of boiling water) _____

Temperature of water in calorimeter at start _____

Final highest temperature of metal and water in calorimeter _____

Specific heat of water in J/g-°C (from a reference source) _____

Calculations

Total heat absorbed by water in calorimeter ($q = m_{H_2O}c_{H_2O}\Delta T_{H_2O}$)

_____

Total heat given off by aluminum (equal to heat absorbed by water)

_____

The specific heat of aluminum ($q = m_{Al}\,c_{Al}\,\Delta T_{Al}$) from **your** data

_____

    Other group results _____ · _____ _____

Average of all trials _____

Literature value of the specific heat of aluminum _____

Percent error in the average _____

PART B. THE SPECIFIC HEAT OF COPPER

Mass of metal   _____

Mass of calorimeter   _____

Mass of calorimeter + water   _____

Mass of water used   _____

Temperature of hot metal
(equal to the temperature of boiling water)   _____

Temperature of water in calorimeter at start   _____

Final temperature of metal and
water in calorimeter   _____

Specific heat of water in J/g·°C   _____
(from a reference source)

Calculations

Total heat absorbed by water in calorimeter
($q = m_{H_2O} c_{H_2O} \Delta T_{H_2O}$)   _____

Total heat given off by copper
(equal to heat absorbed by water)   _____

The specific heat of copper
($q = m_{Cu} c_{Cu} \Delta T_{Cu}$) from your data   _____

Other group results   _____   _____   _____

Average of all trials   _____

Literature value of the specific heat of copper   _____

Percent error in the average   _____

PART C. EXERCISES

1. For the structural stone marble, the specific heat is 0.858 J/g·°C. How much heat is absorbed by a 35.0 g sample of marble if its temperature is raised from 27.0 °C to 55.0 °C?

2. In a calorimetry experiment, the heat capacity of lead was investigated. A 48.0-g sample of lead was heated to 100.0 °C and then poured into 50.0 g of water with an initial temperature of 25.2 °C. The final temperature of the mixture was 27.0 °C, What was the experimentally determined specific heat of lead? You may assume that the heat absorbed by the calorimeter was negligible.

3. Assume you have available to you a 50.0-g sample of glass (c = 0.837 J/g·°C), and a 50.0-g sample of iron (c = 0.452 J/g·°C). Both samples start at room temperature, 24.3 °C. If 500. joules of heat were added to each of these samples, which one would reach the higher temperature? Justify your answer with a calculation.

4. Name at least two potential sources of error in your determination of the specific heat of a metal. For each, decide if it would cause the result to be too high or too low.

EXPERIMENT 6

ELEMENTS AND THE PERIODIC TABLE

GOALS

1. To examine and describe some metallic and nonmetallic elements
2. To understand the organization of the periodic table
3. To identify some elements, given samples to observe and information about their properties

INTRODUCTION

You have been studying the structure of the atom with reference in particular to a number of the smaller elements. The nucleus holds most of the mass of the atom and consists of tightly packed protons and neutrons, with the number of protons determining the particular element under consideration. The electrons are found outside the nucleus, arranged in a relatively small number of shells that are at allowed positions; the number of electrons that can occupy each shell and subshell is strictly limited. You have also learned to represent the electron configuration of an atom in a way that shows which shells and subshells are occupied. For instance, the electron configuration for the neon atom is shown below.

$$Ne: 1s^2 2s^2 2p^6$$

But what do samples of these atoms look like when the sample is big enough to see? In this exercise, you will have an opportunity to examine samples of many elements. Take particular note of how each shows the physical charactistics of a metal or nonmetal and fits into its place in the periodic table. You will then practice writing the electron configuration for each element you examine.

The electron configuration you have learned to write is the ground state, most stable configuration. If an atom is subjected to extra energy, perhaps heat or light or electrical energy, some of its electrons can be pushed into higher, normally empty energy levels to an "excited state." This is usually a very unstable situation, and the excited electron tends to return to its preferred ground state energy level very soon, often with the emission of light. When the light emitted is visible light, sometimes spectacular visual effects can happen, as in the colors you see in fireworks. Among the atoms whose excited states can emit visible light, different colors are possible. In fact, it is sometimes possible to determine the presence of a specific element (especially a metal) in a sample by examining the color of light the sample gives off when its atoms are excited. For several labeled samples of metal salts, you will be able to observe the flame color produced and then use your observations to determine the metal content of an unknown sample.

Electron configurations are the basis for the periodic law, and the organization of the periodic table is based on the periodic law. The periodic table groups elements according to their chemical characteristics, which are a function of the arrangement of their electrons. For instance, Group 1 (Group IA) elements all have one electron in their outer shells, and although they are very different in size they all behave in a similar manner in chemical reactions. The periodic table is a storehouse of information of that sort.

Elements can be classified in many ways depending on the characteristic you are examining. Most elements are metals, and they are found on the left of the table. The nonmetals are on the right, and there are a few elements (the metalloids) that show some characteristics of both metals and nonmetals. On the other hand, you might classify the elements by electronic properties, and then you can think of representative elements, noble gases, or transition elements. In addition, there are several element groups within the periodic table that have their own special names, such as the alkali metals (Group 1), the alkaline earth metals (Group 2 or IIA) and the halogens (Group 17, or VIIA). Check your textbook for more detailed information if you need it to complete the exercises in this experiment.

The third part of the experiment is a puzzle that requires you to identify several elements on the basis of their physical characteristics. If you need to look up some of the necessary data, your textbook is a good source of information, and there are several interactive periodic tables available through the Internet.

REFERENCES

Stoker, H. S. *General, Organic, and Biological Chemistry;* Cengage Learning; Chapter 3.
   Review topics:
   - Atomic structure
   - The periodic table
   - Electron configurations
   - Color and atomic excited states

WEB RESOURCES

Chemistry: WebElements Periodic Table. *http://www.webelements.com/*

How Lasers Work. *http://www.howstuffworks.com/laser.htm*

Gemstone Colors and Defects. *http://webexhibits.org/causesofcolor/12.html*

SAFETY NOTES

The element samples observed in this experiment are sealed and should not be a safety hazard. Do not discard the samples you observe.

Use care handling the hydrochloric acid solution; it is corrosive. Follow your instructor's laboratory regulations about the use of chemical splash goggles and gloves; their use may be governed by state law in your area.

EXPERIMENTAL PROCEDURE

PART A. ELEMENTS AND THE PERIODIC TABLE

Step 1. In your lab you will find several labeled bottles containing some of the naturally occurring elements. Describe each element's appearance. Use sources available to you in the lab to fill in the chart for Part A of the report form.

Step 2. Write the electron configuration for each of these elements, and answer the questions on the report form.

PART B. EXCITED STATES AND COLOR

This exercise is best performed in a fume hood. Available in the lab are four solutions of metal compounds in water. Your task is to determine the color produced by one of the burning metals in its metal salt, and then identify an unlabeled sample. The four metal salts are potassium chloride, strontium chloride, copper chloride, and cobalt chloride.

Step 1. Light a Bunsen burner and adjust the flame for moderate heat.

Step 2. Place about 10 mL of dilute hydrochloric acid solution in a small beaker, and use it for cleaning your spatula. To do this, dip a spatula first in a solution of hydrochloric acid, and then burn off the acid in the burner flame.

> ♥ Be careful about handling the hydrochloric acid. Although it is a dilute solution, it can cause burns.

Step 3. Place about 10 drops of each of the test compound solutions in a well in a spot plate. Then dip the spatula in the metal salt solution you want to test, and observe the color you obtain when the spatula is held in the flame. Before you go on to the next solution, clean the spatula with hydrochloric acid again, as described in Step 2. Test all the solutions and record what you see.

Step 4. One of the solutions available to you is an unidentified sample with only an identifying number. Test your unknown sample for the presence of one of these four metals, and identify the metal in the solution.

PART C. IDENTIFICATION OF ELEMENTS BY PHYSICAL PROPERTIES.

Below are some physical characteristics of a number of elements. From the given information, and information in your textbook, identify the individual elements, and place them correctly on the blank periodic table on the report form.

Element 1. This element does not conduct electricity. It is a colorless gas at room temperature and pressure, has a density of about 90 mg/L at room temperature, and changes from a gas to a liquid at approximately –250 °C.

Element 2. This element, a solid at room temperature, is one of the major elements necessary for life. It has more than one physical state, one black and amorphous, one clear and crystalline, and one in a molecule shaped like a soccer ball. It is not a conductor of electricity. The density of the crystalline form is 3.5 g/cm^3.

Element 3. This metallic element is a transition element with a density of 7.1 g/cm^3. It is in Period 4 (IVA) and has only one major isotope.

Element 4. This is a light green-yellow gas, heavier than air, in Group VIIA (Group 17).

Element 5. Group 1 elements react violently with water to give solutions that turn litmus paper blue. This Group 1 element has a density of 0.9 g/cm^3 and will float on water as it reacts. It burns with a faint purple flame.

Element 6. This nonmetal is a solid, the lightest solid in its family, with a density of 1.96 g/cm^3. A crystalline sample is the color of a dandelion.

Element 7. This element has some of the properties of metals and some of the properties of nonmetals. For instance, it has a somewhat metallic luster, but does not conduct electricity well. This selective conductivity allows it to be used extensively in the production of integrated circuits. Density = 2.33 g/cm^3.

Element 8. Alkaline earth metals react only slowly with water. This one has the usual outer electron shell with two electrons, and an atomic number of 20.

Element 9. This transition element conducts electricity. It is bluish-gray and lustrous. Look in Group IIB (Group 12) for it. It is one of the minor essential elements necessary for good growth and development. The density of the pure element is 7 g/cm^3.

Element 10. This naturally occurring radioactive element is an actinide. It is one of the heaviest elements known, with a density of 19 g/cm^3, and has many commercial and military uses.

EXPERIMENT 6 PRELAB EXERCISES:

ELEMENTS AND THE PERIODIC TABLE

Name_____ Partner _____ Date _____

1. State at least three physical differences that can be used to distinguish between metallic and nonmetallic elements.

2. Write the electron configuration for the potassium atom.

3. Lithium produces a red color when it is placed in a flame. Explain why this is so, using your knowledge of electronic shells and the transitions that are possible for electrons between different electron shells.

4. Atoms of this element have one electron in the outer shell, produce a faint purple color when excited, and have one more neutron than proton in the nucleus. Identify this element.

5. Why are elements in Group 18 (Group VIIIA) sometimes called the "noble gases"?

EXPERIMENT 6 REPORT: ELEMENTS AND THE PERIODIC TABLE

Name _____ Partner _____ Date _____

PART A. ELEMENTS AND THE PERIODIC TABLE

Properties of Some Common Elements

Element	Atomic number	Atomic mass	Appearance	Classify (metal, nonmetal or metalloid)	Will it conduct electricity?	Melting point (°C)	Boiling point (°C)	Density (g/cm^3, or g/mL of a gas)

Write the electron configurations for the elements you have examined.

Element_____ configuration _____

Element_____ configuration _____

Element_____ configuration _____

Element_____ configuration _____

Element_____ configuration _____

Element_____ configuration _____

Element_____ configuration _____

Element_____ configuration _____

PART B. EXCITED STATES AND COLOR

Solution	Color obtained
Potassium chloride	
Strontium chloride	

Flame Tests on Compounds

Solution	Color obtained
Copper chloride	
Cobalt chloride	

What color does chloride ion produce? _____

What compound is in your unknown solution? _____ Identification number _____ Compound present _____

PART C. IDENTIFICATION OF ELEMENTS BY PHYSICAL PROPERTIES.

EXPERIMENT 7

THE SIX-BOTTLE PROBLEM

GOALS

1. To practice observation of chemical reactions
2. To draw conclusions from these observations
3. To write formulas for compounds in a chemical reaction

INTRODUCTION

Imagine this...
What if a puzzled high school teacher brought six unlabeled bottles into your lab with the following story. "The guy who just retired left these bottles on the shelf, but all the labels had fallen off the bottles by the time I saw them. I want to get rid of them, but a couple have heavy metals in them, so I need to dispose of them as hazardous waste. Naturally, I don't want my budget to pay for any more of that kind of stuff than necessary, so I need to know which compound is in which bottle. Can you help me?"

It would then be your job to determine which solution is in each bottle. From the labels, you could deduce which compounds are possible, but you would have to match the label to the bottle. You then would need to recommend which solutions needed the special treatment.

That is the background for this exercise. There are two phases to this activity. First, you will need to experiment with known samples of the six possible solutions, in order to figure out how each one behaves. The Reference Table in the report form will help you organize your observations for this step. Next, you will have to do the same procedure with the unidentified solutions, and deduce which numbered bottle contains each chemical.

Once you have decided the identities of the chemicals in the various unlabeled bottles, you will then need to write appropriate balanced equations for what **might** happen when pairs of solutions are mixed. All the possible mixtures are double replacement reactions, so to balance the equations, you need only follow these usual rules for balancing.

Rules for Balancing Simple Chemical Equations
1. Find the correct formulas for the possible reactants and products. Once you have decided what these are, **do not change the subscripts.** After this, work with **coefficients only**.
2. Find the correct coefficients that ensure that atoms are conserved in the reaction. For example, there must be the same number of oxygen atoms on both sides of the equation. These are relatively simple equations; if you are having trouble balancing one of them, then you may have decided on the wrong formula for a starting reagent or a product.
3. Check to see if a reaction actually occurred by considering what your observations can tell you. If no reaction occurred, write "N. R." instead of products. Your textbook may also be of help in deducing formulas, and on the identity of any solids formed.

A note of caution—sometimes a gas is formed in a chemical reaction. As you write equations, watch for the production of carbonic acid, H_2CO_3, in a reaction. This substance is not your final product when it occurs; it goes on to form the products you actually see, H_2O and the gas CO_2. Another reaction to watch out for is the formation of ammonia gas. If ammonium hydroxide, NH_4OH, is formed in a chemical reaction, a significant amount of it goes on to produce the final products in that reaction, H_2O and the gas NH_3.

ⓘ Watch for these as products, and be sure you don't stop writing too soon!

REFERENCES

Stoker, H. S. *General, Organic, and Biological Chemistry;* Cengage Learning; Chapters 4 and 8.
   Review topics:
   - Ionic compounds
   - Names of ionic compounds
   - Solubility rules
   - Writing chemical formulas
   - Writing chemical equations
   - Chemical and physical changes
   - Observations and conclusions

WEB RESOURCES

How to remember Solubility Rules. *http://scienceray.com/chemistry/how-to-remember-solubility-rules/*

Information Internet: Chemistry's Solubility Rules.
http://www.sparknotes.com/testprep/books/sat2/chemistry/chapter6section3.rhtml

SAFETY NOTES

Some of the unidentified solutions you are handling may contain acids or bases. Acids and bases, especially concentrated ones, can cause severe chemical burns, and some can irritate the linings of the nasal passages. Be extremely careful when handling these chemicals. You must wear chemical splash goggles, and may prefer to wear gloves throughout this experiment. Follow your instructor's recommendations about safety precautions.

♥ Always handle unknown chemicals with caution!

   Mixed solutions should be poured into a labeled waste container provided by the instructor.

EXPERIMENTAL PROCEDURE

PART A. OBSERVATIONS ON KNOWN SOLUTIONS

Step 1. At your station there should be small dropper bottles labeled with the names of the reference solutions. In the spot plates provided, mix 5 drops of any reference solution with 5 drops of another reference solution, and write in the table what you see.

> ⓘ Observe carefully, because a few of the reactions are very difficult to see!

Step 2. Repeat this process until you have made all possible combinations, and record what you see on the report form. Repeat these steps as necessary until you are quite sure what you can observe when any one of the reference solutions is mixed with any other solution.

When you are finished with Part A, dispose of the solutions as your instructor directs.

PART B. OBSERVATIONS ON UNIDENTIFIED SOLUTIONS

Step 1. Perform the same operations on the solutions labeled SB-1, SB-2, and so on. You already know these are the same six solutions you worked with in Part A, but you need to determine which is which. Fill in the chart for Part B with your observations.

Step 2. In the last row of the table for Part B, write the correct formula for each identified solute. That is, in the column headed SB-1, write the formula for SB-1 on the last line, etc.

Step 3. On the last line of the table for Part B, circle the solution(s) that need to be treated as "hazardous waste."

PART C. BALANCED CHEMICAL EQUATIONS

On the third page of the report form, write a **balanced** chemical equation for each of the possible mixtures that could occur when each of the known solutions is mixed with every other one. If a reaction occurs, give the products. If no reaction occurs, write "N.R." after the arrow (for "no reaction").

EXPERIMENT 7 PRELAB EXERCISES:

THE SIX-BOTTLE PROBLEM

Name_____**Partner** _____ **Date** _____

1. Silver, lead, and mercury are considered "heavy metals," and need to be treated with special consideration when you dispose of solutions containing them. What are the chemicals in the reference set that you will need to keep track of as "heavy metal" solutions? Give both the name and the formula for each of the chemicals involved.

2. When sodium sulfate solution is mixed with potassium nitrate solution, a clear solution is formed. Does a reaction occur in this case? Justify your answer.

3. When barium chloride solution is mixed with a solution of sodium sulfate, a white solid forms. Write an equation for this reaction. What is the formula for this white solid?

4. On the Internet, look for information about hydrochloric acid, and explain why it is necessary to wear chemical splash goggles when working with it.

5. Write a balanced equation for the reaction of hydrochloric acid (HCl) with sodium carbonate (Na_2CO_3).

EXPERIMENT 7 REPORT: THE SIX-BOTTLE PROBLEM

Name _____ Partner _____ Date _____

PART A. OBSERVATIONS ON KNOWN SOLUTIONS

Reference Table: Observations of Known Solutions

	Sodium chloride	Hydrochloric acid	Potassium iodide	Sodium carbonate	Silver nitrate	Lead nitrate
Lead nitrate						
Silver nitrate						
Sodium carbonate						
Potassium iodide						
Hydrochloric acid						
Sodium chloride						

Table of Unidentified Solutions

	SB-1	SB-2	SB-3	SB-4	SB-5	SB-6
SB-6						
SB-5						
SB-4						
SB-3						
SB-2						
SB-1						
Identity of solution (Give both name and formula.)	SB-1 =	SB-2 =	SB-3 =	SB-4 =	SB-5 =	SB-6 =

PART C. BALANCED CHEMICAL EQUATIONS

sodium chloride plus hydrochloric acid

sodium chloride plus potassium iodide

sodium chloride plus sodium carbonate

sodium chloride plus silver nitrate

sodium chloride plus lead nitrate

hydrochloric acid plus potassium iodide

hydrochloric acid plus sodium carbonate

hydrochloric acid plus silver nitrate

hydrochloric acid plus lead nitrate

potassium iodide plus sodium carbonate

potassium iodide plus silver nitrate

potassium iodide plus lead nitrate

sodium carbonate plus silver nitrate

sodium carbonate plus lead nitrate

silver nitrate plus lead nitrate

EXPERIMENT 8

THE CHEMISTRY OF WATER

GOALS

1. To investigate the unusual properties of water as a chemical
2. To examine the effect of dissolved substances on the properties of water
3. To examine one method of removing dissolved substances from water

INTRODUCTION

Water is a relatively unusual chemical. We think of it as being very ordinary because we see it everywhere and take its presence for granted, but if you examine its chemical and physical properties and compare them to those of other molecular liquids, you find that in many ways they don't fit the usual pattern. Water will dissolve both ionic and molecular compounds, it will (sometimes) conduct electricity when most other liquids don't, and it has an unusually high boiling point and freezing point for its molecular mass. It has so strong a surface tension that small creatures can skate across the surface of a pond, and you can even float some small metal objects in it (Did you ever try to make a compass by floating a needle on a cup of water?). All three physical states, solid, liquid, and gas, can be reached readily at temperatures near room temperature. And ice floats, whereas with most other substances, the solid form sinks in a bath of the liquid form. All these facts make water essential to the type of chemistry that makes life possible on Earth.

In the first part of this exercise, you will be working with distilled water and a saturated solution of sodium chloride in water. In each step of the procedure you will examine a particular physical property, and compare it in the two solutions. On occasion, to further understand the peculiarities of water you may be asked to compare samples of the water to some other chemical.

The first property you will examine is density. An earlier experiment dealt with the density of liquid water, but this time you will compare the densities of liquid water, salt water, and ice. You can work with the two liquid samples in the same manner as you worked with a sample in Experiment 4, but finding the density of ice requires some planning. Ice is a solid; in order to perform this experiment, you will need to find a way to measure the volume of an ice cube and a way to measure its mass. As in Experiment 4, the volume of the solid (ice) can be obtained by difference, provided the cube can be immersed in a liquid completely, but ice floats in liquid water, so the procedure you have already learned has to be modified a bit. You need to perform the experiment using a liquid that will allow an ice cube to settle to the bottom. The density of isopropyl alcohol (rubbing alcohol) is lower than the density of ice, so this should be a suitable liquid for the standard procedure.

In your earlier density determinations, you gathered information on three trials of each substance and reported an average density. You can repeat each trial to try for greater accuracy in your work for this experiment also, but if you work carefully a single trial with each will be sufficient for you to compare the relative densities of liquid water, ice, and salt water.

You will also have a look at the ability of water to conduct electricity. We are all familiar with warnings about using electric appliances where they might come in contact with water, but **how** does water conduct? Your experiment will examine the conductivity of pure water, water with an ionic solute, and water with a molecular solute, so you can see what components are necessary for water to carry an electric current.

Water has an unusually strong surface tension as well. All liquids have a tendency to form spherical droplets because the molecules on the surface are attracted more toward the molecules of the interior of the sample than to the surrounding air. This is what makes water bead up on a window. This strong surface tension is helpful if you are an insect trying to float on water, but less helpful when you are trying to force water under a patch of dirt on your jeans to clean them. Your experiment will examine the effect of two different types of solutes, salt and detergent, on the surface tension of water. Both are ionic but may not affect surface tension equally. Watch carefully.

Your final experiment with physical properties concerns the effect of solutes on bulk properties of water. "Bulk properties" are those that require a large sample to observe them because an individual molecule won't show them. The usual technical term for these types of properties is "colligative properties," and you will study the colligative properties of water in more detail later in the course. The two you will examine here are freezing point and boiling point: At what temperature does water freeze, and at what temperature does water come to a boil? In general, the magnitude of any colligative property depends on how many particles of solute are dissolved in a liquid, so ionic compounds often have a greater effect than molecular compounds as solutes. That's why you'll be examining these properties using salt water rather than sugar water.

The second part of the experiment deals with "hard" vs. "soft" water. In most parts of the United States, ground water contains significant amounts of dissolved minerals, usually compounds of iron, calcium, and magnesium, that affect how the water tastes, how it looks, and how it interacts with soap. Your exercises here will look at the sudsing ability of a soap (different chemically from a detergent) in hard and soft water, and the effect of using a specialized filter (in this case a Brita® filter) to exchange those metal ions for others that don't have the unwanted effects on the water we use.

In this case, you will be testing hard and purified water for the presence of calcium ions. If calcium is present and you add a solution of sodium oxalate, you will see the formation of insoluble calcium oxalate according to the following equation.

$$Ca^{2+}_{(aq)} + C_2O_4^{2-}_{(aq)} \longrightarrow CaC_2O_{4(s)}$$

REFERENCES

Stoker, H. S. *General, Organic, and Biological Chemistry;* Cengage Learning; Chapters 4 and **8**.
   Review topics:
   - Water
   - Conductivity
   - Colligative properties
   - Density of ice and water
   - Ionic compounds

WEB RESOURCES

There are thousands of web resources about the chemistry of that peculiar compound, water. Here are only a few.

The Chemistry of Water. *http://witcombe.sbc.edu/water/chemistry.html*

Dihydrogen monoxide Research Division-dihydrogen monoxide info. *http://www.dhmo.org/*

Wondernet November 2003-the Wonders of Water.
http://www.nsf.gov/news/special_reports/water/index_low.jsp

SAFETY NOTES

The substances used in this experiment are household substances and are generally considered safe to work with when used as directed. Isopropyl alcohol is flammable, so no open flames should be present in the laboratory during its use. Follow your instructor's laboratory regulations about the use of chemical splash goggles and gloves; their use may be governed by state law in your area.

Materials from Parts A and B of this experiment may be discarded in the waste basket or sink, or as your instructor directs.

EXPERIMENTAL PROCEDURE

PART A. PHYSICAL PROPERTIES OF WATER

A comparison of the densities of water, ice, and salt water

Distilled water
Step 1. Find the mass of a clean, dry 25-mL graduated cylinder and record it.
Step 2. Place in the cylinder about 15-20 mL of distilled water and record the exact volume, using as many significant figures as possible.
Step 3. Find the combined mass of the water and the cylinder, and use your data to determine the density of distilled water.

Ice
Step 1. Find the mass of a clean, dry 25-mL graduated cylinder and record it.
Step 2. Place in the cylinder about 15-20 mL of isopropyl alcohol, and record the exact volume using as many significant figures as possible.
Step 3. Obtain an ice cube that will fit into the graduated cylinder. Working quickly, so it has little opportunity to melt, place it in the isopropyl alcohol, and record the combined volumes. (See Figure 8-1.)

> ⓘ There is room for error here. If the ice melts much, its volume reading will be too low.

Figure 8-1. Using an ice cube in isopropyl alcohol.

Step 4. Record the combined mass of cylinder, alcohol, and ice, and calculate the density of ice.

Salt water
Step 1. Find the mass of a clean, dry 25-mL graduated cylinder and record it.
Step 2. Place in the cylinder about 15-20 mL of saturated salt water and record the exact volume, using as many significant figures as possible.
Step 3. Find the combined mass of the solution and the cylinder, and use your data to determine the density of distilled water.

The effect of solutes on conductivity
Using the conductivity apparatus available to you, check 0.5-mL samples of each of the waters to determine if either can carry an electric current.
Step 1. Place about 10-20 drops (0.5-1 mL) of distilled water in one well of a spot plate, and the same amount of saturated sodium chloride solution in another well.

Step 2. Obtain a conductivity tester from the stock materials, and rinse the electrodes carefully with distilled water (into a waste beaker). Blot the electrodes gently, and immerse them in the sample you are testing.

Step 3. Observe the light on the tester. Does it light up at all? How bright is it? Does it blink? Record your result as

−	No current: You see no light.
+	Weak current: You see a faint steady light.
++++	Strong current: You see a strong, blinking light.

> ⓘ Observe carefully, because in a bright room the dim light is sometimes hard to see.

Step 4. Rinse the electrodes of the tester with distilled water, test the saturated salt solution in the same way, and record your results.

Step 5. Repeat the test using a solution of sucrose (table sugar) in water. Before returning the tester to the stock area, rinse the electrodes again so the next users can be confident of their results.

> ⓘ **Don't** contaminate your solution with the previous one!

Surface tension comparisons

This exercise can be carried out on any flat surface that does not absorb water, but it is most convenient to use a coin as the surface.

Step 1. Lay a clean, dry penny on a flat surface. Using a dropper full of distilled water, count the number of drops of water that can be mounded on the surface of the coin, until the surface "skin" of the water breaks and the coin overflows.

Step 2. On a different penny, try the same observations with a dropper full of salt water. Do you see any difference?

Step 3. Try the same exercise with soapy water.

Step 4. For comparison purposes, repeat the experiment using a dropper full of rubbing alcohol.

Colligative properties
The effect of solutes on freezing point

Step 1. Fill a 400-mL beaker about half full with ice, and add about 100 mL of water to make a cooling bath. Add to this bath about 50 g of solid calcium chloride, and stir well (the calcium chloride does not have to dissolve completely).

Step 2. Into this bath, place two test tubes, one containing 10 mL of distilled water and one containing 10 mL of saturated salt water. Allow these two solutions to cool for 10-15 minutes in the bath, checking the temperature of each solution every 2 minutes or so with a clean thermometer.

Step 3. Record the lowest temperature reached by each solution before it freezes to a solid. If one of the solutions does not freeze, you may discontinue the experiment as soon as you have established which solution has the lower freezing point, pure water or salt water.

The effect of solutes on boiling point

Step 1. Set up a ring stand and Bunsen burner, or a hot plate. Fill one 50-mL beaker half full with distilled water, and another half full with saturated salt water.

Step 2 Add a few boiling chips to each, and bring them to a boil. Determine the boiling point of each solution, and record it.

PART B. WATER HARDNESS

Gauging the hardness of water

Step 1. As a demonstration, your instructor will prepare one sample of clarified water for general use by allowing 1-2 L of hard water to flow through a fresh Brita® filter. Use samples of this water for experiments on "filtered water."

Your instructor will also prepare a stock solution using a spent filter, one that has been used so much that it probably can no longer remove calcium ions from the hard water. Label any samples you use from this stock solution "spent filter."

A third set of filtered samples can be prepared to demonstrate the regeneration of a Brita® filter by passing through it a solution of hydrochloric acid. The effluent from this demonstration should be used for samples labeled "regenerated filter" water.

Step 2. Obtain 10-mL samples of "distilled water," "hard water," "filtered water," "spent filter water," and "regenerated filter water" for your portion of the experiment.

Step 3. Label 5 test tubes hard, distilled, etc., and place 5 mL of the appropriate type of water in each one. Add 1 mL of soap solution to each, cover the mouth of the tube, and shake each tube with a vertical motion for about 15 seconds. Record the height of the suds reached in each tube.

Step 4. Prepare another set of test tubes in the same manner, but without the detergent. If spot plates are available, 10 drops of each solution in a spot plate well is sufficient for Steps 4 and 5. Check each for the presence of calcium ions using a flame test. (Refer to Experiment 6 for the method and the observation to make.)

Step 5. Using the same samples prepared in Step 4, check each for the presence of calcium ions by adding a few drops of sodium oxalate solution. If significant calcium is present, calcium oxalate will precipitate as a white solid.

PART C. EXERCISES.

Refer to the report sheet, and answer the questions for Part C.

EXPERIMENT 8 PRELAB EXERCISES:

THE CHEMISTRY OF WATER

Name_____ **Partner** _____ **Date** _____

1. Ammonia is a gas at room temperature, but it can become a liquid at temperatures lower than −78 °C and even a solid at temperatures far below −78 °C. The density of liquid ammonia is about 0.63 g/cm^3, and the density of frozen ammonia is about 0.7 g/cm^3. Sketch a container of liquid ammonia, to which a cube of frozen ammonia has been added.

2. Tap water may conduct electricity, depending on how it has been purified, but distilled water definitely does not. Would you expect spring water to conduct electricity? Why or why not?

3. What property of water allows you to make a compass by floating a steel needle on the surface of a cup of water?

4. Name one practical use of the fact that water with dissolved ions in it will freeze at a lower temperature than tap water.

5. Why does soap scum form when soap is dissolved in hard water containing calcium and magnesium carbonate?

6. Is calcium oxalate, CaC_2O_4, soluble in water?

EXPERIMENT 8 REPORT: THE CHEMISTRY OF WATER

Name_____ Partner _____ Date _____

PART A. PHYSICAL PROPERTIES OF WATER

A comparison of the densities of water, ice, and salt water

	Water	Ice	Salt water
Mass of cylinder	_____	_____	_____
Volume of liquid	_____	_____	_____
Mass, cylinder + liquid	_____	_____	_____
Volume of liquid and ice		_____	
Mass of cylinder, liquid and ice		_____	
Mass of substance	_____	_____	_____
Density of substance	_____	_____	_____

Rank the substances, liquid water, ice, and salt water, in order of increasing density.

Conductivity observations (-, + or ++++)

Distilled water _____   Salt water _____   Sucrose _____

Surface tension observations (drops used to cover a penny)

Distilled water _____   Salt water _____

Soapy water _____   Alcohol _____

Freezing point observations (°C)

Distilled water _____   Salt water _____

Boiling point

Distilled water _____   Salt water _____

PART B. WATER HARDNESS

Results of Several Tests for Water Hardness

Sample	Sudsing	Flame test for Ca^{2+}	Oxalate test	Hard?
Distilled				
Hard				
Filtered				
Spent filtered				
Regenerated filtered				

PART C. EXERCISES

1. The density of the water in the Great Salt Lake of Utah varies between about 1.06 g/cm^3 and 1.2 g/cm^3, depending on where the sample is collected. In general, it is easier for a person to float in the Great Salt Lake than in freshwater lakes. Why?

2. Given your results in the conductivity experiment, what can you conclude about the importance of presence and type of solutes in the ability of water to conduct electricity?

3. Does the presence of sodium chloride or soap affect the surface tension of water? Justify your answer with your experimental results.

4. Describe the effect of dissolved solutes on the freezing point and boiling point of water.

5. Use the Internet to determine how a water softener works, and relate this to what you have observed in this exercise.

EXPERIMENT 9

IONIC AND MOLECULAR COMPOUNDS

GOALS

1. To review some differences between ionic and covalently bonded compounds
1. To explore the effect of the different types of bonding on a compound's chemical properties
2. To explore the shapes of covalent and ionic compounds

INTRODUCTION

In order to understand how the molecules of life are put together and how they behave, you need an understanding of the types of bonding that can occur. This exercise will use models to illustrate the different types of ionic bonds and covalent bonds, and will allow you to observe a few properties of compounds that are directly affected by the types of bonds present.

Ionic bonds are the kind that occur when ions, the electrically charged forms of atoms, interact with each other to form a formula unit. As in all electric phenomena, opposites attract, so positively charged ions can bond with negatively charged ions, but positive with positive and negative with negative matches do not occur. The most familiar example of an ionic compound is probably sodium chloride, in which one positively charged sodium ion forms a formula unit with one negatively charged chloride ion. Notice that the formula unit, NaCl, has to be neutral overall—the 1+ and the 1- charges add up to a charge of 0 in the complete formula unit. You can look at many other examples, but that general rule will hold in all cases: positive ions combine with negative ions, and the complete formula for one unit of an ionic compound must have no leftover charges.

The formation of ions occurs best if it is relatively easy for one atom to give away an electron or more than one electron, and for the other atom to take on one or more extra electrons, as in compound formation between a metal and a nonmetal. When the elements forming a compound are closer together in the periodic table, it becomes less easy to decide which should "donate" and which should "take on" an electron, and the sharing of electrons becomes more important.

Covalent bonds are formed when sharing of electrons between atoms takes place. The simplest kind of covalent bond to consider is the bond between two atoms of the same type, such as two chlorine atoms forming the Cl_2 molecule.

The electron configuration of chlorine is $1s^2\, 2s^2\, 2p^6\, 3s^2\, 3p^5$. A Lewis dot structure set for two chlorine atoms would look like this, showing only the seven outer shell electrons.

$$: \overset{\displaystyle ..}{\underset{\displaystyle ..}{Cl}} \cdot \qquad \cdot \overset{\displaystyle ..}{\underset{\displaystyle ..}{Cl}} :$$

There is no obvious reason why one should donate an electron and the other gain one because both atoms hold their electrons equally well. However, if they could each share the unpaired ones, those two would then be paired, and each atom would have its octet rule satisfied and a covalent bond would be formed.

$$: \overset{\displaystyle\cdot\cdot}{\underset{\displaystyle\cdot\cdot}{Cl}} : \overset{\displaystyle\cdot\cdot}{\underset{\displaystyle\cdot\cdot}{Cl}} :$$

Lewis structures like this are very helpful when you are trying to figure out bonding, but they become cumbersome when you need to work with a lot of molecular structures, so for drawing actual covalent bonds (as opposed to lone pairs), it is customary to use a single line to mean one shared pair of electrons, or one covalent bond. In this conventional way of drawing structures, the chlorine molecule looks like this.

$$Cl{-}Cl$$

The chlorine molecule is held together by one shared pair of electrons, or a single covalent bond. Other types of covalent bonds are possible as well—double, triple, and coordinate.

Double covalent bonds hold together the oxygen molecule O_2. Each atom is two valence electrons short of an octet, and they share equally in this way.

$$O::O \text{ or } O{=}O$$

The extreme case of multiple bonding comes with nitrogen gas, N_2, in which the covalent bond is a triple bond in which three pairs of electrons are shared.

$$:N:::N:$$

Coordinate covalent bonds are usually single bonds, but both of the electrons for a coordinate bond come from the same atom. You will see in the exercises of this experiment how the ammonium ion forms from an ammonia molecule and a hydrogen ion by the formation of a coordinate bond.

As compounds get more complex, the types of covalent bonds present can affect the overall shape of the molecule. In water, for example, the two lone pairs of the oxygen atom force the molecule into a bent shape, so that they can be as far as possible from each other. Double bonds in large compounds generally produce flat spots in a molecule, and triple bonds produce regions of linear or pencil-like geometries. As you progress in your study of biochemistry, you will see that an understanding of the shape of a molecule is very important to understanding how it functions in the human body.

There are relatively few instances of 100% ionic or 100% covalent bonds; there is really a continuum, with some clearly ionic and some clearly covalent cases, and many others that are somewhere in between. For covalent bonds, the most common "imperfect" case you need to be aware of is the polar covalent bond, such as is found in water. Because oxygen holds electrons more tightly than hydrogen, a covalent bond between oxygen and water can be thought of as lopsided, with the oxygen being relatively (but not completely) negative and the hydrogen being relatively positive. The polar covalent bond has a profound effect on water's ability to act as a solvent for both ionic and covalent compounds.

Keep all these ideas in mind while you explore water chemistry in this exercise.

REFERENCES

Stoker, H. S. *General, Organic, and Biological Chemistry;* Cengage Learning; Chapter 5.
    Review topics:
    - The ionic bond model
    - The covalent bond model
    - Lewis structures of covalent compounds

WEB RESOURCES

Chemical Bonding.
http://www.visionlearning.com/library/module_viewer.php?mid=55

Covalent Bond Formation. *http://intro.chem.okstate.edu/1314F97/Chapter9/CovBond.mov*

SAFETY NOTES

The substances used in this experiment are common substances and are generally considered safe to work with. However, follow your instructor's laboratory regulations about the use of chemical splash goggles and gloves; their use may be governed by state law in your area.

Materials from Part A of this experiment may be discarded in the waste basket, in the sink with lots of water, or as your instructor directs.

EXPERIMENTAL PROCEDURE

PART A. SOME PROPERTIES OF IONIC AND COVALENT COMPOUNDS

Flammability
Your instructor may prefer to do this exercise as a demonstration, because good ventilation is needed.

On two separate watch glasses, in the hood, place small piles of sodium chloride, an ionic compound, and sucrose (table sugar) a covalent compound. A few drops of a second covalent compound, the liquid ethanol, can also be observed on a third watch glass. Try setting each substance on fire using a match, and record what happens.

Solubility
Step 1. Place a small amount (about 0.10 grams, or a spatula-tip full) of sodium chloride in a test tube, and add 5 mL of water. Mix well and observe what happens.

Step 2. Repeat this experiment, using about the same amount of benzoic acid, $C_7H_6O_2$, a covalent molecule, and observe.

Step 3. Try the experiment again, this time using sucrose, $C_{12}H_{22}O_{11}$. Record your observations and try to make a general statement about the solubility of ionic vs covalent compounds. Do you have sufficient data to make a flat statement about the relative solubilities of the classes?

Conductivity in solution
For each of the solutions you made in the solubility test, place about 10 drops in a well of a spot plate, and test the solution for its ability to conduct an electric current.

PART B. SHAPES OF MOLECULES WITH COVALENT BONDS

Covalent bonds in simple compounds
You have available to you a set of molecular models. Your instructor will go over the uses of the different pieces and the color coding used in your particular set.

Step 1. For each of the compounds below make a model, and on the report form draw a picture of the model, making it three-dimensional to the best of your ability.

- **H_2O** water With the oxygen as the central atom, make a model of water and draw it. Attach to your water model two sticks to represent the two lone pairs of the oxygen atom, and draw the shape that includes the electrons. Note that these lone pairs will be extremely important in determining the solvent properties of water in biological systems. A picture of a ball and stick model of water is shown in Figure 9-1.

- **NH_3** ammonia Nitrogen is the central atom in this molecule, and each hydrogen atom is associated with it by a single bond. Note that the nitrogen atom has an unshared pair of electrons. Draw the model and then add to the completed model a stick that represents the lone pair. Which way does this pair point, relative to the geometry formed by the nitrogen-hydrogen bonds? This pair becomes important in the formation of the ammonium ion. Draw this model as well.

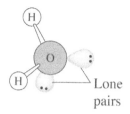

Figure 9-1. The shape of a water molecule.

- **NH₄⁺**   ammonium ion   If you have not already done so, attach to your ammonia model a stick representing the lone pair of electrons. Bring to the model an additional hydrogen ion model, consisting of a hydrogen atom model **without** its attached stick representing a bond. Remember that a hydrogen atom has one electron, but a hydrogen ion (what you have here with no bond attached at present) has no electrons, and has a positive charge. However, the ammonia nitrogen has a complete pair to work with and can use both of them to make an additional bond. When this is done, you have a model of an ammonium ion, still retaining the 1+ charge contributed by the hydrogen ion, but completely covalently bonded within itself as a polyatomic ion.
- **CH₂O**   formaldehyde   Use the carbon atom as the central atom, and attach two hydrogen atoms by single bonds and an oxygen atom by a double bond.
- **HCN**   hydrogen cyanide   The carbon atom is the central one, held to the hydrogen atom by a single bond and to the nitrogen atom by a triple bond. Note that the nitrogen has an unbonded or lone pair of electrons. Which way does the pair point?

Step 2. Describe in words the shape of each model and the types of bonding you observe (single, double, triple, coordinate). Add to your report any comments asked for as well.

Covalent bonds in complex carbon compounds

A carbon compound can contain many atoms, each with its own covalent geometry that contributes to the overall shape of the molecule. The six carbon compounds shown on the next page are all hydrocarbons; that means that they contain only carbon and hydrogen. In making the models of these compounds, remember that carbon can make four covalent bonds, and hydrogen can make only one. It is easiest to chain the carbon atom models together first, making any double or triple bonds as you go, and then add the hydrogen atom models.

For each model you make, draw the structure in three dimensions to the best of your ability, and note both the geometries of the individual bond areas and the overall shape of the molecule you are examining. Also draw the Lewis structures in the space provided.

Models to Make

Name	Molecular formula	Connectivity
Methane	CH_4	H | H−C−H | H
Butane	C_4H_8	H H H H | | | | H−C−C−C−C−H | | | | H H H H
Ethylene	C_2H_4	H H \ / C=C / \ H H
1-Butene	C_4H_8	H H \ / C=C / \ H H C−C−H \ \ H H H
Acetylene	C_2H_2	HC≡CH
1-Butyne	C_4H_6	H H | | H−C≡C−C−C−H | | H H

PART B. (ALTERNATE)

If you have molecular modeling software available in your laboratory, Part B can be done as a computer exercise with virtual models. Make or view computer-generated models of the compounds listed in Part B, and answer the questions concerning them as you would with ball and stick models. Don't forget to take the opportunity to move, turn, and view the models from all directions.

PART C. EXERCISES

Complete the exercises in the report section.

EXPERIMENT 9 PRELAB EXERCISES:

IONIC AND MOLECULAR COMPOUNDS

Name_____ Partner _____ Date _____

1. Draw electron dot structures (Lewis strictures) for lithium chloride, LiCl, and sulfur dioxide, SO_2.

2. Define differences between ionic and covalent bonds.

3. Would you expect magnesium chloride, $MgCl_2$, to be ionic or covalently bonded? Draw a Lewis structure for magnesium chloride.

4. Carbon tetrachloride, CCl_4, is a covalent molecule in which the carbon atom is joined to all four of the chlorine atoms. Draw a 3-dimensional picture of the carbon tetrachloride molecule.

5. For the following compounds, decide if each is ionic or covalent, and predict its solubility in water and the ability of a solution of it to conduct electricity.

Some Physical Properties of Several Compounds

Compound	Solubility in water	Conducts electricity?
Toluene, C_7H_8	_____	_____
Calcium chloride, $CaCl_2$	_____	_____
Potassium bromide, KBr	_____	_____

EXPERIMENT 9 REPORT: IONIC AND MOLECULAR COMPOUNDS

Name_____ Partner _____ Date _____

PART A. SOME PROPERTIES OF IONIC AND COVALENT COMPOUNDS

Compounds Investigated

Name	Ease of melting	Flammability	Solubility	Conductivity
Sodium chloride				
Benzoic acid				
Sucrose				

Can you make any generalized statements about some differences between covalent and ionic compounds?

PART B. SHAPES OF MOLECULES WITH COVALENT BONDS

Covalent Bonds in Simple Compounds

Name	Drawing	Lewis structure	Shape and bonding
Ammonia			
Ammonium ion			
Water			
Water with lone pairs			
Formaldehyde			
Hydrogen cyanide			

Covalent Bonds in Complex Carbon Compounds

Name	Drawing	Lewis structure	Shape of molecule
Methane			
Butane			
Ethylene			
1-Butene			
Acetylene			
1-Butyne			

PART C. EXERCISES

1. Sodium chloride is not very soluble in gasoline, but it is very soluble in water. Suggest an explanation.

2. Acetic acid, a covalently bonded compound, allows a very weak electric current to flow when tested with a conductivity apparatus. Most covalent compounds do not conduct electricity. Account for the observation.

EXPERIMENT 10

INORGANIC REACTIONS

GOALS

1. To synthesize and isolate one or more inorganic compounds
2. To examine the use of the mole concept in describing chemical reactions
3. To use the mole concept in balancing chemical equations
4. To make use of the mole concept in chemical calculations

INTRODUCTION

In this exercise, you will make and isolate two different ionic compounds. One is a simple binary compound, copper(II) oxide. You will take advantage of its relative insolubility in water, mixing a solution of a soluble salt of copper with a solution of sodium hydroxide to cause insoluble copper(II) hydroxide to precipitate from solution. This compound is easily decomposed with gentle heating to the very insoluble compound copper(II) oxide, which you will collect, dry, and weigh. You will then compare the quantity of product copper oxide you obtained to the amount you calculated you should have obtained (more on the calculations later). You might recognize these two reactions—you tried them out in Experiment 2.

As part of your calculations, you will need to learn to do a new type of molecular mass (sometimes called gram-formula weight) for a *hydrate*. Many compounds in crystalline form include specific numbers of water molecules in each formula unit. These types of crystals are called hydrates. The water has to be counted as part of the formula unit and as part of the gram-formula weight, but does not take part in any reaction that does not involve water. You will see in the following example that one reagent is solid copper sulfate pentahydrate, $CuSO_4 \cdot 5H_2O$, which has a gram-formula weight of 249.7 g/mol. This type of formula tells you that in the crystalline form, each copper sulfate formula unit has five water molecules associated with it. Notice that the dot in $CuSO_4 \cdot 5H_2O$ is not a multiplication sign, and does **not** mean multiply by the quantity (5 x 18). You **add the mass of** five molecules of water to the gram formula weight of $CuSO_4$ alone.

The second example you will look at is the preparation of calcium oxalate from separate solutions of sodium oxalate and calcium chloride in water. This one is actually somewhat simpler to perform because the calcium oxalate precipitates from the mixture as soon as you mix the two initial solutions. The equation for its formation looks a little more complex, though, because calcium ions have a charge of 2+, and oxalate ions are polyatomic ions having a charge of 2-, so getting the correct formulas for the starting materials has to take this into account. You won't be able to write a balanced equation for this reaction (or the first set either) unless you get the formulas for the starting materials and products correct. This is always the first step in balancing chemical equations.

In both of these reactions, since you are trying to get 100% of the possible product, the more you can force the reaction to go to completion, the better result you will get. One common way to do this is to be sure that one reagent is in excess (there is more used than needed in theory to perform the reaction), so that one of the reagents must be used up completely. The one chosen to use up completely is usually the more expensive or harder-to-get reagent.

You also have to assume that the product you want to isolate is not soluble in water, so that you can actually collect it all by filtration, and then wash it with extra water so you don't isolate a sample contaminated with the excess reagent. With the two compounds you are working with, this is a reasonable assumption; they are very insoluble in water, and you should be able to clean them up this way with no trouble. Other more soluble compounds require different approaches.

To decide how well your reaction actually worked you will need the balanced equation for the reaction you used to prepare the product compound. The example on the next page shows how you might use the equation and the masses of materials to find out how well your experiment worked, but you will have to perform your own calculations on copper oxide and calcium oxalate. You can use the sample calculation on the next page as a guide.

REFERENCES

Stoker, H. S. *General, Organic, and Biological Chemistry;* Cengage Learning; Chapters 6 and 8.
   Review topics:
   - Solubility rules
   - Common laboratory tools and how to use them
   - The mole and chemical calculations
   - Balancing ionic equations

WEB RESOURCES

Solubility Rules. *http://www.csudh.edu/oliver/chemdata/solrules.htm*

Filtration. *http://www.youtube.com/watch?v=P-UBuAFxJiA*

SAFETY NOTES

Sodium hydroxide is a caustic material, even in dilute solutions. Take special care when handling solutions made using this material. Follow your instructor's laboratory regulations about the use of chemical splash goggles and gloves; their use may be governed by state law in your area.

Materials from Parts A and B of this experiment may be discarded in the waste basket, in the sink with lots of rinse water, or as your instructor directs.

SAMPLE CALCULATIONS

In a precipitation experiment, a student mixed 1.5 g of copper sulfate pentahydrate ($CuSO_4 \cdot 5H_2O$, fw 249.7) with 50 mL of water. The student then made a second solution that contained 1.0 g of silver nitrate ($AgNO_3$, fw = 169.2) in 100 mL of water. The two solutions were mixed, and a precipitate of silver sulfate was obtained. After collecting the silver sulfate, the student found that the yield of silver sulfate was 0.75 g.

What is the balanced equation for this reaction?
    The charges on the ions are Cu^{2+}, Ag^{*}, SO_4^{2-}, NO_3^{2-}.
    The only possible reaction that makes use of these ions is the one below, and silver sulfate is the only insoluble material formed. Note that the hydrate water, $\cdot 5H_2O$, does not take part in the reaction.
    $$CuSO_4 + 2AgNO_3 \longrightarrow Cu(NO_3)_2 + Ag_2SO_4$$

Which starting material is in excess?
    To answer this question, you need to calculate how many moles of each starting material you have, and then refer to the balanced equation to decide which one is used up first.

$$1.5 \text{ g } CuSO_4\cdot 5H_2O \times \frac{1 \text{ mole } CuSO_4\cdot 5H_2O}{249.7 \text{ g } CuSO4\cdot 5H_2O} = 6.0 \times 10^{-3} \text{ mol } CuSO_4\cdot 5H_2O$$

$$1.0 \text{ g } AgNO_3 \times \frac{1 \text{ mol } AgNO_3}{169.2 \text{ g } AgNO_3} = 5.9 \times 10^{-3} \text{ mol } AgNO_3$$

    When you compare these numbers to the balanced equation, you see that to use up all the silver you would need half as many moles of copper sulfate, or about 2.5×10^{-3} moles. You have a lot more than that, so the copper sulfate is in excess, and the silver nitrate will be used up first. Silver nitrate is the limiting reagent, and this is the one you will base your theoretical yield calculation on.

What is the theoretical yield of the silver sulfate?
    A useful rule to remember is that you calculate in this order:
    Convert grams of starting material (silver nitrate in this case) to moles of starting material to moles of product to grams of product (silver sulfate) expected.

    Set up mathematically, it looks like this
$$1.0 \text{ g } AgNO_3 \times \frac{1 \text{ mol } AgNO_3}{169.2 \text{ g } AgNO_3} \times \frac{1 \text{ mol } Ag_2SO_4}{2 \text{ mol } AgNO_3} \times \frac{311.8 \text{ g } Ag_2SO_4}{1 \text{ mol } Ag_2SO_4} = 0.92 \text{ g } Ag_2SO_4$$

What is the percent yield for the reaction as actually done?
$$\frac{0.75 \text{ g}}{0.92 \text{ g}} \times 100\% = 82\%, \text{ to two significant figures}$$

EXPERIMENTAL PROCEDURE

PART A. PREPARATION OF COPPER(II) OXIDE

Step 1. You will need to prepare two solutions to make this compound. First, weigh out 1.70 g of copper(II) chloride dihydrate (fw = 170.48), and dissolve it in 50 mL water in a 400-mL beaker. Next, prepare a solution of 2.00 g of sodium hydroxide in 50 mL of water in a separate 250-mL beaker. Stir each mixture gently until the solids have completely dissolved.

Step 2. Pour the sodium hydroxide solution into the copper chloride solution, and stir to mix well. Allow the new mixture to stand, with occasional stirring, for 10 minutes or longer.

Step 3. Set up a heating apparatus, preferably a hot plate, and heat the mixture **gently** for a few minutes. Under the influence of heat, copper(II) hydroxide is converted to copper(II) oxide. The heating may be stopped when all traces of blue copper hydroxide have disappeared. If some crystals of copper(II) hydroxide appear to be clinging to the sides of the beaker, scrape them down into the solution, or rinse them down with a small amount of water.

> ♥ You do **not** want to boil this solution. Gentle heat is sufficient. If you heat it too vigorously, it may splash, and spatter hot sodium hydroxide about the room.

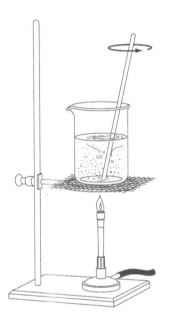

Figure 10-1. An apparatus for heating liquids with a burner.

Step 4. Weigh a piece of filter paper and record the weight; then prepare a gravity filtration setup.

Step 5. Filter the solid copper oxide using good laboratory technique. Remember that for a gravity filtration, the procedure works most conveniently if you filter most of the liquid through the paper first and then scrape the solid onto the filter paper. After most of the solid is on the filter paper, use lots of water to rinse the original beaker, the filter paper, and the solid copper oxide.

> ① Fortunately, copper oxide is not very soluble in water and you probably won't rinse much of your product away with this step.

Step 6. When you are confident that you have rinsed your product completely, place the filter paper with the solid on a dry watch glass, and place this in an oven to dry. Be sure you have identified yours in some way; all the other students will have a product that looks like yours, too!

While your copper(II) oxide product is drying, perhaps for as long as half an hour depending on the temperature of the oven, go on to Part B, and make a sample of calcium oxalate.

Step 7. When your compound and the filter paper are dry, record the combined weights, calculate the mass of copper oxide made, and complete the calculations for Part A. When you are completely finished with all calculations, dispose of your product as your instructor directs.

PART B. PREPARATION OF CALCIUM OXALATE

Step 1. Weigh out 5.0 g of sodium oxalate, $Na_2C_2O_4$, and dissolve in 50 mL water. In a separate beaker, place 1.1 g of calcium chloride and dissolve it in 50 mL of water.

Step 2. Pour the sodium oxalate into the calcium chloride solution, and stir well. Allow the mixture to stand for about 15 minutes, stirring occasionally, then heat the mixture gently (**not** to boiling) for a few minutes, preferably on a hot plate. Allow the solution to cool a bit before going on to the next step.

Step 3. Mark a piece of filter paper with your initials in pencil, and then weigh it. Record the weight, and then filter the solid calcium oxalate, rinsing the beaker and the filter paper well. Remember, in this exercise you want to obtain all the final product possible.

Step 4. Dry the solid calcium oxalate on a watch glass in an oven, and work on the calculations for this experiment while this and the copper(II) oxide are drying.

Answer the questions in the indicated places on the report form. Be sure to show the setup of any calculations needed, as well as your final answer.

EXPERIMENT 10 PRELAB EXERCISES:

INORGANIC REACTIONS

Name_____   Partner: _____   Date _____

In a precipitation experiment, a student mixed 2.0 g of magnesium sulfate heptahydrate ($MgSO_4\cdot7H_2O$) with 50 mL of water. The student then made a second solution, containing 1.0 g of silver nitrate ($AgNO_3$, g-fw = 169.2) in 100 mL of water. After the solutions were mixed, a precipitate of silver sulfate was obtained. After collecting and drying the silver sulfate, the student found that the yield of silver sulfate was 0.75 g.

1. What is the formula weight of magnesium sulfate heptahydrate?

2. What is the balanced equation for this reaction?

3. Which starting material is in excess?

4. What is the theoretical yield of the silver sulfate?

5. What is the percent yield for the reaction as actually done?

EXPERIMENT 10 REPORT: INORGANIC REACTIONS

Name_____   Partner: _____   Date _____

PART A. PREPARATION OF COPPER(II) OXIDE

Mass of copper chloride dihydrate used   _____

Mass of sodium hydroxide used   _____

Mass of the original filter paper   _____

Mass of the filter paper and product   _____

Mass of the product   _____

How many moles of each reagent were used?

    copper chloride   _____

    sodium hydroxide   _____

What is the balanced equation for the reaction?

Which reagent was the limiting reagent?   _____

What is the theoretical yield of the product in moles?   _____

What is the theoretical yield of the product in grams?   _____

What was your actual yield of product in grams?   _____

What is your percent yield for this reaction?   _____

If your actual yield was not between 95% and 105% of the theoretical amount, account for some of the probable sources for difficulties in this experiment.

PART B. PREPARATION OF CALCIUM OXALATE

Mass of sodium oxalate used   _____

Mass of calcium chloride used   _____

Mass of the original filter paper   _____

Mass of the filter paper and product   _____

Mass of the product   _____

How many moles of each reagent were used?

    sodium oxalate   _____

    calcium chloride   _____

What is the balanced equation for the reaction?

Which reagent was the limiting reagent?   _____

What is the theoretical yield of the product in moles?   _____

What is the theoretical yield of the product in grams?   _____

What was your actual yield of product in grams?   _____

What is your percent yield for this reaction?   _____

If your actual yield was not between 95% and 105% of the theoretical amount, account for some of the probable sources for difficulties in this experiment.

EXPERIMENT 11

EMPIRICAL FORMULA OF A COMPOUND

GOALS

1. To practice the experimental determination of a chemical formula
2. To use the mole concept in chemical calculations
3. To apply chemical calculations to experimental data

INTRODUCTION

Chemists have been investigating the behavior of matter for hundreds of years. Because of that, we already know many things that we do not have to deduce again from first principles, such as the structure of the atom, the charges that common ions have, the most likely formulas for chemical compounds, and the like. However, we can still use experimental methods to determine such things in order to illustrate how science is done, and to show the power of the mole concept in predicting the behavior of chemicals.

The focus of this experiment is the determination of the correct formula for a metal oxide when the only data you have are the beginning weight of the metal and the final mass of its oxide. These data, and some calculations using the idea of moles of chemicals reacting, are all that are needed to determine the correct formula. In this case, you have two compounds to examine in two separate experiments—the oxide formed from magnesium and oxygen and the oxide formed from copper and oxygen.

Techniques for studying the two compounds are different. Magnesium oxide is readily formed by heating a piece of magnesium in the presence of air. At first, some of the magnesium may be converted to magnesium nitride because of magnesium reacting with nitrogen from the air, but a second heating in the presence of water will convert this compound to the magnesium oxide you need.

The reactions the magnesium undergoes are these:

$$2Mg_{(s)} + O_{2(g)} \longrightarrow 2MgO_{(s)}$$
$$3Mg_{(s)} + N_{2(g)} \longrightarrow Mg_3N_{2(s)}$$
$$Mg_3N_{2(s)} + 3H_2O \longrightarrow 3MgO_{(s)} + 2NH_{3(g)}$$

The weight of the oxide after heating must include the weight of all the original magnesium and the weight of any oxygen with which it has combined. The masses of magnesium and oxygen in the magnesium oxide formed can then be calculated, and when converted to moles show the actual ratio of atoms of the two in the compound formula. Use the stepwise report sheet to help you figure out the calculations; there are enough data given with the appropriate lines on the report to help you calculate the values needed.

Copper oxide requires a different approach. It is most convenient to "digest" a weighed piece of copper in nitric acid to force it into solution. Then you allow the copper ions to react with hydroxide ions, forming the insoluble copper hydroxide. Heating the copper hydroxide converts it to copper oxide. You can collect the copper oxide, dry it, weigh it, and use this information to discover the formula of the compound. Does all this sound familiar? This reaction has been used in two other experiments in this book, but this time you are looking at quantitative relationships within the compound itself. Again, take the report form in a stepwise fashion. If you work methodically through it, each calculation should follow from the one before it until you are able to deduce the formula for copper oxide.

REFERENCES

Stoker, H. S. *General, Organic, and Biological Chemistry;* Cengage Learning; Chapter 6.
    Review topics
    - Balancing chemical equations
    - Percent composition of a compound
    - The mole and chemical calculations
    - Percent yield
    - Limiting reagents

WEB RESOURCES

Stoichiometry. *http://www.brightstorm.com/science/chemistry/chemical-reactions/stoichiometry*

Stoichiometry and Reactions. *http://chemistry.about.com/od/stoichiometry/Stoichiometry.htm*

General Chemistry/Stoichiometry. *http://en.wikibooks.org/wiki/General_Chemistry/Stoichiometry*

SAFETY NOTES

The usual precautions for handling hot materials apply in this experiment. In addition, take care not to look directly at the magnesium if it flares brightly, because the light can be harmful to the eyes. Wear chemical splash goggles at all times while working on Part A.

    For Part B, you will be handling concentrated nitric acid and sodium hydroxide solutions, both very corrosive materials. You must wear chemical splash goggles throughout this experiment, and gloves are recommended.

EXPERIMENTAL PROCEDURE

PART A. THE FORMULA OF MAGNESIUM OXIDE

Step 1. Wash and dry a crucible and its cover. Set up a ring stand with ring and clay triangle as shown in Figure 11-1, and adjust the height of the ring so that the crucible, when set in the ring, can be heated strongly by the hottest part of the Bunsen burner flame. Set the crucible on the clay triangle, and set the lid on the crucible, but slightly ajar.

> ⓘ The lid is unstable in this position. Be careful not to drop and break it or you will need to start over!

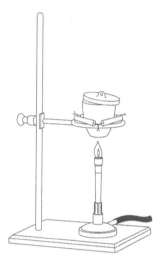

Figure 11-1. The apparatus for heating a crucible.

Step 2. Heat the crucible, at first gently, then strongly for at least 5 minutes. Turn off the burner and allow the crucible to cool to room temperature. After this point, do not pick up the crucible with your fingers. Always use the tongs, even after it is cool. If you need to decide if the crucible and lid have cooled completely to room temperature, touch a **knuckle only** to the side of the crucible to check.

Step 3. When the crucible has reached room temperature, weigh it, together with the lid, and record the weight.

Step 4. Select a piece of magnesium, polish it with steel wool, and roll it into a loose roll that will fit in the bottom of the crucible.

Step 5. Place the magnesium in the crucible, cover it with the crucible lid, and weigh the crucible with the magnesium.

Step 6. Heat the crucible again, at first gently and then strongly. You would like the magnesium to heat to a dull red heat as it turns slowly to ash while combining with oxygen in the air. When the conversion of the magnesium to ash appears to be complete, turn off the burner and allow the crucible and its contents to cool to room temperature.

> ⓘ You would rather the magnesium did **not** catch fire, as flyaway ash can cause your results to be incorrect.

Step 7. Add to the crucible 5 drops of water, and allow it to soak the ash in the crucible. Heat the crucible and its contents again for 15 minutes, first with gentle heat and then more strongly. Cool again to room temperature.

Step 8. Determine the mass of the crucible and its contents and the final mass of magnesium oxide made. Use the starting and ending masses to determine the simplest formula for magnesium oxide. Refer to the report form for help with the calculations.

PART B. THE FORMULA OF COPPER OXIDE

The first part of this experiment must be done in the hood!

Step 1. Polish a piece of copper wire with steel wool, and then weigh it. Place it in a 100-mL beaker.

Step 2. Take the sample to the hood, and add to it 5 mL of concentrated nitric acid. You will immediately see a brown gas start to form. This is nitric oxide, NO_2. Keep the sample in the hood until the formation of nitric acid has stopped and the solution is a deep blue (**not** green, which means there is still some nitric oxide dissolved in it).

> ⓘ ♥ **Really** keep it in the hood—remember Experiment 2?

Step 3. Add 20 mL of 6 M (this is a concentration unit) sodium hydroxide solution to this solution and stir it well. You should see a precipitate at this point. The solution may be removed from the hood now.

Step 4. Place the solution on a warm hot plate, and stir it with heating for a few minutes until the light blue precipitate has been converted completely to a black granular precipitate. This is the copper oxide you are trying to make.

> ⓘ ♥ Don't let the solution boil or it may spatter hot lye and your copper oxide about the room!

Step 5. Place an identifying mark on a piece of filter paper and weigh it. Filter the solution, doing your best to put all the black solid on the filter paper and leave none of it in the beaker. Rinse the beaker and the filter paper as many times as necessary to make the transfer complete and leave no residue of unused sodium hydroxide solution in the filter paper.

Step 6. Dry the sample on a watch glass until it is completely dry, and weigh it again.

Step 7. From the combined mass of paper and solid, find the mass of copper oxide you have made.

Step 8. Use the initial weight of copper and the final weight of copper oxide to determine the formula of copper oxide.

PART C. EXERCISES

Complete the exercises on the report form.

EXPERIMENT 11 PRELAB EXERCISES:

EMPIRICAL FORMULA OF A COMPOUND

Name_____ **Partner** _____ **Date** _____

A sample of calcium hydroxide, $Ca(OH)_2$, was weighed and found to have a mass of 0.5000 g. In an effort to discover the formula of calcium oxide, this sample was converted to calcium oxide by heating. The sample of calcium hydroxide obtained had a mass of 0.3772 g. Follow the steps below to determine the formula of calcium oxide.

1. Find the number of moles of calcium hydroxide in the initial sample.

2. How many moles of calcium ions were in this initial sample? How many millimoles?

3. How many grams of calcium ions were in the initial sample?

4. How many mmoles of calcium ions were in the final sample?

5. How many grams of calcium ions were in the final sample?

6. How many grams of oxygen were in the final sample?

7. How many mmoles of oxygen were in the final sample?

8. What is the exact ratio of mmoles of calcium to mmoles of oxygen in the final sample?

9. To the nearest whole numbers, what is this ratio?

10. The correct formula for calcium oxide is _____ .

EXPERIMENT 11 REPORT: EMPIRICAL FORMULA OF A COMPOUND

Name_____ **Partner** _____ **Date** _____

PART A. THE FORMULA OF MAGNESIUM OXIDE

Raw data

Mass of crucible, lid and magnesium _____

Mass of crucible and lid _____

Mass of magnesium before the reaction _____

Mass of crucible and contents after heating _____

Mass of magnesium oxide formed _____

Mass of oxygen that combined with Mg _____

Calculations

Moles of magnesium used (from mass of Mg at beginning) _____

Moles of oxygen that combined
(from mass of oxygen combined with Mg) _____

Ratio of moles of magnesium to moles of oxygen
(use appropriate significant figures) _____

Closest whole number ratio _____

Best formula for magnesium oxide (using your data) _____

1. If a student loses magnesium oxide ash during the heating process, what effect will that have on the calculated ratio of magnesium to oxygen?

2. Write a balanced equation for the formation of magnesium nitride from magnesium and nitrogen gas.

3. Write a balanced equation for the formation of magnesium oxide from magnesium and oxygen.

PART B. THE FORMULA OF COPPER OXIDE

Raw data

Mass of copper used   _____

Mass of filter paper used   _____

Mass of filter paper plus copper oxide   _____

Mass of copper oxide   _____

Mass of oxygen combined   _____

Calculations

Moles of copper at start (from mass used)   _____

Moles of oxygen combined (from mass of oxygen combined)   _____

Ratio of copper moles used to oxygen moles used
(use correct significant figures)   _____

Closest whole number ratio   _____

Best formula for copper oxide, using your data   _____

PART C. EXERCISES

1. If a student does not completely dry the copper oxide and filter paper, what effect will that have on the calculated ratio of copper to oxygen?

2. Write a balanced equation for the formation of copper(II) hydroxide from copper(II) nitrate and sodium hydroxide, using a metathesis (double replacement) reaction.

3. Write a balanced equation for the formation of copper(II) oxide from copper(II) hydroxide by a decomposition reaction.

EXPERIMENT 12

EQUATION PRACTICE

GOALS

1. To distinguish various ways to classify chemical reactions
2. To balance simple chemical equations
3. To identify redox reactions and the substances oxidized and reduced
4. To identify reducing and oxidizing agents in a chemical reaction

INTRODUCTION

As students begin their study of chemistry, one of the most frustrating hurdles is writing chemical formulas and using them to write balanced equations. Nevertheless, the concept of a balanced equation is extremely important in understanding chemistry. A balanced equation contains a lot of information about a reaction.

Consider the following reaction.

$$2AgNO_3 + K_2Cr_2O_7 \longrightarrow \underline{Ag_2Cr_2O_7} + 2KNO_3$$

It tells you that silver nitrate and potassium dichromate react to make silver dichromate and potassium nitrate. It also tells you that silver dichromate is a solid (underlined product), and that it takes two moles of silver nitrate but only one mole of potassium dichromate to make one mole of silver dichromate. You can also calculate the formula weight of each reactant and product, which can be useful. With some additional information, you can determine that the charge on a silver ion is 1+, and the charge on a dichromate ion is 2+.

The rules for balancing equations are easy to memorize, but application of them takes practice. For this exercise, you will not need to mix any chemicals. Instead, you will concentrate on developing correct equations and predicting the identity of any substance you might observe when you carry out the experiment. Some solutions are available for you to use to try out the reactions for some of the equations you are working with, but you may be able to answer the questions in this exercise without doing anything with the solutions.

Rules governing classification of chemical reactions are rarely black-and-white rules. They are at best guidelines for classifying reactions, and many reactions fit more than one classification depending on what aspect of the reaction you would like to emphasize. Still, for a beginning, the following classifications of chemical reactions are useful.

Combination reactions. A combination reaction is just what it sounds like: two or more substances form new bonds to make a single new substance. Look for this pattern.

$$A + B \longrightarrow C$$

$$2S + 3O_2 \longrightarrow 2SO_3$$

Decomposition reactions. These are the opposite of combinations. Look for a single reactant that falls apart into two or more molecules. Carbonic acid is an example. It is a product of the reaction of carbonate salts with acids, but is not very stable in water; most of it decomposes to carbon dioxide and water, with bubbling of the solution (or solid) as carbon dioxide gas is formed.

$$H_2CO_3 \longrightarrow H_2O + CO_2$$

Single replacement reactions. These are common reactions in which an element reacts with a compound. The final products are a different element and a new compound. The sample pattern to keep in mind is the reaction of aluminum with nickel sulfate.

$$2Al + 3NiSO_4 \longrightarrow Al_2(SO_4)_3 + \underline{3Ni}$$

Double replacement reactions. These reactions happen between two compounds as in the reaction between hydrochloric acid and sodium hydroxide. The positive and negative ions trade partners, and then coefficients can be added to balance the equation. In this particular equation, all the coefficients are 1.

$$HCl + NaOH \longrightarrow NaCl + H_2O$$

Combustion reactions. Combustion reactions are a special example of oxidation-reduction reactions. They are reactions that usually occur using oxygen from the air and that produce heat and light as well as one or more new products.

$$C + O_2 \longrightarrow CO_2$$

Oxidation-reduction (redox) reactions. These reactions occur when an element or ion loses electrons during the reaction (is oxidized) or gains electrons during a reaction (is reduced). The reaction of iron with sulfur to make iron sulfide is a redox reaction. In this reaction, the iron goes from an oxidation state of 0 to on oxidation state of 2+, so it is "oxidized". But those electrons had to go somewhere, and they went to the sulfur, which went from an oxidation state of 0 (in the form of the element) to an oxidation state of 2-. Oxidation and reduction must occur in the same reaction, because the electrons you take from one atom must end up in another atom taking part in the reaction.

$$Fe + S \longrightarrow \underline{FeS}$$

It is also often useful to note that the element that is oxidized is a "reducing agent" because it gives its electrons to something else, while the element that is reduced is an "oxidizing agent" that takes electrons away from some other element.

Perhaps the most <u>general</u> distinction you can make is "redox or non-redox?" However, the most <u>convenient</u> classification depends on what you are studying.

One set of rules for balancing simple reactions is listed here. For more complex reactions (which usually turn out to be some type of redox reaction) there are more complicated rules. This set should get you through the concepts you need to know at present.

Rules for balancing ionic equations for reactions in solution:

1. Write the names of the possible reactants and the possible products. In this exercise, this is already done for you.
2. For each compound, determine the correct formula, including all correct subscripts.

3. Insert coefficients in front of the formulas that will allow the same number of atoms of each kind to be on both sides of the equation.

4. Decide if an observable reaction takes place, considering what you know about solubility rules for ionic compounds and gas formation in chemical reactions. If a visible reaction takes place, then a solid, liquid or gas product should be indicated. In these cases, reaction <u>does</u> occur. In any other cases, where only ions in solution are the potential products, then a "<u>No Reaction</u>" statement is correct.

ⓘ In this step, do not change the subscripts already determined in Step 2, or add extra reagents or products.

ⓘ In this exercise, **something** happens in each example. There are no "No Reaction" combinations.

So how do you decide when a solid is formed? It will depend on the solubility of the potential products in water. If the products are sodium nitrate and potassium chloride, for instance, both are soluble in water, and you cannot expect to see any precipitate from the reaction. General rules for the solubility of ionic compounds can be stated in many ways and have a number of exceptions. The set of rules below is a good general guideline, <u>but is not a complete list</u>. If you are uncertain about the solubility of some ionic compound, it is best to refer to a chemistry reference book such as the *Handbook of Chemistry and Physics* for exact data.

Solubility rules

Rule 1. Ionic compounds of all Group IA (Group 1) elements and all ammonium compounds are soluble in water (Na^+, K^+, NH_4^+, etc, with appropriate anions).

Rule 2. Nitrates, (NO_3^-), acetates ($C_2H_3O_2^-$), and chlorides (Cl^-) are usually soluble in water. Other halides (F^-, Br^- and I^-) are also usually soluble. Important exceptions: halides of silver, lead, and mercury are usually <u>insoluble</u>.

Rule 3. Sulfates (SO_4^{2-}) and hydroxides (OH^-) are usually soluble <u>except for</u> the sulfates and hydroxides of barium, strontium, lead, silver, and mercury.

Rule 4. Carbonates (CO_3^{2-}), hydroxides (OH^-), and phosphates (PO_4^{3-}) not covered by Rule 1 are usually insoluble.

REFERENCES

Stoker, H. S. *General, Organic, and Biological Chemistry;* Cengage Learning; Chapter 9.
Review topics:
- Types of chemical reactions
- Balancing equations
- Solubility rules

WEB RESOURCES

CHEMTUTOR REACTIONS. *http://www.chemtutor.com/react.htm*

How can I balance an equation? *http://misterguch.brinkster.net/eqnbalance.html*

If these sites don't help you, try your own search. There are thousands of sites that offer help on balancing chemical equations.

SAFETY NOTES

The paper portions of this exercise can be completed without wearing chemical splash goggles if done outside the laboratory setting, but any use of chemicals or glassware will require the usual chemical splash goggles and (possibly) gloves.

Some of the solutions you have available to work with contain heavy metals, and heavy metals should never be poured down the sink. Solutions containing heavy metals must be collected and disposed of by a special procedure. Follow your instructor's directions for disposal of any liquids used in this exercise.

EXPERIMENTAL PROCEDURE

PART A. BALANCING EQUATIONS

On the report sheet, you will find word equations for chemical reactions classified in different ways. For each equation, do the following:

- Balance each equation.
- If an insoluble solid is produced in the reaction, underline the formula.
- If a gas is produced in the reaction, write an upward-pointing arrow after its formula.
- If the reaction is a redox reaction, write "redox" under the completed equation.
- If the reaction is a redox reaction, identify the oxidized material, the reduced material, the oxidizing agent, and the reducing agent. Use the following pattern for your answer so it will be easier for your instructor to follow your thinking!

$$\text{REDOX} \quad \underset{\substack{\text{oxidized} \\ \text{(reducing} \\ \text{agent)}}}{\text{Fe}} \quad + \quad \underset{\substack{\text{reduced} \\ \text{(oxidizing} \\ \text{agent)}}}{\text{S}} \quad \longrightarrow \quad \underline{\text{FeS}}$$

PART B. QUESTIONS ABOUT THE REACTIONS

On the report sheet are five questions about the reactions you balanced in Part A. You may need to perform the actual reactions to determine the answers; chemicals for these tests are available.

For any chemical test you wish to perform, use minimum amounts of material. If you are mixing two liquids, a few drops of each in a small test tube or a spot plate well will be enough. If you are mixing a solid with a liquid, a small piece of the solid with a small amount of the liquid is adequate.

> ❤ You must wear chemical splash goggles for any chemical tests you do!

EXPERIMENT 12 PRELAB EXERCISES:

EQUATION PRACTICE

Name_____ Partner _____ Date _____

1. For the following chemicals, write the correct formulas.

    sodium chloride   _____

    barium chloride   _____

    zinc sulfate   _____

    copper(II) hydroxide   _____

    potassium oxide   _____

2. State whether each of the following chemicals is soluble or insoluble in water.

    potassium sulfate   _____

    calcium carbonate   _____

    copper(II) oxide   _____

    silver nitrate   _____

    silver chloride   _____

3. Identify the following reactions by type.

    The production of copper oxide and water from copper hydroxide

    The production of magnesium sulfate and hydrogen from sulfuric acid and magnesium metal

    The production of sodium chloride and water from hydrochloric acid and sodium hydroxide

4. Write one example of a balanced oxidation–reduction reaction here. Use an example <u>other than</u> the reaction between iron and sulfur.

EXPERIMENT 12 REPORT: EQUATION PRACTICE

Name_____ **Partner** _____ **Date** _____

PART A. BALANCING EQUATIONS

Combination reactions

sulfur plus oxygen yields sulfur trioxide

carbon plus oxygen yields carbon dioxide

silver plus sulfur yields silver sulfide

sulfur dioxide plus water gives sulfuric acid

Decomposition reactions

sodium carbonate in the presence of heat produces carbon dioxide and sodium oxide

carbonic acid produces carbon dioxide and water

ammonium hydroxide produces ammonia and water

potassium chlorate ($KClO_3$) plus heat gives potassium chloride and oxygen

Single replacement reactions

copper plus silver nitrate gives silver plus copper nitrate

copper(II) oxide plus hydrogen gives copper and water

zinc plus hydrochloric acid gives zinc chloride and hydrogen

iron(III) oxide plus aluminum gives aluminum oxide and iron

chlorine plus potassium bromide gives bromine plus potassium chloride

Double replacement reactions

    sodium hydroxide plus sulfuric acid gives sodium sulfate and water

    copper(II) sulfate plus sodium hydroxide gives copper(II) hydroxide plus sodium sulfate

    lead chloride plus potassium iodide gives lead iodide plus potassium chloride

    potassium carbonate plus sulfuric acid produces carbon dioxide, potassium sulfate, and water

    barium chloride plus sodium sulfate gives barium sulfate plus sodium chloride

Combustion reactions

    methane plus oxygen gives carbon dioxide and water

    propane (C_3H_8) plus oxygen gives carbon dioxide and water

Oxidation-reduction reactions

    magnesium plus nitrogen in the presence of heat gives magnesium nitride

    zinc plus copper sulfate gives zinc sulfate plus copper

    copper plus silver nitrate gives silver plus copper nitrate

PART B. QUESTIONS ABOUT CHEMICAL REACTIONS

1. What color is lead iodide?

2. Is barium sulfate soluble in water?

3. What substance is formed on the surface of copper metal when it is placed in a solution of silver nitrate?

4. Is the reaction of sodium hydroxide with sulfuric acid exothermic or endothermic?

EXPERIMENT 13

GAS LAW EXPERIMENTS

GOALS

1. To develop an appreciation of how gas volume varies with temperature
2. To develop an understanding of how gas volume varies with pressure
3. To learn how to represent quantitative experimental data on a graph
4. To understand the contribution of individual gases to the overall pressure in a mixture
5. To understand how to calculate these relationships

INTRODUCTION

The overall mathematical relationship that describes the behavior of gases relates a gas sample's volume, temperature, pressure, and number of moles as shown below. This relationship is known as the ideal gas law.

$$PV = nRT$$

For any given gas sample,

P = the pressure exerted by the gas molecules on the surroundings, usually a container of some kind
V = the volume of the container
n = the number of moles of gas molecules
R = the gas constant, determined experimentally (now well known, so it can be looked up)
T = the temperature of the gas, always measured in degrees Kelvin.

For practical measurement of gas behavior, various methods of controlling variables can be used so that only a single relationship at a time need be investigated. It can be shown mathematically, for instance, that the ideal gas law can be used to obtain this relationship.

$$\frac{P_1V_1}{T_1} = \frac{P_2V_2}{T_2}$$

In this "combined gas law" form it predicts the variations among three variables in "before" and "after" conditions. This equation describes two sets of conditions, Set 1 and Set 2, and the relationship between them. However, experiments with three variables all changing at the same time are difficult to do; it's easier to see what's happening if you hold one constant and look at the effects when two of the three change. In this experiment, you will do just that, investigating the relationship between pressure and volume when temperature is held constant, and then the relationship between volume and temperature when the pressure is held constant.

However, if you are working with a mixture such as air or one that includes anesthetic gases, for instance, sometimes you need to know what part of the pressure (partial pressure) is exerted by each gas in the mixture. This isn't covered by the ideal gas law. Another part of this exercise will allow you to use some given experimental data to investigate partial pressure.

Boyle's law: the relationships between pressure and volume in a gas sample

Four variables can affect the behavior of a gas: the number of moles of gas present (how many molecules in the sample), the temperature, the pressure (<u>on</u> the sample, exerted from the outside, or <u>in</u> the sample, exerted by the sample) and the volume of the sample. If any two of these factors are held constant, a simple relationship, called Boyle's law, can be found between the other two.

If the quantity of gas (number of moles, n) is held constant and so is the temperature of the gas, then P (pressure) and V (volume) will vary inversely with one another. This means that the higher the pressure on the sample, the smaller the volume of the sample. Mathematically, Boyle's law is as follows:

$$P_1V_1 = P_2V_2 \qquad \begin{array}{l} P_1, V_1 \text{ are the initial conditions} \\ P_2, V_2 \text{ are the final conditions} \end{array}$$

If three of these are known, the fourth can be calculated. Or, as in your experiment, if one variable is changed in a measured manner, you can see its effect on the other. You will have a closed container to work with, and will change the pressure on it while watching what happens to the volume of the container as the pressure changes.

You will use air as the gas to study, and the variable container of your study sample will be a large plastic syringe with the needle end blocked so the syringe can be made airtight. Air is not really an "ideal" gas, but its variations from ideality are not readily observed in this experiment; much more accurate observations must be made for the discrepancies to appear. For this experiment, the two factors in the ideal gas law that are held constant are n and T, the quantity of gas and the temperature of the gas. Pressure will be exerted on the gas sample by paving bricks piled onto the plunger one at a time.

Once you obtain the data, you will have to graph the relationship. You'll see more on that later.

Charles's law: the variation of a gas volume with its temperature

Charles's law concerns the relationship between the temperature of a gas sample and its volume; this time n and P (atmospheric pressure in this case) of the ideal gas law are kept constant. Again, your variable-size container is a large syringe. You will find that volume is directly proportional to temperature—the higher the temperature the bigger the volume of the sample. If two specific temperatures, T_1 and T_2, are considered, the volumes V_1 and V_2 are related by this form of Charles's law:

$$\frac{V_1}{T_1} = \frac{V_2}{T_2}$$

Your graph of T vs V should yield a straight line of positive slope. It will use only a small part of the full Kelvin temperature scale, but if your data are good you can use the graph to estimate absolute zero or, perhaps, the volume the air sample would reach at 500 K.

Dalton's Law: the partial pressure of oxygen and nitrogen in air

When you work with gases it is frequently necessary to work with a mixture of gases rather than a sample of a single gas. In such cases it is critical to know how to calculate the contributing pressure from each component of the mixture. Dalton's law, the law of partial pressures, assumes your gas sample behaves like an ideal gas, but does not include calculations based on the ideal gas law; the partial pressure relationship requires addition, not multiplication and division. It shows that in any given mixture of gases the total pressure is equal to the sum of the pressures exerted by all the individual components of the mixture.

$$P_T = P_A + P_B + P_C + \ldots$$

Suppose you are working with a mixture of oxygen and helium at a total pressure of 500 torr. If the oxygen pressure is at 100 torr, the helium pressure can be found from this equation.

$$P_T = P_{O_2} + P_{He}$$
$$P_{He} = P_T - P_{O_2} = 500 \text{ torr} - 100 \text{ torr}$$
$$P_{He} = 400 \text{ torr}$$

If you know the total pressure and the percent composition of the mixture, you can also find partial pressures. Suppose for your He-O$_2$ mixture you knew only that it was 20% oxygen, and the total pressure was 500 torr.

$$20\% \text{ of } 500 \text{ torr} = .20 \times 500 \text{ torr} = 100 \text{ torr } P_{O_2}$$
$$80\% \text{ of } 500 \text{ torr} = .80 \times 500 \text{ torr} = 400 \text{ torr } P_{He}$$

A much more convenient mixture of gases to work with is air, a mixture of nitrogen (78%), oxygen (21%), carbon dioxide (.03%), water vapor (variable quantities) and other gases (0.9%). During the course of this experiment you will use some given measurements on air samples to find the concentrations of O$_2$ and CO$_2$. You can calculate the nitrogen concentration then, if you assume that the concentration of water equals its vapor pressure at the temperature of the experiment, and that the concentration of all other gases is about 0.9% of the total exclusive of water.

$$P_{atm} = P_{O_2} + P_{N_2} + P_{CO_2} + P_{H_2O} + P_{other\ gases}$$
$$P_{N_2} = P_{atm} - P_{O_2} - P_{CO_2} - P_{H_2O} - P_{other\ gases}$$

If $P_{other\ gases}$ = 0.9% of all the gases except water, then you can rearrange the equation to this:

$$P_{N_2} = P_{atm} - P_{O_2} - P_{CO_2} - P_{H_2O} - .009(P_{atm} - P_{H_2O})$$

P_{atm} is the barometric pressure for the day the experiment is done, and P_{H_2O} is the vapor pressure of water at the temperature of the experiment; that is a number that can be looked up.

Using graphs to explore experimental data

A reminder about correct graphing procedures is in order. A correctly constructed graph always has a title, so you can readily pick the one of interest from a pile of graphs that all look more or less alike. The date the experiment was done should also be included. Complete labeling of both the x-axis and the y-axis is important; this includes both the name of the variable (in this case P, T, or V) and the units in which it was measured (atmospheres, degrees, bricks, mL, etc.). In general, the x-axis should show the independent variable, the one that you are deliberately varying, and the y-axis should show the dependent variable, the one changed by the carefully measured change in x. For these two experiments, pressure and temperature are the independent variables and volume the dependent one. The names of the experimenters should also be shown on each graph.

All this information reminds you how to label your graph correctly. There are a few other considerations.

Size of the graph can be important. In general, you should construct a graph so that it occupies at least half a page of paper, and a full page is better, because if it is too small, variability in the data is too hard to see. Of course if it's too big, you waste paper, so you need to compromise a bit—a half to one page per graph is about right.

Drawing the line showing the relationship requires judgment. Experimental data are rarely perfect because minor variations in experimental conditions can make a mathematically straight relationship look like a zig-zag line. This is not what you want to show, however. When judging the line for a graph of experimental data, you are going to try to draw a **smooth** line that comes as close as possible (makes a "best fit") to the data points you have. Possibly only a few points will be "on the line," but the smooth variation of the dependent variable with the independent variable will be clear. So watch for that when you graph your experimentally obtained data! And note that relationships between physical quantities do not always yield straight lines—sometimes they are curved lines.

Reproducible "smooth curves" can be estimated this way, or calculated using a computer program. Follow your instructor's directions on this matter.

REFERENCES

Stoker, H. S. *General, Organic, and Biological Chemistry;* Cengage Learning; Chapter 7.
Review topics:
- Kinetic Molecular Theory
- Gas laws: ideal gas law, Boyle's law, Charles's law, Dalton's law, combined law
- Respiration
- Graphing experimental data (refer if necessary to a book on linear algebra for a review)

WEB RESOURCES

Graphing Experimental Data. *http://www.maths.mq.edu.au/numeracy/tutorial/graph.htm*

lesson 1-graphing. *http://www.wellesley.edu/Chemistry/stats/lesson1.html*

CHEMTUTOR GASES. *http://www.chemtutor.com/gases.htm*

Pharmacology II Inhalation Anesthesia-page 1 of 2.
http://mna2001.tripod.com/pharmacology/pharm2/exam1/pharm2.inhalation.1of2.htm

Physics of Diving. *http://www.scuba-doc.com/physicsdive.html*

SAFETY NOTES

For Part A of this experiment, the primary concern is dropping a pile of bricks on the benchtop or on your toes. Depending on the type of bricks available, flying chips also might be a problem in the event you drop something, so be careful to support the bricks you use. You must wear eye protection during this exercise. For Part B, boiling water may splash if sufficient boiling chips are not used.

EXPERIMENTAL PROCEDURE

PART A. INVESTIGATION OF BOYLE'S LAW

Step 1. Obtain a prepared 50-mL syringe, five to seven bricks, and a heavy wire from the general supplies. Check the syringe to be sure the needle end is closed and then insert the plunger. You will find that the air in the syringe resists this, so follow the directions in Step 2 to set the plunger at the right place.

Step 2. Position the syringe plunger at 50.0 mL, at atmospheric pressure. To do this, lay a copper wire alongside the plunger while the plunger is out of the barrel, then insert the wire plus plunger to the desired depth. The presence of the wire prevents the plunger from sealing to the barrel, so excess air can escape rather than be compressed. When the desired plunger position has been reached (50.0 mL for this experiment), remove the copper wire, allowing an airtight seal to form in the syringe. Figure 13-1 shows a diagram of how to do this.

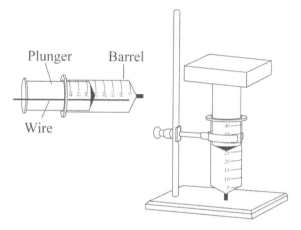

Figure 13-1. The apparatus for investigation of Boyle's law. The first figure shows you how to set the syringe at 50 mL, and the second how to set the apparatus so that pressure can be applied.

Step 3. Clamp the syringe vertically to a ring stand, plunger up, and record the first pressure and volume readings: 0 bricks, 50. mL.

Step 4. Now balance a brick on the plunger and record the new pressure (1 brick) and new volume. It will be necessary for one partner to steady the brick while one quickly reads the volume. Try to use the lightest touch possible. Also try to avoid dropping the bricks on the table. They are messy, noisy, and may damage the table top. Or toes. Repeat the exercise with your remaining bricks and record your data. You should be able to get data for six to eight bricks before the pile gets too big to be manageable.

Step 5. Plot your data on the graph paper provided. Label the x-axis "Pressure (bricks)" and the y-axis "Volume (mL)." Be sure to label the graph "Experiment 13 Part A: Boyle's Law" and follow the other recommendations for a good graphical representation of data.

PART B. INVESTIGATION OF CHARLES'S LAW

Step 1. Place about 700 mL of water in a 1-liter beaker on a ring stand or hot plate, and add a few boiling chips. Place a thermometer in the water, and record the water temperature.

Step 2. Set your syringe to contain about 30 mL of air at atmospheric pressure using the method described for Part A, Step 2. Then place the syringe in the water bath, and record both the volume of air in the syringe and its temperature. Figure 13-2 shows the apparatus to use.

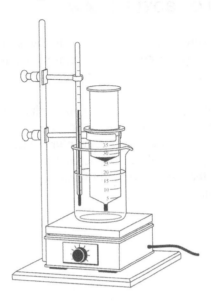

Figure 13-2. The apparatus for the investigation of Charles's law.

Step 3. Heat the water gently to about 40 °C and record the new temperature and volume. For best results it will be necessary to stir the water gently to get even heating of the gas sample. Record your data.

Step 4. Continue this procedure, heating to 50°, then 60°, 70°, and 80°, and record both temperature and volume for each new measurement. You can go as high as 100 degrees in a boiling water bath; be sure you have at least five data points before stopping.

ⓘ Be sure that 0° goes in the correct place in the data table!

Step 5. Find the volume of the air sample in the syringe at 0 °C by placing the syringe in a properly prepared ice bath and recording the new temperature and volume.

Step 6. Plot your data on graph paper, labeling the graph and the axes (including units), and drawing the best smooth line (is it a curve or a straight line?) that approximates your data points.

Step 7. From your graph, predict the volume of the sample at 150 °C, and the volume at -273 °C.

PART C: OXYGEN, CARBON DIOXIDE, AND NITROGEN IN AIR

Simple experiments can determine the partial pressure of carbon dioxide in the air, but they do not always provide accurate data. Instead of asking you to perform one of these unsatisfactory experiments, we have provided you with some experimental data you can use to calculate the value of the partial pressures and then the percentages of oxygen, carbon dioxide, and nitrogen in air.

Percent Nitrogen in the Air by Calculation

In this calculation exercise, you are given the total atmospheric pressure for a certain day, as well as the partial pressure of water vapor for the day's temperature and the partial pressure of carbon dioxide measured by a fairly accurate method. You also have a total volume of air (containing all the gases) and a volume of the oxygen measured experimentally. There are only a couple of steps you need to take to obtain the percent of nitrogen in the air.

You already know this relationship.

$$P_{atm} = P_{O_2} + P_{CO_2} + P_{N_2} + P_{H_2O} + P_{other\ gases}$$

You can find the partial pressure of oxygen because you can find the percent of oxygen in the air as a decimal fraction of the total. For instance, if 35.0 mL of oxygen were recorded with a total gas volume of 83.0 mL, 35.0/83.0, or 42.2%, is the decimal fraction of oxygen in the air. This is a rather high value, but perhaps the student's data were poorly taken. The partial pressure of oxygen is then $0.422P_{atm}$, or 315 mm.

At this point, you have all the necessary data to find P_{N_2}, as long as you make this assumption:

Assume $P_{all\ other\ gases}$ = 0.9% of atmospheric pressure exclusive of water vapor

Therefore,
$$P_{atm} = P_{O_2} + P_{CO_2} + P_{N_2} + P_{H_2O} + 0.009(P_{atm}\text{-}P_{O_2}\text{-}P_{CO_2}\text{-}P_{N_2})\text{-}P_{H2O}$$

This equation is rearranged on the report sheet, with P_{N_2} isolated, to show you the calculation.

Step 1. Calculate the partial pressure of N_2 in your air sample, and record your results.
Step 2. Calculate the percent of nitrogen in air, and complete the report.

EXPERIMENT 13 PRELAB EXERCISES:

GAS LAW EXPERIMENTS

Name_____ Partner _____ Date _____

1. A small gas tank containing nitrogen had a volume of 25 L and an internal pressure of 200.0 atm. The tank developed a small leak, and over time emptied into the closed room it was stored in, to a final pressure of 0.975 atm (the atmospheric pressure in the room). What was the volume of the room (including the tank and assuming there was nothing else stored in the room)?

2. Using the data and the graph below, find the temperature at which the volume of a gas sample reaches 105 mL. Label the graph appropriately.

Volume (mL)	Temperature (K)
45.0	273
65.0	300
75.0	325
85.0	350

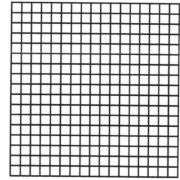

3. In a mixture of gases in a scuba tank, the total pressure in the tank is about 150 atm. Nitrox I is a standard gas mixture used for some types of scuba diving and is composed of 32% oxygen and 68% nitrogen. Find the partial pressure in atmospheres of each of these two components in a full tank of Nitrox I.

4. In a carbon dioxide fire extinguisher at room temperature the carbon dioxide is held at a pressure much higher than atmospheric pressure. When the nozzle is opened the carbon dioxide is released to the atmosphere and then is in an environment where the pressure is equal to the atmospheric pressure. Does the temperature of the carbon dioxide undergoing this change become higher than room temperature, lower than room temperature, or remain the same?

EXPERIMENT 13. REPORT: GAS LAW EXPERIMENTS

Name_____ Partner _____ Date _____

PART A. INVESTIGATION OF BOYLE'S LAW

Pressure (bricks)	Volume (mL)
0	50.0
_____	_____
_____	_____
_____	_____
_____	_____
_____	_____
_____	_____

PART B. INVESTIGATION OF CHARLES'S LAW

Temperature (K)	Volume (mL)
_____	_____
_____	_____
_____	_____
_____	_____
_____	_____
_____	_____
423	_____ (Estimate from graph!)

1. What conclusion can you draw from your first graph concerning the relationship between the volume of a gas sample and the pressure exerted on it?

2. Draw a conclusion from your second graph concerning how the volume of a gas sample varies with temperature, provided the pressure stays constant.

PART C: OXYGEN, CARBON DIOXIDE, AND NITROGEN IN AIR

Determination of the partial pressure of oxygen in air

The day's atmospheric (barometric) pressure	<u>747.5 mm</u>
Vapor pressure of water at about 1 atm, and room temperature	<u>19.5 mm</u>
V_1, <u>total</u> volume of air observed	<u>125.5 mL</u>
V_2, volume of oxygen observed	<u>36.1 mL</u>

What percent of the total gas is O_2?   _____

What is the partial pressure of oxygen (in mm Hg)?   _____

Measured partial pressure of carbon dioxide in air   <u>0.22 mm Hg</u>

Percent CO_2 in air   _____

Percent nitrogen in air—all values are available except P_{N_2}!

$$P_{N_2} = P_{atm} - P_{O_2} - P_{CO_2} - P_{H_2O} - 0.009\,(P_{atm} - P_{H_2O})$$
$$P_{N_2} =$$ _____

$$\%\,N_2 =$$ _____

Do your results agree with "accepted" values for O_2, N_2, and CO_2 in air?

EXERCISES

1. At a constant temperature of 75 °C, a 2.00-L container of oxygen gas at 10 atm was opened into an evacuated container with a volume of 400 L. When the pressure inside the system reached a stable point, what was the pressure of the oxygen in the larger container?

2. A 3.00-L balloon at 50.0 °C was placed in a freezer at a temperature of –40.0 °C. Did the balloon shrink or expand in the new setting? What was its new volume?

3. A 6.00-L gas tank contained a mixture of helium and oxygen at a pressure of 20.1 atm. If the mixture was 75% helium, what were the partial pressures of oxygen and helium?

EXPERIMENT 14

CHANGES OF STATE

GOALS

1. To observe transformation of a substance from its solid to its liquid form and back again
2. To observe the temperature changes as these changes of state to occur
3. To explain the different regions seen on the heating-cooling curve for the change of state of a substance

INTRODUCTION

You've already learned that matter commonly can exist in three different states: a solid, a liquid, or a gas. From personal experience with water, you also know that adding heat to a solid can cause it to melt to its liquid state, and more heat can transform a liquid into its gaseous state. You are going to use this exercise to examine the heat changes that occur in a substance as it goes from one state to another.

There are usually three steps to a complete phase-change cycle. A solid will exhibit a steady rise in temperature as it is heated, up to the point at which it starts to melt. At that point, and until the solid has completely transformed into the liquid, most of the energy put into the substance as heat is used to make the solid-to-liquid transition. The actual amount of heat required per gram of the substance is called the heat of fusion. and differs for different substances. Water has a relatively high heat of fusion, 80 cal (334 Joules) to convert one gram of ice to one gram of water. Overall, you don't see any steady rise in temperature at this stage until the substance is completely in liquid form; all the heat is going to the melting, or fusion, process.

Once the sample is completely liquefied, the liquid can go on absorbing heat and rising in temperature until it comes to a boil. At the boiling point, again, the change in state requires a significant amount of energy, and the liquid stays at boiling temperature until all of it has been converted to the gas form. The heat of vaporization of water, 540 cal/g (2260 Joules/g), is relatively high. The steam, of course, is now a gas, and conforms to Charles's law in how it takes on heat.

Other substances follow the same general pattern, but their physical constants are different, so the numerical values of heat of fusion, heat of vaporization, and actual melting and boiling temperatures are different.

If, instead of a heating curve, you try cooling a liquid to form a solid, you should see the same pattern in reverse: a steady decrease in temperature until the liquid is at its freezing point, then a steady state for a time, while the heat being removed from the liquid allows it to form the solid, then a decrease in the temperature of the solid as fast as heat is being removed from it.

Your exercise will track these changes in two different substances, water and stearic acid. Water is unusual because the solid, liquid, and gaseous phases can be reached with relatively little effort, so you will look at all three phases of water. Stearic acid is a solid at room temperature and liquefies easily, but the vapor phase is harder to reach, so you will look only at fusion of stearic acid.

REFERENCES

Stoker, H. S. *General, Organic, and Biological Chemistry;* Cengage Learning; Chapter 7.
    Review topics:
- Changes of state
- Vaporization
- Fusion (melting, not nuclear fusion)

WEB RESOURCES

Phase Changes. *http://hyperphysics.phy-astr.gsu.edu/hbase/thermo/phase.html*

States of Matter. *http://www.footprints-science.co.uk/states.htm*

CHEMTUTOR STATES OF MATTER. *http://www.chemtutor.com/sta.htm*

SAFETY NOTES

Although you will be working with substances generally recognized as safe, you should expect to wear chemical splash goggles throughout this experiment and exercise usual caution when working with hot materials. Follow your instructor's directions for disposal of the non-water materials you use.

Dry ice (solid carbon dioxide) must be handled with tongs or insulated gloves, since it is at a temperature of -78.5 °C.

Acetone is sometimes used to improve the performance of dry ice baths. It is a flammable liquid with a strong odor and extreme volatility, so it should be used in the hood and in the absence of open heat sources.

EXPERIMENTAL PROCEDURE

PART A. CHANGES OF STATE IN WATER

Step 1. To observe the change in heat content as water goes from solid to liquid to gas, you will need to begin with a sample of ice in a 50-mL beaker. This sample should already have been prepared for you and stored at a temperature below the anticipated 0 °C melting point. Place the beaker of ice on a hot plate at room temperature, and clamp the thermometer so that the whole apparatus does not tip over. Record the temperature of the ice sample at a time of 0 minutes. Add a small stirring rod to the ice to allow you to stir the mixture once it starts to melt.

Step 2. Record the temperature of the ice sample as you begin to heat it with the hot plate. You should record the temperature and the time observed about every minute, until the sample has completely melted and come to a boil and for about an additional 5 minutes. Stirring the mixture helps you obtain more uniform readings.

Step 3. Make a graph of temperature vs. time for this series of changes of state in water. Label the regions of the graph that show melting and boiling. This graph should take only about a half-page of graph paper, but be sure to follow proper procedures for labeling and constructing the graph.

Step 4. Now take the same sample of liquid water in a beaker, and place it in a cooling bath containing dry ice (solid carbon dioxide).

> ● Be sure you know the safest ways to handle dry ice!!

Step 5. Again watch the temperature change with time, and record the temperature of the sample about every two minutes until the sample becomes solid, and for an additional 5 minutes after that.

Step 6. Graph this data with time as the independent variable (x-axis) and temperature as the dependent variable (y-axis), using about a half-page. Label the graph completely, and mark the region that shows the change of state from liquid to solid.

PART B. MELTING AND FREEZING OF A SOLID

You will use another solid to examine how general this heating curve phenomenon is. The substance you will be using is stearic acid, $C_{18}H_{36}O_2$, a major component of animal fats. It is a white solid with a melting point of 69 °C and a boiling point of 361 °C. You will look only at the change of state from solid to liquid since boiling carbon compounds often have a strong odor and significant flammability.

Step 1. Fill a 50-mL beaker about half full with stearic acid, and set it on a cooled hot plate. Arrange a thermometer in the solid, being careful to clamp the thermometer so that it cannot tip your experimental setup over. Take the temperature of the solid, and record this temperature for t = 0 minutes.

Step 2. Heat the stearic acid gently, recording temperature and time about every minute as you did for the water sample, until you have seen the entire sample melt, and for an additional 5 minutes after that.

Step 3. When the heating part of the experiment is finished, remove the stearic acid and thermometer from the hot plate. Continue to record the temperature until the sample has solidified again, and for an additional few minutes after that.

Step 4. Plot the heating and cooling curve on a graph of time vs. temperature, and label the regions where change of state takes place.

EXPERIMENT 14 PRELAB EXERCISES:

CHANGES OF STATE

Name_____ **Partner** _____ **Date** _____

1. Define the following terms.

   fusion

   evaporation

   change of state

2. Identify the following processes as endothermic or exothermic.

   Melting of ice   _____

   Solidification of cooking fat after the pan cools down   _____

   Heating water to bring it to a boil   _____

   Water changing from the liquid to the gaseous state   _____

   Change of lava from liquid to solid rock   _____

3. Sketch a heating/cooling curve for the change of state that occurs when naphthalene, melting point 81 °C, changes from a solid to a liquid. Don't forget to label the axes.

4. Sketch a heating/cooling curve for the change of state that occurs when ammonia changes from a gas to a liquid at -33 °C. (Label the axes!)

EXPERIMENT 14 REPORT: CHANGES OF STATE

Name_____ **Partner** _____ **Date** _____

PART A. CHANGES OF STATE IN WATER

Cooling of Water to Freezing

Time (min)	Temperature of H_2O (°C)	Time (min)	Temperature of H_2O (°C)
0			

1. Sketch the cooling curve for water going from a temperature of 100. °C (as steam) to –10. °C (as ice). Attach the graph of your data to this report.

2. Predict the shape of the heating curve for steam as its temperature changes from 101 °C to 150 °C.

PART B. MELTING AND FREEZING OF A SOLID

Changes in State with Temperature

Time (min)	Temperature of substance (°C)	Time (min)	Temperature of substance (°C)
0			

1. Attach the graph of your data to this report.

2. Using your graph as a guide for what to expect, predict the change-of-state curve for benzoic acid (mp 122 °C, boiling point 249 °C) as the temperature of a sample increases from 50 °C to 250 °C. Be sure to label the graph axes.

3. Make a general statement about the shapes of heating and cooling curves.

EXPERIMENT 15

SOLUTION PREPARATION

GOALS

1. To provide experience in two types of solution preparation
2. To improve laboratory technique with attention to detail
3. To provide practice in calculations associated with the preparation of solutions

INTRODUCTION

In many medical laboratory situations, it is first necessary to dilute a biological sample to obtain a solution of the right concentration to analyze. The more carefully and accurately a worker can do these dilutions, the better the lab results and the better the interpretation for the patient. This exercise will allow you to become familiar with some standard tools for solution making and dilution.

It can be inconvenient to keep large quantities of solutions around, and anyway, some chemicals don't have a long shelf life. Often the best strategy for solution storage is to keep one concentrated solution on hand and dilute some of it to the required concentration as needed. For example, a lab might keep a stock solution of 1.00 M sodium chloride, and a worker would take an appropriate quantity, dilute it to the needed concentration, and continue with the job at hand.

But how do you decide how much to use, or how much to make? The formula you have already learned for dilution calculations is the necessary relationship. C_s and V_s refer to the stock solution, while C_d and V_d refer to the new, diluted solution.

$$C_s \times V_s = C_d \times V_d$$

In order to use this equation you need to know any three of the four variables. Usually you know the concentration of the stock solution and the desired concentration and volume, and only need to decide how much of the stock solution (V_s) to use.

Suppose you had on the shelf 1.00 L of 2.00 M sodium chloride solution, and needed to make 0.500 L of 0.0100 M sodium chloride solution. With V_s as the unknown quantity, you can use the equation to find that you will need to measure out 0.00500 L, or 5.00 mL, of the original solution, and then dilute it to a final volume of 500. mL. Common sense tells you that once you have poured together the water and measured solution, you will need to stir the entire mixture well to insure that it is completely homogeneous.

This dilution equation works for any kind of concentration units, not just molarity, so you will be able to use it in the first part of the exercise as you make a diluted sucrose solution from a stock concentrated solution that uses the % concentration unit. Ordinary laboratory tools will be accurate enough for this determination, but you still should measure as accurately as you can. Try to use a small graduated cylinder for the measurements (they usually have more measuring divisions marked on them than large ones do).

Your instructor will check the accuracy of your work by determining the density of your final solution with a hydrometer, an instrument that measures the density of a solution relative to the density of water (1.00 g/cm^3). A hydrometer usually measures density to three significant figures, so if the relationship between density and sucrose concentration is known, your instructor can calculate the actual sucrose concentration.

The second part of the experiment requires you to use more accurate tools to prepare a dilute solution of potassium permanganate (KMnO$_4$) in water. First you will use a balance to weigh a specific amount of KMnO$_4$, and then use it to make a concentrated solution. You will then calculate its exact concentration from the mass of solute, the solute's gram-formula weight, and the volume of the flask used. This takes advantage of the other relationship between solution and solute when you are working with a single solution,

$$V \times M = \text{moles present in that volume}$$

Once you have found the molarity of this concentrated solution using the equation, you will calculate how much of it you will need to make a more dilute solution. After that, you will make the new, less concentrated solution whose actual concentration will be determined experimentally by your instructor.

The tools you need for accurate solution making are volumetric flasks, pipets, and pipet bulbs. (See Figure 15-1.) The flasks and pipets are carefully manufactured to contain exactly (usually to three or more significant figures) the volume they are supposed to contain, and they usually have only one fill mark instead of a lot of graduations. Refer to your web resources for tips on how to use these tools well. Your instructor may also demonstrate their use.

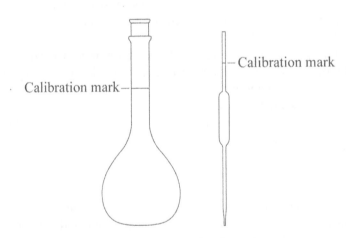

Figure 15-1. A volumetric flask and a volumetric pipet.

Potassium permanganate is a highly colored material, and your instructor will take advantage of that to determine the exact concentration of the solution you prepared. A small amount of your diluted solution will be placed in a spectrometer, a device for measuring how much light is absorbed by a sample at a particular wavelength. This measurement can be related to the concentration of the solution by a simple calculation, and you can then compare the supposed concentration of your solution to the actual concentration to evaluate the quality of your work.

For this exercise, each person can work individually to make up solutions of known concentration. This will allow each of you to have experience with the tools for dilutions and the calculations needed for such work. You will be graded on the accuracy of your lab manipulations, so attention to detail and careful work will help you achieve success!

REFERENCES

Stoker, H. S. *General, Organic, and Biological Chemistry;* Cengage Learning; Chapter 7.
Review topics:
- Per cent concentration
- Molarity
- Dilution calculations
- Pipet use
- Volumetric glassware

WEB RESOURCES

Helpful Hints on the Use of a Volumetric Pipet.
http://www.csudh.edu/oliver/demos/pipetuse/pipetuse.htm

Chemware-Glassware-Volumetric Flasks.
http://www.dartmouth.edu/~chemlab/techniques/vol_flasks.html

Dilution of Solutions Calculations. *http://www.ausetute.com.au/dilucalc.html*

CHEMTUTOR SOLUTIONS. *http://www.chemtutor.com/solution.htm#probs*

SAFETY NOTES

You should wear chemical splash goggles throughout this experiment. Sucrose is a substance 'generally recognized as safe' (GRAS) by the US Food and Drug Administration, but potassium permanganate is somewhat corrosive and can stain skin and clothing. In very small concentrations, potassium permanganate is used for water purification; it can generally be discarded in the drain provided copious amounts of water are used to dilute it, but follow local regulations for disposal.

EXPERIMENTAL PROCEDURE

PART A. PREPARATION OF A SOLUTION FROM A CONCENTRATED STOCK SOLUTION

Available to you in the laboratory is a solution of sucrose in water that is 25% (w/v) sucrose. Your task is to make 25 mL of a solution that is 10% sucrose in water. This solution needs to be prepared accurately to only three significant figures, so volume measurements can be made using a small graduated cylinder.

Step 1. Calculate the amount of the concentrated sucrose solution you will need to use to prepare your final solution, and measure that quantity. Place it in another graduated cylinder big enough to contain all the necessary liquid for your final solution.

Step 2. Add the necessary amount of water to bring the final volume to 25.0 mL and mix well. Be sure your final volume is as close to 25.0 mL as you can manage, and that **all** of the sucrose solution you measured is in it.

Step 3. Bring the solution to your instructor for a concentration determination, and when you have obtained it, calculate your percent error.

PART B. PREPARATION OF A SOLUTION USING A SOLID SOLUTE

Step 1. Using a weighing tray, weigh out about 0.40 grams of potassium permanganate. **Record the exact amount used** and dissolve it in about 50 mL of water. The potassium permanganate can be deceptive. It gives such a dark solution that it is difficult to tell if all the solid is dissolved, so be careful!

Step 2. Pour this solution into your 100.0-mL volumetric flask, being careful to make a complete transfer with no spills and drips. Extra water may be used; many tiny rinses are better than one big one. A funnel is a good idea too. When all the solution has been transferred from beaker to flask, dilute to the mark on the volumetric

> ⓘ Don't use more water than the volumetric flask can hold.

flask, using an eye dropper as shown in Figure 15-2. With gloves on, mix the contents of the flask well. This solution is far too concentrated for your use, so you will need to go on to the next step.

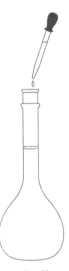

Figure 15-2. Filling the volumetric flask to the line.

Step 3. Perform a dilution to get the concentration needed. Using a 2.00-mL volumetric pipet and a bulb, transfer 2.00 mL of your dark solution to a 250-mL volumetric flask, and dilute it to the mark with water. Mix the solution well, and bring it to your instructor to determine the exact concentration. Your instructor will tell you the exact concentration of the solution you have made.

ⓘ Do not get solution into your pipet bulb, because this can ruin the bulb!

Step 4. Perform the necessary calculations to determine the concentration you were <u>trying</u> to make. Determine your percent error. If it is too high, you may wish to repeat the dilution part of the experiment to obtain better results. Students usually have little trouble making the concentrated solution, but often need practice using the volumetric pipet.

ⓘ You will be graded on the accuracy of your results.

PART C. EXERCISES

Complete the exercises at the end of the report form.

EXPERIMENT 15 PRELAB EXERCISES:

SOLUTION PREPARATION

Name_____ **Partner** _____ **Date** _____

1. How many grams of potassium chloride are necessary to make 100.0 mL of a 0.200 M solution?

2. How many milliliters of that solution would be required to make 250 mL of a 0.0100 M solution?

3. Describe the exact procedure used to make the first solution, including a sketch of the volumetric flask needed.

4. Describe the exact procedure used to make the second solution, and include a sketch of the measuring devices (the pipet and the flask) used.

5. A student's solution of KCl made in this manner was analyzed and found to have a concentration of 0.00950 M. What was the student's percent error for this exercise?

EXPERIMENT 15 REPORT: SOLUTION PREPARATION

Name_____ **Partner** _____ **Date** _____

PART A. PREPARATION OF A SUGAR SOLUTION OF KNOWN CONCENTRATION

Concentration of the stock sugar solution   _____

Calculation of the amount needed for your dilution

                    Amount needed   _____

Expected final concentration of sucrose in water   <u>10.0%</u>

Actual concentration determined by instructor   Trial 1_____

                                                           Trial 2_____

                                                           Trial 3_____

Percent error determination

$\% \text{ error} = \left| \dfrac{\text{true value - experimental value}}{\text{true value}} \right| \times 100\%$ _____

PART B. PREPARATION OF A SOLUTION USING A SOLID SOLUTE

What is the correct formula for potassium permanganate?   _____

How many grams of potassium permanganate did you begin with?   _____

How many moles of potassium permanganate was that?   _____
Show your calculation.

What was the volume of your first volumetric flask?   _____

What was the concentration in moles/liter of your first solution? _____
Show your calculation.

How much of this first solution did you use to make the
second solution?   _____

You diluted 2.00 mL (or 1.00 mL) of the first potassium
permanganate solution to 250.0 mL.

   What was this final concentration in moles/liter?   _____

   What is the true concentration of your diluted solution?   Trial 1 _____

                                                     Trial 2 _____

                                                     Trial 3 _____

   Calculate your percent error.

   % error = $\dfrac{\text{true value-experimental value}}{\text{true value}}$ x 100%   _____

PART C. EXERCISES

1. A student needed to make 5.0 L of a solution of sodium chloride with a concentration of 0.100 M.
Describe how to make this solution using solid sodium chloride and water.

2. How many moles of sodium chloride would be present in the 5.0 L of 0.100 M sodium chloride
solution?

3. The same 0.100 M solution could be made using a 2-L bottle of 1.00 M sodium chloride solution
that the student found on a shelf. What would the procedure be in this case?

EXPERIMENT 16

OSMOTIC PROCESSES

GOALS

1. To introduce the students to osmosis and osmotic pressure
2. To distinguish between osmosis and dialysis
3. To practice the calculation of osmolarity and relate it to the osmotic process

INTRODUCTION

Colligative properties of a solution are properties that depend on the concentration of dissolved particles in a solution: the more dissolved particles, the more that particular property is affected. Colligative properties that can be of biological and commercial interest are boiling point elevation, freezing point depression, and osmotic processes. Although any solvent can demonstrate these properties, the one of most biochemical interest is water, and we will confine our examinations to water solutions. You have looked at boiling point elevation and freezing point depression for water in Experiment 8; in this experiment, you will be looking at the osmotic processes of osmosis, osmotic pressure, and dialysis.

Osmotic processes require a semipermeable membrane (like a cell membrane) that will allow water and small particles to pass through it but prohibit the passage of colloids and large particles such as protein molecules.

The passage of water through a semipermeable membrane from a region where it is concentrated to a region where it is in short supply is termed osmosis. Dialysis is similar, but this term also includes processes in which other small particles (ions and small molecules) pass back and forth through a membrane until a balance, or equilibrium, occurs in which the concentrations of solutes are the same on both sides of the membrane. Cell membranes in plants and animals usually make use of both of these "osmotic processes" to regulate concentrations of ions and molecular nutrients in the cell.

Osmotic pressure is a measure of the driving force for the formation of the osmotic equilibrium. As water (solvent) flows through a membrane such as a cell membrane to equalize its concentration both inside and outside the cell, it can cause the cell to swell up as it takes on all this new water, until the membrane actually bursts from the pressure inside it. This is the explanation of lysis of red blood cells in a bath of distilled water, and the reason that most solutions for injection are *isotonic*, having the same concentration of dissolved particles as plasma.

Since the concentration of dissolved particles is such an important part of understanding osmosis, a new concentration unit is used, *osmolarity*. Calculating the osmolarity of a solution is similar to calculating a molarity; in fact, you can start with a solution's molarity to calculate osmolarity. The calculation rule is to begin with the molarity of a solute, and multiply by the number of particles that solute produces per mole to get the osmolarity.

For instance, a 0.1 M solution of glucose produces 6.02×10^{22} particles per liter of solution (a tenth of a mole of particles for every tenth of a mole of the compound) since it doesn't ionize. But a 0.10 M solution of calcium chloride produces 1.806×10^{24} particles per liter of solution—2 moles of chloride ions and one mole of calcium ions for every mole of $CaCl_2$ that dissolves. The *molarity* of the calcium chloride solution is 0.10 M, but the *osmolarity* is 0.30 osmol/L.

$$\frac{0.10 \text{ moles glucose}}{\text{liter}} \times \frac{1 \text{ mole particles}}{1 \text{ mol glucose}} = \frac{0.10 \text{ moles particles}}{\text{liter}} = 0.10 \text{ osmoles/L}$$

$$\frac{0.10 \text{ moles CaCl}_2}{\text{liter}} \times \frac{3 \text{ mole particles}}{1 \text{ mol CaCl}_2} = \frac{0.30 \text{ moles particles}}{\text{liter}} = 0.30 \text{ osmoles/L}$$

You can calculate the total osmolarity of a solution with many different solutes if you know the molarity of each dissolved substance and the number of particles a mole of it produces when it dissolves. Calculate the osmolarity of each kind of particle, and add the osmolarities of each particle to get the total osmolarity. For example, suppose you had one liter of a solution that has 0.10 osmoles of glucose per liter and 0.30 osmoles of calcium chloride in the same liter of solution. The total osmolarity (moles of particles dissolved per liter of solution) is the sum of the two individual ones, or 0.40 osmoles/L.

In this experiment, you will concentrate on osmotic processes. The first part of the experiment looks at osmosis with a real cell membrane surrounding an egg. You will observe what happens to the egg in a hypotonic, an isotonic, and a hypertonic solution. You will also examine an artificially prepared "cell" made from semipermeable tubing containing a concentrated solution and see what happens to it when it is exposed to distilled water.

REFERENCES

Stoker, H. S. *General, Organic, and Biological Chemistry;* Cengage Learning; Chapter 8.
   Review topics:
   - Colligative properties
   - Molarity
   - Osmolarity
   - Dialysis
   - Osmosis
   - Osmotic pressure

WEB RESOURCES

SparkNotes: Colligative Properties of Solutions: Introduction and Summary.
http://www.sparknotes.com/chemistry/solutions/colligative/summary.html

Colligative properties-wikipedia, the free encyclopedia.
http://en.wikipedia.org/wiki/Colligative_properties

CHEMTUTOR SOLUTIONS. *http://www.chemtutor.com/solution.htm#collig*

SAFETY NOTES

The materials you will be working with are 'generally recognized as safe' (GRAS) but you must follow local regulations about the wearing of safety chemical splash goggles and disposal of the chemicals you will be using.

EXPERIMENTAL PROCEDURE

PART A. OSMOSIS

Osmosis through an egg membrane

Your group needs three eggs that have been soaked for several days in a solution of acetic acid. The effect of the acetic acid is to remove the calcium-containing components of the shell so that all that is left is the semipermeable membrane part of the shell. You will be handling all the eggs several times, and without the hard portion of the shell the eggs are even more fragile than usual, so handle them carefully. For each egg, choose a beaker or weighing boat to weigh it in as the experiment progresses, and a beaker to conduct the experiment in (a total of six beakers needed).

Step 1. Find the weight of each weighing boat and egg at the start of the experiment, and label the weighing boats with mass and an identifier for the egg. You will need the weighing boat weights as well because you will be using them several times throughout this experiment.

> ⓘ It would **not** be a good idea to break a raw egg onto the balance pan, even by accident.

Step 2. Fill three beakers about 3/4 full; one with distilled water, one with normal saline solution (0.9% sodium chloride) and one with saturated sodium chloride solution.

Step 3. Working carefully, place one egg in the distilled water solution, one egg in the normal saline solution, and one egg in the saturated sodium chloride solution. If necessary, add more solution to any beaker in which the egg is not completely covered. Allow the experimental materials to stand for at least 30 minutes while you go on to set up the next experiment.

Step 4. At the end of 30 minutes, carefully lift each egg out of its solution, dry it gently on a paper towel, and reweigh it using the same weighing boat you originally used to determine that egg's weight. Find the mass of each egg.

Step 5. Return the eggs to their osmotic baths and allow them to stand again for 30 min. Refill the beakers if necessary, so the eggs are covered with liquid at all times. At the end of the time, dry and weigh each egg again. Continue this process until you have at least four data points for each egg.

Step 6. Plot the data for each egg on the graph paper provided, using time as the x-axis and the weight of each egg as the y-axis. All three graphs can be plotted on the same piece of graph paper, although it would be convenient if you used different colors for each.

Osmotic pressure

Step 1. Soak a semipermeable membrane in distilled water for at least 10 minutes. Your instructor may already have done this step.

Step 2. Tie one end of the bag in a knot, open the other end, and insert a funnel.

Step 3. Fill the bag about 2/3 full with concentrated sugar solution, and tie off the top of the bag. Try to get rid of as much of the air out of the pocket at the top of the bag as possible, so that the bag is entirely filled with sugar solution. Figure 16-1 shows the procedure. Rinse the outside of the bag thoroughly with distilled water, and lay it in a beaker. Cover it with distilled water and note the time. Observe the bag occasionally for about 2 hours, noting what you see.

PART B. DIALYSIS THROUGH A SEMIPERMEABLE MEMBRANE

Step 1. Prepare another semipermeable membrane as you did above in Step 1.

Step 2. Fill the bag with 10 mL of starch solution and 10 mL of saturated sodium chloride solution.

Step 3. Remove a few drops of the mixed solution to three labeled wells in a spot plate, and tie the bag closed. Rinse the outside of the bag well with distilled water, as shown in Figure 16-1.

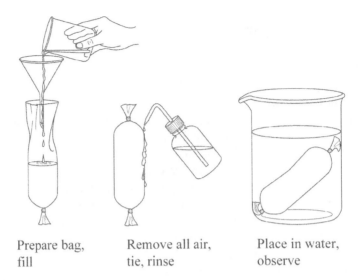

Prepare bag,   Remove all air,   Place in water,
fill   tie, rinse   observe

Figure 16-1. Preparing the dialysis bag.

Step 4. Fill a beaker about 3/4 full of 0.10 M sodium sulfate solution, and place a few drops of this solution in three more wells in your spot plate.

Step 5. Place the dialysis bag into the beaker of sodium sulfate solution, and allow the bag's contents to dialyze for about 30-60 minutes before continuing with Step 6. While you are waiting, continue with the osmosis experiments or with Part C.

Step 6. Swirl the contents of the beaker, and take a few drops of the solution for testing in the spot plate in three other labeled wells. Then break the bag into another beaker, and place a few drops of this solution in each of three spot plate wells.

When you are ready to do the tests, you should have 12 wells of the spot plate filled, three wells for each stage of each solution.

Step 7. Test one well of each solution for the presence of starch by mixing in a few drops of Lugol solution. A positive test for the presence of starch is the formation of a dark blue-black color.

Step 8. Test the second well of each group for the presence of chloride ions by adding a few drops of silver nitrate. A positive test for chloride ions is the formation of a white precipitate. Contamination of these wells by sodium sulfate could also give a white precipitate at this stage because silver sulfate is also an insoluble solid, so draw conclusions carefully.

Step 9. Test the third well of each set for the presence of sulfate ions by mixing its solution with a few drops of barium chloride solution. A white precipitate in this case will indicate the presence of sulfate ions.

PART C. OSMOLARITY EXERCISES

The report sheet has several exercises dealing with the calculation of osmolarity. Work these exercises in the space provided, being careful to keep track of significant figures and units.

EXPERIMENT 16 PRELAB EXERCISES:

OSMOTIC PROCESSES

Name_____ Partner _____ Date _____

1. Explain the difference between osmosis and dialysis.

2. Define these terms:

   isotonic

   hypertonic

   hypotonic

   crenation

   hemolysis

3. Would red blood cells exhibit hemolysis or crenation in a 5.0% solution of glucose?

4. If a solution of starch, glucose, and sodium chloride is placed in a dialysis bag, and the bag is placed in a beaker of distilled water for 24 hours, where would you expect to find the different solutes?

Substance	Inside the bag?	Outside the bag?
Starch		
Glucose		
Sodium ions		
Chloride ions		

5. Predict whether the egg in Part A of this experiment will enlarge or shrink when placed in distilled water. Explain your reasoning.

EXPERIMENT 16 REPORT: OSMOTIC PROCESSES

Name_____ **Partner** _____ **Date** _____

PART A. OSMOSIS THROUGH AN EGG MEMBRANE

Weighing boat masses

	Boat 1	Boat 2	Boat 3
	_____	_____	_____

Observations on Osmosis

Time (min)	Mass of egg 1 (g)	Mass of egg 2 (g)	Mass of egg 3 (g)
0			
30			
60			
90			
120			
150			

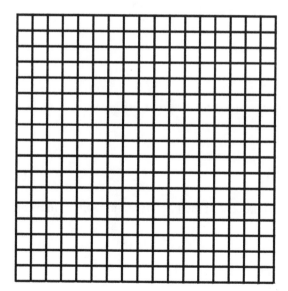

1. Which solution caused an egg to gain the most weight, and which solution caused loss of weight?

2. Explain the results with each egg on the basis of osmosis (water movement through a semipermeable membrane).

Observations on osmotic pressure

How does this exercise demonstrate osmotic pressure? Compare these results with the results from the experiment with the eggs. Are the two exercises consistent?

PART B. DIALYSIS THROUGH A SEMIPERMEABLE MEMBRANE

Tests of Dialysis Solutions

Solution	Test for starch	Test for chloride	Test for sulfate
Distilled water, t = 0 min			
Solution in bag, t = 0 min			
Distilled water, t = 60 min			
Bag solution, t = 60 min			

1. What substances were able to move through the walls of the bag? There was more than one, but you did not test for all of them. Nevertheless, name them all.

2. Using complete sentences, describe how osmosis differs from dialysis.

3. Why are salt and starch still present in the bag? Are they still at their original concentrations? Justify your answer.

PART C. OSMOLARITY EXERCISES

1. Calculate the osmolarity of "normal saline" 0.89% (w/v) sodium chloride in water.

2. Compare the osmolarity of separate 0.010 M solutions of sucrose, calcium chloride, and lithium chloride.

3. A rehydrating solution can be made for intravenous use with the following concentrations of electrolytes: 0.050 mol/L sodium chloride, 0.020 mol/L potassium bicarbonate, 0.092 mol/L of glucose, 0.094 mol/L of sucrose, and 0.009 mol/L of citrate ions. What is the total osmolarity of this solution?

EXPERIMENT 17

OXIDATION-REDUCTION REACTIONS

GOALS

1. To understand the meaning of oxidation and reduction
2. To recognize that both must take place simultaneously on two different substances
3. To rank several metals in a series from most to least reactive
4. To place hydrogen in that activity series
5. To relate the concept of oxidation-reduction to biochemical processes

INTRODUCTION

You have learned one classification system for chemical reactions already. By now you should be familiar with the idea of combinations, decompositions, single and double displacements, and combustion reactions. The other significant classification of reactions considers them as either oxidation-reduction reactions, or not. Although for convenience they were listed in a previous experiment as separate from the other classes, redox reactions happen in several of the classes.

This stems from the fact that the definition of a redox reaction is "a reaction in which there is a transfer of electrons from one atom to another." That makes this a somewhat more general classification system than the one you have learned already. In fact, you can make a table of reaction classes in which redox and non-redox is a super-class for the others.

Classification of Chemical Reactions

Redox	Non-redox
single replacement	double displacement
some combinations	some combinations
some decompositions	some decompositions
combustion	

For instance, in the reaction of hydrogen and oxygen to produce water, combustion of the hydrogen definitely takes place since there is an explosive release of energy with heat and light. It can also be regarded as a combination of two or more substances to form a new substance, but it is an oxidation-reduction reaction as well. The oxidation number of hydrogen goes from 0 to 1+ (in oxidation the oxidation number goes **up**) and the oxidation number of oxygen atoms goes from 0 to -2 (in reduction the oxidation number goes **down**).

$$2H_2 + O_2 \longrightarrow 2 H_2O$$

Notice that oxidation always takes place in combination with reduction, because the electrons given up by one atom have to be taken on by another atom in the same reaction system.

The simplest redox reactions to study are the single replacement reactions of metals with acids. Experiments with different metals and an acid lead to the conclusion that some react readily with acids and some don't, and that metals can actually be ranked in order of how easily they react. Zinc is one example of a very reactive metal.

$$Zn + 2HCl \longrightarrow ZnCl_2 + H_2 \uparrow$$

In this equation, the zinc goes from an oxidation number of 0 to 2+, giving up two electrons to hydrogen atoms (one each) so that the hydrogen atoms go from an oxidation number of 1+ to 0. The zinc is oxidized by the oxidizing agent hydrogen ion, and the hydrogen ions are reduced by the reducing agent zinc. This ranking of metals in their ability to give up electrons is known as the "activity series" or sometimes the "electromotive series" (since electrons moving implies the flow of electric current).

In this exercise you will try to rank several metals in the activity series from most reactive to least reactive. Your strategy for this portion of the experiment is to place a piece of a solid metal (oxidation state in the elemental form = 0) in a solution of another metal's nitrate salt (the dissolved metal's oxidation state will be positive because it is in ionic form). If the dissolved metal plates out on the original solid sample, you know that the dissolved metal must have accepted electrons from the original solid sample. For this to happen, the original solid metal must be **above** the dissolved metal in the activity series. If you waited long enough, you would see the complete disappearance of solid reactive Metal 1 as it gives up electrons and dissolves, and a complete precipitation of Metal 2 as it takes on electrons and precipitates as elemental metal.

For an example of how this might work, consider what happens if a strip of magnesium (Metal 1) is placed in a solution of gold chloride ($AuCl_3$, Metal 2). Gold is at the bottom of the activity series. You would expect to see solid gold plate out onto the strip of magnesium as magnesium (Mg^0) gave electrons to gold (Au^{3+}) according to the following equation. Magnesium metal would slowly dissolve as it became Mg^{2+}. To keep the electrons balanced, the equation must be balanced this way.

$$3\ Mg^0_{(s)} + 2\ Au^{3+}_{(aq)} \longrightarrow 2\ Au^0_{(s)} + 3\ Mg^{2+}_{(aq)}$$

In the second part of the experiment, you will observe the actual single replacement reactions of metals in order to determine the place of hydrogen in the activity series. Some metals, like zinc, will react with hydrogen and some, like gold, will not. To place hydrogen in the activity series, you will have to test various metals from Part A for their reaction with hydrochloric acid. If a reaction occurs, one product is elemental hydrogen, that is, the metal will give electrons to hydrogen ions to make hydrogen with a charge of 0. The metal will be oxidized, and the hydrogen ions will be reduced. The metal is more reactive than H_2, and the metal goes above hydrogen in the series. If no reaction occurs, then hydrogen must be more reactive than the metal, and the metal goes below hydrogen in the series.

This part of the experiment is an exercise in observation and deduction. To help you with the rankings, a good rule of thumb is that "electrons go <u>down</u> the series from most reactive metal to least reactive metal."

In the third part of this experiment, you will have the opportunity to observe a biological application of redox reactions. Metal redox reactions play a significant part in biochemistry, but there is a narrower application of the terms "oxidation" and "reduction" that you will see as well. In biological oxidations of carbon compounds, oxidation most often takes place by the actual addition of oxygen to a molecule, usually with the reduction taking place in some other molecule. The oxidation of glucose to carbon dioxide is a case in point. The formula for glucose is $C_6H_{12}O_6$; glucose contains six oxygen atoms and six carbon atoms. The resulting six molecules of carbon dioxide contain a total of six carbon atoms but 12 oxygen atoms—the carbon atoms have been oxidized. A different molecule in the biological oxidation-reduction system is reduced.

In this experiment, you will observe a simpler reaction, the production of oxygen from hydrogen peroxide in the presence of yeast. This reaction occurs slowly at room temperature in the absence of a catalyst, but readily in the presence of enzymes from yeast. The oxygen from hydrogen peroxide is reduced while another atom in a molecule in the yeast is oxidized. The enzyme molecules in the yeast act as catalysts that make the reaction easier, and do not themselves get changed in the reaction.

REFERENCES

Stoker, H. S. *General, Organic, and Biological Chemistry;* Cengage Learning; Chapter 9.
  Review topics
  - Balancing chemical equations
  - Oxidation reduction reactions

WEB RESOURCES

CHEMTUTOR REDOX AND ELECTROPLATING. *http://www.chemtutor.com/redox.htm*

Reactivity Series. *http://en.wikipedia.org/wiki/Reactivity_series*

Metal Activity Series. *http://www.unr.edu/sb204/geology/mas.html*

SAFETY NOTES

Some of the solutions you will be working with are significantly acidic, and some may stain your skin. Wear chemical splash goggles at all times during this exercise, and follow your instructor's directions about the use of gloves and disposal of the materials you are working with.

Discard the small pieces of metal in the wastebasket or as your instructor directs (**not** in the sink). Solutions should be discarded in the waste container provided by your instructor or in accordance with local regulations.

EXPERIMENTAL PROCEDURE

PART A. AN ACTIVITY SERIES FOR METALS

You have available to you the following materials.

Solid Samples of These Metallic Elements	Solutions of These Metal Salts
tin	tin(II) nitrate
copper	copper(II) nitrate
iron	iron(III) nitrate
aluminum	aluminum nitrate
magnesium	magnesium nitrate
zinc	zinc nitrate
lead	lead nitrate

Step 1. Place about 1 mL (20 drops) of one of the salt solutions in a small test tube and label it.

Step 2. Drop into it a piece of the metal you are testing; be sure to label the tube with the metal name also. Wait about 5-15 minutes before assessing reactivity.

Step 3. Repeat with a new metal and a new salt solution as necessary until all combinations have been tested. To save time, observe as many different metals as possible (six tubes at once) with each salt solution.

Step 4. Observe all your solution/metal mixtures, and record whether a reaction occurred. If it did, write a + in the grid on the report form, and write the equation for the redox reaction (really a half-reaction) that occurred. If reactions seemed to be exceptionally strong, use multiple + signs to indicate strong response, from + to ++++, depending on the vigor of the reaction.

> ⓘ **Hint:** If a metal plates out of solution onto the solid that is already there, the original solid metal has given electrons to dissolved metal ions so that they could become atoms of solid metal.

Step 5. Rank the metals in your list in order of decreasing activity, from **most** easily oxidized first to **least** easily oxidized (most inert).

PART B. PLACING HYDROGEN IN THE ACTIVITY SERIES

Step 1. Place 1 mL of 6 M hydrochloric acid in each of seven test tubes. Label each tube with the metal's name or symbol.

Step 2. Into each tube, place a small piece of the metal to be tested. Observe each sample for several minutes, and rank the metal's ability to react with hydrogen ions.

Step 3. Rank hydrogen in the activity series as best you can with the information available.

PART C. OXIDATION-REDUCTION IN BIOLOGICAL CHEMISTRY

Step 1. Obtain a test tube with a cork and tubing cap from your instructor. Your instructor will show you how to make a gas collection apparatus like the one shown in Figure 17-1.

Step 2. Dissolve a spatula-tip-full of baker's yeast (about 0.10 g) in 5 mL of warm water in the test tube.

Step 3. Add 1 mL of 3% hydrogen peroxide, mix well, and place the cork on the tube.

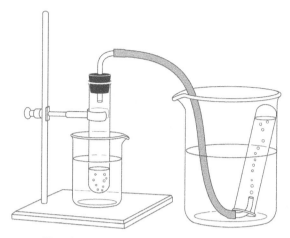

Figure 17.1 A simple gas collection apparatus.

Step 4. Place the tube containing the yeast mixture in a beaker containing warm (about body heat) water, and lead the piece of rubber tubing from the cap to the gas collection test tube. Observe for several minutes, up to 1/2 hour. When a significant amount of gas has been collected, you may stop the experiment.

Step 4. Using a wooden splint, test the gas collected to see if it supports combustion. To do this, light one end of the splint, and then shake it to put out the flame so that only an ember remains. Keeping the gas containing tube upside down, remove it from the beaker of water it has been standing in, and insert the glowing splint into the gas-filled tube.

Step 5. Record your observations.

PART D. EXERCISES

Complete this part of the report.

EXPERIMENT 17 PRELAB EXERCISES:

OXIDATION REDUCTION REACTIONS

Name_____ **Partner** _____ **Date** _____

In a reaction test, a piece of copper metal dropped into a solution of silver nitrate formed a precipitate of silver metal, and the copper metal diminished in size. The solution turned blue.

1. Is this reaction an oxidation-reduction reaction? If so, identify the oxidized and reduced species. If not, identify what other classification this reaction belongs to.

2. Write the equation for the reaction that did take place.

3. Why did the solution turn blue?

4. In the reaction of zinc with sulfuric acid, significant bubbling takes place. What substance causes the production of bubbles?

5. In the biological oxidation-reduction in this experiment, hydrogen peroxide forms oxygen gas and water. Write a balanced equation for this reaction.

6. In the reaction you wrote for Question 5, identify the atom that is reduced.

7. What is the role of the yeast in this reaction?

Name _____ Partner _____ Date _____

PART A. AN ACTIVITY SERIES FOR METALS

Reactions of Metals with Metal Salt Solutions

Metal salt / Metal	Tin(II) nitrate	Copper(II) nitrate	Iron(III) nitrate	Aluminum nitrate	Magnesium nitrate	Zinc nitrate	Lead(II) nitrate
Tin	—						
Copper		—					
Iron			—				
Aluminum				—			
Magnesium					—		
Zinc						—	
Lead							—

Rank the metals in the left-hand column according to increasing activity (ability to be oxidized, or give up electrons).

PART B. PLACING HYDROGEN IN THE ACTIVITY SERIES.

Reaction of Metals with Hydrochloric Acid

Metal	Observations	Ranking from most reactive to least reactive
Tin		
Copper		
Iron		
Aluminum		
Magnesium		
Zinc		
Lead		

Write your complete activity series, inserting hydrogen in the correct place.

PART C. OXIDATION-REDUCTION IN BIOLOGICAL CHEMISTRY.

1. Note your observations on the yeast-catalyzed reaction.

2. Write a balanced equation for the decomposition of hydrogen peroxide.

3. In the decomposition of hydrogen peroxide, which substance is oxidized and which is reduced?

PART D. EXERCISES

1. Gold metal is always at the bottom of the activity series. Explain why this means that gold is a good substance for dental fillings.

2. Use the Internet or book sources to discover the function of iron in heme. Is the iron in ionic or atomic form, and if ionic, what is its charge (oxidation state) ? Is it always in the same oxidation state in heme?

EXPERIMENT 18

EQUILIBRIUM AND RATE IN CHEMICAL REACTIONS

GOALS

1. To clarify the distinction between rate and equilibrium in a chemical reaction
2. To examine the factors that affect equilibrium
3. To examine the factors that affect rates of chemical reactions

INTRODUCTION

At first, when students begin to study chemistry, the discussion centers around what *kinds* of chemical changes can occur, but to understand the biochemistry of the human body it is also necessary to look closely at *how* chemical reactions occur. This means that, instead of looking at many kinds of reactions and large quantities of materials, it's necessary to look at how individual molecules and ions interact with each other in a reaction. To understand a reaction at this level you need to consider two new ideas, equilibrium and rate of reaction.

Equilibrium

When a set of reagents is mixed, the reagents will undergo a chemical reaction provided other conditions are favorable. A reaction system can reach a balance point (equilibrium) at which all the reagents that can react have reacted. It may be reached when only 10% of the reagents have formed products, or there may be some reason why the equilibrium is reached only when 100% of the reagents have reacted. It depends on the nature of the reactants and the conditions of the experiment.

An equilibrium is not static. If you could label individual molecules in an equilibrium reaction, you would find there is a constant exchange of materials. Some reagents are constantly combining to form products while some product molecules are constantly reverting back to the original reactants. That is why the equation for an equilibrium reaction is usually written like this:

$$A + B \rightleftharpoons C + D$$

The two arrows pointing in forward and reverse directions remind you that equilibrium is a dynamic process, with molecules constantly moving back and forth between the product and reactant.

A measure of how well a set of reactants will actually produce products is the equilibrium constant, K_{eq}, which can be derived from the balanced equation for a reaction. A general form for a chemical reaction with its equilibrium constant looks like this.

$$aA + bB \rightleftharpoons cC + dD$$

$$K_{eq} = \frac{[C]^c [D]^d}{[A]^a [B]^b}$$

If K_{eq} is greater than 1, there will be more products than reactants at equilibrium, and the larger the K_{eq}, the better the conversion of reactants to products.

Notice the use in this equation of square brackets, [ ], which in chemistry are always used to mean "the concentration of this substance in moles per liter."

To look at a specific example, consider the reaction of hydrogen and oxygen to form water:

$$2H_2 \ + \ O_2 \ \rightleftharpoons \ 2H_2O$$

To write the K_{eq} for this reaction, you need concentrations of each substance at equilibrium and the coefficients for the materials. For this reaction

$$K_{eq} = \ \frac{[H_2O]^2}{[H_2]^2 \, [O_2]}$$

The equilibrium constant for this reaction is very large, so when the reaction finally reaches equilibrium, essentially all of the reactants have produced products. We say that a reaction of this type "goes to completion."

At any given temperature, the equilibrium constant is just that, a constant that depends on concentrations of all the substances in the equilibrium. But suppose you don't have exactly perfect concentrations of starting materials in the reaction vessel. And of course, as the reaction proceeds the concentrations of both reactants and products will change. Since K_{eq} is constant, deliberately altering the concentration of a reactant or product will change the ratios of all the other materials as well, in order to "maintain the equilibrium."

This idea is expressed in LeChâtelier's principle, which states that any stress on an equilibrium will cause a shift in concentrations of reactants or products in order to maintain the equilibrium. If you add extra reactant, more product will be made. If you reduce the quantity of reactant, more products will convert back to starting materials. If you take away products (by precipitation from the solution, for instance) more products will be made.

In this experiment, you will study several types of equilibrium reactions, applying "stresses" to see how the equilibria act in accordance with LeChâtelier's principle.

You have learned in your study of chemistry that a solution that can dissolve no more of a compound is called a saturated solution. For some compounds, a saturated solution is quite concentrated, while for others it is extremely dilute; it depends on the solubility of the solute under consideration. In any of these saturated solutions, however, there is a dynamic equilibrium. Silver chloride is very insoluble in water, but at any given time some silver ions and chloride ions are recombining as solid silver chloride, and some solid silver chloride particles are redissolving into the separated ions. Even though this is a very unfavorable equilibrium (balanced largely at the reactant side), LeChâtelier's principle still holds.

$$AgCl_{(s)} \ \rightleftharpoons \ Ag^+_{(aq)} \ + \ Cl^-_{(aq)}$$

Your first investigation of equilibrium involves a solution equilibrium, the dynamic equilibrium in a saturated solution of sodium chloride.

Another type of equilibrium that you will study in great detail later is an equilibrium between an acid and a base. So far, we have used only compounds that are strongly acidic, strongly basic, or neither (neutral). However, there are many that are only weakly one or the other, and one class of these compounds is called "pH indicators." They can take part in an ionic equilibrium according to the following equation. In this equation, In is not an element symbol, but an abbreviation for a complex molecular formula.

$$HIn_{(aq)} \ \rightleftharpoons \ H^+_{(aq)} \ + \ In^-_{(aq)}$$
$$\text{Color 1} \qquad\qquad \text{Acid} \quad \text{Color 2}$$

In an acidic solution, Color 1 is the color you see, but in a basic solution, Color 2 predominates. If you have ever checked the pH of a soil sample or a water sample with a paper test strip, you have made use of an indicator like this. Since the color change here is due to an equilibrium reaction, any pH indicator should be subject to the requirements of Le Châtelier's principle. Addition of acid provides H^+ as a product that drives the equilibrium to the left. Addition of a base uses up the acid H^+ ions, and thus removes a product from this equilibrium, allowing more of the HIn to break into ions to maintain the balance. You will look at only one of many examples, the indicator Congo red.

Congo red is a complicated carbon compound with the chemical formula $Na_2C_{32}H_{22}N_6Na_2O_6S_2$ in non-acidic conditions. This is the form of the compound you should represent by In$^-$. The form found in acidic conditions is $NaC_{32}H_{23}N_6Na_2O_6S_2$ (note one less sodium ion and one more hydrogen atom), but this is more easily represented by the abbreviation HIn. These abbreviations help you focus on the important change that takes place, the color change between the acidic and the basic forms.

The third type of equilibrium you will investigate is a "complex ion" equilibrium. Many elements, especially transition metals, can form complex ions that consist of an ion plus another substance, using coordinate bonding. You will look at the complex ion that forms between iron(III) and the thiocyanate ion, SCN$^-$. Of course, there are other ions in the solution, in this case potassium ions and chloride ions, that make the solution neutral overall, but only these two ions take part in the equilibrium you will observe.

$$Fe^{3+}_{(aq)} + SCN^-_{(aq)} \rightleftharpoons FeSCN^{2+}_{(aq)}$$

$$K_{eq} = \frac{[FeSCN^{2+}]}{[Fe^{3+}][SCN^-]}$$

Rate

Notice that the idea of equilibrium says nothing about how long the reaction will take to get there. It may take nanoseconds or megayears, because this also depends on the reaction conditions. This is part of the concept of the *rate* of a chemical reaction.

The rate of a reaction reflects how fast a set of reagents reaches its equilibrium point. There are four or five factors that determine a rate:

- nature of reactants
- surface area of reactants
- temperature at which the reaction takes place
- concentration of reactants
- catalyst to decrease the activation energy of the reaction

Rate can be measured in various ways. One way is to measure how many molecules of reactant form products per second; this measures the rate constant, k, for that particular reaction. Another is to use some visual signal to tell you when the reaction is finished. This second method is the one that you will use to look at temperature and concentration effects.

You have available two solutions, A and B. Solution A contains 1.5% (v/v) hydrogen peroxide. Solution B contains potassium iodide and all the other chemicals necessary to cause the formation of iodine from the reaction of hydrogen peroxide and potassium iodide, and to indicate when the reaction is complete. The reaction that takes place in this solution between peroxide and ions in the solution is this one.

$$H_2O_{2\,(aq)} + 3I^-_{(aq)} + 2H^+_{(aq)} \longrightarrow I_3^-_{(aq)} + 2\,H_2O$$

A second reaction takes place with another chemical in the solution, potassium thiosulfate, which uses up the I_3^- ions until the very end of the timing period, and then a third reaction with starch causes the I_3^- ion to give a deep blue-black color to signal the end. Although the chemistry of the system is complex, the observations to make are simple. You will make up a series of different concentrations of hydrogen peroxide using the 1.5% stock solution and then time their reactions with solution B. The reaction time in each case is finished when the blue-black starch-iodine color appears.

A second set of reactions with the 1.5% peroxide solution, but performed at different temperatures, should help you understand how temperature affects reaction rate.

It is important for good results to be sure you mix your solutions well and time them carefully from beginning to end of each experiment. If you are not careful about these things, your data will be hard to interpret. If necessary, repeat the series until you are certain of what you are looking at. And be sure that you always start with dry glassware, because residual wash water can affect concentrations, and this reaction is extremely sensitive to concentration changes.

The effect of a catalyst on the decomposition of hydrogen peroxide

Catalysts help reactions reach equilibrium faster by lowering the activation energy. You can see the effect of a catalyst on a reaction by looking at a different reaction of hydrogen peroxide, its decomposition to oxygen and water. In this case, you will observe the effect of catalysis by adding powdered manganese dioxide to hydrogen peroxide and observing the evolution of oxygen gas.

Catalytic effects are of paramount importance in biological systems where the reactions of life, by virtue of their large K_{eq}'s, are favored but slow because of the conditions inside a living cell.

REFERENCES

Stoker, H. S. *General, Organic, and Biological Chemistry;* Cengage Learning; Chapter 9.
  Review topics
  - Equilibrium constants
  - LeChâtelier's principle
  - Catalysts
  - Rates of chemical reactions

WEB RESOURCES

HENRI LOUIS LE CHATELIER.
http://www.woodrow.org/teachers/chemistry/institutes/1992/LeChatelier.html

Equilibrium and LeChatelier's Principle. *http://www.capital.net/com/vcl/equil/equil.htm*

Kinetics: Factors Affecting Reaction Rate.
http://library.thinkquest.org/C006669/data/Chem/kinetics/factors.html

Equilibrium Java Applet. *http://mc2.cchem.berkeley.edu/Java/equilibrium/index.html*

SAFETY NOTES

Wear chemical splash goggles when handling the chemicals used in this experiment, and you may wish to wear gloves as well. Disposable pipets, if used, are usually regarded as "sharps" and should be disposed of appropriately. Dispose of used reagents according to local ordinances.

In general, if you are a small-volume user the organic solvents can be discarded in an appropriately labeled waste container in the hood, and water solutions or suspensions of the unknown substances can be discarded in the sink, flushed down the drain with copious amounts of water. However, local ordinances govern your procedure.

EXPERIMENTAL PROCEDURE

PART A. EQUILIBRIUM IN CHEMICAL REACTIONS

A solubility equilibrium

Step 1. Obtain a 1.0-mL sample of saturated sodium chloride solution in a test tube. Describe what you see in the current equilibrium situation, and then perturb the equilibrium by adding 1 drop of concentrated hydrochloric acid, and note any change that occurs.

> ♥ Concentrated HCl is a strong acid. Use in the hood and use care!

Step 2. For a second 1.0-mL sample of saturated sodium chloride solution, predict what will happen when you add about 5 drops of saturated sodium nitrate solution, and then carry out that experiment.

An acid-base equilibrium

Step 1. In a small test tube, place 10 drops (about 0.5 mL) of 0.10 M hydrochloric acid. Add to it 2 drops of Congo red indicator, and note the appearance of the solution.

> ⓘ Note that this is a different concentration of HCl!

Step 2. Now add to this mixture 1 drop of basic 1.0 M sodium hydroxide solution and observe the change.

Step 3. Recycle the solution to the acidic form by adding 10-20 drops of the hydrochloric acid solution, just barely enough to return the solution to its acidic form.

Step 4. Again add a few drops of the sodium hydroxide solution and observe the cycle again.

A complex ion equilibrium

Step 1. First set up three reference solutions to show you the colors of the iron(III) ion, the thiocyanate ion, and the iron thiocyanate complex ion (potassium ions and chloride ions are colorless in solution). To do this, place 10 drops of iron chloride solution in one labeled test tube, 10 drops of potassium thiocyanate solution in a second labeled test tube, and a mixture of 8 drops of potassium thiocyanate and 2 drops of iron chloride solution in the third tube.

Step 2. Set up another tube like the third one above to observe the effect of adding extra reactant. To do this, place 8 drops of the potassium thiocyanate solution and 2 drops of the iron chloride solution in the tube, and mix well. This is similar to your observation tube.

Now add to this tube a few crystals of solid potassium thiocyanate, mix well, and compare colors with your reference tubes. Record your observations.

Step 3. Set up another test tube to observe the effect of removing a reactant. Place 8 drops of the potassium thiocyanate solution and 2 drops of the iron chloride solution in the tube, and mix well.

To this solution add a few crystals of sodium monohydrogen phosphate solution, and mix well. Some of the phosphate ions react with iron(III) ions and remove them from the equilibrium reaction. Record and explain what you observe here.

PART B. RATES OF CHEMICAL REACTIONS

The effect of concentration on reaction rate

Step 1. Obtain from your instructor 15 mL of solution A, the hydrogen peroxide solution. Divide it into three portions in three different test beakers. Leave the first beaker alone; this will be used for your reaction in which $[H_2O_2]$ is 1.5%.

Step 2. To the second sample, add 5 mL water and mix. Label this one 0.75% H_2O_2.

Step 3. To the third tube add 10 mL water, and label it 0.5% H_2O_2. Use a stopwatch for timing.

Step 4. Set up three small beakers, each containing 5 mL of solution B. Working quickly, pour all of your 1.5 % peroxide solution into one of these beakers, mix well, and start the stopwatch. Stop it when the reaction mixture turns blue/black, and record the time for this reaction.

Step 5. Measure 5 mL of your 0.75% peroxide solution in a dry graduated cylinder. As you did in Step 4, add this sample all at once, with mixing, to the second beaker of solution B, and record the time for the reaction to occur.

Step 6. Repeat again, using 5 mL of the 0.5% peroxide solution.

The effect of temperature on reaction rates

Plan to do three reactions; one at ice temperature, one at room temperature, and one at 50 °C.

Step 1. Prepare a cooling bath for the cold reaction by filling a 250-mL beaker half-full of an ice-water mixture.

Step 2. Place in this mixture, in separate containers, 5 mL of solution A and 5 mL of solution B. Allow them to cool (pre-incubate) for 10 minutes before beginning the reaction.

Step 3. Take the temperature of the ice bath just before you begin the reaction.

Step 4. Working quickly, pour all of your peroxide solution A into solution B, mix well, and start the stopwatch. Stop the stopwatch when the reaction mixture turns blue/black, and record the time for this reaction.

Step 5. Record the room temperature, and time the reaction of a new set of chemicals A and B at room temperature.

Step 6. In a 250-mL beaker, heat a sample of water to about 50 °C on a hot plate. Prepare separate 5-mL samples of solutions A and B, and pre-incubate them for 10 minutes in the hot water bath (off the hot plate).

Step 7. Just before mixing solutions A and B, take the temperature of the water. Then mix the solutions, start the stopwatch, and time the reaction at the elevated temperature.

The effect of a catalyst on the decomposition of hydrogen peroxide

This reaction may be done as a demonstration or by student groups. It can be started before the other experiments.

Step 1. Prepare two gas collection devices like the one in figure 18-1, and fill the reaction well of each with 3% H_2O_2.

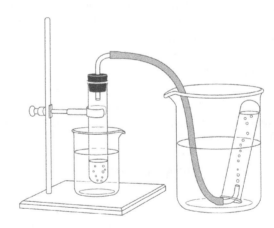

Figure 18-1. A small gas-collection device.

Step 2. To the first gas reaction well, add a pinch of sand, an inert solid substance. To the second, add (using a spatula!) about the same amount of solid MnO_2. Fit the corks on each apparatus.

Step 3. Observe these reactions over the course of the lab period to determine if the decomposition of hydrogen peroxide proceeds faster in the presence or absence of the catalyst MnO_2.

EXPERIMENT 18 PRELAB EXERCISES:

EQUILIBRIUM AND RATE IN CHEMICAL REACTIONS

Name_____ **Partner** _____ **Date** _____

1. Describe the difference between equilibrium and reaction rate.

2. Write an equilibrium expression for K_{eq} for the reaction shown below.

$$2NO + O_2 \;\rightleftharpoons\; 2NO_2$$

3. The substances in Question 2 are all gases. What would be the effect on the equilibrium concentration of NO if you removed NO_2 as it is formed?

4. What factors can be altered if you wish to increase the rate of a chemical reaction?

5. Does a catalyst increase the **rate** of a chemical reaction??

6. Does a catalyst change the K_{eq} of a reaction?

EXPERIMENT 18 REPORT: EQUILIBRIUM AND RATE IN CHEMICAL REACTIONS

Name_____ Partner _____ Date _____

PART A. EQUILIBRIUM IN CHEMICAL REACTIONS

A solubility equilibrium

1. Write the equilibrium expression for the solubility of sodium chloride in water.

2. Describe what is happening at the molecular level in the saturated sodium chloride solution.

3. What did you see when you added HCl? How is this consistent with Le Châtelier's principle?

4. Predict what you should see when you add sodium nitrate solution. Explain.

5. Did your experiment confirm your prediction?

An acid base equilibrium

1. What did you see in acidic solution?

2. What did you see in basic solution?

3. Write the equilibrium for Congo red, showing the color changes that accompany it.

A complex ion equilibrium

1. The visible effect of adding extra thiocyanate ion was _____ .

2. This means that the equilibrium shifted to produce _____ when extra reactant was added.

3. The visible effect of adding sodium phosphate was _____ .

4. This means the equilibrium shifted to produce _____ when reactant was removed.

5. When zinc is mixed with hydrochloric acid, the reaction is as shown below.

$$Zn_{(s)} + 2HCl_{(l)} \longrightarrow ZnCl_2 + H_{2(g)}$$

If sufficient HCl is present, the reaction will go to completion, and all the zinc will be used up. Explain how this is an illustration of LeChâtelier's principle.

PART B. RATES OF CHEMICAL REACTIONS

The Effect of Concentration on Reaction Rate

Concentration of H_2O_2	Time to completion
1.5%	
0.75%	
0.5%	

1. Describe your results in a sentence or two, and reach a conclusion about the effect of concentration on the rate of the reaction of hydrogen peroxide with potassium iodide.

The Effect of Temperature on Reaction Rate

Temperature	Time to completion

2. Describe your results in a sentence or two, and reach a conclusion about the effect of temperature on the rate of the reaction of hydrogen peroxide with potassium iodide.

The effect of a catalyst on the decomposition of hydrogen peroxide

1. Describe your observations on the decomposition of hydrogen peroxide in the presence and absence of manganese dioxide.

2. Would you expect large chunks or smaller chunks of a catalyst to be more effective? Why or why not?

EXPERIMENT 19

ACIDS, BASES, AND pH

GOALS

1. To define acids and bases
2. To examine the concept of weak acids and bases
3. To define the pH scale
4. To use the pH scale to determine the acidity of household and environmental mixtures

INTRODUCTION

The simplest way to understand acids and bases is to recognize that acids give hydrogen ions in solution, and bases give hydroxide ions. This is really an over-simplification. It works fine when you are studying acids and bases that completely ionize in water, but is insufficient to explain what goes on in biological systems in which reactions with favorable but weak equilibrium constants are the norm. The more general Brønsted-Lowry definition helps more: An acid is a proton donor, and a base is a proton acceptor. This definition paves the way to understanding the behavior of many compounds that act as acids or bases but don't look the part.

A weak acid or base is one that only partly ionizes in solution. One example of a weak acid is acetic acid, $HC_2H_3O_2$. It has one acidic proton (the others are tied up in fully covalent bonds), but even that one will dissociate only part of the time. You can write an equilibrium equation that shows this.

$$HC_2H_3O_{2(l)} \rightleftharpoons H^+_{(aq)} + C_2H_3O^-_{(aq)}$$

The easiest weak base to understand is ammonia, NH_3. It uses the lone pair of electrons on the nitrogen to take on a proton from water (act as a proton acceptor) and make a solution that does have hydroxide ions in solution.

$$NH_{3(g)} + H_2O_{(l)} \rightleftharpoons NH_4^+_{(aq)} + OH^-_{(aq)}$$

For solutions of weak acids and bases, such as most biological systems, one extremely important question is, "How acidic is it?" To determine relative acidity and basicity, the pH scale was developed. It describes solutions with a hydrogen ion concentration $[H^+]$ of 1 M to 1×10^{-14} M, the range that is possible in water. The equation for the relationship between pH and $[H^+]$ is this.

$$pH = -\log [H^+]$$

You should refer to your textbook for a more detailed discussion, but the general rule to remember is that any pH below 7.0 is the pH of an acidic solution, and any pH above 7.0 is the pH of a basic solution. The farther away from 7.0, the more acidic or basic the solution is. Of course, many substances are neutral. In this exercise, you will be looking at a number of common household and environmental materials and examining the question "Is this substance acidic, basic, or neutral?"

There are two common ways to do this. One you may already be familiar with is the use of pH test paper. These papers are simply pieces of filter paper soaked in one or more indicator solutions so that the paper turns one color in the presence of an acid, and another color in the presence of a base. The pH paper takes advantage of the weak acid equilibrium that was discussed in Experiment 18. In an acidic solution, LeChâtelier's principle requires that the extra H^+ of the solution shift the concentrations to the left (acid) color, while in basic solution, the H^+ of this equilibrium is used up in combination with the base to make water, allowing more of the In^- (basic color) to form.

$$HIn \rightleftharpoons H^+_{(aq)} + In^-_{(aq)}$$
$$\text{Color 1} \qquad\qquad\qquad \text{Color 2}$$
$$\text{in acidic solution} \qquad\qquad \text{in basic solution}$$

The simplest form of pH paper is litmus paper, which shows you only if a solution is acidic or basic. If a paper is soaked in several indicators that change color at different pH, then papers that show actual pH values can be made. You will be using both wide-range paper that will give whole pH values between 1 and 12 and narrow-range pH paper that can help you determine pH to the nearest tenth of a pH unit.

Be aware that the color in a pH test strip can be washed out of it, so the wrong way to test a solution is to dip a paper into the liquid. The best way to proceed is to lay a small piece of the pH paper on a paper towel and barely dampen it with water. Then touch a stirring rod first to the solution to be tested and then to the damp paper, and compare the color you see with the color chart that comes with the paper.

An instrumental method of determining pH makes use of a pH meter that might look something like the picture in Figure 19-1.

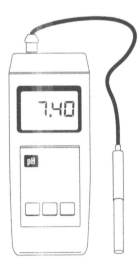

Figure 19-1. A standard pH meter.

There are a number of different brands, and the operating instructions are similar but not identical for each brand, so your instructor will explain how the kind available to you operates. You will check several solutions with both paper and a pH meter.

Among the solutions you will test are solutions of salts made from weak acids and weak bases as well as some made from strong acids and strong bases. Not all of these salt solutions will be neutral even though the salts themselves have neither H^+ nor OH^- ions in them.

In the environmental department, you will look at how to check the pH of a soil sample and a water sample. In many parts of the country, the soil is extremely acidic, and the addition of lime (calcium carbonate) is recommended to bring it to a more neutral state. In other parts of the country, the soil is rather alkaline (basic), and soil acidification is needed. Several chemicals can be added to the soil to acidify it, but you will look only at one soil sample of this type, that has been treated with iron sulfate. Since pH is only meaningful in a water solution, soil samples are usually suspended in water, and the pH of the resulting water is checked.

Water samples are easy to check for pH. If acid rain is a problem in your area, you may find that open water samples collected around campus are significantly acidic.

REFERENCES

Stoker, H. S. *General, Organic, and Biological Chemistry;* Cengage Learning; Chapters 8 and 9.
  Review topics
  - Equilibrium
  - Definitions of acids and bases
  - Reactions of acids and bases
  - The pH scale
  - Hydrolysis of salts

WEB RESOURCES

Miami Museum of Science-pH Panel/Ammonia. *http://www.miamisci.org/ph/phammonia.html*

Miami Museum of Science-The pH Factor. *http://www.miamisci.org/ph/index.html*

CHEMTUTOR ACIDS AND BASES. *http://www.chemtutor.com/acid.htm*

Id-84 IRON DEFICIENCY OF LANDSCAPE PLANTS.
http://www.ca.uky.edu/agc/pubs/id/id84/id84.htm

SAFETY NOTES

Follow local regulations for the disposal of these chemicals. Some are strongly acidic or strongly basic, and the usual precautions for handling such materials must be followed. Wear chemical splash goggles and perhaps gloves throughout this experiment.

EXPERIMENTAL PROCEDURE

PART A. pH OF COMMON CHEMICALS

PH in our households

Step 1. For each of the household substances in the list below, place 5 mL of the substance in a small beaker, and find the pH of the substance using first wide-range and then narrow-range pH paper.

chlorine bleach	cola
liquid soap	milk
baking soda	vinegar
baking powder	beer
lemon juice	liquid drain cleaner
spring water (bottled)	mouthwash

Step 2. Find the pH of each substance in the table using a pH meter as directed by your instructor.

Step 3. Calculate the [H$^+$] for each, and identify each as acidic, basic or neutral.

> ⓘ Be sure to rinse the electrode between readings!

Indicator color and pH

There are five indicators available in dropper bottles. For each indicator, follow the following steps.

Step 1. Place 5 mL of 0.1 M hydrochloric acid in a small beaker. To it add 3 drops of the indicator to be tested.

> 🖤 Treat HCl with caution!

Step 2. Check the pH of this solution with meter or paper.

Step 3. To the solution, add 1.0 M sodium hydroxide solution dropwise until the indicator just turns to its basic color. Record this pH as the pH at which the indicator is fully converted to its basic form.

Step 4. Using your experimentally determined pH in Step 3, alculate the [H$^+$] concentration at which you observed the transition.

pH of salt solutions

Step 1. In labeled wells in a spot plate, place 5 drops of each of the following solutions: ammonium chloride, sodium chloride, sodium acetate, ammonium acetate, and iron chloride.

Step 2. Check the pH of each solution first with wide-range pH paper to find the approximate pH and then with narrow-range paper, and record the pH of each sample to the nearest tenth of a pH unit.

Step 3. Perform the calculations necessary to determine the [H$^+$] concentration in each of these solutions.

Alternate Step 1: Use 5-mL samples of each solution, and determine the pH with a pH meter.

PART B. pH IN OUR ENVIRONMENT

Soil pH

There are three soil samples available for you to examine. One is a native soil specimen from your campus or nearby. One has been "limed," and one has been "acidified" with standard garden treatments for this purpose. For each sample, follow the directions below to determine its pH.

Step 1. Scoop about 50 mL of the sample from the stock, and place it in a 250-mL Erlenmeyer flask. Add 100 mL of distilled water, and swirl the mixture for 15 minutes. If a stirring motor and stir bar are available, these will be the most efficient tools to use.

Step 2. Allow the bulk of the soil to settle. Dampen a piece of narrow-range pH paper (pH range 4-8) with water. Drop one drop of the water from your soil sample onto the paper, and record the pH.

Water pH

Determine the pH of your local tap water, distilled water, and a sample collected from a local pond or stream. For each sample, use either pH paper or a pH meter as your instructor directs.

PART C. EXERCISES

Complete the exercises on the report sheet.

EXPERIMENT 19 PRELAB EXERCISES:

ACIDS, BASES, AND pH

Name_____ Partner _____ Date _____

1. What is the function of an indicator?

2. Write the equation for the general form of an indicator equilibrium. Underline the form of the indicator in acidic solution, and circle the form of the indicator that is in largest concentration in basic solution.

3. Find the pH of a solution that has a concentration of hydrochloric acid equal to 0.001 M.

4. Using the Internet, find the normal pH of rainfall (not acid rain). Would a rain sample with a pH of 5.0 be considered "acid rain?"

5. Predict whether the pH of a solution of sodium sulfate should be acidic, basic, or neutral. Justify your prediction.

EXPERIMENT 19 REPORT: ACIDS, BASES, AND pH

Name_____ **Partner** _____ **Date** _____

PART A. pH IN OUR HOUSEHOLDS

pH of Common Household Materials

Substance	pH by paper	pH by meter	$[H^+]$ by paper	$[H^+]$ by meter
Bottled water				
Mouthwash				
Drain cleaner				
Lemon juice				
Beer				
Baking powder				
Vinegar				
Baking soda				
Milk				
Soda				
Liquid soap				
Chlorine bleach				

Do the hydrogen ion concentrations as measured by both methods agree with each other? If not, what might be the reason?

pH of Indicator Solutions

Substance	Color in acid	Color in base	pH of color change
Phenolphthalein			
Congo red			
Bromthymol blue			
Alizarin yellow			
Methyl red			

Why do the indicators change color at different pH values?

pH of Salt Solutions

Substance	pH	[H$^+$]
Ammonium chloride		
Sodium chloride		
Sodium acetate		
Ammonium acetate		

Can you determine a general rule of thumb for how to decide if a salt solution is acidic, basic, or neutral?

PART B. pH IN OUR ENVIRONMENT

pH of Local Soil Samples

Test sample	pH
Normal soil	
Limed soil	
Iron sulfate treated soil	
Tap water	
Distilled water	
Open water	

1. How did liming the soil sample affect the soil's pH?

2. How did the application of iron sulfate affect the soil's pH?

3. Explain why lime is used to make acidic soil more neutral.

4. Why is iron sulfate useful to acidify soil?

5. Is the local pond water significantly acidic?

PART C. EXERCISES

1. Oak leaves lying in water can cause the water to have a pH of 3 or less. Is the water acidic, basic, or neutral?

2. When soap is made in a pre-industrial process, animal fat is boiled with water that has been steeped with wood ash. Use the Internet or book sources to find out why the wood ashes are needed, and whether the pH of the water is acidic, basic, or neutral.

3. Suggest a way to determine the pH of a swimming pool.

EXPERIMENT 20

BUFFERS

GOALS

1. To prepare a buffer that holds an acidic pH
2. To prepare a buffer that holds a basic pH
3. To investigate buffer capacity as a function of concentration

INTRODUCTION

A buffer is a solution that resists changes in pH. To do that, it must contain a weak acid and its conjugate base in equilibrium with each other. For example, if a solution is buffered at pH = 4.0, then reactions that produce or consume protons can take place without changing the pH much. If some H^+ ions are released into the solution, then the basic part of the buffer reacts with them. If a reaction takes up hydrogen ions, then the acidic part of the buffer releases more to maintain the original pH.

There are four steps to preparing a good buffer after you decide on the pH you need and the appropriate compound set to maintain it. In this exercise, you will use three different salts obtained from phosphoric acid (H_3PO_4) to make two different buffers and test them.

First you must calculate the correct amount of the acidic component and the correct amount of the basic component of the solution. Suppose you needed 1.0 liter of a solution buffered at pH 2.7. An appropriate weak acid and its conjugate base would be H_3PO_4 and NaH_2PO_4, respectively. Notice that the "weak base" is the dihydrogen phosphate ion, $H_2PO_4^-$, but you won't find a bottle of dihydrogen phosphate ions alone on the shelf; they are available to you as the sodium salt. The calculation uses the Henderson-Hasselbach equation in which A^- is the basiccomponent, and HA is the acidic component.

$$pH = pK_a + \log \frac{[A^-]}{[HA]}$$

Temperature and exact concentration can also be a factor, so for the buffers you are making up, you are provided with the number of moles of each substance you need to use. For a 0.10 M buffer, you would need .0178 mol of H_3PO_4 and 0.0821 mol of sodium dihydrogen phosphate. This works out to 1.74 g of phosphoric acid, but the sodium salt needs consideration. It is available as the hydrate, $NaH_2PO_4 \cdot H_2O$, so the formula weight you use needs to take that into account. The mass of the salt needed is 9.85 g.

To make up the solution, use the techniques you studied earlier in the course: dissolve the materials in slightly <u>less than</u> the needed amount of water first. At this point, you have a buffer solution, but it may not be at the precise pH you want. You will need to check it to see what pH you actually obtained, and if it is not quite right, add a little acid or base to make it right. If you use a strong acid or base, you can force the buffer pH in the right direction, and then you can dilute to the final volume and begin to work with it.

Once you have made your buffers, you will observe their ability to resist changes in acidity by comparing their actions to the action of pure water.

Another aspect of buffers that needs consideration is capacity. The molarity of the buffer will determine how much acid shift it can actually absorb. In the body, for instance, the carbonate buffer is not very concentrated, and major shifts in blood pH can be more than it can handle, leading to acidosis or alkalosis that can be life-threatening. You will make a diluted buffer from one of your 0.10 M buffers and compare the capacity of each buffer to absorb protons with no change in pH.

How do you make a diluted buffer of the desired concentration? You will be working with two solutions, so the usual dilution equation applies.

$$C_s V_s = C_d V_d$$

Refer to your text for a refresher on solution calculations if you need it.

REFERENCES

Stoker, H. S. *General, Organic, and Biological Chemistry;* Cengage Learning; Chapters 8 and 10. Review topics

- Equilibrium
- pH
- The Henderson-Hasselbach relationship
- Neutralization reactions
- Solution dilutions

WEB RESOURCES

CHEMTUTOR ACIDS AND BASES. *http://www.chemtutor.com/acid.htm*

Miami Museum of Science-The pH Factor. *http://www.miamisci.org/ph/*

Buffers for pH control (c) Rob Beynon, University of Liverpool. http://www.liv.ac.uk/buffers/buffercalc.html

SAFETY NOTES

Follow local regulations for the disposal of these chemicals. Some are strongly acidic or strongly basic, and the usual precautions for handling such materials must be followed. Wear chemical splash goggles and perhaps gloves throughout this experiment.

EXPERIMENTAL PROCEDURE

PART A. THE PREPARATION OF A BUFFER OF pH 8.0

Step 1. To make 100 mL of a 0.10 M (100 mM) phosphate buffer, you will need to weigh out 0.0094 moles of the basic component and 0.0005 moles of the acidic component. The two substances you need to use are sodium monohydrogen phosphate ($Na_2HPO_4 \cdot 7H_2O$, molecular mass = 268.07) and sodium dihydrogen phosphate ($NaH_2PO_4 \cdot H_2O$, g-fw = 119.98). Before doing the calculation, identify which one of these is the basic component and which one is the acidic component. Then calculate the required mass of each compound.

Step 2. Using the amount of each salt that you calculated, dissolve the salts together in about 50-75 mL water in a 400-mL beaker. When both are completely dissolved, check the pH of the solution with a pH meter following your instructor's directions for the use of the type of meter you have.

Step 3. If the pH of this solution is not 8.0, adjust it to the right pH using one of the following solutions. If the pH is below 8.0, use 6.0 M sodium hydroxide. Add a drop at a time, checking the pH after each addition, until 8.0 is reached. If the pH is above 8.0, follow the same procedure, but use 6.0 M hydrochloric acid to adjust the pH.

> ♥ Treat HCl and NaOH with caution!

Step 4. Dilute the solution to 100 mL and mix well. Take the solution to your instructor for a final reading of the pH you obtained.

Step 5. Place about 20 mL of your prepared buffer in each of two beakers. Add 20 mL of distilled water to a third. Set one of the beakers of prepared buffer aside for use in Step 6. Using the pH meter and the first beaker of buffer solution, observe the pH of the solution when 3 drops of 0.10 M HCl is added and mixed in. Compare this with the pH of distilled water (Beaker 3) when 3 drops of HCl are added.

> ⓘ Be sure to rinse the electrode between readings!

Step 6. Use the second beaker of buffer and a new sample of distilled water to check the buffer's resistance to addition of base. Check the pH of the buffer before and after the addition of 3 drops of 0.10 M sodium hydroxide solution. Compare this with the reaction of distilled water when sodium hydroxide is added.

PART B. THE PREPARATION OF A BUFFER OF pH 12.3

The procedure for this section of the exercise is exactly the same as the procedure for Part A. However, the two salts you need to use are $Na_2HPO_4 \cdot 7H_2O$, molecular mass = 268.07, and $Na_3PO_4 \cdot 12H_2O$, molecular mass = 380.13. Abbreviated forms of the steps are shown here for reference. Refer to Part A if you need more detail.

Step 1. Identify the acidic and basic components of the buffer mixture. Calculate the amount of each needed. You will need 0.0054 moles of the acidic component and 0.0045 moles of the basic component.

Step 2. Dissolve the salts in water as you did in Step 2 of Part A, and check the pH of the solution.

Step 3. Adjust the pH to exactly 12.3 using either acid or base as needed.

Step 4. Make up the solution volume to 100 mL, mix, and take it to your instructor for checking.

Step 5. Check the ability of the buffer to resist changes in pH as you did in Part A.

PART C. BUFFER CAPACITY

Step 1. Calculate the number of milliliters of the 0.10 M phosphate buffer needed to prepare 100 mL of a 0.010 M buffer. Prepare 100 mL of a 0.01 M phosphate buffer by diluting the correct quantity of the 0.10 M buffer appropriately. Divide it into two portions in separate beakers.

    Step 2. Check the pH of one container of diluted solution, and then add three drops of 0.10 M HCl. Observe the pH using a pH meter.

    Step 3. Record the pH of the second sample of diluted buffer, and add three drops of 0.10 M sodium hydroxide. Observe the pH.

    Step 4. Compare your results with the resistance to pH change you observed in Part A.

PART D. EXERCISES

Complete the exercises at the end of the report form.

EXPERIMENT 20 PRELAB EXERCISES:

BUFFERS

Name_____ Partner _____ Date _____

1. What is the function of a buffer?

2. How does the capacity of a buffer relate to its concentration?

3. Calculate the quantity, in grams, of formic acid and sodium formate necessary to make 500. mL of a buffer that has a formic acid concentration of 0.75 M and a sodium formate concentration of 0.75 M. You will need to look up or calculate the molar masses of these two compounds as your first step. See question 5 below for their chemical formulas.

4. A student tried to make a buffer of pH 4.0 using sodium acetate and acetic acid. The initial pH of the buffer solution was 3.9. How should the student adjust the pH to the desired value?

5. What is the pH of the buffer described in Question 3? The formic acid ($HCHO_2$) concentration is 0.75 M, the sodium formate ($NaCHO_2$) concentration is 0.75 M, and the pK_a of formic acid is 3.74.

EXPERIMENT 20 REPORT: BUFFERS

Name_____ Partner _____ Date _____

PART A. THE PREPARATION OF A BUFFER OF pH 8.0

Which phosphate salt is the acidic component?   _____

Calculate the number of grams of this salt needed
to make 0.000500 moles.   _____

Which phosphate salt is the basic component?   _____

Calculate the number of grams of this salt needed
to make 0.00940 moles.   _____

Initial pH reading, before adjustment   _____

Did you use HCl or NaOH to adjust the pH?   _____

Final pH reading   _____

Instructor's check   _____

pH of Solution with Addition of Acid or Base

Solution	pH with added acid	pH with added base	Change in pH
pH 8.0 buffer			
Distilled water			

1. Did the buffer solution resist changes in pH?

2. Explain this observation in terms of weak acid-base equilibrium.

PART B. THE PREPARATION OF A BUFFER OF pH 12.3

Which phosphate salt is the acidic component?   _____

Calculate the number of grams of this salt needed
to make 0.000500 moles.   _____

Which phosphate salt is the basic component?   _____

Calculate the number of grams of this salt needed
to make 0.00940 moles.   _____

Initial pH reading before adjustment   _____

Did you use HCl or NaOH to adjust the pH?   _____

Final pH reading   _____

Instructor's check   _____

pH of Solution with Addition of Acid or Base

Solution	pH with added acid	pH with added base	Change in pH
pH 12.3 buffer			
Distilled water			

1. Did the buffer solution resist changes in pH?

2. Explain this observation in terms of weak acid-base equilibrium.

PART C. BUFFER CAPACITY

Calculate and record the number of milliliters of the 0.10 M
phosphate buffer needed to prepare 100 mL of a 0.010 M buffer.   _____

Check of Capacity of Buffers of Different Molarity

Solution	pH	pH after acid addition	pH after base addition
0.01 M buffer			
0.10 M buffer (from Part A)			

Did the 0.01 M buffer resist changes in pH? How well, compared to the 0.10 M buffer?

PART D. EXERCISES

1. A student needed 1000 mL of a 0.10 M buffer of pH 4.0 Calculate the amount of sodium acetate
and acetic acid needed for this solution.

2. What should the student do if the actual pH of the prepared buffer is 4.1?

3. What could the student do if the actual pH is 3.9?

EXPERIMENT 21

TITRATION

GOALS

1. To demonstrate the analysis by titration of an acidic sample and a basic sample
2. To determine the molar concentration and (weight/volume) percent of acetic acid in vinegar
3. To practice the calculations associated with titrations

INTRODUCTION

Simple tests of acidity like the use of pH paper can only help to a certain extent in determining the acid concentration of a solution. Often more precise analysis is needed, and for this a titration is required.

A titration uses a solution of known concentration to react completely with a precisely measured quantity of another substance. When a titration is done, there are two stages: determining the concentration of an acid or base solution extremely accurately and then using that solution to determine the concentration of the sample of interest. This process uses both of the usual dilution equations.

$$V \times M = \text{moles present}$$
$$C_s V_s = C_d V_d$$

In this experiment, you will first determine the concentration of a solution of sodium hydroxide by using it to titrate a very accurately weighed sample of a solid acidic substance, potassium hydrogen phthalate (KHP for abbreviation, but actual formula $C_8H_5KO_4$, molar mass 204.2151g/mole).

To determine the concentration of the sodium hydroxide solution, you will find out how much of the solution will react with a very exactly known quantity of KHP. The neutralization equation for the reaction of NaOH with KHP is this.

$$\text{NaOH} + \text{KHP} \longrightarrow \text{KNaP} + \text{H}_2\text{O}$$

From the equation you can see that one mole of sodium hydroxide reacts with 1 mole of KHP, so if you know the volume of sodium hydroxide solution used and the number of moles of KHP it reacted with, you can determine the molarity of the sodium hydroxide solution to four or more significant figures. The steps go like this.

1) From the mass of KHP, find the number of moles used. Use the correct formula weight!
2) Moles KHP = moles NaOH in volume of NaOH used. This volume of NaOH solution used is the experimentally determined number.

3) Then use the usual relationship

V_{NaOH} x M_{NaOH} =   moles of NaOH in that volume
(from titration)   (unknown)   (from moles of KHP used)

Rearranged to isolate the unknown value you are trying to find, this equation becomes

$$M_{NaOH} = \frac{\text{moles of NaOH in that volume}}{V_{NaOH}}$$

The new tool you will be using for this experiment to find the volume of the sodium hydroxide is a *buret*. This is a device that is designed to deliver precisely controlled volumes of liquid in a dropwise fashion into another solution. However, it is only as good as your technique in using it, so be sure you understand how to use it correctly and how many significant figures to record when you use it. Your instructor will give you a short lesson on this, and several of the suggested Internet resources show you how to use a buret correctly.

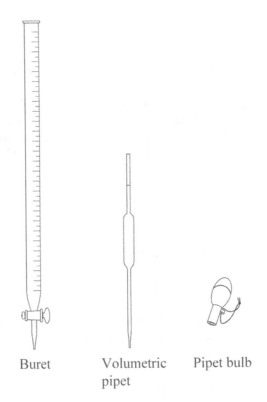

Buret   Volumetric   Pipet bulb
                   pipet

Figure 21-1. Tools for a titration: a buret, a volumetric pipet, and a pipet bulb.

How do you know when the neutralization reaction is finished? End points can be determined in various ways. You will use the most non-technological way, adding an indicator to the reagent flask. Phenolphthalein indicator is colorless in acidic solutions and turns pink in basic solutions, so at the end point, when acid exactly matches base, the solution should be a very faint pink. A dark pink means you have gone too far, and the volume of base you record may be far more than was actually needed, leading you to an erroneous conclusion about the molarity of your base.

Once you know the molarity of the sodium hydroxide solution accurately, you are finished with the KHP and are ready to do the analysis you expected to do in the first place, the concentration of acetic acid in vinegar. The procedure is similar to the first one except that the analyte (the material you want to analyze) is a liquid and is best measured by volume. This means using a volumetric pipet, this time a 5.00-mL pipet. Since you are using two *solutions* this time, not a solution and a solid as in Part A, the relationship between the acid and basic solutions is most easily calculated using the neutralization equation.

$$C_a \ \times \ V_a \ = \ C_b \ \times \ V_b$$

	acid	5.00 mL	known	read from
	concentration		from	titration
			Part A	

On your report form, there is an equation for converting molarity of acetic acid to %(w/v), so you can also see if your vinegar sample conforms to expected concentrations as reported on its label.

You will be evaluated on the accuracy of your results, so be sure to do your best.

REFERENCES

Stoker, H. S. *General, Organic, and Biological Chemistry;* Cengage Learning; Chapters 8 and 10.
Review topics
- Solution calculations
- Neutralization reactions
- Common laboratory tools and techniques

WEB RESOURCES

How to Perform a Titration. *http://www.wikihow.com/Perform-a-Titration*

ChemLab-Techniques-Titration. *http://www.dartmouth.edu/~chemlab/techniques/titration.html*

CHEMTUTOR ACIDS AND BASES. *http://www.chemtutor.com/acid.htm#titr*

SAFETY NOTES

Follow local regulations for the disposal of these chemicals. Some are significantly acidic or basic, and the usual precautions for handling such materials must be followed. Wear chemical splash goggles and perhaps gloves throughout this experiment.

EXPERIMENTAL PROCEDURE

PART A. CONCENTRATION OF A SODIUM HYDROXIDE SOLUTION

Step 1. Obtain from your instructor three pre-weighed samples of potassium hydrogen phthalate, $C_8H_5KO_4$, fw 204.2151. Label these samples so you can tell which is which, and record their sample numbers and weights.

Step 2. Place one of the samples in a 125-mL Erlenmeyer flask, and dissolve it as much as possible in 25 mL of water. Most of it should go into solution readily. If a bit remains undissolved, this is not a problem; it will dissolve as the analysis proceeds. Label this solution with the sample number. If you have enough flasks available, all three samples can be prepared at once.

> ⓘ Don't lose any of the powder. This is the primary standard!

Step 3. Add 5 drops of phenolphthalein indicator to the solution.

Step 4. Prepare a buret for use as your instructor directs. In general, this involves cleaning the buret and then rinsing it with three small volumes of the sodium hydroxide solution (the base) you are going to analyze. Then mount the buret on the ring stand, and fill it with the base to be analyzed. Run some of the base through the tip of the buret and be certain that any air bubbles have been removed.

Step 5. Record the buret reading just before beginning the titration.

Step 6. Titrate the acid sample with the sodium hydroxide to a very **faint** pink end point. The fainter the better—this is easiest to see if the flask is sitting on a piece of white paper or card stock. You can use large (1-5 mL) volumes at first, but as you get nearer the end point you will need to go drop by drop, or even by half-drops, with continual swirling.

Step 7. Record the final buret reading, and find the volume of sodium hydroxide actually used.

Step 8. Repeat the titration for the other two potassium hydrogen phthalate samples, recording the data as you perform each one.

Step 9. Calculate the molarity of the sodium hydroxide solution as determined in all three samples and the average of the three trials. You will need this average in Part B.

PART B. THE ACID CONCENTRATION OF VINEGAR

Step 1. Using a volumetric pipet, measure 5.00 mL of the vinegar solution into a 125-mL Erlenmeyer flask. Dilute it with about 25 mL of water so the titration will be easier to see, and add 4 drops of phenolphthalein indicator solution.

Step 2. Refill your buret with fresh sodium hydroxide solution, and titrate the vinegar sample to a faint pink end point. Record the volume of sodium hydroxide solution used.

Step 3. Repeat this analysis twice more, with two more 5.00-mL samples of vinegar, and record the volume of sodium hydroxide used for each one. If you are obtaining reproducible results, all three should agree very closely.

Step 4. Using the titration relationship, the known molarity of the sodium hydroxide solution and the volumes of sodium hydroxide solution and vinegar used, calculate the molarity of vinegar as determined for each sample. Report this and the average of the three trials.

Step 5. Following the model calculation provided, find the percent of acetic acid in the vinegar.

PART C. EXERCISES

Complete Part C of the report.

EXPERIMENT 21 PRELAB EXERCISES:

TITRATION

Name_____ **Partner** _____ **Date** _____

To discover the molarity of a sodium hydroxide solution, a student used it to titrate a 1.0056-g sample of potassium hydrogen phthalate. The titration required 24.52 mL of the sodium hydroxide solution.

The same sodium hydroxide solution was used to titrate 10.00 mL of a solution of hydrochloric acid, HCl, of unknown concentration. The titration required 23.05 mL of the sodium hydroxide solution.

Use this information to answer the questions below.

1. How many moles of potassium hydrogen phthalate (KHP) were used?

2. How many moles of sodium hydroxide were used in the titration?

3. What was the molarity of the sodium hydroxide solution?

4. What was the molarity of the hydrochloric acid solution?

5. What was the percent concentration (w/v) of the hydrochloric acid solution? (Hint: From your answer in Question 4, determine the number of grams of HCl in 1000 mL of solution, and use that information to find the number of grams, then the %, in 100 mL of solution.)

EXPERIMENT 21 REPORT: TITRATION

Name_____ Partner _____ Date _____

PART A. CONCENTRATION OF A SODIUM HYDROXIDE SOLUTION

Standardization of a Sodium Hydroxide Solution

KHP sample number			
Mass of sample of KHP			
Moles of KHP			
Buret reading, start			
Buret reading, end			
Volume of NaOH solution used			
Moles of NaOH in that vol.			
Calculated molarity of NaOH (Use V x M = moles)			

1. Sample calculation

2. The average molarity of the sodium hydroxide determinations is _____ .

PART B. THE ACID CONCENTRATION OF VINEGAR

Titration of a Vinegar Solution

Vinegar sample number			
Volume of vinegar used			
Buret reading, start			
Buret reading.end			
Volume of sodium hydroxide used			
Molarity of vinegar Use $C_aV_a = C_bV_b$			

Average molarity of vinegar in three trials   _____

Sample calculation (density of vinegar 1.005 g/mL, Molar mass of acetic acid 60.05 g)

$$\frac{\text{Moles acetic acid}}{\text{Liter}} \times \frac{60.05\ \text{g}}{\text{mol}} \times 100\% \quad = \quad \%\ (\text{w/v})\ \text{acetic acid}$$

Experimentally determined acetic acid in vinegar _____

PART C. EXERCISES

1. How does your result for the concentration of vinegar compare to the stated concentration of acetic acid on the vinegar bottle label?

2. What were three major sources of difficulty in the reproducibility of your titrations, and how might you avoid them another time?

EXPERIMENT 22

HOUSEHOLD CHEMISTRY IN THE SPOTLIGHT

GOALS

1. To learn about some of the chemicals used in the home
2. To examine some chemical reactions possible with these chemicals
3. To review solubility rules and balanced chemical equations

INTRODUCTION

Chemistry is not confined to the laboratory. Every day you deal with chemicals, even internally; that's what your study of general, organic, and biological chemistry is all about. Sometimes it's helpful to take stock of the chemicals you use daily to remind yourself just how important chemistry is in your daily life.

In this exercise, you will be experimenting with 15 different white solids, each labeled with only a sample number. All of them are pure chemicals that can be found readily in the pharmacy, hardware store or grocery store. You will observe their solubility, reactivity to various other chemicals, and color changes in order to identify each. Correctly identifying each one is your major goal.

Success in this exercise requires you to draw on many previous observations of chemical reactions and to observe carefully the results of the ones you try here. For example, there are five kitchen substances in the list: sodium chloride, sucrose, cream of tartar, cornstarch, and MSG (monosodium glutamate). Three of these are soluble in water (cream of tartar is not soluble and cornstarch is not very soluble), and three of the five would taste salty, so these household tests are not sufficient to classify them. However, sodium chloride reacts with silver nitrate, and MSG dissolves in hydrochloric acid, while sodium chloride and sucrose do not, so a little deduction will allow you to sort the set into correctly labeled bottles. Cornstarch is the easiest, since it is the only one that will react with iodine solution.

Because there are 15 materials to deal with in this exercise, not just five, the steps involved in the complete set of identifications are somewhat more complicated than this example. You will need to try all the reactions <u>in the correct order</u> and observe the results carefully in order to correctly identify the chemicals you are dealing with.

The flow chart for this experiment should be especially helpful. It shows a general progression of the steps for identification of the solids along with a brief note on what to expect at each step for a positive result. Refer to it frequently as you work through the exercise in order to keep your thoughts organized.

REFERENCES

Stoker, H. S. *General, Organic, and Biological Chemistry;* Cengage Learning; Chapters 8 and 10.
    Review topics
    - Balancing chemical equations
    - Solubility rules

Other references:

Oliver-Hoyo, M.; Allen, D.; Solomon, S.; Brook, B.; Ciraolo, J.; Daly, S.; Jackson, L. *J. Chem. Educ.*
2001, *78*, 1477-1478.

WEB RESOURCES

Chemical & Engineering News: What's That Stuff? *http://pubs.acs.org/cen/whatstuff/stuff.html*

A Drachm of This and a Tincture of That-Household Chemistry.
http://www.victoriancanada.com/household_chemistry.html

Chemistry Experiments You Can Do at Home.
http://chemistry.about.com/od/homeexperiments/ChemistryExperiments_You_Can_Do_at_Home.html

SAFETY NOTES

Sodium hydroxide is a caustic material, and hydrochloric acid is corrosive, even in dilute solutions. Take special care when handling solutions made using these materials. Follow your instructor's laboratory regulations about the use of chemical splash goggles and gloves; they may be governed by state law in your area.

Materials from Part A and B of this experiment may be discarded in the waste basket, in the sink with lots of rinse water, or as your instructor directs.

The white solids are all relatively non-hazardous, and most of them are generally recognized as safe, but not all of them are! Treat them with the respect you would treat any unknown substance. **Do not taste any of them for the identification!**

EXPERIMENTAL PROCEDURE

PART A. CHEMICALS IN THE HOME

The fifteen household chemicals you will identify are listed on the report form. For each one, find the formula, a common name, and a household use. Suitable sources are available in the library and on the Internet.

ⓘ Do Part A before coming to class.

PART B. IDENTIFICATION OF INDIVIDUAL HOUSEHOLD CHEMICALS

Use the guidelines below to fill in the chart on the report sheet for Part B of this exercise. You need to do the cold water solubility test for all 15 chemicals **first**, but for other tests, use the flow chart to guide you concerning which tests are suitable for each chemical. If you do not do a specific test, write N. A. for "not applicable" in the box.

ⓘ You do NOT have to do all tests for all substances!

Step 1. Solubility in cold water. For each of the 15 substances available, do a solubility test with water. Use about 0.1 g (about a spatula-tip-full) of the solid, and add to it 3 mL of water. Stir or shake well, and decide if the substance is soluble in water. Some may only be partially soluble.

ⓘ The solubility tests are crucial to success—observe carefully.

Step 2. Reaction to iodine. For each chemical that is **insoluble** in water, place a few crystals in a well of a spot plate, and add a drop of iodine test solution. A blue-black color indicates the chemical is cornstarch. For any chemical that is insoluble in water, **and** unreactive to iodine solution, continue to Step 3.

Step 3. Reaction with hydrochloric acid. Place a small amount of the insoluble substance to be tested in the well of a spot plate. Add to it 10 drops of 6 M HCl and observe for any reaction. Calcium carbonate will produce bubbles, potassium tartrate will dissolve, and calcium sulfate will remain unchanged.

☻ HCl is corrosive, so be careful.

NOTE: For Step 4 and later, you should need to work with only those materials that **dissolved** in cold water.

Step 4. pH of water solutions. Find the pH of the water solution for each substance to be tested. To do this, first use wide-range pH paper and then narrow-range pH paper to determine the pH of the solution to the nearest tenth of a unit. Refer to the table in the Flow Chart to determine what pH is appropriate for water solutions of several of the unknown samples.

NOTE: Only those water-soluble substances with a nearly neutral pH need be tested in the next step.

Step 5. Reaction with sodium hydroxide. Check the reaction of the substance to sodium hydroxide solution. Dissolve a small amount of it in water, and mix this solution with an equal amount of 6 M sodium hydroxide. If a precipitate forms, the numbered substance is magnesium sulfate.

Step 6. Benedict's test (copper(II) test) for reducing sugars.

a) For the remaining substances, set up a series of test tubes so you can test their reactions to copper ions all at once.

b) Dissolve 0.1 g of each substance in a separate test tube using 3 mL of water. If you have leftover solution from earlier tests, you can use that, because only about 1 mL is really needed.

c) Add to each test tube 1 mL of Benedict's reagent, and mix well.

d) Heat the solutions in a warm water bath on a hot plate for 5 minutes, and observe any color changes. A positive test for glucose and fructose is the formation of a green solution from the blue one, followed by precipitation of red-orange copper(I) oxide. Sucrose and sodium chloride will not give a positive test.

Step 7. Clinistix test. Glucose and fructose can be identified using a test strip such as Clinistix. For any substance that tested positive in Step 6, test a water solution with a glucose test strip, using the package directions. Fructose should not give a positive reaction to the strip or will at most have a very slow reaction, while glucose will react immediately.

Step 8. Reaction of soluble substances with HCl. Place about 1 mL of the water solution of the substance to be tested in a test tube. Add about 10 drops of 6 M hydrochloric acid and observe after mixing. MSG will precipitate as glutamic acid, but the HCl will have no visible effect on sucrose or salt.

Step 9. Reaction with silver nitrate. The only unidentified substances at this point should be sucrose and sodium chloride. There are many ways to distinguish between these two. Sucrose is more soluble in hot water than sodium chloride, and sodium chloride will react with silver nitrate while sucrose will not. Choose one or more of these reactions to test the last two unidentified chemicals.

PART C. SOME REACTIONS OF HOUSEHOLD SUBSTANCES.

Complete the relevant equations on the third page of the report form.

FLOW CHART
15 UNIDENTIFIED HOUSEHOLD CHEMICALS
substance examined

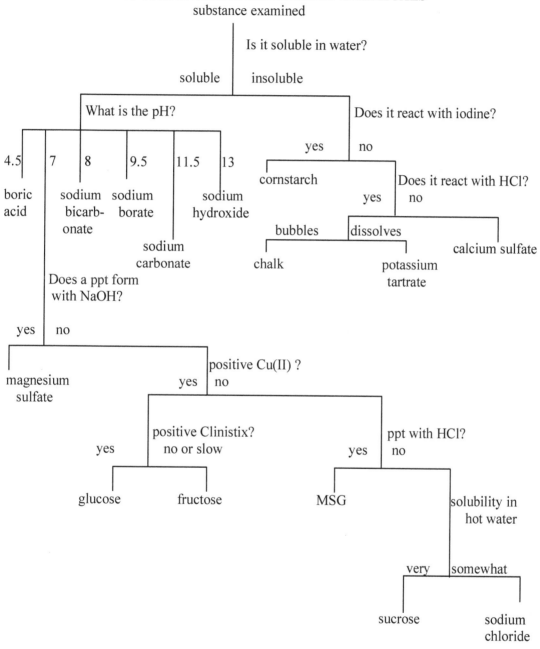

Is it soluble in water?

soluble · insoluble

What is the pH?

Does it react with iodine?

4.5 | 7 | 8 | 9.5 | 11.5 | 13

boric acid

sodium bicarb-onate

sodium borate

sodium hydroxide

sodium carbonate

yes · no

cornstarch

Does it react with HCl?

yes · no

bubbles · dissolves

chalk

potassium tartrate

calcium sulfate

Does a ppt form with NaOH?

yes | no

magnesium sulfate

positive Cu(II) ?

yes | no

positive Clinistix?
no or slow

ppt with HCl?

yes | no

yes · no or slow

glucose · fructose

MSG

solubility in hot water

very | somewhat

sucrose

sodium chloride

EXPERIMENT 22 PRELAB EXERCISES:

HOUSEHOLD CHEMISTRY IN THE SPOTLIGHT

Name_____ **Partner** _____ **Date** _____

1. For the following chemicals, use book or Internet sources to find the formula and predict whether each will be soluble in distilled water, acid, or base.

	Formula	Soluble in water?	Soluble in acid?	Soluble in base?
Calcium carbonate				
Magnesium sulfate				
Potassium tartrate				
Boric acid				
Calcium sulfate				

2. Which substances in the list of 15 household chemicals in this experiment will react with Benedict's solution? Refer to the chart in Part B of the report for guidance. What will a positive test look like?

3. Which of the 15 chemicals should produce carbon dioxide when treated with acid solutions?

4. Suggest one way to distinguish between sucrose and sodium chloride.

5. How could you distinguish between calcium sulfate and magnesium sulfate if these were the only two unknown samples to examine?

6. Name at least one of the household substances in this experiment that is caustic (strongly basic) and potentially hazardous, and describe how it should be handled.

EXPERIMENT 22 REPORT: HOUSEHOLD CHEMISTRY IN THE SPOTLIGHT

Name_____ Partner _____ Date _____

PART A. CHEMICALS IN THE HOME

Library Research on Household Chemicals

Substance name	Formula	Common name	Household use
Cornstarch			
Calcium sulfate			
Sodium chloride			
Sodium bicarbonate			
Calcium carbonate			
Potassium hydrogen tartrate			
Sucrose			
Sodium hydroxide			
Fructose			
Glucose			
MSG			
Magnesium sulfate			
Boric acid			
Sodium borate			
Sodium carbonate			

PART B. IDENTIFICATION OF INDIVIDUAL HOUSEHOLD CHEMICALS

Test Results for Individual Chemicals

Compound number	Solubility in cold water	Reaction to iodine	Reaction to 6 M HCl	pH of water solution	Reaction with NaOH	Reaction with copper(II)	Reaction to Clinistix®	Reaction to AgNO₃	Compound name
1									
2									
3									
4									
5									
6									
7									
8									
9									
10									
11									
12									
13									
14									
15									

PART C. SOME REACTIONS OF HOUSEHOLD SUBSTANCES

Write balanced equations for these reactions, which were some of those observed in the experiment you just completed.

1. The reaction of calcium carbonate with HCl

2. The reaction of sodium borate ($Na_2B_4O_7$) with water to produce a basic solution

3. The reaction of magnesium sulfate with sodium hydroxide

4. The reaction of MSG with HCl to produce insoluble glutamic acid

5. The reaction of sodium bicarbonate with HCl

6. The reaction of sodium carbonate with water to produce a basic solution

EXPERIMENT 23

TECHNIQUES IN RADIOCHEMISTRY

GOALS

1. To demonstrate half-life in radioactive materials
2. To observe the effects of shielding on radioactivity
3. To practice techniques for handling radioactive materials
4. To practice writing nuclear decay equations

INTRODUCTION

The three most common types of radioactivity found in nature are alpha particles, beta particles, and gamma rays. They are emitted from the nuclei of unstable atoms and can be harnessed for the service of medicine, but there are potential hazards involved. The intensity of the radiation, its penetrating power and potential for cellular damage, and the half-life of a radioactive material are all important in the safe use of radioactive substances. In any actual work with radioactive material, you would also need to be aware of necessary safety precautions to avoid unnecessary spread of radioactivity in your work area. This exercise will help you examine these concepts using models instead of significantly radioactive substances.

Radioactivity can be measured in a variety of ways, but the most basic way is to count the number of atoms that decay per unit of time by measuring the *intensity* of the radioactive emissions. Usually this quantitative measure is referred to as "counts per minute" (cpm) or "disintegrations per minute" (dpm). Cpm and dpm are related but slightly different quantities. The unit of measurement is the Curie, or 3.7×10^{10} dpm/sec. A good tool for rough measurement of radioactivity intensity is a Geiger counter, which is sensitive to γ-, β-, and x-rays. It takes advantage of the fact that radioiactive particles can ionize atoms they strike, which in turn can be detected by the instrument—the more decompositions, the more ions registered by the counter as cpm. In a laboratory setting, other more sensitive tools can be used as well, that you may have an opportunity to use as your studies continue.

Naturally, ways to shield people from radiation have received a lot of attention. Alpha particles and most beta particles are easily stopped by lightweight shielding, even by skin, although inhalation or ingestion of either can cause serious problems in large doses. Gamma rays are significantly more energetic and require more massive shielding. Part A of this exercise allows you to examine shielding of a radioactive source by various substances.

The *half-life* of a radioactive substance (or radionuclide) is the amount of time it takes for half of a given sample to decay into smaller nuclides called daughter products. Half-lives can be very short, as for P-32 ($t_{1/2}$ = 14.3 days) or extremely long, as for Pu-244 ($t_{1/2}$= 76,000,000 years). The half-life of a diagnostic radionuclide is of great importance; short half-lives are preferred for these in order to minimize unnecessary exposure of the patient and the environment to radiation. On the other hand, if the half-life is too short, the substance will decay to non-radioactie status before it arrives at the medical center. In this experiment, you will look at a simulated radioactive decay to observe how half-life works.

For research and diagnostic studies, radioactive materials are often purchased in relatively concentrated solution. That is, they have a defined concentration of radioactive material that is probably much stronger than the amount needed for any procedure. How does the worker obtain the right level of radioactivity for the procedure? One common way is by serial dilution. You have already learned the necessary techniques for diluting a sample of a concentrated solution to make one of a lower concentration. The techniques and calculations are the same for radioactive solutions except that concentrations are measured in dpm instead of % (w/v) or molarity, and you need to take even more than usual care to have no spills or accidents with the material you are working with.

When a lab worker or medical professional works with low-level radioactive materials, such as are commonly used in diagnostic or metabolic studies, there are a number of important safety precautions to take to ensure that the radioactive substance is not spread over the workplace. Well-fitting disposable gloves are a must, as are tools that can be cleaned easily. The wash water can be collected, if necessary, for disposal. All work takes place on an absorbent mat similar to a diaper, and in a fume hood that will not recirculate any radioactive material that escapes into the atmosphere. The worker also wears a coverall or lab coat that can be discarded if needed. When the experiment is finished, all the contaminated materials, including the mat, are gathered together for disposal in a special container that eventually is taken to an appropriate waste disposal facility. Some nuclides with very short half-lives can simply be stored until they have decayed to 1% or less of their original activity, but others require special disposal procedures.

In Part C of this experiment, using techniques similar to those that might be used in a research laboratory, you will try a serial dilution to obtain a solution that is 1/1000 as concentrated as the original. However, you will be working with a substance that glows strongly (fluoresces) when exposed to an ultraviolet (UV) or "black light." It is not radioactive, but using it will allow you to assess how well you could handle a radioactive sample if you needed to.

REFERENCES

Stoker, H. S. *General, Organic, and Biological Chemistry;* Cengage Learning; Chapters 8 and 11.
   Review topics
   • Radioactive half-life
   • Nuclear equations
   • Radiochemical techniques
   • Dilution of solutions

WEB RESOURCES

Radioisotope techniques. *http://www.darvill.clara.net/nucrad/hlife.htm*

Nuclear Energy. *http://www.aisp.net/vster/nuclear10.htm*

Re: Halflife source: Bi-212 demo. *http://einstein.byu.edu/~masong/HTMstuff/Radioactive2.html*

SAFETY NOTES

The ores recommended for the demonstration of shielding effects are all low-level sources and may be handled safely. The fluorescent compound in this exercise (calcofluor white M2R) is non-hazardous at the levels used. However, spills will fluoresce under a black light. Avoid looking directly at the UV light since UV rays can harm the eyes.

EXPERIMENTAL PROCEDURE

PART A. SHIELDING OF RADIOACTIVE SOURCES

This exercise can be done either as a demonstration by your instructor or by individuals. The source of radioactivity is a naturally occurring uranium ore that emits a mixture of alpha, beta, and gamma radiation at very low levels.

Step 1. Wearing disposable gloves, place the ore sample on a paper towel on the bench top. Position a Geiger counter close to the sample, and take a reading of decompositions per minute from the uncovered sample.

Step 2. Examine the effect of paper as a shield by placing an index card between the source and the counter. Also try a stack of 25 cards.

Step 3. Try different materials as potential radiation shields. Available to you are water, glass, and lead sheets.

> Of course you must wear chemical splash goggles, too.

PART B. MODELING RADIOACTIVE DECAY

Obtain from your instructor a set of 100 objects with either printing on one side only or different markings on each side. At your instructor's discretion, you may have 100 pennies to work with or 100 wooden disks or perhaps even 100 candies. You will also have two cloth bags for storage; one labeled "radioactive" and one labeled "decayed." These tokens act as a model for a sample of radioactive material at a time equal to zero half-lives. You can assume that all of these tokens are "radioactive" at this point.

Step 1. Shake your tokens in the "radioactive" bag for exactly 30 seconds (one half-life), being sure that they are well mixed. Pour them out onto the bench top, and separate all those that have printing or "heads" up. Count these models of "decayed" radioactive atoms, record how many there are, and place these in the "decayed" container. The number of "undecayed" tokens represents the amount of radioactive material left after one half-life has passed.

> ⓘ Be sure the items are **well** mixed. This can be surprisingly hard to accomplish.

Step 2. Return the "undecayed" tokens to the "radioactive" bag, shake them well for another half-life, and again pour them out onto the bench. Count and record the number of "decayed" tokens, discard them in the "decayed" container, and count and return the still "radioactive" ones to the "radioactive" bag.

Step 3. Continue this procedure until you have only one "undecayed" token left.

Step 4. Make a graph of your results, with "half-lives" as the x-axis and number of "radioactive" tokens as the y-axis. Draw the best smooth curve that fits your data.

> ⓘ Be sure to label the graph completely!

PART C. LABORATORY TECHNIQUES IN RADIOCHEMISTRY

Step 1. Using your best "clean room" technique, obtain 10 mL of calcofluor white solution in a graduated cylinder. Place it on an absorbent mat on your benchtop along with the three volumetric flasks, three pipets, pipet bulbs, and tissues you will need. Wear disposable gloves and (preferably) an apron or lab coat throughout this experiment.

Step 2. Do a serial dilution of this solution to obtain a solution that is 1000 times less concentrated than the one you started with.

a) To begin this procedure, make a 1:10 dilution in the following manner. Use a 1.00-mL volumetric pipet and pipet bulb, and withdraw 1.00 mL from the starting solution.

b) Transfer this to a 10.0-mL volumetric flask, and discard your used pipet in the "contaminated" bin provided.

c) Dilute the solution in the 10.0-mL flask to the mark, cap, and mix well. Use this solution in Step 3.

d) Wash and dry your volumetric pipet before continuing to Step 3. Remember, all the washings are "contaminated" and must be contained for mock disposal procedures. If your lab is well supplied with volumetric glassware, you may be permitted to use a new pipet for each dilution.

Step 3. Make a 1:100 dilution of the original solution by starting with your 1:10 dilution. Again, take 1.00 mL of this material (use a new pipet at this stage), transfer it to a new 10.0-mL volumetric flask, and fill to the mark. Mix well, and use in Step 4.

Step 4. Make a 1:1000 dilution of the original solution, starting with the 1:100 made in Step 3. The same process applies.

Step 5. Observe the solutions you made under an ultraviolet (UV) light.

Step 6. Wash all your glassware, including pipets, into a beaker using minimal amounts of water, and lay your glassware, pipet bulbs, and disposable gloves on the absorbent mat along with your washed glassware.

Step 7. Check your technique (or have your instructor do this) by observing your work area and your clothing under a UV light. Calcofluor white has a strong blue fluorescence under a UV light, and any spills will show up readily. When your work area has been checked, you may discard all solutions into the sink or as your instructor directs. As usual, if the materials are washed into the sink, wash them down the drain with lots of water.

PART D. EXERCISES

Refer to the report form for some sample exercises on half-life and nuclear equations.

EXPERIMENT 23 PRELAB EXERCISES

RADIOCHEMICAL TECHNIQUES

Name_____ **Partner** _____ **Date** _____

1. Define these terms:
    cpm

    half-life

    dpm

2. Why is it <u>not</u> reasonable to use a synthetic (man-made) radioactive substance with a half-life of one millisecond as a diagnostic tool?

3. A student measured the intensity of emission from a radioactive mineral at various distances from the piece of rock. The results are shown in the table. Graph the result (distance is the independent or x-axis variable), and state a conclusion about the variation of intensity with distance from the radiation source.

cpm	Distance (cm)
62,500	1
13,100	2
6900	3
400	4

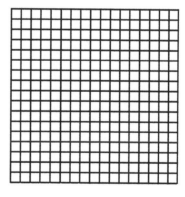

Conclusion:

4. If a medical worker has available 10 mL of a solution of a diagnostic radionuclide with a total activity of 1 mCi/10 mL, how would that worker prepare 10 mL of a solution with an activity of 0.01 mCi/mL?

EXPERIMENT 23 REPORT: RADIOCHEMICAL TECHNIQUES

Name_____ **Partner** _____ **Date** _____

PART A. SHIELDING OF RADIOACTIVE SOURCES

Counts per Minute with Various Shielding Materials

cpm, no shielding	cpm, 1 card	cpm, 25 cards	cpm water	cpm 1 glass	cpm, 2 glass	cpm, 1 lead	cpm, 2 lead

Describe the shielding ability of the various materials. Which types of radioactivity (alpha, beta, or gamma) can be stopped by each kind of material?

PART B. MODELING RADIOACTIVE DECAY

Radioactive Decay as a Function of Half-life

Half-lives	Radioactive	Decayed

1. Graph your decay results here. Describe in a sentence the visual appearance of this line. Don't forget to label the graph.

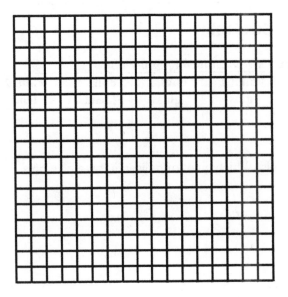

2. How would the line look if you used a half-life of 90 seconds instead of a half-life of 30 seconds?

PART C. CLEAN LABORATORY TECHNIQUES IN RADIOCHEMISTRY

1. Were your solutions fluorescent? What color were they?   _____

2. Did the fluorescence weaken as the concentration of calcofluor
white decreased?   _____

3. How many spills did your instructor find in your work area
and on your person?   _____

4. What is your instructor's assessment of your handling technique?   _____

5. Discuss how you could improve your handling skills.

PART D. EXERCISES

1. Phosphorus-32 has a half-life of 14.3 days. Once a sample has been used in a diagnostic test, how many days must it be stored to reach 1% of its original activity so that it can be discarded as non-hazardous waste?

2. Uranium-235 is a beta emitter. Write a nuclear equation for its decay.

3. A radioactive solution contained a nuclide that gave 2×10^8 counts per minute per mL of solution. That is, in every mL of the solution, 2×10^8 radioactive atoms decayed every minute. How much of this solution would a lab worker need to measure to obtain 2×10^7 counts per minute?

4. Using the same solution as in Question 3, how would a worker prepare 10 mL of a solution containing 2.7×10^7 cpm/mL? (Hint: This is a dilution question using two solutions.)

5. Beryllium-10 decays into boron-10 and a radioactive particle. Write the complete nuclear equation for this decay.

EXPERIMENT 24

PROPERTIES OF SATURATED HYDROCARBONS

GOALS

1. To understand the structure of alkanes
2. To investigate shapes of saturated hydrocarbons
3. To explore the reactions of saturated hydrocarbons

INTRODUCTION

The simplest organic compounds to understand are alkanes. They are hydrocarbons and contain only carbon-carbon and carbon-hydrogen single bonds. They are relatively inert, undergoing few reactions under normal conditions, but are of particular importance because understanding them is the basis for understanding more complex molecules that are of prime importance in biological systems.

Alkanes are also the simplest organic substances to name. The naming of all organic molecules is governed by strict rules, and the root words for many names are derived from the names of the first 10 to 20 alkanes in the series. In order to progress in your study of organic and biological chemistry, you will need to be familiar with the conventions for naming compounds. You should refer to your textbook for an overview of the rules.

Alkanes can only undergo a few chemical reactions without using some kind of forcing conditions. The most important of these is combustion, the combination with oxygen to generate carbon dioxide, water, and energy in the form of heat and light. We use this reaction to cook on a gas stove, heat homes with oil or gas heat, and run internal combustion engines to drive cars. Under some conditions, alkanes can be converted into haloalkanes, which have somewhat different properties, among them a lack of flammability. They are often studied together in courses like this, because both classes contain only single carbon-carbon bonds.

This exercise will examine physical properties of alkanes and haloalkanes using hexane and dichloromethane, CH_2Cl_2, as representative examples. Consider what you see, and interpret your observations in the light of what you know about polarity and density.

For the second part of the experiment, you will be using molecular models. Depending on the tools available in your lab, these may be physical ball-and-stick models or computer-generated 3-dimensional pictures. Molecular modeling exercises should give you a feel for the way in which the shape of a molecule is governed by the types of bonds present and by the size of the molecule. A static picture on a piece of paper cannot show the twisting and tumbling of long chains of carbon atoms in the same way a physical or computer-generated model can.

REFERENCES

Stoker, H. S. *General, Organic, and Biological Chemistry;* Cengage Learning; Chapter 12.
   Review topics
   - Structural formulas
   - Isomers
   - Conformations
   - Covalent bonding

WEB RESOURCES

These resources are best viewed using the Accelrys Chime 2.6 Chime plugin from *http://accelrys.com/products/informatics/cheminformatics/chime/no-fee.php*

Life's Molecules. *http://biology.wsc.ma.edu/biology/courses/molecules/*

Cyclic Alkanes and Alkenes. *http://www.elmhurst.edu/~chm/vchembook/506cyclic.html*

SAFETY NOTES

Organic chemicals often have strong odors and are usually volatile and flammable. Treat them with respect, and avoid inhaling the vapors, using them in the hood whenever possible. Use small amounts of materials, and discard them into appropriately labeled waste jugs when finished, not down the sink. Some chemicals in this experiment may be strongly acidic—treat all the chemicals used with respect. You may wish to wear gloves during this experiment, and you must wear chemical splash goggles.

EXPERIMENTAL PROCEDURE

PART A. PROPERTIES OF SATURATED HYDROCARBONS

Step 1. In each of two test tubes place 1 mL of water. Add to the first tube 5 drops of hexane and to the second 5 drops of dichloromethane, mix each well, and record what you see.

Step 2. In two other test tubes place 1 mL (about 20 drops) of toluene (another hydrocarbon). Into one sample mix 5 drops of dichloromethane, and into the other sample mix 5 drops of hexane. Mix each well, and record what you see.

Step 3. Place 10 drops of hexane in a well of a spot plate, and 10 drops of dichloromethane in a second well. Test each with the conductivity tester to determine if they will conduct electricity.

Step 4. Place 1 mL of hexane in an evaporating dish in the hood, and touch a lighted match to it. Record what you see. Repeat this exercise with 1 mL of dichloromethane.

Step 5. Place 10 drops of hexane in each of two test tubes. To one tube, add 5 drops of potassium dichromate-sulfuric acid solution (a strong oxidizing agent). To the second tube, add 1 drop of iodine solution and observe.

> ♥ This reagent is strongly acidic. Treat it carefully!

Step 6. Repeat Step 5, but this time use dichloromethane as the reactant.

PART B. STRUCTURES OF ALKANES

Using molecular models or a computer-based molecular modeling program, make models of the following compounds. Draw each model, and answer the questions relating to it.

n-hexane	*cis*-1,2-dimethylcyclopentane
2,3-dimethylbutane	*trans*-1,2-dimethylcyclopentane
3-chlorohexane	*trans*-1-3-cyclopentane ~~dimethyl~~
methycyclopentane	cyclohexane
3-methylpentane	2-methylpentane

PART C. EXERCISES

Answer the questions on the report form concerning the properties of alkanes and haloalkanes.

EXPERIMENT 25

UNSATURATED HYDROCARBONS

GOALS

1. To investigate the physical and chemical properties of alkenes and alkynes
2. To practice naming unsaturated hydrocarbons
3. To draw structures of unsaturated hydrocarbons and their reaction products

INTRODUCTION

"Unsaturated" in reference to organic compounds refers to any compound that has at least one double or triple bond in it. Many compounds of biological importance have other functional groups as well as multiple carbon-carbon bonds, but they still exhibit the properties of the alkenes as well as the properties of their more complex functional groups.

Alkenes have one or more double carbon-carbon bonds, while alkynes have one or more triple bonds. They may be as simple as acetylene, C_2H_2 (actually an alkyne, in spite of its –ene ending), or as complex as vitamin A, with twenty carbon atoms and five double bonds. The reactions they undergo are governed by the relative reactivity of their multiple bonds. Alkynes generally have very similar chemistry to alkenes, which is one reason they are often studied together.

Outside of combustion, addition is the main reaction of unsaturated compounds. Halogens such as bromine can react with either alkenes or alkynes to give haloalkanes.

The strong acids HCl and HBr can also react by addition to the multiple bond. In the case of an alkene the product looks like this.

In larger molecules, though, you need to be sure that you use Markovnikov's rule to predict which isomer you get.

One other class of unsaturated compound is worth mentioning, and that is the class of aromatic compounds. Because of the special nature of the alternating single-double bond arrangement in six-membered rings, they are especially stable. They don't really act like alkanes, but they are very inert relative to alkenes and alkynes and are usually considered a separate class of unsaturated compound.

You will have a chance to look at the reactivity and structure of all three classes of unsaturated compounds in this exercise.

REFERENCES

Stoker, H. S. *General, Organic, and Biological Chemistry;* Cengage Learning; Chapter 13.
    Review topics
    • Nomenclature of alkenes and alkynes
    • Geometry of alkenes and alkynes
    • Reactions of unsaturated hydrocarbons

WEB RESOURCES

This resource is best viewed using the Accelrys Chime 2.6 plugin from
    http://accelrys.com/products/informatics/cheminformatics/chime/no-fee.php

Cyclic Alkanes and Alkenes. *http://www.elmhurst.edu/~chm/vchembook/500Bhydrocarbons.html*

SAFETY NOTES

Organic chemicals often have strong odors and are usually volatile and flammable. Treat them with respect, and avoid inhaling the vapors by using them in the hood whenever possible. Use small amounts of materials, and discard them into appropriately labeled waste jugs when finished, not down the sink. Some chemicals in this experiment may be strongly acidic—treat all the chemicals used with respect. You may wish to wear gloves during this experiment, and you must wear chemical splash goggles.

EXPERIMENTAL PROCEDURE

PART A. PROPERTIES OF UNSATURATED HYDROCARBONS

Step 1. To observe some of the physical properties of unsaturated hydrocarbons, set up four test tubes, each with 1 mL of water in it. Add to one 5 drops of hexane in order to compare unsaturated hydrocarbons with a saturated hydrocarbon. To the second, add 5 drops of cyclohexene; to the third, 5 drops of 1-pentyne; and to the fourth, 5 drops of toluene.

Step 2. Mix the solutions well, and draw conclusions about the density and polarity of these compounds.

Step 3. To observe some of the chemical properties of unsaturated compounds, set up four test tubes, each containing 10 drops of one of these compounds: hexane (for comparison purposes), cyclohexene, 1-pentyne, and toluene.

Step 4. To each, add 3 drops of a solution of iodine in hexane, and record your observations. Iodine itself has a strong color. If that color changes, a reaction occurs. Be sure you understand what you see!

Step 5. This test is best done as a group activity, unless your lab has large areas of hood space. Place 10 drops of your test compounds (hexane, cyclohexene, 1-pentyne, and toluene) in four evaporating dishes in the hood. Touch a match to each one in turn, and note how each one burns.

Step 6. Styrene is an alkene that polymerizes readily. Place 1 mL of styrene in a disposable medicine cup, and add 2 drops of benzoyl peroxide solution (a catalyst). Stir the mixture, and observe over the course of 1/2 hour.

Step 7. Dissolve a drop of cooking oil in 1 mL of hexane, and add 1 drop of the iodine solution. Record your observations.

PART B. SHAPES OF ALKENES AND ALKYNES

Step 1. Using ball-and-stick molecular models or a computer-based modeling program, make models of the following compounds, and answer questions about them on the report form.

cis-2 pentene	toluene (methylbenzene)
trans-2-pentene	naphthalene ($C_{10}H_8$)
2-pentyne	1,3-pentadiene
cyclohexene	*n*-pentane
cyclooctene	

Step 2. Using the plug-in Chime or another web-based modeling program, find models on the internet of the following polycyclic aromatic compounds, and answer the questions about them.
phenanthrene
anthracene
benzanthracene

PART C. EXERCISES

Complete the exercises in Part C of the report form.

EXPERIMENT 26

ALCOHOLS, PHENOLS, AND ETHERS

GOALS

1. To observe the differences among the structures of alcohols, phenols, and ethers
2. To examine some physical properties of these classes of compounds
3. To learn to distinguish among primary, secondary, and tertiary alcohols
4. To observe some common reactions of alcohols

INTRODUCTION

Alcohols, phenols, and ethers all contain singly-bonded oxygen as the major functional group.

an alcohol   a phenol   an ether

Alcohols can be further subdivided into primary, secondary, and tertiary alcohols, each with slightly different reactivity requirements.

a primary alcohol   a secondary alcohol   a tertiary alcohol

Primary alcohols, which contain the group R-CH$_2$OH, will oxidize to form aldehydes, which themselves can further oxidize to carboxylic acids. Secondary alcohols have the general structure RCHOHR'; they have only one hydrogen atom on the carbon with the -OH group. They will oxidize to ketones but stop there. Cyclohexanol, your model compound for this experiment, is one such secondary alcohol. Tertiary alcohols, R$_3$COH, do not oxidize readily because the oxidation reaction requires at least one hydrogen atom on the alcohol carbon, and tertiary alcohols have none.

Phenols are really a type of alcohol, with some special properties because the –OH group is on an aromatic ring. Their main additional property is the ability to lose the hydrogen atom of the alcohol in a weak acid equilibrium, as shown on the next page.

All three classes, including ethers, are somewhat polar, leading to partial or complete solubility of the smaller members of each series in water. However, ethers are the least polar of the lot. In a large molecule, an ether functional group contributes to the overall polarity of the molecule, but does not undergo many reactions.

To summarize the major differences among these functional groups, we can say this:

- The major biochemical reactions of alcohols are oxidation, usually to a ketone, aldehyde, or acid, and elimination of water (for primary and secondary alcohols) to form a double bond. Substitution of other atoms for the alcohol -OH can also occur.
- Phenol reactivity is governed by its aromaticity and by its acidity. The aromatic ring stabilizes the carbon-oxygen bond and allows the OH to act sometimes like an alcohol, as in the formation of an ester, and sometimes like an acid, in the release of hydrogen ions in solution.
- Ethers are generally unreactive because of the strength of the two oxygen-carbon bonds in the ether functional group. However, they make good solvents for many organic compounds, and ether groupings are often found in biochemical molecules.

This exercise will show you some examples of each of these reactions. Pay particular attention to the reactions that can be used to identify an alcohol or phenol or alkyl halide in the presence of compounds containing other functional groups because these "spot tests" will be useful later.

REFERENCES

Stoker, H. S. *General, Organic, and Biological Chemistry;* Cengage Learning; Chapter 14.
   Review topics
- Structures of primary, secondary, and tertiary alcohols
- Structures of phenols and ethers
- Chemical reactions of alcohols
- Chemical reactions of phenols

WEB RESOURCES

Alcohols, Phenols, and Ethers. *http://www.hcc.mnscu.edu/chem/V.20/*

Functional Groups-Organic Chemistry. *http://chemistry.about.com/library/weekly/aa062703a.htm*

SAFETY NOTES

Organic chemicals often have strong odors and are usually volatile and flammable. Treat them with respect, and avoid inhaling the vapors by using them in the hood whenever possible. Use small amounts of materials, and discard them into appropriately labeled waste jugs when finished, not down the sink. Some chemicals in this experiment may be strongly acidic. Treat all the chemicals used with respect. You may wish to wear gloves during this experiment, and you must wear chemical splash goggles.

EXPERIMENTAL PROCEDURE

PART A. STRUCTURES OF ALCOHOLS, PHENOLS, AND ETHERS

Step 1. Using either physical models or a modeling program provided by your instructor, make models of cyclohexanol, phenol, and diethyl ether.

> ⓘ This can be done with ball and stick models, space-filling models, or a modeling program.

Step 2. Draw each on your report form, and answer the questions about them.

PART B. PHYSICAL AND CHEMICAL PROPERTIES OF ALCOHOLS, PHENOLS, AND ETHERS

Solubility

Step 1. Place 1 mL of water in each of three test tubes.

Step 2. Place 5 drops of cyclohexanol, phenol solution, and diethyl ether in separate test tubes. Shake each well, and observe the solubility of the compounds.

Step 3. To discover the limits of solubility of alcohols as a function of carbon number, set up seven test tubes, each containing 1 mL of water. In one tube, place 2 drops of methanol, mix well, and observe. Repeat this procedure with ethanol, propanol, butanol, pentanol, hexanol, and heptanol.

Step 4. Set up another test tube with 1 mL of water, and repeat the solubility test with 0.10 g of glucose, a polyol.

Oxidation

Step 1. In the hood, set up three evaporating dishes, each containing 10 drops of one of the following compounds: ethanol, phenol (phenol is a solid, so use 0.1 g instead) and diethyl ether. Touch each sample with a match and observe.

Step 2. Set up three test tubes, each containing 10 drops of one of the following compounds: ethanol, phenol, diethyl ether. To each tube, add 1 drop of the oxidizing agent sodium dichromate in sulfuric acid, and mix.

> ☙ The sodium dichromate solution is strongly acidic.

Step 3. You have available, in dropper bottles, three samples marked Sample A, Sample B, and Sample C. Determine which sample is an alcohol.

Substitution

Step 1. To 10 drops of *tertiary*-butanol in a test tube, add 1 mL of the Lucas reagent and mix well. Allow the solution to settle. A layer of *t*-butyl chloride forms on the surface of the mixture.

Step 2. After about 1 minute, withdraw a few drops of <u>only</u> the upper liquid to a new test tube. Use this sample to test for the presence of halogens in the Beilstein test described in Step 3.

> ☙ Lucas reagent contains hydrochloric acid, so be cautious!

Step 3. The Beilstein test for halogens. Clean a heavy-gauge copper wire by holding it in a Bunsen burner flame in the hood until the flame glows yellow. Allow the wire to cool. Dip the cleaned wire in the sample to be tested, and insert it into the flame again. A green flame indicates the presence of a halogen in the compound being tested.

PART C. EXERCISES

Answer the questions on the report form for Part C.

EXPERIMENT 27

ALDEHYDES AND KETONES

GOALS

1. To review the naming and structures of aldehydes and ketones
2. To investigate identification tests for aldehydes and ketones
3. To examine important reactions of aldehydes and ketones
4. To point out how the reactivity of aldehydes and ketones differs

INTRODUCTION

The C=O group is very important in many organic molecules. It forms part of several different functional groups such as carboxylic acids and amides, but it can also act as a functional group in its own right. This "carbonyl group" by itself is the important part of the ketone functional group and the aldehyde functional group.

ketone   aldehyde

An aldehyde is simply a ketone bonded to an end carbon atom so that one of the carbon atom's R groups is a hydrogen atom. However, that hydrogen atom alters the reactivity of the molecule, so aldehydes are really a different functional group—they do all the things ketones do and more besides.

Your study of alcohols has already shown you one way to make a ketone, from a secondary alcohol by oxidation with sodium dichromate in sulfuric acid. Aldehydes can be made in the same way from primary alcohols, but they are still quite reactive, and in the presence of sodium dichromate would react further to give a carboxylic acid. To make an aldehyde from an alcohol, a much milder oxidizing agent would be used in the lab. In biochemical systems, this reaction is controlled by specific enzymes, so that oxidation of a primary alcohol stops at the aldehyde stage when necessary.

Both ketones and aldehydes will undergo reduction, adding hydrogen atoms to the carbonyl group to form an alcohol. A ketone will give a secondary alcohol, and an aldehyde will give a primary alcohol. A mild reducing agent for lab use is sodium borohydride, $NaBH_4$, which you will use in this experiment to observe a reduction; this reaction is shown on the next page. Biochemical reductions are enzyme controlled, and the hydrogen atom source is often NADH, which you will study later in this course.

In this experiment, you will reduce the ketone benzophenone to benzhydrol; one of the exercises in the report requires you to look up the structure of benzophenone and benzhydrol. (Hint: doing this ahead of time will save you lab time!)

One other reaction that ketones and aldehydes can undergo is acetal or hemiactal formation. This is a reaction that happens when a ketone or aldehyde is allowed to react with an alcohol in the presence of an acidic catalyst. An acetal is formed when two moles of the alcohol add to the oxygen of the carbonyl group, and a hemiacetal is just what it sounds like, half an acetal, formed when only one mole of alcohol reacts. Below is a sample of each type of compound. We will not do any reactions with them in this lab exercise, but they will become important when you study the chemistry of carbohydrates. Note that hemiacetals can sometimes form rings.

an acetal   a hemiacetal   another hemiacetal

REFERENCES

Stoker, H. S. *General, Organic, and Biological Chemistry;* Cengage Learning; Chapter 15.
    Review topics
    • Structures and nomenclature of aldehydes and ketones
    • Reactions of aldehydes
    • Reactions of ketones
    • Spot tests for identification of aldehydes and ketones

WEB RESOURCES

Chemistry Molecular Models. *http://www.uwsp.edu/chemistry/pdbs*

Hangover. *http://en.wikipedia.org/wiki/Hangover#Causes*

Oxygen Containing Compounds-Aldehydes and Ketones. *http://mcat-review.org/aldehydes-ketones.php*

SAFETY NOTES

Organic chemicals often have strong odors and are usually volatile and flammable. Treat them with respect, and avoid inhaling the vapors by using them in the hood whenever possible. Use small amounts of materials, and discard them into appropriately labeled waste jugs when finished, not down the sink. Some chemicals in this experiment may be strongly acidic. Treat all the chemicals used with respect. Be especially careful with concentrated sulfuric acid, and be sure you handle it safely. You may wish to wear gloves during this experiment, and you must wear chemical splash goggles.

EXPERIMENTAL PROCEDURE

PART A. STRUCTURES OF ALDEHYDES AND KETONES

Using a molecular modeling program or physical models, produce models of the following aldehydes and ketones. Draw the models on the report form, and circle the functional group in each molecule.

benzaldehyde
acetaldehyde (ethanal)
cyclohexanone

benzophenone
acetone (2-propanone)

| ⓘ Ball and stick models or other hands-on models can be used. |

PART B. CHEMICAL TESTS FOR ALDEHYDES AND KETONES

There are five chemical samples available to test. Determine which are aldehydes, which are ketones, and which are neither on the basis of three spot tests.

Step 1. The 2,4-DNP test. Set up a series of five test tubes labeled A-E. Place a few drops of one of the chemicals in each of the appropriate tubes. To each, add 2 drops of 2,4-DNP reagent, mix well, and allow to stand for 10 minutes. A positive test is the formation of a precipitate, which may range from yellow to deep red. Ketones and aldehydes will both react with 2,4-DNP reagent.

| ☕ The reagent contains sulfuric acid, so be careful. |

Step 2. The sodium dichromate test. Set up an additional set of five samples A-E in test tubes. To each, add 2 drops of sodium dichromate/sulfuric acid reagent, mix, and observe. A positive test is the formation of a green to blue-green solution as the chromium is reduced from orange chromium(VI) to green chromium(III), and the organic substrate is oxidized. Secondary or primary alcohols and aldehydes will react with this reagent, but ketones will not.

| ☕ The reagent is strongly acidic, too. |

Step 3. The Purpald™ test. Prepare a new set of five samples A-E. To each add 2 drops of Purpald™, a reagent that turns purple in the presence of aldehydes, but not with any other functional group. From the data from these tests, deduce which of samples A-E is a ketone, which is an aldehyde, and which is neither.

| ☕ This solution is strongly basic. |

PART C. REDUCTION OF A KETONE:BENZOPHENONE TO BENZHYDROL

Step 1. Weigh 0.15 g of solid sodium borohydride and place it in a test tube, along with 10 mL isopropyl alcohol. Caution: NaBH$_4$ can burn in the presence of water, so be careful!

Step 2. Add to the mixture 1.25 g of benzophenone, and mix well. Place a small beaker upside down over the test tube. The object of this is to keep the isopropyl alcohol vapor in the test tube, but **not** to seal off the test tube, because it should remain at atmospheric pressure throughout the reaction. Heat this mixture in a hot water bath on a hot plate for 30 minutes, swirling occasionally. Because isopropyl alcohol is flammable, this step should be done in an efficient hood. The water bath does not need to boil.

Step 3. Pour the reaction mixture into about 30 mL of a mixture of ice and water, and stir vigorously. The product is a solid, but sometimes it is hard to crystallize. If your product does not come out of solution as a solid, but instead as an oil, ask your instructor for suggestions for forcing the crystals to form.

Step 4. Filter the product using a suction filtration apparatus. Wash the crystals with a 50:50 mixture of cold methanol and water. This should purify your product sufficiently for the final test. Allow the apparatus to pull air through your product on the filter paper for at least an additional 5 minutes.

Step 5. When your crystalline product appears to be dry, weigh it on a filter paper or weighing boat, and record the yield in grams. Benzophenone is the limiting reagent. Calculate the percent yield.

> ⓘ The theoretical yield for this reaction is 1.26 g.

Step 6. Check for the presence of an -OH functional group and leftover ketone. To check for the formation of an alcohol, place a few crystals of your product in a test tube along with a few drops of sodium dichromate/sulfuric acid reagent. Is the spot test positive for alcohol? Test the starting benzophenone in the same manner as a control.

Step 7. Check your product benzhydrol for the presence of a ketone functional group. To check for the presence of a ketone, place a few crystals of your product in a test tube, and use another test tube containing a few crystals of benzophenone as a control. To each tube, add 10 drops of "2,4-DNP reagent." If there is benzophenone in the sample, a red, fine-grained powder will precipitate. A positive test will indicate that you have some leftover starting ketone in your product and that the reaction did not go to completion.

EXPERIMENT 28

CARBOXYLIC ACIDS AND ESTERS

GOALS

1. To review the differences in structure between carboxylic acids and esters
2. To investigate neutralization of carboxylic acids
3. To practice ester formation from an acid derivative and an alcohol
4. To observe saponification and hydrolysis of an ester

INTRODUCTION

Carboxylic acids and esters are usually considered together in a beginning study of organic and biochemistry because their functional groups are closely related. Esters can be made from carboxylic acids and vice versa.

a carboxylic acid
general formula RCOOH

an ester
general formula RCOOR'

Among the most important acids in the body are fatty acids, carboxylic acids with long carbon chains. The triacylglycerols are in fact esters of these acids with glycerol, so a knowledge of the reactions they undergo is critical to understanding the metabolism of dietary fats and oils.

No matter how long the carbon chain, much of the chemistry of a carboxylic acid is governed by the acid group. If an acid dissolves in water, it can weakly ionize to give a solution with an acidic pH, and then it's a weak electrolyte. Carboxylic acids will also react with bases in neutralization reactions, and will also undergo a new type of reaction, the esterification reaction.

Esterification of an acid is a condensation reaction between an acid and an alcohol. The two combine, releasing a molecule of water. A mineral acid catalyst or an enzyme catalyst in the body is usually necessary.

In the lab, it is essential that an <u>anhydrous</u> acid catalyst such as sulfuric acid be used instead of (say) HCl, which is only 37% acid. Water is a product in this reaction, and water present at the beginning will, according to LeChâtelier's principle, discourage the formation of the product ester. When you make an ester in this exercise, be sure the containers are dry until the ester reaction is completely finished.

Esters undergo relatively few reactions. Unless there is some other functional group in the molecule as well, the main reaction is the return to the acid and alcohol parent compounds. This can be accomplished in two ways, hydrolysis or saponification.

Hydrolysis involves the addition of water to the ester linkage, breaking it apart so that the acid, RCOOH, and the alcohol, ROH, are formed. The hydrolysis process is illustrated below with ethyl acetate, but it is a general reaction for any ester. In the lab, a catalyst in the form of a mineral acid is usually used, but in the body hydrolysis of esters is usually mediated by an enzyme catalyst.

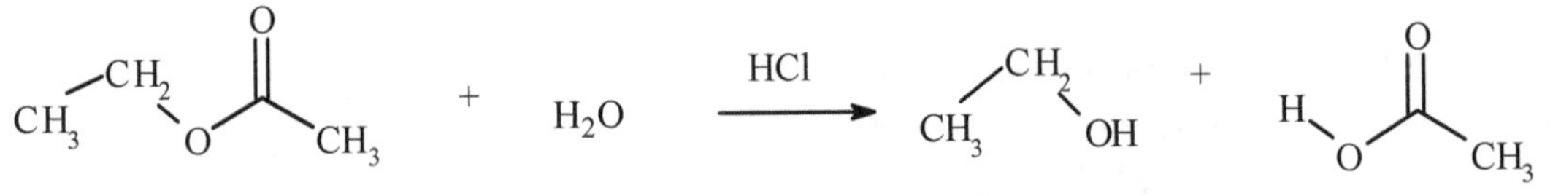

Saponification also breaks apart an ester linkage, but the catalyst is a base, and there is an intermediate step before the complete carboxylic acid can be reached. The alcohol is formed during the reaction, but the acid is formed as the anion, which is soluble in the water solution of the reaction. To obtain the free acid, which can precipitate from the solution if the acid is an insoluble one, it is necessary to add extra mineral acid to protonate the anion. The reaction is shown below for methyl benzoate.

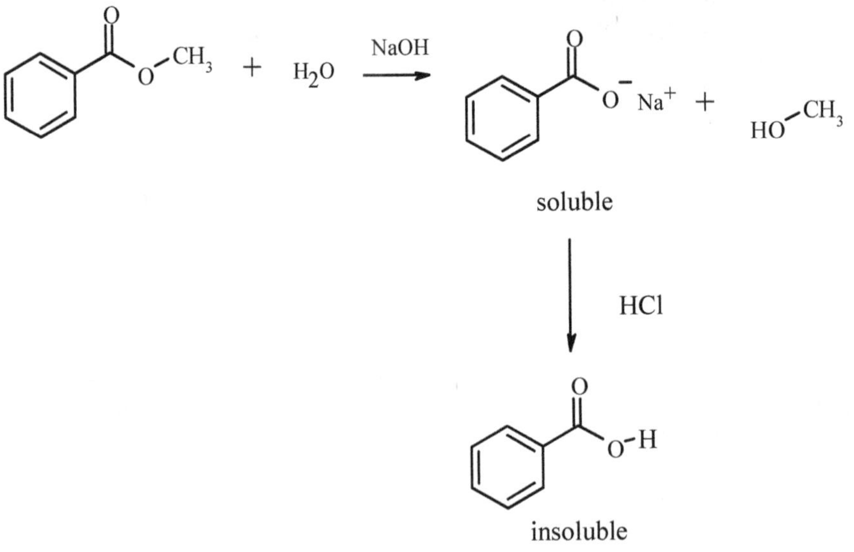

soluble

HCl

insoluble

In this experiment, you will try some simple reactions of carboxylic acids and then move on to examine the hydrolysis and saponification reactions of esters. Read all parts of the experiment carefully before coming to class. For best time management, you may prefer to start on Part B of the experiment before doing the relatively simple reactions in Part A.

REFERENCES

Stoker, H. S. *General, Organic, and Biological Chemistry;* Cengage Learning; Chapter 16.
  Review topics
  - Functional groups containing oxygen
  - Nomenclature of acids and esters
  - Esterification
  - Saponification
  - Hydrolysis

WEB RESOURCES

Simple Aldehydes. *http://www.indigo.com/models/gphmodel/molymod-aldehydes-62009.html*

Carboxylic Acids and Derivatives. *http://www.hcc.mnscu.edu/chem/V.22/*

Carboxylic Acids and Their Derivatives.
http://www.fsj.ualberta.ca/chimie/chem161/bACIDS/sld001.htm

SAFETY NOTES

Organic chemicals often have strong odors and are usually volatile and flammable. Avoid inhaling the vapors by using them in the hood whenever possible. Use small amounts of materials and discard them into appropriately labeled waste jugs when finished, not down the sink. Some chemicals in this experiment may be strongly acidic. Treat all the chemicals used with respect. You may wish to wear gloves during this experiment, and you must wear chemical splash goggles.

EXPERIMENTAL PROCEDURE

PART A. STRUCTURES OF ACIDS AND ESTERS

Use a molecular modeling program, a set of physical models, or an Internet search for this part of the exercise. For each of the following compounds,
   a) draw the structural formula,
   b) draw the ball and stick model, and
   c) circle the functional group in the model drawing.

ⓘ This exercise can be done with ball and stick models or a computer-based modeling program.	

acetic acid   propanoic acid
iso-amyl acetate   ethyl propanoate
ethyl benzoate   3-aminobenzoic acid

PART B. REACTIONS OF ACIDS

Ionization

To test a liquid for the degree of ionization, test for conductivity with the usual battery-operated apparatus. To test, first rinse the electrodes with distilled water, then immerse their ends in the liquid to be tested. The brightness of the light is an indication of the degree of ionization of the substance tested. A blinking light is the strongest indicator.

Step 1. Dissolve 5 drops of acetic acid in 10 drops of water in a spot plate.
Step 2. Repeat the exercise with propanoic acid.
Step 3 Compare with 10 drops of 6 M HCl in another well of the spot plate.

> 🛡 6 M HCl is corrosive.

Neutralization

Step 1. Place 0.1 g of benzoic acid in a test tube and add 2 mL water. Stir well. Does it dissolve?

Step 2. Add to this solution a drop of 6 M NaOH solution, and stir. Continue adding and observing until all the benzoic acid has dissolved. How many drops did it take? Write the equation for what happened.

Step 3. Add to the test tube, dropwise, about 2 mL of 6 M HCl, stirring after each addition. Observe and record what you see. What is the white substance formed? Write an equation for its formation.

> 🛡 6 M NaOH is very caustic, so be careful.

Esterification

Step 1. Set up a warm water bath on a hot plate, using a beaker half full of water and adding a few boiling chips. You will need this bath for about 1/2 hour, so if it begins to boil dry, add more water. Don't let the beaker of water go completely dry.

Step 2. In a test tube, place 0.1 g of benzoic acid, 1 mL of methanol, and 2 drops of concentrated sulfuric acid. Place the test tube in the water bath and heat for 1/2 hour at boiling. Some of the methanol may boil away during this time; add methanol if you see white crystals forming in your test tube.

Step 3. At the end of the heating time, pour the reaction mixture into 50 mL of water in a beaker. Observe the mixture for the presence of droplets floating on the water, and cautiously smell the mixture. Note your observations and write an equation for the reaction that occurred.

> 🛡 Caution here, too! Concentrated sulfuric acid is highly corrosive!

PART C. REACTIONS OF ESTERS

Hydrolysis

Step 1. Place 10 drops of methyl benzoate in a test tube along with 10 drops of 6 M HCl and 2 mL of water. Place this mixture in a hot water bath for 30-45 minutes, and stir occasionally.

Step 2. Cool the mixture in an ice/water bath for 15 minutes, and observe the result.

Saponification

Step 1. Place 10 drops of methyl benzoate in a test tube along with 10 drops of 6 M NaOH and 2 mL of water. Place this mixture in the hot water bath, and allow this reaction to proceed at the same time as the hydrolysis reaction.

Step 2. After 30-45 min, cool the reaction mixture in an ice bath and observe. Do you see any solid forming? Add to this mixture about 2 mL (40 drops) of 6 M HCl and observe.

Step 3. Write the equations for what happened in Step 1 and Step 2.

EXPERIMENT 28 REPORT: ACIDS AND ESTERS

Name_____ Partner _____ Date _____

PART A. STRUCTURES OF ACIDS AND ESTERS

Structure	*Model*

acetic acid

iso-amyl acetate

ethyl benzoate

propanoic acid

ethyl propanoate

3-aminobenzoic acid

PART B. REACTIONS OF ACIDS

Ionization

Extent of Ionization of Carboxylic Acids

Acid	Observation of conductivity	How well ionized?
Acetic acid		
Propanoic acid		
Hydrochloric acid		

Neutralization

Drops of NaOH used to dissolve benzoic acid _____

Equation for the reaction

Reaction of the basic benzoic acid solution with 6 M HCl

The white solid formed is _____

Esterification

    Observations

    Equation

PART C. REACTIONS OF ESTERS

Hydrolysis

    Observations

    Equation

Saponification

    Observations after cooling

    Observations after acidification

    Equation for sequence in Step 1 and Step 2.

EXERCISE

Ibuprofen, a commonly used anti-inflammatory agent, is a carboxylic acid with the following structure. When used to treat people, the sodium salt is used instead of the carboxylic acid. Why?

EXPERIMENT 29

ESTERIFICATION REACTIONS

GOALS

1. To demonstrate esterification reactions
2. To synthesize a common medication that is an ester
3. To demonstrate the spot test for the presence of a phenol

INTRODUCTION

In this exercise, you will actually prepare and isolate two esters, methyl salicylate and acetylsalicylic acid. For both reactions, the limiting reagent is salicylic acid, but because it has both an acidic functional group and an alcohol functional group, two possible esters can be made, depending on what other starting materials you use. Both of these esters are of significant commercial importance.

Methyl salicylate forms when the acidic functional group of salicylic acid reacts with methanol to form the ester. The common name for this chemical is "oil of wintergreen," and it is the main flavoring agent in wintergreen candies and gum. This is a typical esterification, requiring only the two starting materials and an acidic catalyst. However, it is necessary to avoid the presence of extra water in the reaction mixture because an additional product is water, and the presence of extra product molecules would, as you learned earlier, prevent the reaction from proceeding to the right

(LeChâtelier's principle at work). This is the reason that the acid catalyst used is usually sulfuric acid, which has very little water in it, as opposed to (say) hydrochloric acid, which is more than 60% water.

Acetylsalicylic acid is the chemical name for aspirin, a familiar over-the-counter medication. In the reaction to make this compound, the phenolic –OH of the salicylic acid acts as the alcohol and reacts with acetic acid or a derivative of it to produce the ester. The properties of this ester are very different from the properties of methyl salicylate; one is a liquid and the other is a solid, for instance, and the odors are not at all alike. Their physiological actions are very different, as well.

In this reaction, you will take advantage of a well-known trick in organic chemistry, using a derivative of a carboxylic acid instead of the acid itself. In this case, the derivative is acetic anhydride, made from two molecules of acetic acid.

$$CH_3\text{-}\overset{\overset{\displaystyle O}{\|}}{C}\text{-}O\text{-}\overset{\overset{\displaystyle O}{\|}}{C}\text{-}CH_3$$

acetic anhydride

Anhydrides usually will react better in condensation reactions like this one than the free acids will, and you are more likely to get significant amounts of product in a short time this way. Using an anhydride has its drawbacks, though. Most anhydrides belong to a class of chemicals known as "lachrymators"; they will cause your eyes to water. Acetic anhydride is a mild lachrymator as such things go, but you still will want to keep the reaction mixture covered as much as possible, and wear chemical splash goggles to protect your eyes from the vapors. Use a lachrymator in the hood! Excess acetic anhydride will be destroyed by reaction with water during the isolation of your product.

$$CH_3\text{-}\overset{\overset{\displaystyle O}{\|}}{C}\text{-}O\text{-}\overset{\overset{\displaystyle O}{\|}}{C}\text{-}CH_3 \;+\; 2\,H_2O \longrightarrow 2\,CH_3COOH$$

Once you make the two esters, you will need to assess how pure the actual samples are. There are many high-tech ways to do this, but in this case you will observe a well-known reaction of a phenol group with ferric chloride. If there are molecules containing a phenolic –OH in the sample tested, ferric chloride will form a complex compound with them and produce a vivid color, usually red, green or purple depending on what else is in the molecule. Your methyl salicylate should give a strong ferric chloride test since you did not change the phenol part of the molecule, but if your aspirin is pure you should see little or no reaction to the presence of ferric chloride.

You will want to begin the preparation of aspirin while the preparation of methyl salicylate is in progress, because like many endothermic organic reactions, the esterification of salicylic acid with methanol takes significant time. This time can be used well for beginning the next reaction.

REFERENCES

Stoker, H. S. *General, Organic, and Biological Chemistry;* Cengage Learning; Chapter 16.
   Review topics
   - Structures of acids and esters
   - Reactions of acids and alcohols
   - Acid derivatives as synthesis reagents

WEB RESOURCES

Aspirin. *http://www.chemheritage.org/EducationalServices/pharm/asp/asp00.htm*

How do Artificial Flavors work? *http://science.howstuffworks.com/question391.htm*

SAFETY NOTES

Organic chemicals often have strong odors and are usually volatile and flammable, so avoid inhaling the vapors by using them in the hood whenever possible. Use small amounts of materials, and discard them into appropriately labeled waste jugs when finished, not down the sink. Some chemicals in this experiment may be strongly acidic. Treat all the chemicals used with respect. You may wish to wear gloves during this experiment, and you must wear chemical splash goggles.

EXPERIMENTAL PROCEDURE

PART A. FORMATION OF METHYL SALICYLATE

Step 1. Place 0.2 g of salicylic acid in a test tube, and add 30 drops (about 1 mL) of methanol. Add 5 drops of concentrated sulfuric acid, and mix the chemicals.

> ♥ Concentrated sulfuric acid is highly corrosive!

Step 2. Set up a warm water bath on a hot plate, and heat the reaction mixture for about 1/2 hour, replenishing the methanol if it evaporates during the process. You may wish to start the synthesis of aspirin while watching this reaction.

Step 3. At the end of the heating time, cool the reaction mixture and pour it into 5 mL of water in a test tube while stirring the mixture. An insoluble liquid should come out of solution and sink to the bottom of the test tube (methyl salicylate is more dense than water). If a solid forms, there is still some unreacted salicylic acid left. Try to avoid carrying any unreacted solid forward into the next step.

Step 4. Using a micropipet such as a Pasteur pipet or disposable plastic pipet, transfer the bottom layer to a new test tube, being careful not to carry forward any unreacted solid. Cautiously smell the liquid you have isolated.

> ⓘ Try not to carry forward any of the water solution, either.

Step 5. Set up two test tubes, one containing salicylic acid and one containing a few drops of your methyl salicylate. To each add about 10 drops of acetone to dissolve the contents of the tubes.

Step 6. Try a ferric chloride test on each of the samples to determine if there is still a phenolic –OH in the compound represented. To do this, add 5 drops of 5% ferric chloride solution to each sample, and observe for a color change. Salicylic acid has a free phenol group; does methyl salicylate?

PART B. SYNTHESIS OF ASPIRIN

This reaction can be started while the reaction in Part A is going on.

Step 1. In a small Erlenmeyer flask, place about 1 g (recorded to two decimal places) of salicylic acid.

Step 2. Add to this, in the hood, 1.5 mL of acetic anhydride, and 3 drops of concentrated sulfuric acid (caution!). The acetic anhydride should be measured with a calibrated plastic pipet provided for the purpose.

> ♥ Acetic anhydride is a lachrymator— careful! Use a lachrymator in the hood!

Step 3. Set up a hot water bath, and heat the reaction mixture until the solid has dissolved, and then for an additional 5-10 minutes. You can use the same hot water bath you are using for the methyl salicylate reaction.

Step 4. Cool the mixture to room temperature before proceeding. The fastest way to do this is to hold the flask under cold running water for a few seconds.

Step 5. Pour the reaction mixture onto a mixture of ice and water (about 25 mL total) and stir to induce crystallization. Cool for 10 minutes in an ice bath. While you are waiting, chill some distilled water by adding ice to it.

Step 6. Collect the product by suction filtration.

Step 7. With the suction still on, use a little of your chilled water to rinse the crystals free of reaction solution, and allow the suction to continue for an additional 5 minutes to dry your product as much as possible.

Step 8. Find the mass of your product by scraping it into a tared (pre-weighed) flask and re-weighing. Calculate the per cent yield of your reaction (the theoretical yield from 1.00 g of salicylic acid is 1.30 g).

Step 9. Test your product for the presence of a free phenol group.

a) Place a few crystals of salicylic acid in a test tube. In another tube, place an equal amount of your acetylsalicylic acid, and in a third place a small amount of powdered commercial aspirin.

b) Dissolve each sample in about 10 drops of acetone. All of the commercial aspirin may not dissolve because there are some insoluble binders in the commercial tablets.

c) To each tube add 5 drops of ferric chloride solution and mix well. Record your observations. Save these tubes for comparison if you continue with Steps 10-12.

Steps 10-12 may be done if there is sufficient time.

Step 10. You can purify your acetylsalicylic acid by the process known as recrystallization. To the acetylsalicylic acid product in the Erlenmeyer flask, add about 10 mL of water, and heat gently to boiling on a hot plate. All the crystals should go into solution; if they don't, add a bit more water. Set the mixture into an ice bath to cool, and allow the purified product to crystallize for at least 10 minutes before going on to Step 11.

Step 11. Suction filter the product, and wash it with a small amount of ice-cold water. Allow the suction to remain on the sample for 10 minutes before proceeding.

Step 12. Weigh your purified product, and observe the color of it. Place a few crystals of your product in a test tube and add 10 drops of acetone and 5 drops of ferric chloride solution. Record the result.

PART C. EXERCISES

Answer the questions for Part C on the report form.

EXPERIMENT 30

AMINES AND AMIDES

GOALS

1. To examine the difference in structure between amines and amides
2. To review the reactions of amines
3. To emphasize the basic properties of amines
4. To synthesize an amide from an amine

INTRODUCTION

Amines are nitrogen-containing molecules that often have strong fishy or even putrid odors. Many amines, such as morphine or caffeine, have strong biological action, and many prescription drugs contain the amine group.

Because nitrogen normally has a valence of three, there are three classes of amines to be aware of—primary, secondary, and tertiary amines. Their classification depends on how many R groups are attached to the nitrogen of the amine group.

methyl amine
a primary amine

diemthyl amine
a secondary amine

trimethyl amine
a tertiary amine

Simple amines, with up to five or six carbon atoms, are somewhat soluble in water, dissolving as does ammonia to give an amine cation, a hydroxide ion, and a basic solution.

Larger amines, though not soluble in water, can act as Brønsted-Lowry bases, reacting with hydrogen ions in a neutralization reaction to form compounds called amine salts. These salts are usually soluble in water. They can be changed back into the free amine by treating them with a base such as sodium hydroxide.

An amine forms an amide with a carboxylic acid or a derivative of a carboxylic acid. An amide has characteristics different from those of amines. Amides are not very basic, not very soluble in water, and do not form salts with acids. Their major reaction is to hydrolyze in the presence of an acidic or basic catalyst back to the parent amine and carboxylic acid.

Formation of an amide is illustrated here with the reaction you will do in this experiment, the formation of acetaminophen (one brand name is Tylenol®). As in the formation of an ester, the reaction works best if you use the anhydride of the carboxylic acid, in this case acetic anhydride.

| p-amino phenol | acetic anhydride | | acetaminophen | acetic acid |

Amides are of great importance in the chemistry of proteins since it is amide functional groups that hold the amino acids residues together in protein polymers. Amides can also undergo hydrolysis to return to the original carboxylic acid and amine. This is one of the major initial processes in the digestion of meat, for example.

REFERENCES

Stoker, H. S. *General, Organic, and Biological Chemistry;* Cengage Learning; Chapter 17.
Review topics
- Structure of amines and amides
- Reactions of amines
- Basic properties of amines
- Synthesis of amides
- Reactions of amides

WEB RESOURCES

Chemfinder.com. *www.chemfinder.com*

Drug Profiles: Acetaminophen or Tylenol® for Migraines.
http://www.migraines.org/treatment/protylnl.htm

Tylenol (Acetaminophen) Toxicosis in Cats.
http://www.addl.purdue.edu/newsletters/1998/spring/acet.shtml

SAFETY NOTES

Organic chemicals often have strong odors and are usually volatile and flammable. Some amines can be toxic and should be used in the hood. Use small amounts of materials, and discard them into appropriately labeled waste jugs when finished, not down the sink. Some chemicals in this experiment may be strongly acidic. Treat all the chemicals used with respect. You may wish to wear gloves during this experiment, and you must wear chemical splash goggles.

EXPERIMENTAL PROCEDURE

PART A. PROPERTIES OF AMINES AND AMIDES

Step 1. Odor of amines. Place a few drops of triethyl amine, aniline, and benzamide (a few crystals in this case) in three test tubes. Cautiously smell each chemical and describe the odor.

> ⓘ You may wish to begin first on Part B of this exercise because there are several waiting periods involved, and the tests in Part A can be done during those times.

Step 2. Solubility of amines in 10% HCl. Using the same samples you prepared in Step 1, add to each one a few drops of acid, mix and observe. Continue this step until all the chemical has dissolved, or until 2 mL of HCl has been used.

Step 3. Reaction of an amine with acid and base. In a test tube place 1 mL of 10% HCl, and add to it 5 drops of aniline. Does the aniline dissolve? Add to this solution, dropwise, 1 mL of 10% NaOH, and observe the result. Write an equation for the reaction of an amine salt with base.

no dissolve

react heavily

Step 4. Reaction of an amine with water. Add triethyl amine (1 drop) to 2 mL of water, and check the pH of the resulting solution. Write an equation for what happened to the amine in water.

increased pH(10)

Step 5. Change of an amine odor with addition of an acid. Cautiously note the odor of the triethylamine solution. Neutralize the solution by adding 20 drops of 10% HCl solution drop by drop. Smell the mixture. Did the odor change? Write an equation for this reaction.

> 🛡 Amine odors can be very strong and unpleasant. Some amines can be toxic and should be used in the hood.

PART B. SYNTHESIS OF ACETAMINOPHEN, A MEDICINAL AMIDE

Step 1. Weigh about 1.0 g of *p*-aminophenol into a weighing boat, and transfer it to a small Erlenmeyer flask. Be sure to record exactly how much *p*-aminophenol you used, because you will be asked to calculate a percent yield for this reaction.

Step 2. Using the provided disposable plastic measuring pipets add about 3 mL of water and then 3 mL of concentrated acetic acid to the flask. Then add 2 mL of acetic anhydride. Do this step in the hood, because concentrated acetic acid and acetic anhydride can cause chemical burns and tears.

Step 3. Warm the reaction mixture on a hot plate in the hood for about 15 minutes. Do not boil the reaction mixture. If it looks as if the mixture will boil, take it off the hot plate for a short time. Cool to room temperature before proceeding.

> ⓘ The purpose of the heating is a **gentle** warming.

Step 4. Pour the reaction mixture into about 25 mL of a mixture of ice and water in order to precipitate the acetaminophen. Stir and cool in an ice bath for at least 15 minutes before collecting the solid.

Step 5. Collect the solid acetaminophen by suction filtration. Rinse the crystals with a small amount of ice-cold water, and then allow the vacuum to remain on the sample for several minutes (until the funnel has stopped visibly dripping). Weigh your crude product.

Step 6. Calculate the percent yield of your compound. The theoretical yield from 1.0 g of *p*-aminophenol is 1.36 g. Check it for solubility in 10% HCl solution. Starting *p*-aminophenol will dissolve in acid solution, but acetaminophen, an amide, should not.

PART C. EXERCISE

Complete the report with the exercise listed under Part C.

EXPERIMENT 31

IDENTIFICATION OF AN UNKNOWN ORGANIC COMPOUND

GOALS

1. To review the chemistry of organic functional groups
2. To recognize some simple tests for the presence of functional groups
3. To determine the structure of an unidentified organic compound

INTRODUCTION

Imagine this...

You have been hired by the university to help clean out a chemical storeroom. You have found a box full of bottles without labels; the old labels have fallen off. There is an inventory list for the box, so you know which chemicals are possibilities, but you don't know which chemical is in each bottle. In order to keep or dispose properly of these compounds, you need to know what is in each bottle.

Your specific mission is to identify the one compound, either a solid or a liquid, that you have been assigned.

This exercise summarizes the concepts you have been working on for the past several weeks in lab. Each week we have examined one or two functional groups and considered their reactions toward various chemicals. For example, you have used the Tollens reagent and Purpald™ to distinguish aldehydes from ketones and the sodium dichromate reagent to identify alcohols. If chemical substances are truly unknown, however, you might need to use several of these tests and others besides to determine their identities. That's what you'll be doing for this experiment.

You can make these assumptions about your unknown chemical:

- It is a pure organic compound.
- Testing it with the reagents described below will give you sufficient information to identify it.
- You will need a combination of observations on these tests, what you can learn about the physical properties of the possible compounds, and deduction in order to solve your puzzle.

The Inventory List Found with the Unlabeled Bottles

cinnamic acid	cinnamaldehyde
phenol	2-propanol
ethylenediamine	methyl ethyl ketone
diphenylamine	salicylic acid
benzyl alcohol	benzoic acid
hexamethylenediamine	benzophenone
salicylaldehyde	propanoic acid
ethyl acetate	benzaldehyde
toluene	benzhydrol
4-chlorobenzaldehyde	aniline
methyl benzoate	4-chlorobenzoic acid
	cyclohexene

REFERENCES

Stoker, H. S. *General, Organic, and Biological Chemistry;* Cengage Learning; Chapters 12–17.

  Review topics
  • Reaction classes for functional groups
  • Tests for the presence of different functional groups
Other References:

Aldrich Catalog of Fine Chemicals; Aldrich Chemical Company: Milwaukee, 1999. Other editions are fine.

CRC Handbook of Chemistry and Physics, 50th ed.: Boca Raton, 1970. Later editions are fine.

Merck Index, 10th ed.; Merck: Whitehouse Station, 1983. Later editions are fine.

WEB RESOURCES

Chemfinder.com. *http://www.chemfinder.com*

Experiment 8. *http://en.wikipedia.org/wiki/Functional_group*

SAFETY NOTES

Wear chemical splash goggles when handling the chemicals used in this experiment. You should try to work with only a few drops or crystals of your unknown compounds, because several of the substances have strong odors, and some are potentially toxic if used carelessly. (Phenol is a special case in point here).

  Some of the reagents contain strong acids or bases, and you may prefer to wear gloves while working with these chemicals.

  You should always wear gloves when working with bromine in carbon tetrachloride. However, continue to work as neatly and efficiently as possible. Do not assume that lightweight polyethylene or vinyl gloves are totally impervious to all chemicals; they can dissolve or allow some chemicals to pass through to your skin. **Never** put an unprotected finger over an open test tube to mix a solution.

Tollens reagent should not be kept from day to day since it can precipitate a highly explosive solid. At the end of each day, discard according to your instructor's directions any leftover Tollens reagent you have made.

Glass disposable pipets are usually regarded as "sharps" and should be disposed of appropriately.

Dispose of used reagents according to local ordinances. In general, if you are a small-volume user, the organic solvents can be discarded in an appropriately labeled waste container in the hood, and water solutions or suspensions of the unknown substances can be discarded in the sink, flushed down the drain with copious amounts of water. However, local ordinances govern your procedure.

EXPERIMENTAL PROCEDURE

PART A. UNIDENTIFIED COMPOUND X

Below are some general hints for you, but the major part of the work is up to you. Good luck! Step 1 should be completed before class to maximize your work time.

Step 1. Before beginning the experiment, you will need to find out the structures of all the possible unknown compounds in the list on p 302, and record the structures and perhaps some useful physical properties in a notebook or on the report pages. For example, what is the compound's physical state? What is its melting point or boiling point? There are several useful books listed in the References section of this experiment that should be available in your library. You may also wish to investigate these compounds using the Internet; one useful URL is listed in the Web Resources section.

Step 2. Each individual has an unidentified sample for which that individual is solely responsible. To facilitate your deductions, answer all of these questions in writing in a notebook or on these pages before coming to class.

1. The possible compounds can be grouped into functional group classifications. Which ones are amines? Which ones are acids, alcohols, etc?

2. Within those groups, which ones are solids, and which ones are liquids?

3. (You won't have the answer for this question until you begin work in lab.) Is your compound soluble in water? If it is, what is the pH of the solution? What would that tell you about your compound? If your compound is not soluble in water, answers to the following questions will help you identify it.

4. What kind of functional group causes a bromine solution to turn colorless?

5. What kind of functional group reacts with sodium dichromate reagent <u>and</u> 2,4-DNP reagent?

6. What kind of functional group reacts with 2,4-DNP reagent, but not with sodium dichromate reagent?

7. What kind of functional group reacts with ferric chloride?

8. What kind of functional group dissolves in both sodium bicarbonate solution and in sodium hydroxide solution?

9. What kind of functional group dissolves in only sodium hydroxide solution?

10. What kind of functional group burns with a sooty flame?

11. What kind of functional group will react with Benedict's solution?

12. What kind of functional group dissolves only in acid solution?

13. What kind of functional group can dissolve in water to give a basic solution?

14. What kind of functional group can dissolve in water to give an acidic solution?

Step 3. Solubility tests and pH tests. If any of these tests gives you a clear result, you are already almost finished with your investigations! If the compound you are investigating is not soluble in water, however, you have more to do. The solubility tests in acidic and basic solutions are particularly important since they can immediately tell you if your compound is (or is not) a carboxylic acid, an amine, or a phenol.

> ⓘ Pay particular attention to the Flow Chart for this experiment, which summarizes the answers to most of these questions.

In general, a solubility test is carried out with about 1 mL of water, acid, or base, and one or two drops of a liquid compound or a few crystals of a solid compound. "Soluble" is defined as more than three parts in 100 (3%).

> ⓘ Be careful as you do these tests. A mistake at this stage can cause you endless difficulty.

Step 4. Combustion test. This test distinguishes aromatic from non-aromatic compounds. Dip a clean spatula in your compound and hold the sample in the flame of a Bunsen burner until it catches fire (do this in the hood). Take the sample out of the fire and watch it burn. Aromatic compounds burn with lots of dark, sooty smoke; compounds with only a few C=C double bonds give off a small amount of dark smoke; and aliphatic compounds burn cleanly with little or no white smoke.

Step 5. If you have been able to identify a functional group in your compound, you need to refer to the list of possible compounds and see if you have sufficient information to decide on the compound identity. If there are still several possibilities, try some of these further tests to determine what other functional groups might be in your compound.

If at any point in the investigation you have sufficient information to identify your sample, you do not need to do any of the remaining tests. For instance, if your compound is soluble in NaOH but not NaHCO₃, it is not a phenol, so you don't need to do a ferric chloride test.

> ⓘ Don't do any more tests than you need to do.

Step 6. Beilstein test. This test is positive if your compound contains a halogen. Heat a looped copper wire in a burner flame in the hood until it glows red, then allow it to cool. You will want to hold the wire with a forceps. After the wire has cooled, dip it into your unknown sample and hold the wire in the burner flame until the compound ignites. If the compound burns with a distinct green flame, your compound contains chlorine, bromine, or iodine.

Step 7. Bromine in carbon tetrachloride reagent. This reagent tests for the presence of carbon-carbon double bonds. Place a small amount of your compound in a test tube, and dissolve it in acetone. Add a drop of the bromine reagent and observe. A decrease in the color of the bromine reagent

> ☻ Always wear gloves and work in the hood when handling bromine solutions.

indicates the presence of a double bond. Phenols sometimes also give decolorization, but there is a further test for the presence of a phenol to help you determine if this is a competing reaction.

Step 8. Sodium dichromate/sulfuric acid reagent for primary or secondary alcohols. Dissolve a few drops of your sample in acetone, and add a drop of the sodium dichromate reagent. A green color (change from the original orange-red) indicates the presence of an easily oxidized group. Usually either a primary or a secondary alcohol gives this test, but aldehydes can also react.

Step 9. 2,4-DNP test for aldehydes and ketones. Mix 5-10 drops or a small spatula-full of your sample with 10 drops of ethanol in a test tube. Add 10 drops of the 2,4-DNP reagent. Mix well, and allow the mixture to stand for 10 minutes. Record what you see. A yellow, orange, or red precipitate is a positive test for an aldehyde or ketone. Usually yellow precipitates are produced from non-aromatic ketones, and dark orange or red precipitates from aromatic ketones.

> ☻ **Caution**: The reagents for Steps 8 and 9 contain sulfuric acid. Be careful handling them.

Step 10. The Tollens test for aldehydes. First make the Tollens reagent. Mix 2 mL of Tollens Solution A (5% silver nitrate) with 1 drop of 10% sodium hydroxide solution (Tollens B). Stir well; a gray precipitate should form. Add to this sufficient 2% ammonium hydroxide solution to just dissolve the precipitate. More than 20 drops may be necessary, so be patient—the precipitate will dissolve. Use this solution in the next step.

> ☻ Tollens reagent is strongly basic.

In a **very** clean small test tube, place 5 drops of a liquid or a small spatula-full of a solid unknown. Add about 10 drops of dimethoxyethane to improve solubility of this substance in the Tollens reagent. Mix with it 1 mL of freshly prepared Tollens reagent, and allow the mixture to stand for 10 minutes. A positive test is the formation of a silver mirror coating the inside of the test tube. If the result is uncertain, heat the tube in a warm water bath for 2 minutes, but do not boil.

Tollens reagent is a mild oxidizing agent, so occasionally other easily oxidized compounds such as primary or secondary alcohols can react if the heating is too prolonged or vigorous. Be cautious in your interpretation of this test, especially if you obtained a silver mirror only on heating! Other compounds besides aldehydes can oxidize in the presence of Tollens reagent and heat, and these false positives can make it difficult to figure out your results. Follow the directions exactly, and note details of exact times and appearances!

Step 11. The Purpald™ test. Try this if your compound is probably an aldehyde, because the Tollens test can be difficult to interpret. Dissolve a few drops of your compound in acetone, and add about 5 drops of Purpald™ solution. A purple color is a positive test for an aldehyde.

> ♥ This reagent is strongly basic, too.

Step 12. Ferric chloride reagent for phenols. If your solubility tests indicated that your compound might contain a phenol, you should do this test. Dissolve a few drops of your compound in acetone, and add about 5 drops of ferric chloride solution. A phenol will give a deep color, usually purple, red, green or blue.

If you do all of the necessary tests on your unknown sample, you should have sufficient information to allow you to identify it successfully. Remember, if you have already decided that your compound is a carboxylic acid, it is not necessary to test it for the presence of an alcohol functional group. Some of the tests are redundant if you have already identified the major functional group in your compound, and some can give false positive tests if tried on a compound with the wrong functional groups present. You do not have to do any more reactions than necessary to deduce the identity of your unknown sample.

PART B. UNIDENTIFIED COMPOUND Y

If time permits, your instructor may assign you a second unknown sample to identify. The list of possible compounds is the same, and the procedure is the same; you will just have to work with a different compound.

EXP. 31 FLOW CHART

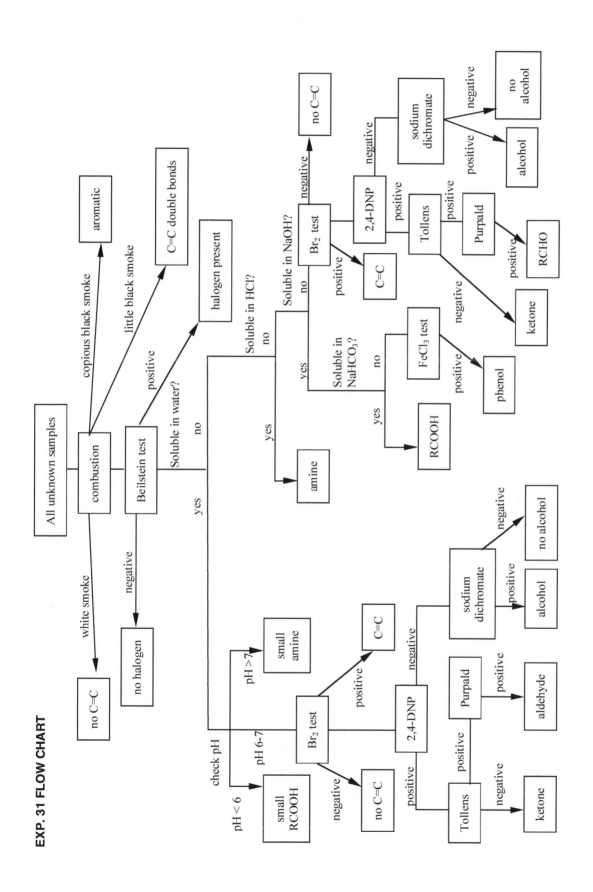

EXPERIMENT 31 PRELAB EXERCISES:

INVESTIGATION OF MIXTURES

Name_____ Partner _____ Date _____

1. For the following test procedures, identify the class of functional group present in the unknown sample.

Test procedure	Result with sample	Functional group present
2,4-DNP test	yellow precipitate	
Tollens test	silver on sides of test tube	
mixed with sodium bicarbonate solution	dissolves readily	
dissolves in water to test pH	pH = 9	
mixed with bromine in carbon tetrachloride	bromine turns colorless	

2. What test could you use to tell the difference between a carboxylic acid and an ester?

3. What class of organic compound will dissolve in 6 M HCl?

4. A student discovered that an unknown liquid sample reacted with the 2,4-DNP reagent, but not the Tollens reagent. The sample was one of the two compounds below. Circle the correct one, and explain your choice.

5. When you are working with unidentified samples, why is it especially important to be careful about wearing goggles and gloves?

6. Complete Step 1, Part A of this experiment.

EXPERIMENT 31 REPORT: IDENTIFICATION OF AN UNKNOWN ORGANIC COMPOUND

Name_____   **Partner** _____   **Date** _____

PART A. UNIDENTIFIED COMPOUND X

Structures and names of possible compounds. Notes of physical characteristics such as solubility, odor, and color may also be useful.

Sample number   _____
Physical description (physical state, odor, other)

Solubility Tests for Sample Number _____

In water	pH if soluble in water	In acid	In NaOH	In NaHCO$_3$

Results of Other Spot Tests for Sample Number _____
If the test was not done, write NA in the table.

Other tests	Observations	Positive or negative?	Possible functional group present?
Combustion			
Beilstein			
Br$_2$/CCl$_4$			
Na$_2$Cr$_2$O$_7$/ H$_2$SO$_4$			
2,4-DNP			
Tollens			
Purpald™			
FeCl$_3$			

This sample number _____ was identified as
    Name

    Structure

PART B. UNIDENTIFIED COMPOUND Y

Sample number _____
Physical description (physical state, odor, other)

Solubility Tests for Sample Number _____

In water	pH if soluble in water	In acid	In NaOH	In NaHCO$_3$

Results of Other Spot Tests for Sample Number _____
(If the test was not done, write NA in the table.)

Other tests	Observations	Positive or negative?	Possible functional group present?
Combustion			
Beilstein			
Br$_2$/CCl$_4$			
Na$_2$Cr$_2$O$_7$/ H$_2$SO$_4$			
2,4-DNP			
Tollens			
Purpald™			
FeCl$_3$			

This sample number _____ was identified as
    Name

    Structure

EXPERIMENT 32

CHROMATOGRAPHY OF MEDICINAL SUBSTANCES

GOALS

1. To introduce the concept of chromatography
2. To investigate the structures of some common pain relievers
3. To reinforce the concept of polarity in organic compounds

INTRODUCTION

The technique of chromatography has been known for nearly a century. It is a valuable method for separating mixtures so that the individual components can be analyzed. In the years since its discovery the technique has been extended, applied in new situations, and automated to such an extent that almost every laboratory that routinely analyzes mixtures uses it, including many medical technology and research laboratories.

Now imagine this...

Imagine that you work in a medical laboratory in a small, remote hospital. A three-year-old female patient, Sally S., was admitted to the emergency room in a comatose state. Apparently Sally needed some sweets for a play tea party, but there were no cookies or candies in the house. For a substitute, she pulled all the pill bottles off the shelf in the bathroom and put those colorful tablets, caplets, and capsules on the plates and then ate all the "candies" herself.

Fortunately, the items on the medicine shelf were limited, consisting of generic brands of the over-the-counter drugs aspirin, acetaminophen, caffeine, ibuprofen, and naproxen. In the emergency room, appropriate therapy for a drug overdose was begun, and because the timing and treatment were favorable, Sally eventually recovered.

To determine what medications the child had ingested, an automated form of the chromatography technique was used. This experiment will use the separation of familiar over-the-counter medicines to demonstrate the power of chromatography as an analytical tool.

The earliest form of chromatography, thin-layer chromatography (TLC) was reported by Tswett in 1906 and is still useful today in some situations. It takes advantage of the differing polarities of pure substances and their relative affinities for a solid substance, the "solid phase" in a TLC analysis. The solid phase is a thin layer of a highly polar substance coated onto an inert backing, hence the "thin-layer" name of the technique. You apply to the solid phase the mixture you want to analyze and allow it to dry in place. You then place the prepared "plate" in a chamber containing a small amount of appropriate solvent, the "mobile phase."

The different substances in the mixture have differing abilities to stick to the surface of the solid phase. When the mobile phase runs past the mixture attached (adsorbed) to the surface of the solid phase, the individual components of the mixture detach from the solid phase at different rates. These detached components then move along with the liquid solvent to different positions on the solid phase. Usually, less polar substances move more readily than highly polar ones. In this way, if all the conditions are right, the unknown substances separate from each other and can be isolated, and then studied further individually.

If the unknown mixture consists of compounds whose identity you suspect, you can often confirm their identity directly on the TLC plate. You can run extra spots of "standards," or known samples, side by side with the unidentified ones, and compare the distances traveled by the various components. If a standard and a compound in the unknown sample run to the same point on the chromatogram, you have strong evidence that the two are identical.

Sometimes you will need to compare results from different TLC plates. In these cases, you need one further piece of information, the R_f or retention factor. This is a calculated number for each individual component of a mixture:

$$R_f = \frac{\text{distance traveled by compound from origin}}{\text{distance traveled by solvent from the origin}}$$

The R_f thus shows what fraction of the possible distance a compound actually went up the plate, expressed as a decimal fraction. This fraction should be approximately the same for a particular compound in a given solvent system, no matter if the two compared plates actually ran to different heights. In the particular samples you will be using, you should calculate the R_f for each compound in a sample even though you may be able to determine the identities of the unknowns by inspection.

If a sample contains more than one compound (often the case) you will need to determine an R_f for each component of the mixture.

You will use TLC to separate and analyze a mixture of common over-the-counter medicines in order to identify the individual components.

REFERENCES

Stoker, H. S. *General, Organic, and Biological Chemistry;* Cengage Learning.
   Review topics
   - Organic functional groups
   - Chromatography
   - Polarity of organic molecules

Drug Facts and Comparisons 2011; Lippincott Williams and Wilkins, 65th ed.: St. Louis, 2010; Many other editions are available for reference.

Elder, J. W. *J. Chem Educ.* **1995** *72* 1049. "Thin-Layer Chromatography of Analgesics—An Update."

Roueche, B. *The Medical Detectives;* Times Books: New York, 1980; pp. 61-80;"$CH_3CO_2C_6H_4CO_2H$ (Aspirin)." A more recent edition of this book is also available.

Tswett, M. *Chemische Berichte* **1906** *24* 316-323. "Physikalish-chemische Studien über das Chlorophyll. Die Adsorptionen."

WEB RESOURCES

Thin Layer Chromatography. *http://orgchem.colorado.edu/hndbksupport/TLC/TLC.html*

TLC Retention Factor. *http://orgchem.colorado.edu/hndbksupport/TLC/TLCrf.html*

TLC Procedure. *http://orgchem.colorado.edu/hndbksupport/TLC/TLCprocedure.html#visualize*

SAFETY NOTES

Organic chemicals often have strong odors and are usually volatile and flammable. Treat them with respect and avoid inhaling the vapors by using them in the hood whenever possible. Use small amounts of materials, and discard them into appropriately labeled waste jugs when finished, not down the sink. Some chemicals in this experiment may be strongly acidic, so treat all the chemicals used with respect. You may wish to wear gloves during this experiment, and you must wear chemical splash goggles.

All chromatography chambers should stay covered since several of the solvents have strong odors.

Do not look directly at the UV light; it can cause retinal damage.

Dispose of used reagents according to local ordinances. In general, if you are a small-volume user the organic solvents can be discarded in an appropriately labeled waste container, and water solutions or suspensions of the analgesics can be discarded in the sink, flushed down the drain with copious amounts of water. However, local ordinances govern your procedure. TLC plates can be discarded in the wastebasket, but don't discard them until your notebook record is complete and you have completely finished this laboratory activity.

Microliter pipets are usually regarded as "sharps" and should be disposed of appropriately.

EXPERIMENTAL PROCEDURE

PART A. CHROMATOGRAPHY OF KNOWN SAMPLES

This activity will introduce you to the technique of thin-layer chromatography and allow you to become familiar with the tools for performing an analysis. You will be using two colors of egg dye for the unknown materials in this analysis. Your instructor will give you some pointers about making and handling satisfactory TLC plates.

Step 1. Prepare the solvent chambers and samples.

a) Prepare 10 mL of a 50:50 mixture of acetone and water.

b) Label three 400-mL beakers "acetone," "50:50 acetone:water," and "water." In each, place a piece of filter paper to line the sides of the beaker, as shown in Figure 32-1.

c) To each beaker, add 10 mL of the appropriate solvent and cover the beaker with a watch glass or plastic wrap held by a rubber band. You now have three separate solvent systems (mobile phases) to try to separate a mixture of dyes. Set the beakers aside while you prepare the TLC plates for use.

Figure 32-1. A beaker set up as a chromatography chamber.

Step 2. Prepare the TLC plates.

a) In a small labeled test tube, place one drop of green food color. Also prepare a separate sample of red, and a sample containing one drop of red and one drop of green. Dilute each one with about 1 mL of acetone.

b) On each plate, make a small mark about 1 cm from the bottom, at the edge of the plate. This marks the origin of the chromatogram, and all your samples should be applied at this distance from the bottom of the plate. Use a 10-μL pipet to apply samples to the plates. See Figure 32-2 for the proper arrangement of spots.

Figure 32-2. A TLC plate with five samples applied. Note the mark for the origin.

Step 3. On each plate, apply three samples: pure green food color, a mixture of green and red, and pure red. At the top of the plate, write in pencil the identity of the spots and the name of the solvent system to be used. Allow all applications to dry before proceeding. Small spots are better than big ones.

Step 4. Run the chromatograms as shown in Figure 32-3.

a) Place each chromatogram in the appropriately labeled beaker with the origin on the bottom. Cover the beaker, and allow the chromatogram to develop until the solvent front, as seen by a damp line on the plate, is about 1 cm from the top of the plate. This may take as long as 45 minutes. During this time you should begin to plan the next step. DO NOT slosh the samples around while the chromatogram is running.

> ⓘ Be sure that the liquid level in the beaker is below the lowest part of all the spots on the plate as you begin, or you will get confusing results. See Figure 32-3.

Figure 32-3. Placement of the TLC plate in the chromatography chamber.

b) While this set of chromatograms is running, begin preparations for Part B. When the chromatography is complete, remove the plate(s) from the beakers with forceps and immediately mark the solvent front, since it is hard to see after it dries.

Step 5. Analyze your results. Using a pencil, circle each spot you can see on the plates. Probably your original sample separated into several spots; this would be expected if you have a mixture of compounds. Record for each chromatogram the distance the solvent traveled past the origin, the distance <u>each</u> spot traveled past the origin, and the color of each spot.

Step 6. Now examine the plate in a dimly lit area under an ultraviolet (UV) light. Because the chromatographic plates have been impregnated with a fluorescent indicator, compounds attached to the silica gel will show up as dark spots. This visualization method is unnecessary if your compounds are colored, but is very useful if they are white or colorless as they will be in Part B.

At this point, you should be able to determine which of the three solvent systems would allow you to separate green from red, and identify the individual components of each sample. Now that you are familiar with the technique, you are ready to try the TLC of a mixture of colorless analgesic compounds.

PART B. SEPARATION AND IDENTIFICATION OF MEDICINAL SUBSTANCES

The goal of this analysis is to determine which of five medicines the patient might have ingested, so that appropriate therapy can be applied. There are six samples for you to analyze by TLC: five standards that are pure samples of the possible medications dissolved in a solvent, and one sample to be analyzed that could have been isolated from such a patient. You also have available two solvents, A and B, which can be used on TLC plates to separate and identify the components of the mixture.

> ⓘ Both solvents are quite odorous, so you should work in the hood using these materials.

Solvent A is relatively non-polar, and solvent B is relatively polar. The sample to be analyzed may contain one, two, or three of the possible medications. You will need to determine how many compounds are in the sample and which ones they are, using TLC.

Step 1. Mix the appropriate solvent system. 95% A and 5% B is a good mixture to try first, but don't make more than you need. A mixture of 9.5 mL A and 0.5 mL B should be sufficient for several chromatograms. Pour this sample into a prepared jar or beaker, and cover it until you are ready to use it.

Step 2. Obtain a TLC plate about 4.5 cm wide by 10 cm tall, and mark the origin using a pencil. Mark on it as well, at the top of each channel you will use, an identifier to help you remember which sample is in which channel. Using the standards and your unknown sample, spot each sample on the plate at the origin, using a 10- or 20-μL capillary pipet. The samples are all colorless, so you may not see any sample buildup. If you need to, overlay an additional spot at each site to build up a significant amount of sample.

> ⓘ Remember, the smaller the diameter the spot, the better the TLC experiment will run.

Step 3. Place the chromatogram gently in the development chamber, cover the chamber, and allow the solvent to run to within 1 cm of the top of the plate.

Step 4. Remove the plate with forceps, and mark the solvent front. Lay the plate in the hood to dry as much as possible.

Step 5. Under ultraviolet light, mark all the components of each solution, using a pencil to circle the visible spots.

Step 6. Remove the plate from the UV source, and calculate the R_f for each of the spots that you see. Draw this plate in the report form, making any observational notes that you need as well.

Step 7. If your chromatogram has run well, you should be able to identify the components of your mixed sample on the basis of their appearance compared to standard samples and on the basis of their R_f values. If you are still unsure, run a second chromatogram using a new plate, improving your technique on this second trial.

It is also possible to use the two solvents, A and B, in different proportions to see if you can improve the separation of the mixture components. If you are still having trouble deciding on the identities of the components of your sample, try a different solvent mixture.

EXPERIMENT 32 PRELAB EXERCISES:

CHROMATOGRAPHY OF MEDICINAL SUBSTANCES

Name_____ Partner _____ Date _____

1. What is the purpose of chromatography?

2. A mixture of compound X and compound Y was chromatographed by TLC along with known samples of X and Y, with the following results. Which compound, X or Y, is likely to be more polar? Why?

3. On the chromatogram, identify the origin and the solvent front.

4. Calculate the R_f of all compounds shown on the figure.

5. On the plate, identify the components of the mixture as either X or Y.

EXPERIMENT 32 REPORT:

CHROMATOGRAPHY OF MEDICINAL SUBSTANCES

Name_____ Partner _____ Date _____

PART A. CHROMATOGRAPHY OF KNOWN SAMPLES

Drawings of the three TLC plates

A: in acetone   B: in 50:50 acetone:water   C: in water

R_f of each component on each plate:

A: In acetone   B: in 50:50 acetone:water   C: in water

Which solvent appears to give the best separation of food color mixtures?

PART B. SEPARATION AND IDENTIFICATION OF MEDICINAL SUBSTANCES

The structures of the five possible compounds

aspirin   caffeine

acetaminophen   ibuprofen   naproxen

R$_f$ data

Channel   A   B   C   D   E   F

Origin

The medicinal substance(s) in sample #_____ are

_____   _____   _____

EXPERIMENT 33

INTRODUCTION TO CARBOHYDRATE CHEMISTRY

GOALS

1. To introduce some structural features of carbohydrates
2. To review reactions of carbohydrates
3. To examine the hydrolysis of di- and polysaccharides

INTRODUCTION

Carbohydrates are complicated organic molecules that contain carbon, hydrogen, and oxygen in the ratio CH_2O. They can also be defined by their functional groups, as polyhydroxy aldehydes and ketones. Monosaccharides such as glucose and fructose are well known as food substances, but larger units built of two or more monosaccharides are also common; sucrose, starch, and cellulose are three examples.

The carbohydrates are the first of a series of biochemical molecules you will need to be familiar with, and the first you will study that exhibit chirality, the ability to rotate polarized light in one direction or another. This ability is due to chiral centers in the carbon chain; any carbon atom that has four different types of substituents will be chiral, and even simple carbohydrates have several chiral centers. This leads to the possibility of myriad carbohydrates, all with the same molecular formula but all isomers of each other. Still, their reactions in general are similar, which simplifies their study somewhat.

A basic understanding of possible reactions of carbohydrates does not initially depend on their chiral nature. Their reactions are the reactions of aldehydes and alcohols; hemiacetal and acetal formation is one of the most important. In very simple aldehydes, the reaction might look like this.

An acid catalyst is usually necessary. An acetal is not very reactive because the aldehyde functional group is blocked, or "protected," by being made into an unreactive ether. A hemiacetal, that still has the original aldehyde oxygen atom with only a hydrogen atom attached, can still undergo some reactions.

Hemiacetals are particularly important in sugar chemistry because the chain of carbons even in a monosaccharide is relatively long, and an alcohol on the end of the chain can attack its own aldehyde and make a hemiacetal. In fact, most aldoses and ketoses with more than five carbons are actually in a cyclic hemiacetal form derived from the original aldehyde most of the time. This is a spontaneous equilibrium between the hemiacetal and the aldehyde.

One more complexity of this equilibrium is that, in the hemiacetal form of an aldose or ketose, a new chiral center has been produced. The aldehyde carbon is now tetrahedral and thus can be optically active. A solution of glucose, as an example, will contain all three species in equilibrium with each other. This important equilibrium is called *mutarotation*, and is illustrated below for glucose.

Disaccharides and larger polysaccharides are made up of individual monosaccharide units joined to each other in an acetal linkage. Acetals are reactive in acidic water solutions and give the individual carbohydrate units back. For instance, hydrolysis of starch, its reaction with water, releases glucose into solution.

Because of the many functional groups in carbohydrates, there are many color tests for their presence; some of them are quite specific for a single sugar. You will learn about five of them in this experiment. Benedict's test, Fehling's test, and the Tollens test are all tests for reducing sugars, that have a free α-hydroxy aldehyde or ketone. For the Benedict's and Fehling's tests, the carbohydrate is oxidized to a carboxylic acid, and Cu^{2+} ions are reduced to Cu^{+} ions. Since the Benedict's and Fehling's tests take place in basic solution, however, the gluconic acid formed is present as the anion and is soluble in water solution. What you see when you do the test is the change in the solution color from the blue of Cu^{2+} ions to the orange color, and eventual precipitation of, Cu^{+} ions as solid Cu_2O. The difference in the Benedict's and Fehling's solutions lies mainly in the other dissolved species that help the reaction go well.

The Tollens reaction is also an oxidation of an aldehyde. However, in this case silver ion is reduced, from Ag^{+} to Ag^{0}. The silver metal will precipitate out of solution, and if the test tube is clean enough it will actually plate onto the glass, forming a "silver mirror" that is quite distinctive and is actually shiny enough to use as a looking glass, although it is a bit small. Again, the reaction takes place in basic solution, and the gluconic acid formed stays in solution as the anion.

The criteria to look for in a reducing sugar are 1) the presence of a free carbonyl group or a potentially free carbonyl group (in a hemiacetal or hemiketal) and 2) a hydroxyl group on the adjacent carbon. The derivative of glucose shown on the next page is a complete acetal and is non-reducing. This complete acetal does not give an equilibrium concentration of the aldehyde.

You will also use a test strip for the detection of glucose. Glucose oxidase is an enzyme that specifically catalyzes the oxidation of glucose. Although it will facilitate the oxidation of other monosaccharides, those reactions are much slower. This property has been used to produce a commercial product diabetics use to test their urine for the presence of glucose. The reaction is coupled in the test strips to a color-producing reaction, so that it is easy for a patient to determine if a glucose problem is developing.

Starch is the only common carbohydrate that will react with a solution of potassium iodide and iodine in water. This solution forms the triiodide ion, I_3^-, which can become trapped in the coils of starch molecules and produce a deep blue-black color.

In this exercise you will first look at some computer-generated models and then some ball-and-stick models that demonstrate the concepts of hemiacetal formation and mutarotation. Some color tests for the presence of various saccharides follow, and then you will examine the hydrolysis of di- and polysaccharides.

REFERENCES

Stoker, H. S. *General, Organic, and Biological Chemistry;* Cengage Learning; Chapters 18.
   Review topics
   - Chirality
   - Monosaccharides and polysaccharides
   - Mutarotation
   - Reactions of carbohydrates

WEB RESOURCES

Absorption of Monosaccharides.
http://www.vivo.colostate.edu/hbooks/pathphys/digestion/smallgut/absorb_sugars.html

Exploring the Molecules of Life: Carbohydrates.
http://academic.pg.cc.md.us/~ssinex/blt/carbohydrates/carbohydrates.htm

Chapter 25: Carbohydrates. *http://www.mhhe.com/physsci/chemistry/carey5e/Ch25/ch25-3-1.html*

Diabetes Education Resource, Medications. *http://www.neha-diabeteseducation.ca/Medications.php*

Sugar Molecules page. *http://www.nyu.edu/pages/mathmol/library/sugars/*

SAFETY NOTES

Organic chemicals often have strong odors and are usually volatile and flammable. Use small amounts of materials and discard them into appropriately labeled waste jugs when finished, not down the sink. Some chemicals in this experiment may be strongly acidic. Treat all the chemicals used with respect. You may wish to wear gloves during this experiment, and you must wear chemical splash goggles.

EXPERIMENTAL PROCEDURE

PART A. STRUCTURES OF CARBOHYDRATES

If you have a computer-based molecular modeling program available, complete Steps 1 and 2. Otherwise, begin with Step 3.

Step 1. Using a computer-based molecular modeling program, generate models of each of the following carbohydrates.

Step 2. Observe the computer-generated models, allowing them to rotate and turning them so that you can describe their shapes. Then answer the questions on the report sheet about these compounds.

> ⓘ Be sure to use several display forms, including ball-and-stick.

β-D-glucose	sucrose
β-D-fructose	cellulose
starch	

Step 3. Using ball-and-stick models, make a model of β-D-glucose using the computer-modeled structure as a guide. Draw this model in three dimensions on the report form.

Step 4. Using that same model, convert it into the Fischer projection formula of the open form of glucose, drawing the model on the report form.

Step 5. Now convert the model to the α-D-glucose form and draw it. On <u>each</u> drawing, circle the hemiacetal or aldehyde, <u>and</u> mark the anomeric carbon with an asterisk.

PART B. SOME REACTIONS OF CARBOHYDRATES

For each test that you do, record what the result looks like and whether this indicates a positive or a negative test.

Spot tests

Step 1. Set up a hot water bath on a hot plate for the Benedict's and Fehling's tests. The beaker should not be more than half-full, and if you heat the water for a long time, it may be necessary to refill the beaker.

Step 2. Benedict's test for reducing sugars. In four labeled test tubes, place 1 mL of solutions of the following carbohydrates: glucose, fructose, sucrose, and starch. To each add 1 mL of Benedict's reagent. Heat in a hot water bath for 5 minutes, and observe the results.

Step 3. Fehling's test for reducing sugars. In four labeled test tubes, place 1 mL of solutions of the following carbohydrates: glucose, fructose, sucrose, and starch. To each sample, add 1 mL of Fehling's solution A **and** 1 mL of Fehling's solution B. Mix well. Heat the samples in the hot water bath for 5 minutes and observe.

> ♥ Benedict's, Fehling's, and Tollens reagent are all strongly basic.

Step 4. Tollens test for reducing sugars. This test is usually most reliable at room temperature; it can be difficult to interpret if the reaction mixture is heated. First make the Tollens reagent if it has not already been prepared for you: mix 10 drops of Tollens Solution A with 10 drops of Tollens Solution B in a large test tube. Stir well; a gray precipitate should form. Add to this suffucient 2.0 M ammonium hydroxide solution to <u>just</u> dissolve the precipitate More than 20 drops may be necessary, so be patient—the precipitate <u>will</u> dissolve. Use this solution in the next step. Alternatively, your instructor may make up a quantity of fresh Tollens reagent before class starts. Follow your instructor's directions here.

In four **very** clean small test tubes, place 5 drops of glucose, fructose, sucrose, and starch. Mix each with 1 mL of freshly prepared Tollens reagent, and allow the mixture to stand for 10 minutes. If the result is difficult to interpret, heat in a warm water bath for 2 minutes, but do not boil. A positive test is the formation of a silver mirror coating the inside of the test tube.

Step 5. Glucose oxidase test for glucose. Using a clinical test strip such as Clinistix® according to the label directions, check each of your four standard samples for the presence of glucose. Note the times at which any changes in the test strips occur, since timing can be critical in distinguishing between fructose and glucose.

Step 6. Iodine test for starch. Again using the four standard samples, and in addition a small sample of cellulose in a fifth tube, check the reaction of iodine with each by adding 1 drop of iodine test solution to each sample.

Hydrolysis of di- and polysaccharides

Step 1. In three new test tubes, place 1-mL samples of starch, sucrose, and cellulose. To each add 1 mL of water and 1 mL of 6 M HCl solution. Heat the samples in a water bath for 30 minutes.

> ⚠ 6 M HCl is strongly acidic—handle with care.

Step 2. At the end of that time, cool the reaction mixtures to room temperature, and test each mixture for the presence of glucose and reducing sugars using test strips and Benedict's test.

Step 3. Test the starch and cellulose hydrolysis solutions with iodine solution. What do the results tell you?

EXPERIMENT 34

IDENTIFICATION OF AN UNKNOWN CARBOHYDRATE

GOALS

1. To review the chemistry of carbohydrates
2. To reinforce the concept of isomerism in carbohydrates
3. To compare the reactivities of monosaccharides and polysaccharides

INTRODUCTION

Carbohydrates are a class of organic compounds found in many places in nature. As their name implies, the simple ones contain only carbon, hydrogen and oxygen. Some of them are polysaccharides as well, being built up into long chains from individual units of small sugar molecules. Your laboratory activity, in which you will identify the carbohydrate present in a urine sample, will require you to be familiar with the structures and major reactions of carbohydrates, so before beginning this experiment review the introductory sections in your text concerning carbohydrates.

The presence of an undue quantity of sugar of any sort in the urine is known as glycosuria. This condition is commonly found in the presence of diabetes, but some other disease states, some types of poisons, and even starvation can produce glycosuria. Sometimes sugar is spilled into the urine simply because a healthy individual has ingested a large quantity of sugar in a short period of time. Thus it can be important to determine which specific sugar is causing the glycosuria. For example, glucosuria, a very specific form of glycosuria, is commonly associated with the presence of diabetes mellitus, so the source of the glucose in the urine needs to be determined without undue delay. On the other hand, fructosuria is usually an alimentary condition; that is, it may occur simply because the person has recently eaten large quantities of fruit.

At one time, wet chemical tests were widely used for the determination of carbohydrates in body fluids such as urine. These tests, however, were relatively slow and required the handling of numerous reagents. In a clinical laboratory setting these tests have largely been replaced by spot tests using enzyme assays, but they can still be useful in a more relaxed setting for practicing laboratory and diagnostic skills. You will be using the tests studied in Experiment 33 as well as several additional tests.

Your sample today is a "patient urine sample" that has tested positive for glycosuria. The differential diagnosis for the patient requires that you find out which specific carbohydrate is present in your sample. All the available carbohydrates—fructose, glucose, sucrose, starch, maltose, galactose, xylose, amd arabinose—have unique chemical properties that allow them to be identified. Some react similarly in some tests; however, if you use several tests you can identify each of the sugars by its combination of reactions. The notes in this Introduction summarize the useful information about the tests you will use.

> ⓘ Remember that "speed is fine, but accuracy is final" when it comes to lab results!

There are eight sugars and polysaccharides of known identity for you to use as standard samples, to see how they behave toward the 10 chemical tests to perform. You will also test your unknown sample with a glucose test strip. Comparing the behavior of your unknown sample with the standards should allow you to determine the identity of the sugar present. You may make these assumptions about your unknown sample:

- Each sample contains only one of the possible carbohydrates.
- There are no interfering substances in the samples, so the tests should be easy to observe.
- Performing the 11 suggested tests on your sample will give you enough information to identify your unknown sample.

Notes on the tests to perform

Most carbohydrates normally are found in the cyclic hemiacetal or hemiketal form, but you should also keep in mind that there is a small equilibrium concentration of the free aldehyde or ketone present and that this form can undergo reactions as well. The equilibrium among the three possible forms is shown below for the simple monosaccharide glucose.

It is, in fact, the free aldehyde (or ketone) that usually reacts with the test reagent in the experiments you will be investigating.

The fermentation test should be positive for any hexose except galactose. The enzyme zymase, present in yeast, causes the decomposition of hexoses into ethanol and carbon dioxide. In addition, other enzymes in yeast, maltase and sucrase, can first split the disaccharides maltose and sucrose into hexoses, which then ferment as well.

$$C_6H_{12}O_6 \longrightarrow 2\ C_2H_6O + 2\ CO_2$$

After you begin the incubation of the samples with yeast, watch for the formation of bubbles in the small tube, which indicates the formation of carbon dioxide and fermentation activity.

Benedict's Reagent, Fehling's Reagent, and Barfoed's Reagent all test for reducing sugars. The reaction in each case is an oxidation-reduction reaction—the aldehyde is oxidized to a carboxylic acid while the blue Cu^{2+} is reduced to Cu^+ in copper(I) oxide. The copper(I) oxide forms a reddish precipitate, and it is the change from a deep blue solution to a greenish one with a red precipitate that signals a positive test for a reducing sugar. The general reaction is shown below.

Any carbohydrate with a potentially free carbonyl group ($R_2C=O$) will reduce the copper ion in these solutions. Fructose reacts very rapidly, glucose and galactose more slowly. The carbonyl groups of carbohydrates such as sucrose, starch, and cellulose are engaged in bonding between the units and will not react under normal conditions. However, even starch and cellulose will give weak tests if the solution is heated intensely, so care is needed in the interpretation of these tests.

The criteria to look for are 1) the presence of a free carbonyl or a potentially free carbonyl (in a hemiacetal or hemiketal) and 2) a hydroxyl group on the adjacent carbon. The derivative of glucose shown on p 327, for example, is a complete acetal, and is non-reducing. This complete acetal does not give an equilibrium concentration of the aldehyde. Compare it to sucrose to see why sucrose is not a reducing sugar.

Barfoed's reagent, Bial's reagent, and Molisch reagent all are very specific in their activity. Barfoed's reagent will give the usual reaction for reducing sugars, but it will react only with monosaccharides that are reducing. Bial's Reagent reacts only with pentoses. The Molisch reagent reacts rapidly with monosaccharides and more slowly with polysaccharides, so for this test the timing of the reaction will be important.

The Tollens reagent distinguishes between lactose and glucose in this group of carbohydrates; lactose will not react. Tollens reagent is a weak oxidizing agent for any substance that can be oxidized readily. Aldehydes or potential aldehydes (hemiacetals) generally will react with the Ag^+ ions of Tollens reagent to produce carboxylic acids and silver metal, which plates out on the test tube as a "silver mirror." The reagent will thus react with any saccharide that contains an aldehyde group, so you may need to be very discriminating in your observations.

> ⓘ It is easy to misinterpret the Tollens test.

Rubner's reagent helps to distinguish between lactose and other monosaccharides. Lactose gives a brick-red color, dextrose gives a coffee-brown color, maltose gives a light yellow color, and levulose gives no color at all. Other monosaccharides give no color change.

Seliwanoff's reagent is a test specific for fructose. Fructose, a ketose, is rarely found in urine, but it may appear in cases of liver disease. When a ketose is heated with a strong acid, hydroxymethyl furfural is formed, and this compound forms a red complex with the organic compound resorcinol. The aldoses will give the red color more slowly. A red color that develops in 20 seconds or less during the test indicates a ketose. If the red color develops after 20 seconds, it is likely that the carbohydrate is an aldose.

Iodine test reagent is specific for starch. The reagent contains both elemental iodine and iodide ion from potassium iodide, which together form the triiodide ion, I_3^-. This ion forms a dark blue complex when treated with starch. Cellulose and the simple sugars produce no color change.

Clinistix® should react only with glucose. Most clinically useful spot tests for glucose now take advantage of an enzyme reaction using glucose oxidase. This is quick and efficient and has the major advantage of being specific for glucose only. Clinistix® test strips operate in this fashion. A few other monosaccharides can give a color change on the test strip if the timing of the reading is longer than recommended, so follow the directions on the bottle carefully.

REFERENCES

Stoker, H. S. *General, Organic, and Biological Chemistry;* Cengage Learning; Chapters 18.
  Review topics
  - Carbohydrate chemistry
  - Mutarotation
  - Spot tests for carbohydrates
  - Hydrolysis of carbohydrates

Other references:

Merck Index, *10ᵗʰ edition*, Merck: Whitehouse Station NJ, 1983. Later editions are fine.

Shriner, R. L. et al., *The Systematic Identification of Organic Compounds,* Wiley: New York, 1998.

WEB RESOURCES

Carbohydrates. *http://www.cem.msu.edu/~reusch/VirtualText/carbhyd.htm*

Chemfinder.com. *http://www.chemfinder.com*

Tests for Carbohydrates. *http://faculty.mansfield.edu/bganong/biochemistry/carbos.htm*

SAFETY NOTES

Organic chemicals often have strong odors and are usually volatile and flammable, so avoid inhaling the vapors by using them in the hood whenever possible. Use small amounts of materials, and discard them into appropriately labeled waste jugs when finished, not down the sink. Some chemicals in this experiment may be strongly acidic. Treat all the chemicals used with respect. You may wish to wear gloves during this experiment, and you must wear chemical splash goggles.

EXPERIMENTAL PROCEDURE

Before beginning the chemical tests, draw on the report form the structures of all the possible carbohydrates you may encounter in this activity. Most will be in your text, or refer to the Merck Index or to a site such as <*www.chemfinder.com*> online.

   Test 1. Fermentation. Use the test tubes provided in the general supplies for this test and save your own test tubes for the other tests.

   In a large test tube held upright in a beaker, place 15.0 mL of yeast suspension and about 4.00 mL of your carbohydrate solution. Mix well. Place the entire mixture in a large test tube. Now place a small test tube upside down in the larger test tube. Cover the opening of the large test tube with plastic film and invert the tube, so that the smaller one is now right-side up. Allow the smaller one to fill with your fermentation mixture as much as possible, then re-invert the assembly so that the full little tube is again upside down with no air bubble in the tip.

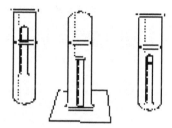

   Place the entire apparatus in a 37 °C water bath. Fermentation tubes may also be used. Observe over a period of time to see if bubbles of carbon dioxide gas collect within the smaller test tube. Allow the fermentation to continue for 30-90 min before deciding if the test is positive or negative.

   General instructions for other tests. While the fermentation is running, set up a test tube rack with eight test tubes, each labeled with the name of one of the known carbohydrates or your sample number. In each test tube, place 1-2 mL of the carbohydrate solution to be tested. Perform Test 2, described below, on all eight of the carbohydrates, and record your observations in the data table provided. Rinse your tubes with distilled water, and refill with the appropriate carbohydrates. Then go on to Test 3. Continue until all the tests have been performed on all the carbohydrates.

   In the report form there is a record sheet to help you organize your results from this activity. There may not be enough room to record all the information you need in the small spaces available; if this is the case, make more detailed observations in a notebook or as your instructor directs. You can never have too much detail in investigations of this sort!

   Test 2. Benedict's Test. Place 2.0 mL of Benedict's solution in a test tube, and heat it in a gently boiling water bath for 2 or 3 minutes. Add 10 drops of your carbohydrate solution to be tested. Mix thoroughly, and heat it in a boiling water bath for an additional 5 minutes. A color change from blue, to green, to yellow, to brick red indicates that the carbohydrate is a reducing sugar. The rate of reaction is very important, so observe carefully and write your observations down.

   Test 3. Fehling's Solution. Place 2.0 mL of Fehling's Solution A <u>and</u> 2.0 mL of Fehling's Solution B in a test tube and mix. Add 10 drops of your carbohydrate solution and mix thoroughly. Heat gently for 2 to 5 minutes; the rate of reaction is very important. Record both the time and the color change.

Test 4. Barfoed's Reagent. Place 5.0 mL of Barfoed's Reagent in a test tube, add 10 drops of your carbohydrate solution and mix thoroughly. Heat gently for 2 to 5 minutes; the rate of reaction is very important. Record both the time and any color change.

Test 5. Bial's Reagent. Place 1.0 mL of your carbohydrate to be tested into a test tube. Add 3.0 mL of Bial's Reagent and mix. Heat gently until the solution just starts to boil. Then let stand for 5 minutes. Record what you see.

Test 6. Tollens test. The Tollens reagent may have been made up already for your use. If not, make it up as follows, and use it the same day you make it. To a thoroughly clean test tube add:

20 drops of 5% silver nitrate

1 drop of 10% sodium hydroxide solution

Stir well; a gray precipitate (Ag_2O) should form. Add to this sufficient 2% ammonium hydroxide solution to just dissolve the precipitate. More than 20 drops may be necessary, so be patient. The precipitate will dissolve. Use this solution in the next step.

Place 2.0 mL of fresh Tollens reagent in a test tube. Add a few drops of the carbohydrate solution and mix well. Allow to stand for 10 minutes and record any results you see. After that, heat in a warm water bath for 1 or 2 minutes. Look at the walls of the test tube and record your observations for a positive test.

Silver mirrors will not dissolve in any reagent at your disposal. Return your silvered test tubes to the stockroom for replacement or deal with them as your instructor suggests.

Discard any unused Tollens reagent in the provided waste jar for disposal by your instructor. Do not store it in your drawer.

Test 7. Rubner's Test. To a test tube containing 1.0 mL of your carbohydrate solution add 5.0 mL of a 30% solution of lead acetate in water. Shake well, and filter if there is any solid in the mixture—you need a clear solution. Place the clear solution in a test tube, and place the test tube in a boiling water bath **in the hood** for several minutes. Add 2.0 mL of concentrated ammonium hydroxide solution and reboil. Describe the color of the solution and precipitate (if any).

The solutions for Rubner's test contain lead and should be discarded in the specially prepared disposal jar, not in the drain. In addition, rinse the test tubes used and add the rinses to the disposal jar, not to the drain.

Test 8. Seliwanoff's Test. To 3.0 mL of freshly prepared Seliwanoff's reagent add 10 drops of your carbohydrate solution and mix well. Boil for less than 20 seconds in a water bath and observe. If a red color develops after a longer heating time, it may be informative as well, so be sure to record both times and colors observed.

Test 9. Iodine Test. To 2.0 mL of your carbohydrate solution add 2 drops of iodine solution and mix well. Record your observations on each carbohydrate.

Extra test. Clinistix® test.

Follow the directions on the bottle to determine if glucose is present in your unknown sample only. The test strips may also react with some of the other saccharides, so interpret your results carefully.

EXPERIMENT 34 PRELAB EXERCISES:

IDENTIFICATION OF AN UNKNOWN CARBOHYDRATE

Name_____   Partner _____   Date _____

1. Write an equation that shows the interconversion of glucose from the α– to the β– form via mutarotation.

2. Would you expect fructose to give a positive test with Seliwanoff reagent?

3. Why is sucrose not considered a reducing sugar?

4. Starch is made up of repeating glucose units, but it does not give a positive Benedict's test. Why?

5. If a carbohydrate does not give a reaction to Tollens reagent, what does that tell you about its structure?

EXPERIMENT 34 REPORT: IDENTIFICATION OF AN UNKNOWN CARBOHYDRATE

Name_____ Partner _____ Date _____

Structures of the Standard Carbohydrates

Structure	Expected behavior
Fructose	
Glucose	
Xylose	
Sucrose	
Maltose	
Galactose	
Arabinose	
Starch	

Results of Carbohydrate Tests

Indicate your results for each test in the table. Be sure to include in the table whether a test was positive or negative, and what it looked like.

	Fructose	Glucose	Sucrose	Galactose	Xylose	Arabinose	Maltose	Starch
Fermentation								
Benedict's								
Fehling's								
Barfoed's								
Bial's								
Tollens								
Rubner's								
Seliwanoff								
Iodine								
Clinistix®								

Unknown sample number _____

Identity of the sample _____

Evidence for this conclusion:

EXPERIMENT 35

PROPERTIES OF LIPIDS

GOALS

1. To examine the physical and chemical properties of lipids
2. To rank some edible lipids for degree of unsaturation
3. To determine the fatty acid content of some edible lipids

INTRODUCTION

The class of biological molecules called lipids has many subclassifications. In general, the biological molecules that are not water soluble fall into this class, but that general statement includes many components of membranes, body fats, cholesterol, and prostaglandins, all with widely differing structures.

This exercise will focus on the chemistry of the simplest lipids, triacylglycerols. These are esters made from fatty acids and the trialcohol glycerol. The triacylglycerol shown below is called *tristearin*, because it is made from glycerol and three molecules of stearic acid. Naturally occurring triacylglycerols may contain all the same fatty acid residue, but most will contain several different fatty acids, the exact composition depending on many biological factors. Tristearin is one of the simplest to consider.

$$CH_2-O-\overset{\overset{O}{\|}}{C}-(CH_2)_{16}CH_3$$

$$H-\overset{|}{C}-O-\overset{\overset{O}{\|}}{C}-(CH_2)_{16}CH_3$$

$$CH_2-O-\overset{\overset{O}{\|}}{C}-(CH_2)_{16}CH_3$$

a triacylglycerol

Like most organic compounds, the triacylglycerols, rich in carbon and hydrogen, will burn to give carbon dioxide and water. This is the basis for the use of organic oils such as whale oil or tallow for lamps in primitive societies, but at a slower pace oxidation of triacylglycerols is also a major source of the energy needed for biochemical reactions. In a less desirable situation, the slow oxidation of triacylglycerols is also responsible for cooking fats going rancid at high temperatures.

Unsaturated compounds, those containing carbon-carbon double or triple bonds, will react with any of the halogens by an addition reaction, breaking the double bond and making instead a haloalkane. The reaction is shown here for chlorine, but iodine and bromine will also cause this reaction. It usually proceeds almost instantly at room temperature with corresponding bleaching of the color of the original halogen as it reacts.

You may be aware that animal fat or vegetable oil is a major raw material for one of our oldest industries, the making of soap. You have already studied this reaction because it is a reaction common to all esters. The ester bonds in triacylglycerols can be broken by boiling the fat with lye, or sodium hydroxide, to make glycerol and the sodium salt of each fatty acid molecule broken loose. The salt of the of the fatty acid is the soap.

triglyceride   glycerol   3 soap molecules

In the second part of this experiment, you will use the halogenation reaction for a practical application. One way to determine the degree of unsaturation of any organic compound, and especially lipids, is to determine the "iodine value" for that compound, the number of moles of iodine (or any other halogen) that will react with one mole of the compound of interest. The higher the iodine value, the more double or triple bonds present in the molecule. So for any given fat, such as lard or butter or corn oil, the more halogenation per gram that occurs, the more unsaturated the fat. In this exercise you will determine the relative iodine values for eight triacylglycerols as they are used commercially, and rank them in order of increased unsaturation. Because an exact iodine value is a bit complex to determine, you will simply count the drops of a bromine solution necessary to react with all the double bonds of the fat you are investigating, and rank your triacylglycerols on that basis.

There are more accurate methods for determining unsaturation number, but this method allows you a rough ranking of fats. In addition, you may be able to use a proportional factor to multiply your result by, in order to arrive at a "true" iodine value as a measure of unsaturation.

REFERENCES

Stoker, H. S. *General, Organic, and Biological Chemistry;* Cengage Learning; Chapters 19.
Review topics
- TLC
- Heat content
- Saponification of esters
- Halogenation of alkenes

WEB RESOURCES

Caveman Chemistry: Hands-on Projects in Chemical Technology. *http://cavemanchemistry.com/*

Chemistry Molecular Models. *http://www.uwsp.edu/chemistry/pdbs* This resource requires the use of the Chemscape Chime plugin.

Lipids. *http://www.cem.msu.edu/~reusch/VirtualText/lipids.htm*

LOTION MAKING 101 FROM SNOWDRIFT FARM.
http://www.snowdriftfarm.com/lotionmaking101_p3.html

SAFETY NOTES

Organic chemicals often have strong odors and are usually volatile and flammable, so avoid inhaling the vapors by using them in the hood whenever possible. Use small amounts of materials, and discard them into appropriately labeled waste jugs when finished, not down the sink. Some chemicals in this experiment may be strongly acidic. Treat all the chemicals used with respect.

Bromine is a particularly corrosive element. You must wear gloves while working with this reagent, and you must wear chemical splash goggles throughout the experiment.

EXPERIMENTAL PROCEDURE

PART A. REACTIONS OF LIPIDS

Step 1. The oxidation of fats. This reaction is best observed by groups of four or more since it should be done in the hood. Place a small amount of each fat to be tested in an evaporating dish in the hood. Using a match, try to light the fat on fire, and observe what happens. The reaction is similar for all fats, so using one solid fat and one liquid fat is sufficient.

Step 2. Halogenation of a fat. Place a small amount of corn oil (0.1 g or so) in a 50-mL beaker, and dissolve it in 2 mL of hexane. Add a drop of bromine/hexane solution and observe the result. Write a general equation for the reaction that occurred using olein as the reactant fat.

Step 3. The saponification of a fat. In a small beaker, place a small piece (0.1 g or so) of butter. Add to this fat 5 mL of 6 M NaOH. Heat this mixture, with constant stirring, to the boiling point. Continue heating gently until the liquid has reduced in volume to about half the original volume or even less, and then set the beaker aside to cool. Take care to avoid spatters of the hot sodium hydroxide solution. Observe the solid made. Does it look like the original fat? Write the equation for what happened during this reaction, using tristearin as the reactant fat. Resuspend the solid in 10-20 mL of water. To this mixture, add with stirring 10 mL of 6 M hydrochloric acid, and observe what happened. Write the equation for this step as well, referring to the reactions of acids and esters if you need a hint.

> ⓘ Do not let the entire amount of liquid evaporate.

PART B. THE RELATIVE IODINE VALUE OF EDIBLE OILS

There are eight oils to be investigated. Your instructor may assign you one or all of the oils. They are

coconut oil	canola
butter	corn oil
lard	sunflower
olive oil	safflower oil

Step 1. Weigh a small (25 mL) Erlenmeyer flask; then place in it 0.2 mL of one of the fats to be investigated, and weigh the flask again. Calculate the mass of the fat used. This flask will be your reaction flask. You will need to clean and dry this flask between uses.

Some of the fats may be solid. In that case, place a small chunk in the flask—a piece about the size of a kidney bean. You will need to get all your sample into the reaction flask, not stuck at the lip. Dissolve your sample in 2.0 mL of hexane, and cork the flask lightly.

Step 2. In a similar flask, place 3.0 mL of hexane, and cork the flask. This will be the control flask. You will be able to use the same control flask for each fat studied. Transfer both control and reaction flasks to an absorbent mat. Add one drop of the bromine/hexane solution to the control flask, and note the color. This is the end color you are looking for; when your reaction flask reaches this color, you are finished with that particular fat analysis.

> ☻ The mat is a safety precaution in case you spill either fat or reagent.

Step 3. Add a drop of the bromine (☻ !) solution to the reaction flask, and swirl the mixture gently. As the reaction between bromine and the double bonds of the fat occurs, the bromine color should disappear. Continue adding bromine reagent, with swirling, counting the drops needed to complete the reaction of the fat (until the color of the reaction flask matches the color of the control flask). Record the number of drops needed.

> ☻ Bromine is a particularly corrosive element. You must wear gloves and chemical splash goggles.

Step 4. Pour the contents of the reaction flask into the waste container provided. Rinse the flask first with water and then with ethanol, and dry it.

Step 5. Repeat the analysis with the other oils assigned to you, being careful to clean and dry the reaction flask between each run.

Step 6. You should make a second trial of each fat in order to determine the reproducibility of your results.

Step 7. Calculate the drops of reagent/gram of fat for each fat studied, and rank the fats in increasing order of iodine number. (1 = most saturated, 8 = most unsaturated)

When you are completely finished with the experiment, fold the mat up neatly and discard it in the waste basket unless there was a substantial spill of bromine. If there was, ask your instructor how to proceed.

PART C. EXERCISES

Answer the questions in Part C of the lab report.

EXPERIMENT 36

SOAPS AND DETERGENTS

GOALS

1. To recognize the chemical differences between soaps and detergents
2. To prepare a sample of a soap
3. To examine the relative sudsing abilities of soaps and detergents in waters of varying hardness

INTRODUCTION

Soaps belong to a general class of chemicals called "surfactants." These are substances that have a long, non-polar carbon chain and a small ionic part on the end of the chain. Because of these polar and non-polar areas within the surfactant molecule, surfactants are good at solubilizing both polar and non-polar materials in water. Since most dirt is oily and non-polar, it is not soluble in water until lifted by some type of surfactant, and this is where soaps and detergents come in. Simple soaps are the salts of fatty acids.

Soaps have been known for hundreds of years. Before there was a major soap industry in developed countries, and even today in undeveloped areas, soap-making was a cottage industry done in the home from ingredients readily available. The source for the fatty acids was whatever animal or vegetable oil was available—tallow or bacon grease, or in warmer countries palm oil. This was cooked in water together with lye obtained by soaking wood ashes in water, or later with purchased lye. The reaction could be carried out readily, although today we would be horrified at the lack of

$$\text{triacylglycerol} \xrightarrow[\text{H}_2\text{O}]{\text{NaOH}} \text{glycerol} + 3\ \text{Na}^+ \ \text{O}^- \text{(CH}_2\text{)}_{16}\text{CH}_3 \quad \text{3 soap molecules}$$

safety precautions the home chemists used.

It has also been known for a long time that soaps don't work very well in hard water. Sodium and potassium salts of fatty acids, the kind you get with a lye hydrolysis of animal or vegetable fat, are soluble in water. They make a fine suds and lift oily dirt readily provided the water is soft. You may have read of using only rain water for washing, and that's why. But hard water contains dissolved ions of calcium, iron, and magnesium, and the iron, calcium and magnesium salts of fatty acids are insoluble in water. Instead of making suds, they make inactive soap scum that redeposits on the clothes, leading to the legendary "tattletale gray" of advertising fame.

$$2 \ (CH_2)_{16} \ CH_2 \overset{\overset{O}{\parallel}}{-\!\!-} O^- \ Na^+ \ + \ Ca^{2+}{}_{(aq)} \longrightarrow \left((CH_2)_{16} \ CH_2 \overset{\overset{O}{\parallel}}{-\!\!-} O^- \right)_2 Ca^{2+}{}_{(s)} \ + \ 2 \ Na^+{}_{(aq)}$$

There are various ways for the hard water problem to be solved, but one of them is by the use of detergents. To a chemist, the words "soap" and "detergent" are not interchangeable; they mean two different kinds of chemicals.

Synthetic detergents, or syndets, play an important role in many industries, including the laundry industry. Syndets are usually sodium or potassium salts of sulfonic acids instead of carboxylic acids. A common one is sodium lauryl sulfate, often called SDS for its proper chemical name, sodium dodecyl sulfate. These compounds retain the ability to form suds in hard water and thus make washing your laundry easier, even without the use of soft water. In addition, syndets have many industrial uses.

$$CH_3 (CH_2)_{11} \!-\! OSO_2^- \ Na^+$$

sodium lauryl sulfate

In this experiment, you will make three types of soap from fats and compare the sudsing ability of your home-made soaps to SDS and to Ivory Snow®, a well-known commercial soap. Your first soap will be a sodium salt. Sodium soaps are hard soaps that usually can be dried to a cake such as you might be familiar with. Soaps made with potassium hydroxide as the lye source, however, are usually soft or liquid, such as you might use in a dispenser for hand-washing. A popular alternative soap is Castille soap which is made with olive oil. Chemically it is very similar to other hard soaps, but the soap is clear, rather than opaque, and this makes it more acceptable to many consumers.

You may be expected to make only one soap and pool your results with other students, or you may make all three for your own comparisons. Follow your instructor's directions on this.

REFERENCES

Stoker, H. S. *General, Organic, and Biological Chemistry;* Cengage Learning; Chapters 19.
  Review topics
  • Reactions of esters
  • Saponification
  • Actions of surfactants

WEB RESOURCES

All Vegetable Soaps and Recipes. *http://millersoap.com/soapallveg.html*

Soap. *http://www.chemistryexplained.com/Ru-Sp/Soap.html*

The Chemistry of Soap. *http://www.chandlersoaps.com/chemistry-of-soap.html*

SAFETY NOTES

Do not plan to use your soap for everyday washing. You have no way to assess the amount of unreacted sodium hydroxide left behind, and the amount may be sufficiently high to make this soap unsafe to use.

Organic chemicals often have strong odors and are usually volatile and flammable, so avoid inhaling the vapors by using them in the hood whenever possible. Use small amounts of materials, and discard them into appropriately labeled waste jugs when finished, not down the sink. Some chemicals in this experiment may be strongly acidic or basic. Treat all the chemicals used with respect. You may wish to wear gloves during this experiment, and you must wear chemical splash goggles.

EXPERIMENTAL PROCEDURE

PART A. PREPARATION OF SOAPS

A hard soap

Step 1. In a 250-mL beaker, place 5 g of lard and 5 mL of 20% sodium hydroxide solution. Add to this mixture 5 mL of 95% ethanol and mix well using a wooden tongue depressor or other flat-bladed object.

Step 2. Heat this mixture gently on a hot plate until most of the liquid has evaporated and the thick mixture briefly retains the imprint of the stirring splint. Remove the soap mixture from the hot plate to cool. Ethanol is flammable. If your mixture ignites, cover it with a watch glass to extinguish the flame, and remove it temporarily from the hot plate using a hot pad.

> ♥ Avoid spattering the hot lye.

Step 3. Before the mixture starts to harden, stir into it 30 mL of saturated salt solution (brine). Allow the soap to settle, and filter it by gravity.

> ⓘ The heating step should take no more than 20 min.

Step 4. Scrape your soap into a small pre-weighed weighing boat to cure and harden. You may be directed to keep it in your drawer until your next class meeting, but you can continue with the rest of this experiment using the fresh soap. For the fresh soap, the weight would be meaningless because of the water incorporated in it. If you keep the soap and your report until the next week, find the weight of the actual soap prepared before you turn in your report.

A soft soap

Follow the directions above using potassium hydroxide instead of sodium hydroxide. The directions are repeated here for convenience.

Step 1. In a 250-mL beaker, place 5 g of lard and 5 mL of 20% potassium hydroxide solution. Add to this mixture 5 mL of 95% ethanol and mix well using a wooden tongue depressor or other flat-bladed object.

Step 2. Heat this mixture gently on a hot plate until most of the liquid has evaporated and the thick mixture briefly retains the imprint of the stirring splint. Remove the soap mixture from the hot plate to cool. Ethanol is flammable. If your mixture ignites, cover it with a watch glass to extinguish the flame, and remove it temporarily from the hot plate using a hot pad. This should take about 20 minutes.

Step 3. Before the mixture starts to harden, stir into it 30 mL of saturated salt solution (brine). Allow the soap to settle, and filter it by gravity. If your soft soap turns out to be more liquid than solid, so that you can't filter it, simply withdraw the water layer with a pipet, and do not try to filter it.

Step 4. Scrape your soap into a small pre-weighed weighing boat to cure. It will probably not completely solidify, even on standing. You be directed to keep it in your drawer until your next class meeting, but you can continue with the rest of this experiment using the fresh soap. If you keep the soap and your report until the next week, weigh the actual soap prepared next week before turning in your report.

A clear soap (Castille soap)

Follow the directions for lard-based hard soap, but use olive oil in place of the lard. The directions are repeated here for convenience.

Step 1. In a 250-mL beaker, place 5 g of olive oil and 5 mL of 20% sodium hydroxide solution. Add to this mixture 5 mL of 95% ethanol and mix well using a wooden tongue depressor or other flat-bladed object.

Step 2. Heat this mixture gently on a hot plate until most of the liquid has evaporated and the thick mixture briefly retains the imprint of the stirring splint. Remove the soap mixture from the hot plate to cool. Ethanol is flammable. If your mixture ignites, cover it with a watch glass to extinguish the flame, and remove it temporarily from the hot plate using a hot pad. This process should take about 20 minutes.

Step 3. Before the mixture starts to harden, stir into it 30 mL of saturated salt solution (brine). Allow the soap to settle, and filter it by gravity.

Step 4. Scrape your soap into a small pre-weighed weighing boat to cure and harden. You may keep it in your drawer until your next class meeting, but you may continue with the rest of this experiment using the fresh soap. If you keep it and your report until the next week, weigh the actual soap prepared before you hand in your report.

Do not plan to use your soap for everyday washing. You have no way to assess the amount of unreacted sodium hydroxide left behind, and the amount may be sufficiently high to make this soap unsafe to use.

PART B. COMPARISON OF SOAPS AND DETERGENTS

You will investigate the three kinds of soaps the class has made, a commercial soap, and a synthetic detergent, sodium lauryl sulfate (SDS).

Step 1. Dissolve a small sample of each of the soaps in 5 mL of water in a large labeled test tube. Each soap sample should be about the size of a pea. Also prepare a similar sample of sodium lauryl sulfate and a sample of Ivory Snow® Check the pH of each solution, either with pH paper or a meter.

Step 2. Cover the tubes with plastic wrap, hold the wrap down, and shake each tube vigorously for 10 seconds. Measure the height of the suds layer formed.

Step 3. Repeat this experiment using all the soaps and SDS, but using the hard water provided for your use. Record the height of the suds obtained in each case.

Step 4. Repeat the experiment using locally obtained muddy water.

Step 5. Set up a new set of five tubes containing the soaps and the detergent. To each sample add 1 mL of 6 M HCl and observe the result.

♥ Avoid contact with the HCl.

PART C. EXERCISES

Answer the questions in Part C of the report form.

EXPERIMENT 37

PROPERTIES OF AMINO ACIDS

GOALS

1. To examine the structural differences and similarities of some amino acids
2. To perform a hydrolysis of a small peptide
3. To analyze a mixture of amino acids
4. To examine the properties of amino acids at different pH levels

INTRODUCTION

Amino acids are the monomer units that make up all proteins, so in order to understand the structure and function of proteins, you need to have a grasp of the chemistry of amino acids as well.

The 20 amino acids that are present in all proteins contain a central carbon that has as substituents an amine, a carboxylic acid, a hydrogen atom, and a side chain R group. All have the L- optical configuration at that central carbon.

Amino acids exhibit the properties of acids and of amines. The amino group acts as a base, reacting with acids to make the nitrogen atom positively charged, and the acid portion reacts with bases to form the negatively charged carboxylate ion. But the situation is complicated by the fact that there are both acidic and basic functional groups within the same molecule. The amino acids are actually more usually found in the "zwitterion" form in the neutral state.

The 20 amino acids are divided roughly into classes based on the chemical nature of the side chains. The nature of the side chain has an enormous effect on the overall polarity of the molecule and its ability to carry positive charges, and thus on the three-dimensional structure of proteins. Some amino acids have basic side chains that consist of some type of amino group in addition to the one on the amino acid end of the molecule, some have a carboxylic acid side chain, and some have neutral side chains that might be either nonpolar or polar.

In the zwitterion form, the amino acids may not be very soluble in water (or blood) at neutral pH, but it really depends on the nature of the side chain and on the overall charge on the molecule. In a neutral amino acid such as glycine, solubility can be encouraged by making the molecule charged. How do you do that? By changing the pH.

At pH 7, glycine, for example, is neutral, but at pH 3, both the amine and the carboxylic acid would be protonated, making the molecule positively charged overall. At pH 10, the acid group would be in the carboxylate form (negatively charged), and the amine would be back in its basic form. The overall charge on the glycine would be negative. In both cases (in acid and in base) the charge on the molecule increases its solubility because water will tend to dissolve ionic compounds better than neutral ones.

glycine at pH 3   glycine at pH 7   glycine at pH 10

The other major reaction of amino acids to review is the formation of an amide by the condensation of the acid functional group of one amino acid with the amine functional group of another. This amide group is the primary linkage of amino acids to make protein polymers. The di-amino acid (dipeptide) glycylglycine is shown below.

glycylglycine at pH 1

The amide group in such a compound is subject to the usual type of chemical activity for amides—it can be hydrolyzed back to the original amino acids by allowing it to react with water in the presence of a catalyst.

In today's experiment, you will take advantage of both behaviors. The first part of the experiment is a chromatography experiment. The samples you need to examine are all dissolved in 0.1 M HCl to encourage the solubility of the amino acids. You will run a chromatogram on paper, using a solvent that is known to separate the various amino acids well.

There are several known amino acid samples to run, as well as three unidentified sample mixtures. Two are artificially mixed, but one is made by the hydrolysis of a dipeptide. This particular amide, as its methyl ester, is available commercially as the artificial sweetener, aspartame. The aspartame was treated with hydrochloric acid in boiling water for a brief time to break apart the two amino acids and convert the methyl ester back to the acid. Your chromatography experiment should allow you to determine the two amino acids present in aspartame, as well as any amino acids present in the other two unidentified sample mixtures labeled A and B.

Most amino acids are white solids and produce colorless spots on a chromatogram. To see them on the chromatogram, you will take advantage of a color reaction of amino acids and proteins, the ninhydrin reaction. Ninhydrin is a yellow chemical that produces dark blue compounds in the presence of proteins or amino acids (except in the presence of proline, which gives a yellow-brown color).

Paper chromatograms often run rather slowly, taking an hour or more to reach a good stopping point. While yours is running, you will be able to examine some other properties of amino acids such as the solubility and structures mentioned earlier.

REFERENCES

Stoker, H. S. *General, Organic, and Biological Chemistry;* Cengage Learning; Chapters 17 and 20.
Review topics
- Amino acid structure
- Zwitterions
- pI
- Chromatographic techniques

WEB RESOURCES

Biochemistry of Amino Acids. *http://themedicalbiochemistrypage.org/amino-acids.html*

Chemfinder.com. *www.chemfinder.com*

Chemistry Molecular Models. *http://www.uwsp.edu/chemistry/pdbs#AMINO ACIDS*
This resource requires the Chime plugin.

SAFETY NOTES

Organic chemicals such as the mixed BAW solvent used in this experiment often have strong odors and are usually volatile and flammable, so avoid inhaling the vapors by using them in the hood whenever possible. Use small amounts of materials, and discard them into appropriately labeled waste jugs when finished, not down the sink. Some chemicals in this experiment may be strongly acidic. Treat all the chemicals used with respect. You may wish to wear gloves during this experiment, and you must wear chemical splash goggles.

EXPERIMENTAL PROCEDURE

PART A. PAPER CHROMATOGRAPHY OF AMINO ACIDS

Before you arrived for class, three samples to investigate were prepared: Unknown Sample A is a sample of aspartame, a dipeptide ester, that was hydrolyzed to the free amino acids. Unknown Sample B and Unknown Sample C are both mixtures of amino acids dissolved in dilute HCl.

Your task is to determine the amino acids present in samples A, B, and C.

Step 1. Before working with your amino acid samples, prepare a chromatography chamber from a 400- or 600-mL beaker.

a) Pour into the beaker about 15 mL of the prepared solvent labeled BAW. It consists of a mixture of 1-butanol, acetic acid, and water in the proportions 60:15:25.

b) Add to the beaker a circle of filter paper that fits on the bottom; this will provide a flat surface for your chromatogram to stand on so it will run evenly. Figures showing the appropriate setup are shown in Experiment 32.

> ⓘ Keep this chamber covered because it stinks.

c) Cover the beaker with aluminum foil or plastic wrap and swirl the solvent gently so that the air within the chamber becomes saturated with the vapor of the solvent.

Step 2. Obtain a piece of filter paper approximately 15 cm x 10 cm, and prepare it for use as the chromatography substrate. Refer to Figure 37.1 to see how to set it up.

a) Mark the origin for the chromatography about 1 cm above the bottom (long) edge, using a faint pencil (not pen!) line.

b) Along that line, mark 12 places for sample spots with a small x. Mark each spot with the abbreviation for one of the samples to be run. These are shown in the table below.

ala	Unknown sample A (from aspartame)
his	gly
pro	ile
phe	tyr
asp	val
Unknown sample B	Unknown sample C

Step 3. Now apply the samples to the appropriate places at the origin using a capillary pipet as demonstrated by your instructor. The amino acid samples have been dissolved in 0.1 M HCl, so be careful not to spill. To get a large enough concentration of each sample to be readily visible at the end, you will want to reapply the samples more than once. Remember, though, that each spot must be dry before reapplying because you want the overall sample spot to have as small a diameter as possible. The prepared chromatogram should look like Figure 37-1. Note that the spots need to be at least 2 cm from the edges of the paper. The three unknown samples should also be in the center of the paper, not on the ends.

Step 4. Roll the paper into a cylinder with the short edges overlapping, and staple the paper at top and bottom so that it can stand upright. Be sure that the overlap does not take in any of the spots, which would then run in an irreproducible manner. Stand the cylinder upright in your chromatography chamber, and cover the chamber again. You will see the solvent begin to climb the paper by

> ⓘ A rubber band around the plastic wrap will help keep it in place.

capillary action, and should leave the chamber untouched (i.e., unshaken) until the chromatogram is finished. This will take about 1 hour. In the meantime, continue with Part B and Part C of this exercise.

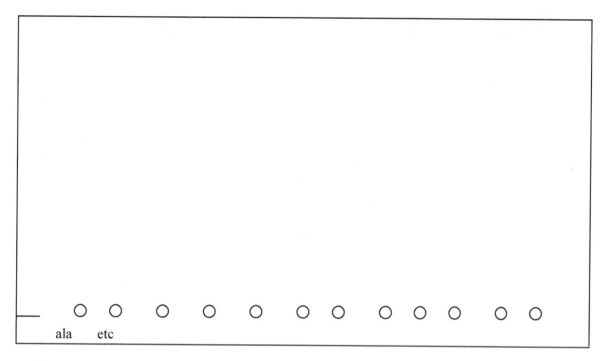

ala   etc

Figure 37.1 Preparation of the chromatogram.

Step 5. Your chromatogram has run far enough when the solvent front has traveled to within 1-2 cm of the top of the paper. Remove the paper from the solvent **and** mark how far the solvent traveled. Allow the paper to dry. Discard the excess solvent in the waste jar provided.

> ① Dry the paper in the hood!

Step 6. When the paper is dry, place it in the chamber provided, and spray it with ninhydrin spray until the paper is damp but not wet. Do your best to control the spray. Warm the paper over (not touching) a hot plate to develop the color of the ninhydrin-amino acid complex.

> ♥ Ninhydrin will stain the proteins in your skin as well. Use gloves for this step.

Step 7. Measure the distance traveled by the solvent front and the distance traveled by each amino acid. Calculate the R_f for each known sample, and for each amino acid present in the unknown samples A-C, and note the colors produced. From these data, identify the components of the samples A-C. Refer to Experiment 32 for a review of R_f and its uses.

Step 8. Be prepared to hand in your chromatogram with your report.

PART B. BEHAVIOR OF AMINO ACIDS IN SOLUTION

Solubility behavior

Step 1. Set up three labeled test tubes, each containing a few crystals of one of the following amino acids: aspartic acid, valine, and histidine. To each add 10 drops of 0.1 M hydrochloric acid, and observe their behavior. Are any soluble or insoluble in acidic solution?

Step 2. Using the same amino acids, observe their solubility behavior in distilled water.

Step 3. Again observe the solubility of these amino acids, using 0.1 M NaOH as the solvent.

Color reactions

For these reactions, shift to using the <u>solutions</u> of amino acids available. Use histidine, aspartic acid, proline, and tyrosine.

> ♥ Use caution with the acids and bases!

Step 1. Set up four test tubes, each containing 5 drops of an amino acid solution. To each add 10 drops of ninhydrin solution and mix well. Observe the color formation and make a note of it. Most amino acids will react to ninhydrin although they don't all give the same color.

Step 2. Set up four test tubes again, each containing a different amino acid. To each add 5 drops of concentrated nitric acid, and warm the sample in a hot water bath on a hot plate for 5 minutes. This is a test for proteins containing tyrosine, which contains a phenol.

PART C. EXERCISES

Complete the questions on the report form.

EXPERIMENT 38

PROPERTIES OF A PROTEIN

GOALS

1. To investigate the structure of proteins
2. To determine the type of environment that can denature a protein
3. To examine some spot tests for the presence of proteins

INTRODUCTION

Proteins are polymers of amino acids held together by the peptide bond, which is really an amide linkage. But there is much more to them than that. This <u>primary</u> structure is covalently linked, but there are other superstructures built on this backbone that make the protein active in the way you see in nature. The <u>secondary</u> structure of a protein is the intramolecular hydrogen bonding that occurs between different amino acid molecules in the same protein—the original chain folds back on itself, either into an α–helix or a β-pleated sheet arrangement. These secondary regions of structure may be only a part of the molecule; it is quite common for any given protein chain to have many helical regions, a number of pleated regions, and some irregular ones as well. Further <u>tertiary</u> structure occurs when the amino acid side chains come into play, making salt bridges, disulfide linkages and other relatively weak but important interactions. To be active, many proteins (hemoglobin is a familiar example) have more than one protein chain tied together in quaternary structure, and prosthetic groups such as metals or light-absorbing molecules to make them active. Any disruption of the entire protein structure can cause temporary or permanent inactivation of the protein. Changes in pH, addition of metals that interact with disulfide bridges, addition of organic poisons, heat and cold can all cause inactivation of a protein due to disruption of the tertiary structure.

One such complex protein is used in this experiment to demonstrate some properties of proteins. The protein phycocyanin is a conjugated globular protein, consisting of several chains of amino acids along with a number of chromophores, or light-harvesting organic molecules. It is part of the Photosystem II complex found in cyanobacteria and plays an important part in the photosynthesis that these green algae can use to convert sunlight to energy.

In its native, properly folded state, phycocyanin is a deep blue. It also has the additional easily observed property of fluorescence—when light that the chromophores can absorb shines on it, part of the absorbed light is given back out as light. In this case, the chromophore molecules absorb some wavelengths of visible light, and the fluorescence is red. So you have two obvious properties to observe to decide if the protein structure has been altered, color and fluorescence. Inactivated protein, especially when completely unfolded and denatured, will be yellow and not show any visible fluorescence.

Phycocyanin is easily isolated in active form as long as the pH of the solution used is approximately neutral. You will isolate it (actually probably a solution of several closely related proteins) by grinding up spores of a blue-green alga, *Spirulina platescens,* and extracting the freed protein into a sodium phosphate buffer at pH 7. This solution will then be available to you to examine as you treat it with various agents that disrupt tertiary structure. Look for changes in both color and fluorescence as a guide that something is happening.

REFERENCES

Stoker, H. S. *General, Organic, and Biological Chemistry;* Cenage Learning; Chapter 20.
 Review topics
  - Primary, secondary and tertiary structure of proteins
  - Denaturation of proteins
  - Tests for proteins

Bowen, R.; Hartung, R.; Gindt, Y.; *J. Chem. Educ.* **2000** *77* 1456.

WEB RESOURCES

Molecular Structure of Phytocyanin.
http://www.wellesley.edu/Chemistry/nhk/ppt_cyano/sl_16_phycocy_molecule.html

Phycocyanin. *http://nist.rcsb.org/pdb/explore.do?structureID=2VJR* This image requires the program Jmol for a 3-dimensional representation. Jmol can be downloaded free at this site.

Botany online: Photosynthesis-Photosynthetic Membrane. *http://www.biologie.uni-hamburg.de/b-online/e24/24d.htm*

Cyanophyta. *http://www.uaex.edu/pperschbacher/Fish/Cyanobacterium.htm*

Life History and Ecology of Cyanobacteria. *http://www.ucmp.berkeley.edu/bacteria/cyanolh.html*

SAFETY NOTES

Organic chemicals often have strong odors and are usually volatile and flammable, so avoid inhaling the vapors by using them in the hood whenever possible. Use small amounts of materials, and discard them into appropriately labeled waste jugs when finished, not down the sink. Some chemicals in this experiment may be strongly acidic. Treat all the chemicals used with respect. You may wish to wear gloves during this experiment, and you must wear chemical splash goggles.

EXPERIMENTAL PROCEDURE

PART A. ISOLATION OF PHYCOCYANIN

Step 1. Obtain 1000 mg (1.0 g) of dried Spirulina spores, and place them in a mortar. Add to the mortar 1000 mg (1.0 g) of fine silica, and grind the two substances vigorously together in the mortar for four minutes or longer.

Step 2. When you have reduced the spores to a very fine flour, divide the mixture into four test tubes that will fit into the centrifuge available, and suspend the solids in 7 mL of 0.10 M phosphate buffer, pH 7. Vortex the mixture well to be sure all the particles are suspended.

Step 3. Centrifuge the samples for 3 minutes, using the centrifuge according to the directions your instructor gives you. At the end of this time, most of the cell debris should be packed in the bottom of the test tubes. Transfer the liquid only to new test tubes using a disposable pipet. The solid may be discarded.

Step 4. Filter all the solutions by gravity into a small flask, and discard the solids. The protein you have isolated is dissolved in the dark blue solution. Proteins can be quite fragile at this stage. If your solution is dark blue, you have isolated an active form of the phycobiliprotein known as phycocyanin.

Step 5. In a dim room, shine a flashlight up through the bottom of the flask as shown in Figure 38-1, and describe what you see. The red fluorescence and blue color of the protein phycocyanin are characteristic of the protein in its native state. It is also soluble in water in its native conformation. Loss of color, loss of fluorescence, or formation of a milky solution will indicate that some unfolding of the protein form has occurred, with partial or complete denaturation.

Figure 38-1. Observe the solution by transmitted light.

PART B. PROPERTIES OF A PLANT PROTEIN

For each step in this sequence, use about 1 mL of your protein solution. Observe what happens, and describe the event as an indicator of denaturation or inactivation.

Effect of denaturants

Step 1. Prepare a control sample by diluting 1 mL of the protein solution with 3 mL of phosphate buffer. Keep this sample throughout the experiment to compare it with various treated samples. At room temperature, was the color or other visible attribute affected by the dilution? For each treated sample, observe the color and fluorescence both before and after the treatment.

Step 2. Bring a water bath to a boil on a hot plate. Dilute 1 mL of the original protein solution with 3 mL of phosphate buffer at pH 7. Heat the mixture in the boiling water bath for 5 minutes, and then compare its visible attributes with those of the untreated sample. Return the tube to room temperature and observe it again

Step 3. Dilute 1 mL of the protein solution with 3 mL of the phosphate buffer. Plunge the tube into an ice bath for 15 minutes, and describe what happens. Return the tube to room temperature and observe.

Step 4. Dilute 1 mL of the protein solution with 3 mL of 1.0 M sodium chloride solution, and describe the effect on the solution.

Step 5. Dilute 1 mL of the protein solution with 3 mL of 95% ethanol and observe.

Step 6. Dilute 1 mL of the protein solution with 3 mL of 0.3 M lead nitrate solution and observe.

Step 7. Place 1 mL of the protein solution in a test tube, and add to it, dropwise, about 1 mL of 0.10 M HCl. Record the observations on this solution, and then neutralize the solution by adding 0.10 M NaOH solution. Is the change in the protein reversible?

Step 7. Place 1 mL of the protein solution in a test tube, and add to it, dropwise, about 1 mL of 0.10 M NaOH. Record the observations on this solution, and then add an equal amount of 0.10 M HCl solution. Is the change in the protein reversible?

The pI of phycocyanin

Place the rest of your protein solution in a small beaker, and dilute it to about 10 mL with distilled water. Add to it dropwise, with shaking, 0.10 M acetic acid until the solution just becomes milky. Using a pH meter, record the pH of the solution as the apparent pI of this protein.

PART C. EXERCISES

Complete the exercises at the end of the report.

EXPERIMENT 38 PRELAB EXERCISES:

PROPERTIES OF PROTEINS

Name_____ **Partner:** _____ **Date** _____

   1. What is a chromophore?

2. Phycocyanin is a conjugated protein. Use appropriate resources to determine what a "conjugated protein" is.

3. What effect does a denaturant have on a solution of a protein?

4. Why is a buffer usually used instead of distilled water to isolate a protein from its native life form?

5. Why does a protein precipitate from solution at its pI?

6. Why is it necessary to balance a centrifuge with test tubes of equal weight in opposite wells?

EXPERIMENT 38 REPORT: PROPERTIES OF PROTEINS

Name_____ Partner _____ Date _____

PART A. ISOLATION OF PHYCOCYANIN

1. Describe the original spores. Observations such as color, odor, transparency, and apparent fluorescence are relevant.

2. Describe the final solution in similar terms, comparing it to the dried spores you began with.

3. How many mL of protein solution did you isolate? _____

PART B. PROPERTIES OF A PLANT PROTEIN

Results of Protein Treatment with Various Agents

Test	Color	Fluorescence	Was the change reversible?
Dilution			
Heat			
Cold			
Sodium chloride			
Ethanol			
Lead nitrate			
HCl			
NaOH			

The pI of phycocyanin

Number of drops of acetic acid required   _____

Final pH by meter   _____

Apparent pI of the protein   _____

PART C. EXERCISES

1. Using phycocyanin as a model, what types of agents can you say will permanently denature a protein?

2. What types of agents will simply inactivate a protein?

EXPERIMENT 39

THE PROTEIN CONTENT OF EGG WHITE

GOALS

1. To practice quantitative laboratory techniques
2. To learn the construction and use of a standard curve
3. To determine the concentration of protein in a common food source
4. To apply a laboratory result to a practical problem

INTRODUCTION

Imagine...

Starry Sky Enterprises, a local company that markets camping equipment, is planning to introduce a line of freeze-dried foods for extended backpacking trips, and their advertising department thinks that they can use nutritional information about the products in their ads. The company is a small one with a very limited budget, so they have turned to the local university for help in determining some of the nutrition information they need. Their reasoning is that students can get some practical experience in analysis while at the same time providing a useful service to a local company at a small cost. The instructors for your class have agreed to allow their students to help.

One of the products that Starry Sky plans to sell is powdered egg, and the company wishes to know, among other things, how their powdered egg protein content compares to that of a fresh egg. Recognizing that egg yolk and egg white may have different nutritional values, Starry Sky has requested a protein analysis on fresh egg white and fresh egg yolk separately rather than on whole fresh eggs. Your class has been assigned the task of determining the amount of protein in egg white. The company would like the report to give the protein concentration in grams of protein per gram of egg white and grams of protein per serving, assuming one serving is one egg white of about 50.0 g.

The approach...

How can you go about this? There are a number of color reactions characteristic of amino acids and proteins; you can use these <u>qualitatively</u> to test for the presence or absence of proteins, or <u>quantitatively</u> to discover the exact concentration of protein (usually in mg/mL) in a given solution. Chemists often use such protein content determinations in biochemical research and in nutrition laboratories, and sometimes in clinical laboratories as well.

In order to do the quantitative analysis, you will prepare a series of protein solutions in which the exact protein concentration is known, and mix each with a reagent (biuret) that turns a deep purple-blue in the presence of proteins. You will see that the final color of the reagent depends on the concentration of protein—the greater the protein concentration, the deeper the color. You will measure the depth of color of each solution and then construct a graph, a "standard curve," that shows the relationship between concentration and color. You'll use the graph to analyze two protein solutions of unknown concentration. One of these solutions is a solution of bovine serum albumin

(BSA), a protein found in the blood of cattle; this analysis is mainly for practice constructing and reading the graph. However, the second unknown sample is the egg white solution for the Starry Sky analysis.

After you learn the concentration of the egg white solution, you will need to use your knowledge of conversion factors to convert the diluted concentration you actually measure to the original egg white concentration and then to the concentration of protein in one serving.

More about the method...

The biuret reagent contains copper sulfate, sodium potassium tartrate, sodium hydroxide, and potassium iodide, and reacts with proteins and other compounds containing -NH$_2$ or amide groups to give a blue-violet solution. The intensity of the color produced is proportional to the quantity of protein in the solution.

So how do we measure the color intensity? We use a type of instrument known as a *spectrometer*. This is a device designed to measure the quantity of light of a specific wavelength (in this case, 550 nm) that a solution absorbs. For most laboratory purposes, this number is converted to a unitless measurement between 0 and 1.0 called the absorbance (abbreviation A) of the solution. The greater the quantity of absorbing material such as a biuret-protein complex, the more intense is the color and the higher the absorbance reported.

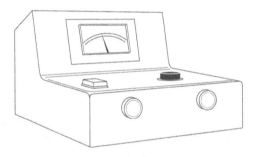

Figure 39-1. A typical small spectrophotometer.

As in most spectrophotometric techniques, you will need to analyze a "blank" sample at the same time as you do your regular analysis. A "blank" is a control sample that contains all the reagents except the one under analysis (in this case the protein). You will subtract the blank absorbance from each of the analytical sample absorbances to compensate for any color due to reagents alone.

The analysis of your data

By preparing a standard curve you can determine the exact quantity of any protein that produces a given absorbance. For your standard curve, you will prepare a series of protein solutions (in this case, BSA) of known concentration, carry out the biuret reaction on each solution, and measure the absorbance of each solution using the spectrometer.

You will need to construct a graph next, showing the variation of corrected absorbance with protein concentration. The relationship should be linear, and if you used careful technique in your analysis you will see a straight line relating protein concentration as the independent variable (on the x-axis) to absorbance as the dependent variable on the y-axis. The diagram on the next page shows an example of a standard curve for determination of the exact protein concentration in an unknown solution. Note that the data are not perfect. However, you can draw a "best" straight line which approximates all the data points, and this "best fit" line is the one you should use to determine the concentration of unknown protein solutions.

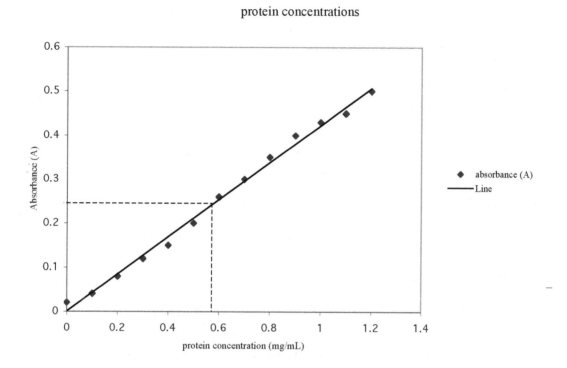

Figure 39-2. Determining the protein content of a solution.

You can then analyze any protein solution of unknown concentration but known color intensity with reference to this graph. In this example, for instance, a sample of unknown concentration with *A* (absorbance) = 0.25 must have a protein concentration of 0.6 mg/mL in order to lie on the line produced by the standard solutions. If you are familiar with using a spreadsheet program for graphing and data analysis, you will find this exercise even simpler.

In this activity, you will construct a standard curve by preparing and analyzing solutions of known concentration of BSA, and will use this curve to determine the protein concentration in two different solutions, a solution containing an unknown concentration of BSA, and a solution of egg white.

REFERENCES

Stoker, H. S. *General, Organic, and Biological Chemistry;* Cengage Learning; Chapters 20 and 21.
Review topics
- Characteristics of proteins
- Amino acids
- Peptide formation

WEB RESOURCES

Guide to Spec 20. *http://faculty.uca.edu/march/bio1/scimethod/Spec20_AbsAnthocyanin.htm*

Helpful Hints on the Use of a Volumetric pipet.
http://www.csudh.edu/oliver/demos/pipetuse/pipetuse.htm

Recording and Graphing Data. *http://www.harpercollege.edu/tm-ps/chm/100/dgodambe/thedisk/labtech/data4.htm*

Spectronic_20. *http://www.chemistry.nmsu.edu/Instrumentation/Spectronic_20.html*

Using Microsoft Excel to Create a Chart or Graph with Pictures in the Graph.
http://www.internet4classrooms.com/excel_picto_chart.htm

SAFETY NOTES

Wear chemical splash goggles when handling the chemicals used in this experiment. The biuret reagent contains sodium hydroxide, a strong base.

Disposable pipets, if used, are usually regarded as "sharps" and should be disposed of appropriately. Dispose of used reagents as your instructor directs, according to local ordinances. In general, if you are a small-volume user, the organic solvents can be discarded in an appropriately labeled waste container in the hood, and water solutions or suspensions of the unknown substances can be discarded in the sink, flushed down the drain with copious amounts of water. However, local ordinances govern your procedure.

> ● You may prefer to wear gloves while working with the biuret reagent.

EXPERIMENTAL PROCEDURE

Step 1. Prepare the solutions for the standard curve and the unknown samples.

a) Label 15 test tubes 1-15.

b) Using measuring pipets and a pipet bulb, place in each tube the appropriate reagents. The table on the next page is your guide for the amount of each reagent needed in each tube. The reagents in Tube #1 are the control solution, and the reagents in Tubes 2-11 are the solutions to use when constructing the standard curve. Solutions 12 and 13 contain the unknown BSA solution to analyze, and Solutions 14 and 15 contain the egg white solution to analyze. Samples 12-15 are deliberately done in duplicate, so you can check the accuracy of your technique.

> ⓘ Be careful using the pipets! If you need to, practice using the bulb so that you do NOT get liquid into the bulb. Check with your instructor on how to proceed if this does happen.

Step 2. Mix well each solution you make, using a vortex mixer if available. Allow the solutions to stand at room temperature to develop color for 15-30 minutes.

Step 3. While you are waiting, prepare a sheet of graph paper for your standard curve. You will need to calculate the concentration of BSA in each of the Tubes #2-11, but you can do that from the initial known concentration of the solution, which is 10 mg/mL.

Prepare the graph so that these calculated concentrations of BSA protein in Tubes 2-11 are the x-axis numbers, and the corrected absorbances (to be determined in Step 4) for these samples are the y-axis values. Protein concentrations should range from 0 to 10 mg/mL, and absorbance values should range from $A = 0$ to $A = 1.0$.

> ⓘ The entire graph should take up at least half a page—don't make it too small to read! Be sure to label your graph completely and correctly and you **must** use a ruler, to draw the axes and, eventually, the "best fit" straight line.

Step 4. Following the directions of your instructor, read the absorbance of each sample using the spectrometer and the special tubes provided. Record these absorbance numbers in appropriate columns in the table on the <u>second</u> page of the report form (p 392). You will need to read the absorbance of the solution in each of the 15 tubes.

Step 5. Calculate the "corrected absorbance" for your sample tubes #2-11 by subtracting the absorbance of Tube #1, the blank, from each of the values in the "absorbance" column of the table (values for Tubes 2-15). Record these numbers in the appropriate column.

Step 6. Now construct the standard curve showing absorbance of the known BSA solution vs. its concentration, using the corrected absorbances for samples 2-11 as the y-axis values and the protein concentrations in these tubes as the x-axis values.

Step 7. For each unknown protein sample, average the two absorbance determinations to get your best result, and record the average.

Step 8. Using the average BSA value, determine from the graph the best x-value that gives the concentration of protein in the unknown BSA solution.

Step 9. Do the same for the egg white solution, using the average egg white protein absorbance to determine its corresponding protein concentration.

Step 10. Calculations. Now you are ready to approach the real question. What is the protein content, in various units, of an egg white? Use the following questions as a guide as you answer the questions on the first page of the report form.

- What is the mass of the egg white used to make the solution you analyzed? This number should be on the bottle.
- What is the concentration of the solution of egg white in mg of egg white per mL of solution? This number should also be reported on the bottle.
- Use this information to determine the protein content of a raw egg white in units of mg/g.

- Once you have determined the mg of protein per gram of egg white, determine the total mg of protein in the egg white.

Your lab report will consist of the data sheet, the graph, and your calculations giving your results in the units requested. If you have access to a computer program that draws graphs and does least-squares line fits or linear trend lines, you may wish to use that instead of a hand-drawn graph. Follow the advice of your instructor.

EXPERIMENT 39 PRELAB EXERCISES:

THE PROTEIN CONTENT OF EGG WHITE

Name_____ **Partner** _____ **Date** _____

1. Use the Internet or print sources available to you to determine the composition of biuret solution.

2. Given the data below, circle the sample number of the solution that has the largest concentration of colored materials (all the solutes are not necessarily colored, so concentration of solutes is not useful here).

Absorbance of Protein Solutions

Sample number	Absorbance (550 nm)	Concentration of solutes (total)
100	0.10	1.0 M
200	0.20	2.0 M
300	0.30	0.15 M
400	0.40	0.60 M
500	0.50	0.80 M

3. Given the data below, graph the data with absorbance as the y-axis and concentration as the x-axis variable.

Sample number	Absorbance	Concentration (mmol/L)
100	0.10	0.20
200	0.20	0.30
300	0.30	0.40
400	0.40	0.60
500	0.50	0.70

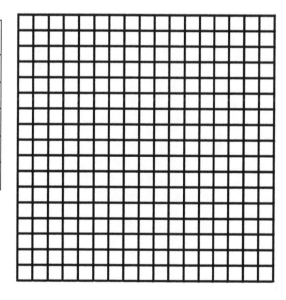

4. From the graph, what would be the expected absorbance of a solution with a concentration of 0.25 mmol/L?

EXPERIMENT 39 REPORT: THE PROTEIN CONTENT OF EGG WHITE

Name _____ **Partner** _____ **Date** _____

Use the graph you have constructed from the data on page 392 to answer these questions.

1. From the standard curve and the data you obtained from the unknown BSA samples, what is the BSA concentration in the unknown solution in Tubes 12 and 13? _____

2. What is the mass of the egg white used to make the solution you analyzed (given)? _____

3. What is the concentration of the solution of egg white in mg of egg white per mL of solution? This should be on the bottle label. _____

4. From the standard curve and the data you obtained from the unknown egg white solution (Tubes 14 and 15), what is the protein concentration of the egg white solution? _____

5. Find the protein content of egg white in mg protein per g of egg white. Show your work. A suitable setup is given here.

$$\frac{?? \text{ mg protein}}{\text{mL solution}} \times \frac{1000 \text{ mL solution}}{???\text{g egg white}} = \frac{??\text{mg protein}}{\text{g egg white}}$$

6. Find the protein content of an egg white in percent protein. Show your work.

7. From the data above, calculate the number of grams of protein in a 50.0-g portion of egg white.

Data for the Determination of Protein in Egg White

tube #	Protein (mL)	H$_2$O (mL)	Biuret reagent (mL)	[Protein] in mg/mL	Raw absorbance	Corrected absorbance	
1	0.0	1.0	4.00			0.00	
2	.10	.90	4.00				
3	.20	.80	4.00				
4	.30	.70	4.00				
5	.40	.60	4.00				
6	.50	.50	4.00				
7	.60	.40	4.00				
8	.70	.30	4.00				
9	.80	.20	4.00				
10	.90	.10	4.00				
1	1.00	0.00	4.00				
12	1.00 BSA unknown	0.00	4.00				AVERAGE [BSA] =
13	1.00 BSA unknown	0.00	4.00				
14	1.00 mL egg white solution	0.00	4.00				AVERAGE [egg white] =
15	1.00 mL egg white solution	0.00	4.00				

EXPERIMENT 40

PROTEINS FROM MILK

GOALS

1. To isolate casein from milk
2. To examine the behavior of casein as a buffer
3. To examine casein as a representative protein
4. To determine the pI of casein

INTRODUCTION

Milk is a food familiar to all of us. It contains an amazing number of nutrients: protein, carbohydrate, fat, and many trace minerals and vitamins. It's not surprising that it has been advertised as "nature's most nearly perfect food." The milk product most available in the United States is cow's milk, in dried or liquid form, full-fat or reduced fat form, and many flavored forms. Many Internet references will give you sources for detailed information on the composition of cow and other kinds of milk, including human milk. Protein in cow's milk is the subject of this exercise.

The primary milk protein is casein, which is suspended in milk in colloidal form. It is held in solution in part by the fact that the pH of milk is several units above the pI of casein, so the casein is present in charged form. Reducing the pH of the milk allows the overall charge on casein to be neutralized, and the neutral protein will then precipitate from the whey. There are actually at least five casein proteins, that act similarly in the isolation exercise you will perform, so we really ought to refer to "caseins" not "casein." The other major milk protein, lactalbumin, is soluble in whey and is not precipitated by treatment with acid in this fashion.

This exercise has three parts. In the first part of the exercise, you will use liquid skim milk treated with dilute acetic acid to precipitate casein and isolate it. The casein you isolate will be rinsed with a mixture of alcohol and ether, which will dissolve any fats that may have co-precipitated with the casein sample and leave you a sample of protein to dry, weigh, and examine. This sample will be used to determine the percent of caseins in a sample of liquid skim milk.

In the second part of this exercise, you will do a simple titration of a new milk sample to estimate the approximate pI for casein in cow's milk. For this part of the exercise, you will need to use a pH meter to determine the pH at which the casein appears to be completely isolated. Your instructor should be able to review tips on the best methods for operating the brand of pH meter you have available, but you will need to rely on your own judgment to determine when the casein has completely precipitated. The casein from this exercise and the whey formed will be used in the third part of the experiment.

Once you have isolated casein and whey, you will need to redissolve a small sample of the casein in water, using a sodium hydroxide solution to bring the pH back above the pI of casein. You can then test the casein solution and the whey for the presence of protein and sugars such as milk sugar.

REFERENCES

Stoker, H. S. *General, Organic, and Biological Chemistry;* Cengage Learning; Chapters 20, 21, and 26.

Review topics
- Protein structure
- Peptide bonds
- Titration
- pI

WEB RESOURCES

Chem 2O06-97/98-Experiment 11.
http://www.chemistry.mcmaster.ca/~chem2o6/labmanual/expt11/2o6exp11.html

Healthy Weight with Dairy.
http://www.nationaldairycouncil.org/Research/DairyCouncilDigestArchives/Pages/dcd78-6Page1.aspx

Nat. Academies Press, Nutrition during Lactation (1991).
http://www.nap.edu/books/0309043913/html/297.html

SAFETY NOTES

Organic chemicals often have strong odors and are usually volatile and flammable, so avoid inhaling the vapors by using them in the hood whenever possible. Use small amounts of materials, and discard them into appropriately labeled waste jugs when finished, not down the sink. Some chemicals in this experiment may be strongly acidic. Treat all the chemicals used with respect. You may wish to wear gloves during this experiment, and you must wear chemical splash goggles.

EXPERIMENTAL PROCEDURE

PART A. WHAT PROPORTION OF MILK IS THE PROTEIN CASEIN?

Step 1. Weigh a graduated cylinder (or tare it) and measure into it 25 mL of skim milk. Record (or calculate if necessary) the mass of the milk. Place the milk in a 100-mL beaker. Add to this sample 5 mL of 10% acetic acid, and stir well.

Step 2. When the precipitation of the protein appears to be complete, warm the mixture gently on a hot plate (to about 50-60 °C) for about 5 minutes, stirring occasionally. Do not bring the solution to a boil. Allow the solid to settle and decant (pour off) as much liquid as possible, being careful not to lose any solidified protein. The liquid from this portion of the experiment can be discarded in the sink.

Step 3. Place the protein in a small beaker and, in the hood, stir it with 10 mL of a 50% ethyl alcohol-50% ethyl acetate mixture to dissolve fats that may have precipitated with the protein. Discard the liquid by decanting it into the waste jar provided in the hood.

Step 4. Add to the protein 10 mL of distilled water, and stir the mixture well. Discard this liquid into the provided waste jar.

Step 5. Find the mass of a piece of filter paper that will fit into a suction filtration apparatus, and record its weight. Filter the solid protein from the liquid, and rinse it with two 10-mL portions of distilled water.

Step 6. Remove the paper and solidified protein from the apparatus, and place them together in a beaker or on a watch glass labeled with your name, lab number, and date. Place the sample in a warm (<70 °C) oven to dry, and continue with Part B of this exercise. Plan to complete Part A toward the end of the lab time, after the sample is dry. Discard the liquid from this portion of the exercise. Go on to Part B while you are waiting to do Step 7.

Step 7. When your sample is dry, weigh the filter paper and protein together, and calculate the mass of protein obtained from 25 mL of skim milk.

Step 8. Calculate the percent of casein in the sample. Share your results with at least three other groups, and calculate the average casein content of milk using all those results.

PART B. DETERMINATION OF THE pI OF CASEIN

Step 1. Place 25 mL of skim milk in a 50 mL beaker.

Step 2. Prepare a pH meter for use according to the instructions provided by your instructor, and measure the pH of the milk, recording it on the report sheet.

Step 3. Add 10 drops of 10% acetic acid to the milk, stir it well, and after 30 seconds, record the pH of the mixture as well as the appearance.

Step 4. Repeat Step 3 until it appears that the liquid in the beaker is as clear as possible and that no more protein is precipitating. Record your observations at each step (appearance and pH)—it should not take more than about 10-15 mL of acetic acid in all. When you are finished, be sure to rinse your electrode with distilled water and store it as directed by your instructor.

Step 5. Identify the approximate pI of casein. This will be the pH at which the solution becomes as clear as possible, and no further clarification of it takes place with additional acid.

Step 6. Decant the whey into another beaker and save it for Part C, and then purify and isolate the casein as you did in Part A, Steps 3-5.

When you are finished with Part B, you should have a portion of damp casein to work with and a portion of whey to examine as well in Part C.

PART C. CHEMICAL TESTS ON CASEIN AND WHEY

Step 1. Redissolve the casein for the tests. Place a small piece of casein (about the size of a pea) in a small beaker and add to it 25 mL of water and 1 mL of 0.5 M NaOH solution. This step should neutralize any acetic acid left in the sample and redissolve at least some of the proteins. Use the liquid portion of this sample for the tests below. The whey can be used directly.

Step 2. Test for protein. Using ninhydrin, test the isolated samples (casein and whey) for the presence of protein in this step. Place 1 mL of the casein solution from Step 1 and 1 mL of the whey into separate, labeled test tubes. For controls, prepare an additional labeled test tube containing only 1 mL of distilled water and another containing 1 mL of 1% egg albumin solution.

Add to each solution 10 drops of ninhydrin solution, and heat all the samples in a hot water bath on a hot plate for 5 minutes. Record the results that you see.

Step 3. Test for reducing sugars. Set up five labeled test tubes to contain casein solution, whey, albumin, water, and glucose respectively. The last three are controls that should show distinct positive or negative results for the presence of reducing sugars.

To each of the five tubes add 3 mL of Benedict's reagent, and warm the samples in a hot water bath on a hot plate for 5-10 minutes. Record the result of each test.

DON'T FORGET TO FINISH PART A!

EXPERIMENT 40 REPORT: PROTEINS FROM MILK

Name_____ Partner _____ Date _____

PART A. WHAT PROPORTION OF MILK IS THE PROTEIN CASEIN?

Mass of cylinder if needed   _____

Combined mass of cylinder and milk   _____

Mass of 25 mL of milk   _____

Mass of filter paper   _____

Mass of dried paper and protein   _____

Mass of protein recovered   _____

Mass of protein recovered by three other groups   _____

_____

_____

Average protein mass recovered (all groups)   _____

Percent of casein in milk (show calculation below.) _____

1. Why was it better to use the average of several trials of the isolation process in Part A than to rely on your results alone?

2. The expected concentration of casein in milk is about 2.7%. How does your value from Part A compare to this number? If it is significantly different from the accepted value, give at least two reasons why this could be the case.

PART B. DETERMINATION OF THE pI OF CASEIN

Treatment of Milk with 10% Acetic Acid

Amount of acetic acid added (drops)	Appearance	pH of milk
0.0		

1. From your data, determine the pH of milk.   _____

2. From a reference source, record the pH of milk.   _____

3. According to your data, what is the approximate pI for casein? _____

4. How does this compare with the accepted value of pI = 4.6? Give at least one reason why your data may not agree with the accepted value.

PART C. CHEMICAL TESTS ON CASEIN AND WHEY

Test for Protein Using Ninhydrin

Sample	Color	Result (-, or + to ++++)
Casein		
Whey		
Water		
Albumin		

1. Did the whey test positive for protein? If it did, what might account for this?

Test for Reducing Sugars

Sample	Color	Result (-, or + to ++++)
Casein		
Whey		
Water		
Albumin		
Glucose		

2. Did either the casein or the whey give a positive test for reducing sugars?

3. What sugar in milk would be expected to give a positive Benedict's test?

4. Would you expect either the casein or the whey to be positive for a test for cholesterol (you did not use this test)? Why or why not?

EXPERIMENT 41

ENZYME ACTION: AN INVESTIGATION OF LACTASE ACTIVITY

GOALS

1. To investigate the specificity of an enzyme
2. To examine the factors affecting the activity of an enzyme as a catalyst
3. To consider the relationship between protein structure and enzyme function

INTRODUCTION

Enzymes are naturally occurring protein catalysts that help the biological reactions necessary for life to occur at a meaningful rate. They are usually extremely specific in their action; only a small number of substrates will undergo reaction with any given enzyme, and the pH range and temperature range for maximum catalyst activity is often quite narrow. In addition, like other catalysts, enzyme catalysts can be inhibited from acting if a foreign molecule is present to block or otherwise affect the active site for substrate attachment. In this exercise you will observe some of these effects as you study the enzyme lactase.

Lactase operates on the milk sugar lactose. Lactose is a disaccharide consisting of a galactose unit and a glucose unit, and is found in all types of milk. The first step in the digestion of lactose is the lactase-mediated cleavage of the glycoside bond between the galactose and glucose, which releases the two monosaccharides for further digestion and energy production.

Because lactase is a relatively sturdy enzyme, it is possible to isolate a solution of it and use it to test the various factors that affect enzyme activity.

In fact, there is a major commercial use for isolated lactase. Many people have a condition known as "lactose intolerance" in which they are unable to digest lactose readily. The unchanged lactose is instead fermented by intestinal bacteria, and this fermentation can cause gas, nausea, bloating, and diarrhea for the unfortunate individual. To avoid these distressing symptoms, lactase-deficient persons can take advantage of a number of commercial products containing lactase. One of these products is Lactaid®, a readily available preparation of lactase that you will be using in this experiment.

You will study the effects of substrate type and concentration, as well as enzyme concentration, on the production of glucose from lactose. In addition, you will study temperature effects, pH effects, and inhibitor effects on the activity of lactase. The basic experiment consists of incubating a sample of lactose with a small amount of lactase at an optimum temperature (about 30 °C), and testing the solution for the presence of glucose after a set time. For the glucose test you will use a Tes Tape®, a piece of paper impregnated with a reagent that signals the presence of glucose in a solution. In each subsequent run of the experiment, you will alter a single variable such as temperature or pH and examine the lactase activity under each set of conditions.

Although each individual experiment is simple, there are seven sets of experiments to complete, so be sure to organize your time well!

REFERENCES

Stoker, H. S. *General, Organic, and Biological Chemistry;* Cengage Learning; Chapters 21 and 24.
    Review topics:
- Equilibrium and rate
- Protein structure and function
- Catalysis
- Carbohydrate metabolism

WEB RESOURCES

Enzyme Action. *http://www.hillstrath.on.ca/moffatt/bio3a/digestive/enzanim.htm*

Lactose Intolerance. *http://digestive.niddk.nih.gov/ddiseases/pubs/lactoseintolerance/#whatis*

New Enzyme section. *http://web.fccj.org/~dbyres/enzyme1.html*

SAFETY NOTES

Some of the reagents contain strong acids or bases, and you may prefer to wear gloves while working with these chemicals. **Never** put an unprotected finger over an open test tube to mix a solution.

Glass disposable pipets are usually regarded as "sharps" and should be disposed of appropriately.

Dispose of used reagents according to local ordinances. In general, if you are a small-volume user, the organic solvents can be discarded in an appropriately labeled waste container in the hood, and water solutions or suspensions of the unknown substances can be discarded in the sink, flushed down the drain with copious amounts of water. However, local ordinances govern your procedure.

EXPERIMENTAL PROCEDURE

PART A. NORMAL ACTIVITY OF LACTASE

Step 1. Prepare three test tubes labeled "lactose," "treated lactose," and "glucose." In the first two tubes, place 1 mL of 1% lactose solution, and in the third place 1 mL of a 1% glucose solution. Preincubate the three tubes in a warm water bath at approximately 30 °C for 5 minutes before proceeding to Step 2.

Water at 30 °C is easily obtained from the hot tap water available in the lab, but heated water in the building water system is usually at a higher temperature than that, so you will need to adjust the water temperature using cold water and a thermometer before continuing with the experiment. The first tube is the control for this experiment, the second is the experimental sample, and the third is the control for the eventual test for lactase activity. Watch the temperature of the water bath, and warm it up as necessary to insure that the samples are all incubated at a temperature around 30-35 °C.

Step 2. Add to the first control tube 3 drops (about .05 mL) of distilled water. Add to the second tube ("treated lactose") 3 drops of liquid Lactaid®, and add to the third tube 3 drops of water. Mix each well. Allow the samples to stand in the warm water bath for 5 minutes before proceeding to Step 3.

Step 3. Check each sample for the production of glucose from lactose using Tes Tape®, which gives a specific reaction with glucose. To do this, remove about 2 drops of the sample you want to test, and place it on a small piece of Tes Tape®. Record your results as – (negative) or + to ++++ depending on the strength of the reaction you observe.

PART B. THE EFFECT OF SUBSTRATE ON ENZYME ACTIVITY

Step 1. Again label four test tubes with the name of the substrate or control to be investigated. Tube 1 should contain 1 mL of distilled water; Tube 2 should contain 1 mL of a 1% lactose solution; Tube 3, 1 mL of 1% sucrose; Tube 4, 1 mL of 1% glucose solution. Preincubate all samples at 30 °C for 5 minutes.

Step 2. Add to each of the first three samples 3 drops of Lactaid®, and to the glucose solution 3 drops of distilled water. Incubate all the samples in the warm water (30 °C) bath for 5 minutes.

Step 3. Test each sample for the presence of glucose as described in Part A, and record your results.

PART C. THE EFFECT OF ENZYME CONCENTRATION

Step 1. Prepare a diluted solution of the enzyme solution Lactaid® by mixing 1 drop of Lactaid® with 20 drops of distilled water.

Step 2. Prepare four test tubes labeled "lactose," "Lactaid®," "diluted Lactaid®," and glucose. In the first three of these, place 1 mL of the 1% lactose solution and in the fourth place 1 mL of glucose.

Step 3. Add 3 drops of water to the lactose-only tube, 3 drops of Lactaid® to the second tube, 3 drops of diluted Lactaid® to the third, and three drops of water to the fourth. Mix the solutions, and incubate all the samples for five minutes in the warm water bath.

Step 4. Test each solution for the presence of glucose, and record your results.

PART D. EFFECT OF SUBSTRATE CONCENTRATION

Step 1. Prepare a diluted solution of the lactose by mixing 1 drop of the lactose solution with 20 drops of distilled water.

Step 2. Prepare four test tubes labeled "untreated lactose," "lactose," "diluted lactose," and "glucose." In the first two of these, place 1 mL of the 1% lactose solution; in the third, place the diluted lactose solution; and in the fourth place 1 mL of glucose. Preincubate these solutions at 30 °C for 5 minutes.

Step 3. To the controls (the first and fourth tubes) add 3 drops of water. To the second and third tubes, add 3 drops of undiluted Lactaid®. Mix the solutions, and incubate at 30 °C for 5 minutes.

Step 4. Test each sample for the presence of glucose, and record your results.

PART E. EFFECT OF TEMPERATURE ON ENZYME ACTIVITY

In this section, you will need to study standard Lactaid® activity on the standard lactose solution, but at different temperatures. You will need, in addition to the usual 30 °C bath, an ice bath, and a water bath above 70 °C for the incubations. To minimize confusion for this phase of the experiment, plan to test only one temperature at a time. Complete directions are given for the 30 °C experiment and abbreviated ones for the other two temperatures, since the only variable is the temperature of incubation.

Step 1. Prepare three test tubes labeled "lactose," "treated lactose," and "glucose." In the first two tubes, place 1 mL of 1% lactose solution, and in the third, place 1 mL of a 1% glucose solution. Preincubate the three tubes in a warm water bath at approximately 30 °C for 5 minutes before proceeding to Step 2.

Step 2. Add to the first control tube 3 drops (about .05 mL) of distilled water. Add to the second tube ("treated lactose") 3 drops of liquid Lactaid®, and add to the third tube 3 drops of water. Mix each well. Allow the samples to stand in the warm water bath for 5 minutes before proceeding to Step 3.

Step 3. Check each sample for the production of glucose from lactose. Record your results as − (negative) or + to ++++ depending on the strength of the reaction you observe.

Step 4. Repeat the experiment, but instead of using a 30 °C bath, preincubate the samples in a beaker containing about 50 mL of ice and 50 mL of water. Record the temperature of the bath and the results for the presence of glucose after 5 minutes of incubation of the samples with the Lactaid®.

Step 5. Repeat the experiment, but instead of using a 30 °C bath, preincubate the samples in a beaker containing about 50 mL of water at a temperature between 70 and 90 °C. Record the temperature of the bath and the results for the presence of glucose after 5 minutes of incubation of the samples with the Lactaid®.

PART F. EFFECT OF pH ON ENZYME ACTIVITY

In this portion of the exercise, you will test the activity of lactase at high and low pH. All the samples will be incubated at 30 °C so that there will be only one variable (pH) to be considered.

Step 1. Prepare five test tubes labeled "lactose," "pH 7," "pH 2," "pH 10," and "glucose." In the first four tubes, place 1 mL of 1% lactose solution, and in the fifth place 1 mL of a 1% glucose solution. Preincubate the five tubes in a warm water bath at approximately 30 °C for 5 minutes before proceeding to Step 2.

Step 2. To Tubes 1, 2, and 5, add 1 drop of distilled water. To the third tube, add 1 drop of 1 M HCl solution, and to the fourth tube, add 1 drop of 1 M NaOH solution. Continue to incubate at 30 °C while you do Step 3.

Step 3. Add to the first control tube 3 drops (about .05 mL) of distilled water. Add 3 drops of liquid Lactaid® to the second, third, and fourth tubes, and to the fifth tube add 3 drops of water. Mix each well. Allow the samples to stand in the warm water bath for 5 minutes before proceeding to Step 4.

Step 4. Check each sample for the production of glucose, and record your results.

PART G. EFFECT OF AN INHIBITOR ON ENZYME ACTIVITY

For this exercise, ethanol will be used as a model inhibitor of enzyme action. You will need four tubes for this exercise, labeled "lactose," "Lactaid®," "ethanol," and "glucose." All samples will be incubated at 30 ˚C.

Step 1. Prepare the four tubes. In the first three, place 1 mL of the substrate lactose solution. In the fourth, place 1 mL of the glucose solution. Preincubate the solutions in the water bath for 5 minutes.

Step 2. In the first, second, and fourth tubes, place 5 drops of distilled water. In the third tube, place 5 drops of 95% ethanol. To Tubes 2 and 3 add 1 drop of undiluted Lactaid® solution. Mix each solution and incubate for 5 minutes in the warm water bath.

Step 3. Test each sample for the presence of glucose, and record your results.

no change
brownish
green
brown

EXPERIMENT 42

URINALYSIS

GOALS

1. To review some normal and abnormal components of human urine
2. To examine urine samples for evidence of metabolic errors
3. To examine the end product of the urea cycle

INTRODUCTION

Imagine...

A number of patients have come into the clinic this morning. As usual, some of them are in normal health, and some have some fairly severe conditions that will need treatment. Several of the patients exhibit gross symptoms of untreated diabetes, and several may have urinary tract infections (UTI). At least one appears to be passing significant amounts of blood in his urine. One may be a laxative abuser although she refuses to admit it. It's a busy morning, and some of the patients exhibit symptoms that imply they have several of these conditions all at once.

You are already aware that the composition of body fluids can be a guide to the health of a patient. There are many tests that can be performed on urine, and you will investigate several of them during this laboratory exercise. Your job is to analyze the urine samples from three of the patients and report the results so that treatment can begin.

There will be two parts to this activity. In Part A, you will examine one normal and five abnormal samples using some typical tests that can be performed on urine samples, and observe outcomes of those tests. In Part B of this exercise, you will have three unknown samples to examine, and each of these samples will exhibit one or more pathological conditions that can be identified by means of some of the tests you have studied as well as some additional tests, which are described in the directions for Part B.

Exercises of this nature have a two-fold purpose: to allow you to observe chemical tests of clinical significance, and to encourage you to begin to rely on your own observations and judgment when faced with a clinical situation of unknown scope. You should begin each part of the experiment with a preliminary physical assessment of the sample, noting specific gravity, turbidity, pH, and odor.

The relative concentration of salts in a urine sample is assessed by the sample's specific gravity, which is measured with a urinometer. The normal specific gravity of urine is about 1.015-1.020, or slightly more dense than water, due to the dissolved salts present. Note that this figure is accurate to four significant figures. Any number appreciably outside this range is important, and in disease states, the specific gravity may vary from a low of 1.001 to a high of about 1.06. Urines with higher specific gravity are more concentrated than normal and may be an indication that the patient is not eliminating urine well. A pale, low-specific-gravity urine may indicate that the patient is not eliminating salts well. Very dark urine may be very concentrated or may indicate the presence of abnormal substances that require further investigation. Turbid or cloudy urine, if fresh, often contains excess protein; this may come from a variety of sources but is rarely normal.

The odor of a urine sample is worthy of note; it can be recorded as normal, aromatic, medicated, ammoniacal, putrid, or strong. The pH of normal urine is about 4.8-7.5, or slightly acidic to neutral. If it is excessively acid or alkaline, further tests are in order.

None of these preliminary examinations is sufficient to diagnose an ailment; they are merely preliminary indicators to be noted. A red sample, for example, may merely mean that the patient has been eating beets.

Specific dipstick tests, such as Clinistix®, Albustix®, etc., are very commonly used because they are specific for the substances of concern. Clinistix®, for example, rely on an enzyme-mediated reaction using glucose oxidase to detect glucose, so that other substances won't interfere.

Your unknown samples for Part B are a little more complicated. They have been set up so that each sample may exhibit one or more of four possible pathological conditions for which further tests may be needed. In addition, it is possible that some unknown samples are perfectly normal, so you will need to use care in interpreting your results. The four conditions to consider can be diagnosed with the help of five more tests in conjunction with those you learned in Part A.

The four conditions to consider in your testing are laxative abuse, hematuria, albuminuria, and glucosuria.

Laxative abuse is not unusual. Individuals may begin to use an over-the-counter laxative during a bout of constipation and then become dependent on it for daily bowel movements. Chemical laxatives can in time affect intestinal tissues adversely, and a patient who is in fact laxative dependent will need to be weaned from its use. Laxative abuse is not always reported by the patient, since the patient is often embarrassed about it. However, a once commonly used over-the-counter laxative is phenolphthalein, a phenolic chemical that is also useful as an indicator of basic solutions—it is deep fuschia pink in the presence of bases at a pH above about 8. Thus the presence of phenolphthalein can easily be detected in a urine or feces sample by adding a bit of sodium hydroxide to the liquid. Although a stool sample is the best substance to check, phenolphthalein also spills into the urine, so a urine sample will suffice.

As its name suggests, phenolphthalein is a phenol. Phenols will react with ferric chloride, $FeCl_3$, to give a strong blue or green color, so a positive ferric chloride test on a urine sample is one method for distinguishing the presence of phenolphthalein. This way, you have an extra check on the presence of a laxative abuser among your patients.

Blood in the urine, or hematuria, can indicate many problems, from a simple urinary tract infection to prostate trouble or to major kidney dysfunction. Guaiac is a natural product that has the ability to react with hemoglobin and produce a blue color, so a solution of it is a useful diagnostic tool for the presence of blood in urine. Guaiac has also been used for many years to test for occult (non-visible) blood in stool samples.

The presence of protein, or proteinuria, can be determined by the use of the biuret reagent, which reacts with various amides and amide-like groups to produce a strong blue-violet color. You will need to be careful in your interpretation, though, because the reagent itself is blue; only a change from the initial color will indicate the presence of protein. Hemoglobin may also react, so you will need to interpret your results carefully!

SAFETY NOTES

Wear chemical splash goggles when handling the chemicals used in this experiment. Some of the reagents contain strong acids or bases, and you may prefer to wear gloves while working with these chemicals.

The "urine" samples used are synthetic and are not really urine. Nevertheless, treat them with the same precautions you would use for handling any body fluid.

Glass disposable pipets are usually regarded as "sharps" and should be disposed of appropriately.

Dispose of used reagents according to local ordinances. In general, if you are a small-volume user, the organic solvents can be discarded in an appropriately labeled waste container in the hood, and water solutions or suspensions of the unknown substances can be discarded in the sink, flushed down the drain with copious amounts of water. However, local ordinances govern your procedure.

REFERENCES

Stoker, H. S. *General, Organic, and Biological Chemistry;* Cengage Learning; Chapters 24-26.
Review topics
- The urea cycle
- Composition of normal urine
- Common tests for urine samples
- Density and specific gravity

Other reference:

S. Frankel and S. Reitman, in *Gradwohl's Clinical Laboratory Methods and Diagnosis,* 6th ed. or later; Mosby: St. Louis, 1963; pp 1808-1836.

WEB RESOURCES

Allrefer Health-Urinaylsis: Normal Values (Routine Urine Test, Urine Appearance and Color). *http://health.allrefer.com/health/urinalysis-results.html*

Urinalysis. *http://www-medlib.med.utah.edu/WebPath/TUTORIAL/URINE/URINE.html*

Urinalysis:" The Test. *http://www.labtestsonline.org/understanding/analytes/urinalysis/test.html*

Urine Reagent Strip Expected Values. *http://www.craigmedical.com/urinalysis_techs.htm*

EXPERIMENTAL PROCEDURE

PART A. OBSERVATIONS ON NORMAL AND ABNORMAL SAMPLES OF KNOWN COMPOSITION

Step 1. Dipstick tests. Your instructor will provide several types of test strips for you to use on normal and abnormal urine samples. The brand names and functions vary from manufacturer to manufacturer, so follow your instructor's guidelines for the test strips to use. In order not to waste these relatively expensive diagnostic tools, only test the normal sample and the appropriate abnormal sample with each test strip. Report your results as – (negative) or + to ++++ for positives.

Step 2. Sodium hydroxide test. Place 0.5-1.0 mL of the urine sample in a small test tube. Add 1-2 drops of 10% sodium hydroxide solution, shake gently, and observe. The presence of phenolphthalein (a common over-the-counter laxative ingredient) in the urine sample will cause the formation of a pink color.

Step 3. Guaiac test. Mix together 1.0 mL of guaiac solution and 1.0 mL of "old oil of turpentine." Layer on this solution about 0.5 mL of the urine sample to be tested. A blue-green ring at the interface, followed by a light or dark blue ring will appear at the point of contact of the two solutions. If you now shake the mixture well, the entire solution will turn blue if the test is positive. Record your results for both a positive and a negative sample of urine.

Step 4. Ferric chloride test. To 1.0 mL of the urine sample in a small test tube, add 0.5 mL 5% ferric chloride solution, and mix well. The presence of phenolic substances such as salicylates will cause a dramatic color change, usually to blue, green or brown. Record the results for both positive and negative samples.

Step 5. Biuret test. Place about 1.0 mL of the urine sample and 1.0 mL of a control sample in separate test tubes. To each add about 10 drops of 10% sodium hydroxide solution and mix. Add a few drops of 10% copper sulfate solution to each, mix, and observe the results. The presence of albumin or related materials in the urine will cause a rose to violet tint depending on the concentration of the albumin. Be careful not to misinterpret a negative result—unreacted copper sulfate is blue and might confuse you!

PART B. EXAMINATION OF URINE SAMPLES OF UNKNOWN COMPOSITION

You have between one and three numbered patient urine samples to analyze—your instructor will notify you of the samples you must investigate. Your patients may have **any combination** of the following problems **or may be perfectly normal.**

diabetes	UTI	hemoglobinuria	salicylate poisoning	laxative abuse

Your task is to determine which problem(s), if any, each patient has and report the results to your instructor. You may use any combination of wet chemical tests and rapid dip tests to reach your conclusions. You may not need to perform all the tests described described here. Perform only those needed to determine if any of the five above pathological conditions is present.

EXPERIMENT 42 PRELAB EXERCISES:

URINALYSIS

Name_____ **Partner** _____ **Date** _____

1. What are appropriate precautions to take when handling samples of body fluids?

2. Would a specific gravity of 1.2 for a urine sample be considered high or low? What might it indicate in terms of the health state of the patient?

3. Why is it important to time your observation of test strips carefully after dipping them in the sample to be tested?

4. An individual's urine sample exhibited a normal specific gravity, and the addition of sodium hydroxide solution to the sample produced a deep pink solution. What might this finding indicate concerning the patient's health state?

5. If a urine sample tested positive with the ferric chloride test but negative with the sodium hydroxide test, what would that tell you about the contents of the sample?

EXPERIMENT 42 REPORT: URINALYSIS

Name_____ Partner _____ Date _____

PART A. OBSERVATIONS ON NORMAL AND ABNORMAL SAMPLES OF KNOWN COMPOSITION

Results of Various Tests

	Test strip type and result	Sodium hydroxide test	Guaiac test	Ferric chloride test	Biuret test
Normal sample					
Laxative abuse sample					
Hematuria sample					
Albuminuria sample					
Glucosuria sample					

State in your own words what you would expect to see in a sample that was positive for glucose and positive for laxative abuse.

PART B. EXAMINATION OF URINE SAMPLES OF UNKNOWN COMPOSITION

Results for Several Samples

Sample number			
Test strip(s) used and results			
Sodium hydroxide test			
Guaiac test			
Ferric chloride test			
Biuret test			

What pathological conditions, if any, are indicated by each sample? For each sample you analyzed, report your conclusions in a paragraph, giving your reasons for reaching those conclusions.

EXPERIMENT 43

WATER SOLUBLE VITAMINS

GOALS

1. To review the structures of water- and oil-soluble vitamins
2. To identify the components of a mixture of water-soluble vitamins
3. To determine quantitatively the vitamin C content of a vitamin tablet

INTRODUCTION

We are all aware that a healthy diet requires the intake of "vitamins and minerals". It can be easy to understand what the minerals are needed for: sodium and potassium, calcium and iron all have their recognizable functions in the body. Vitamins are less easy to classify, and have broad functions. They are relatively small organic compounds, containing many different kinds of functional groups, and usually are required as cofactors or coenzymes for the proper functioning of the catalysts needed for biochemical reactions.

Vitamins are not classified by their functional groups, because they are all so different in structure. Instead, they are broken into two classes, water-soluble vitamins and oil-soluble vitamins. As organic molecules, they follow the general rule that "like dissolves like," and non-polar vitamins dissolve readily in body fat and may be stored, while more polar water-soluble vitamins dissolve in the blood, and are readily excreted in the urine.

In the body, water-soluble vitamins are lost easily, and must be made up constantly from foods taken in. And because they are soluble in water, cooking a vitamin-rich food can decrease its vitamin content significantly. This is one reason raw leafy vegetables, for example, are often recommended by dietitians over cooked vegetables of the same kind.

Analysis of vitamin content in foodstuffs can be an important tool for deciding on the food value of a substance, and you will be trying two different kinds of analysis today. The first exercise uses thin-layer chromatography (TLC) to separate a mixture of water-soluble vitamins and allows you to identify the components of the mixture, while the second exercise is the analysis of the vitamin C content of a nutritive supplement commonly available in the grocery store.

Thin layer chromatography as a technique has been covered before in this manual, in Experiment 32, and in the essay on common laboratory tools in the front. Refer to those two documents for a refresher on how the technique works. A small quantity of the sample to analyze (the analyte) is placed in small quantity near the bottom of a chromatography plate, a thin plastic sheet that has been coated with a layer the solid phase (silica gel, in this case) of the chromatography system. The initial spots of analyte may be invisible to the naked eye, but can be seen as dark spots on a light ground if the silica gel is viewed under an ultraviolet light.

Running the chromatogram requires that you place it in a chamber of suitable solvent and allow the solvent to slowly move up the plate by capillary action. As it moves past the spots of samples, the different solubilities of the individual compounds in the samples allow them to move along with the solvent, but not necessarily at the same rate. The distance each compound can move, as a fraction of the distance the solvent actually moved (the R_f of each compound), is a physical constant that can be used to help identify an unknown analyte.

You will have a set of standard solutions of vitamins B_2, B_3, B_6, and C, and a sample to analyze that could contain one of them or any combination of them. After you have prepared a TLC plate and placed it in the development chamber you will then begin on the vitamin C analysis, and eventually go back to the TLC analysis at the end of the period.

Since the vitamins you will be testing are all colorless, analysis of the TLC plates will require you first to make the location of each compound visible, either with an ultraviolet light or with iodine vapor. When you have found all the spots on the plate, you will need to calculate the R_f for each, in both the standard samples and the unknown sample, and then by comparison of these values determine the identity or identities of the vitamins in your unknown sample.

For the vitamin C analysis, you will perform a titration. Titration is a technique that you are already familiar with, but you will be using it in a new fashion this time, as a "redox" titration, to analyze a sample of vitamin C from a commercial vitamin tablet. After making up a solution of the vitamin C to be tested, you will allow it to react with a compound called dichloroindophenol (DCIP), a compound that is colored in its oxidized form and colorless in its reduced form. The color of the oxidized form depends on the solvent—it is blue in a neutral or basic solution, and red in an acidic solution.

What you actually see in the titration is based on these facts. The DCIP solution is initially blue, which is expected of this compound in a neutral solution. As it drops into the flask containing the (acidic) ascorbic acid solution, an oxidation-reduction reaction takes place, and the DCIP takes on two electrons and two hydrogen atoms (the DCIP is reduced) to form a colorless compound. As the DCIP is reduced, the ascorbic acid is oxidized. So as long as you see the blue solution dropping into the colorless ascorbic acid solution and the contents of the flask remaining colorless, ascorbic acid is still present and you should continue the titration. The end point occurs when there is no more ascorbic acid to react with the DCIP. However, you will see a color change still, not from blue to colorless, but from blue to amber-pink, the color of unreacted DCIP in acidic solutions. The number of milliliters of the DCIP solution can be used to calculate the vitamin C content of the original solution and from there you can obtain the actual vitamin C content of the tablet you were analyzing. As for most color-producing reactions, you will also need to do a titration on the solvent alone, so that if it produces any color at all, you can compensate for this in the calculation that you do.

Your actual raw data for the calculation will consist of the milliliters of DCIP solution needed to produce the pink color in the ascorbic acid solution, and the number of milliliters of the DCIP needed to produce a pink color in the blank solution. You will need to do the titration three times, and take an average of those to report.

The reaction for the redox titration is this.

1 ascorbic acid + 1 DCIP $\longrightarrow$ 1 dehydroascorbic acid + 1 reduced DCIP

A calculation sample is set up on the report sheet for you that takes into account the known concentration of the DCIP solution and the volume of stock vitamin C solution you used in each titration. Your first step will be to subtract the blank value of milliliters of DCIP used from each raw DCIP volume, to obtain a corrected volume of DCIP used in the titration. The second and third conversion factors use the known molarity of the DCIP (in mmol/liter) and the known stoichiometry of the reaction (one DCIP oxidizes one ascorbic acid molecule) to determine how many milligrams of ascorbic acid were in the sample you actually titrated.

$$\frac{mg_{aa}}{sample} = mL\ DCIP_{(corr)} \times \frac{1\ L\ DCIP}{1000mL} \times \frac{0.862\ mmol\ DCIP}{1\ L\ DCIP} \times \frac{1\ mmol_{aa}}{1\ mmol_{DCIP}} \times \frac{176\ mg_{aa}}{1\ mmol_{aa}}$$

There is one further calculation step to take. How many milligrams of vitamin C were present in the initial full tablet? The sample used in each titration is 1/500th of the total in a single tablet of vitamin C, so multiplying the mg/sample by 500 will give you a final answer concerning the actual vitamin C content of the tablet you investigated.

REFERENCES

Stoker, H. S. *General, Organic, and Biological Chemistry;* Cengage Learning; Chapter 21.
Review topics:
- Vitamins and minerals
- Oil and water soluble vitamin structures
- Titration
- Use of indicators

Ponder, E. L.; Fried, B.; Sherma, J. Acta Chromatographica 81 (14) 2004. See also *onlinelibrary.wiley.com/doi/10.1002/jps.2600800417/pdf*

WEB RESOURCES

Analyzing the Vitamin C in Fruit Juices. www.cerlabs.com/experiments/1087540622X.pdf

Vitamin. http://www.britannica.com/ebc/article?tocId=9382099&query=vitamins&ct

Vitamin C. http://www.britannica.com/eb/article?tocId=9075556&query=vitamins&ct

SAFETY NOTES

Some of the reagents contain strong acids or bases, and you may prefer to wear gloves while working with these chemicals. You should always wear gloves when working with bromine in carbon tetrachloride. **NEVER** put an unprotected finger over an open test tube to mix a solution.

Glass disposable pipets are usually regarded as "sharps" and should be disposed of appropriately. Dispose of used reagents according to local ordinances. In general, if you are a small-volume user, the organic solvents can be discarded in an appropriately labeled waste container in the hood, and water solutions or suspensions of the unknown substances can be discarded in the sink, flushed down the drain with copious amounts of water. However, local ordinances govern your procedure.

EXPERIMENTAL PROCEDURE

PART A. TLC OF WATER SOLUBLE VITAMINS

Step 1. In the hood, prepare a chromatography chamber by lining a tall (at least 20 cm tall) beaker or jar with a piece of filter paper. Place in the chamber about 10 mL of the solvent system to be used, and cover the chamber. Allow the vapors inside the chamber to come to equilibrium while you prepare the plates to analyze.

> ● The solvent contains benzene, so be sure to keep the vapors in the hood!

Step 2. Obtain a 5 x 15 cm thin layer chromatographic plate, and using a pencil, mark it about 1 cm from the bottom for 5 analytical channels, as the diagram shows.

Step 3. Spot your plate with standard, known samples of vitamins B_2, B_3, B_6, and C, and with an unknown sample mixture, that may contain one, two, or three of these vitamins. It may be necessary to build up a strong concentration of these compounds at the origin by spotting the plate multiple times in the same place. If you need to do that, it will be important to allow each spotting to dry before adding another layer.

Step 4. When your samples are completely dry, lower the plate into the TLC chamber in the hood, being careful to allow the solvent to contact the bottom of the plate evenly. Cover the chamber and allow the chromatogram to run until the solvent line is at least 10 cm from the origin. This may take some time. Begin to work on Part B while the chromatogram is running.

Step 5. When the chromatogram is complete, remove it from the chamber using forceps, and mark the solvent front. Allow the chromatogram to dry in the hood before developing the color of the spots.

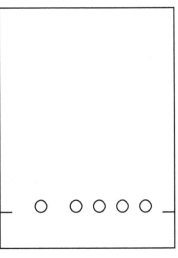

Step 6. To visualize the samples, first view under a UV light. Some of the samples should glow against a darker background, and others will show as darker spots on a lighter background. Mark any spots you see by circling them with pencil. Then place the plate in an iodine chamber for a few minutes to be sure you have seen all the compounds present.

Step 7. Measure the distance the solvent moved, and the distance to the center of each observed spot. Calculate the R_f for each of the standard samples of vitamins, and for each spot observed in the unknown sample. Record your data and determine the identity of the vitamins in the unidentified sample.

PART B. TITRATION: QUANTITATIVE DETERMINATION OF ASCORBIC ACID

Steps 1-2 may be done in advance by your instructor, or groups of several students may combine to make their own stock solution of ascorbic acid. Each student performing the titration analysis will need about 15 mL of the stock solution.

Step 1. Weigh one 500 mg tablet of vitamin C, record the mass, and transfer it to a mortar. Using the pestle provided, grind the tablet to a fine powder. The powder will contain vitamin C, but also various binders and starches used in manufacturing, Transfer the powder as completely as possible to a 100 mL beaker and add 50 mL of water. Stir the solution for 1 minute, and then allow it to stand for 10 minutes, so than any undissolved materials settle to the bottom of the beaker.

Step 2. In the hood, set up a 100 mL volumetric flask with a funnel that contains a loosely packed plug of glass wool. Transfer the vitamin C solution as completely as possible to the 100 mL volumetric flask, using several rinses of the beaker to complete the transfer. You do not need to transfer the undissolved powder. Dilute the solution to the mark on the flask with distilled water. Mix the vitamin C solution well, by capping the flask and inverting it several times to insure that the solution is uniform throughout. This stock solution of ascorbic acid should contain about 5 mg of ascorbic acid per mL.

Step 3. Using a pipet and a suction bulb, transfer 0.2 mL of the stock solution to an Erlenmeyer flask, and add to it 20 mL of distilled water. Mix well.

Step 4. Set up a clean buret, and rinse it with a few mL of the indicator solution (DCIP), then fill it with DCIP solution and remove any air bubbles still remaining in the tip. Take an initial reading for the liquid level in the buret.

Step 5. Titrate the ascorbic acid solution by adding the DCIP solution to the Erlenmeyer flask about 1 mL at a time, and swirling the resulting mixture. Continue until an addition of DCIP produces a faint amber to pink color, indicating that all the ascorbic acid has reacted with DCIP. Record the final buret reading. Be patient after each addition; as you get closer to the end point, the reaction may take longer to occur.

Step 6. Repeat the titration twice more, using a cleaned Erlenmeyer flask and a new 0.2 mL sample of ascorbic acid for each trial. It may be necessary to refill your buret before beginning one of these—don't try to read the buret past the calibrated area.

Step 7. Now determine a blank in this manner. Place 20 mL of water (no ascorbic acid at this point!) in a clean Erlenmeyer flask, and titrate it with DCIP solution until a faint end point is reached. The volume of DCIP that is needed to reach this point will have to be subtracted from each raw experimental volume before the calculation is finished.

Step 8. Perform the calculations for each of the titrations, following the steps shown in the sample calculations. Your result at the end should give you the milligrams of ascorbic acid in 0.002 tablet. Multiplying this mass by 500 should tell you the amount of ascorbic acid in one tablet.

EXPERIMENT 43 PRELAB EXERCISES: WATER SOLUBLE VITAMINS

Name_____ **Partner:** _____ **Date** _____

1. Find and draw the structures of the water-soluble vitamins used in this exercise.

Vitamin C   Vitamin B_2

Vitamin B_3 Vitamin B_6

2. Below is a thin-layer chromatogram (TLC) for a sample that contained ascorbic acid and vitamin B_6. Calculate the R_f for each substance in the chromatogram. Which one is most likely to be the ascorbic acid (the most polar)?

Solvent front

○

○

origin

3. Why do water-soluble vitamins need replenishing in the body?

EXPERIMENT 43 REPORT: WATER SOLUBLE VITAMINS

Name_____   **Partner:** _____   **Date** _____

PART A. TLC OF WATER SOLUBLE VITAMINS

Unknown sample number _____

Draw your TLC plate here, noting all spots observed and the solvent front. Note the measurement of each distance traveled.	Record here the R_f of each pure vitamin, and of each component of your unknown sample.
	B_2 B_3 B_6 C Unknown (s)

Identity of the material(s) in your unknown sample number _____

Did your plate give clear results? If not, what might you do to improve the result?

Which of the four vitamins analyzed appears to be the **most** polar?

PART B. TITRATION: QUANTITATIVE DETERMINATION OF ASCORBIC ACID

Mass of vitamin C tablet   _____

Titration of ascorbic acid with DCIP

	Trial 1	Trial 2	Trial 3	Blank
Initial buret reading (mL)				
Final buret reading (mL)				
Volume of DCIP used (mL)				
Corrected volume, DCIP$_{(corr)}$ (mL)				

CALCULATION OF THE ASCORBIC ACID IN THE TABLET

Abbreviations: aa = ascorbic acid; DCIP$_{(corr)}$ = (titration volume – blank)

$$\frac{mg_{aa}}{0.2 \text{ mL sample}} = mL\ DCIP_{(corr)} \times \frac{1 \text{ L DCIP}}{1000 mL\ DCIP} \times \frac{0.862 \text{ mmol DCIP}}{1 \text{ L DCIP}} \times \frac{1 \text{ mmol}_{aa}}{1 \text{ mmol}_{DCIP}} \times \frac{176 \text{ mg}_{aa}}{1 \text{ mmol}_{aa}}$$

Record this value in the table below.

mg ascorbic acid in the tablet = 500 x mg of ascorbic acid per sample titrated.

Record your results in the table below.

Calculation of the ascorbic acid content of a Vitamin C tablet

	Trial 1	Trial 2	Trial 3
Ascorbic acid in the 0.20 mL sample (mg)			
Ascorbic acid in the tablet (in mg)			

Average of the three trials: (mg of ascorbic acid per tablet)   _____

Do your results agree with the stated value on the label of the bottle? If not, what might be some reasons that account for the discrepancy?

Other Kaplan Books on Graduate Admissions

Graduate School Admissions Adviser
GRE/GMAT Math Workbook
The Yale Daily News Guide to Fellowships and Grants
Guide to Distance Learning

GRE*

1999–2000

By the Staff of Kaplan Educational Centers

Simon & Schuster

*GRE is a registered trademark of the Educational Testing Service, which is not affiliated with this product.

Kaplan Books
Published by Kaplan Educational Centers and Simon & Schuster
1230 Avenue of the Americas
New York, NY 10020

The material in this book is up-to-date at the time of publication. Educational Testing Service may have instituted changes after this book was published. Please read all material you receive regarding the GRE test carefully.

Project Editor: Richard Christiano
Cover Design: Cheung Tai
Interior Page Design: Krista Pfeiffer
Production Editor: Maude Spekes
Managing Editor: David Chipps
Executive Editor: Del Franz

Special thanks to Ben Paris, David Stuart, Judi Knott, and Linda Volpano.

Manufactured in the United States of America
Published Simultaneously in Canada

March 1999

10 9 8 7 6 5 4 3 2 1

ISSN: 1090-9117
ISBN: 0-684-85667-0

CONTENTS

This book was designed for self-study only and is not authorized for classroom use. For information on Kaplan courses, which expand on the techniques offered here and are taught only by highly trained Kaplan instructors, please call 1-800-KAP-TEST.

PREFACE

You've probably heard the good news: According to recent surveys, Americans with a graduate degree earn, on average, 35 to 50 percent more than do those with just a bachelor's degree. No, that's not a misprint: 35 to 50 percent more.

Maybe that's one reason there are more people than ever in the United States taking the GRE and applying to graduate schools. In fact, the number of GRE takers has virtually *doubled* in the last decade. No, that's not a misprint either. Only about 293,000 took the test back in 1987, but about half a million took it in 1998.

What do these remarkable statistics mean for you, the prospective grad-school applicant? Well, they mean that while the rewards of advanced study are lucrative, the competition for getting into a good graduate school is as keen as it's ever been. There are a lot of people out there thinking about going to graduate school. Meanwhile, the variety of graduate programs offered by graduate institutions is also growing. And the degree to which they're all keeping pace with the dizzying changes in technology varies widely as well.

That's why, now more than ever, you've got to find a graduate school with a program that's exactly right for you. And not only do you have to find that school, you've got to get into it—at a time when there are more people than ever pounding on the grad school door.

And that's where Kaplan comes in. We at Kaplan have had decades of experience getting people into great grad schools, and one thing we've learned is that *you must have a comprehensive strategy*. You can't approach graduate school admission in a casual, piecemeal way. If you want to maximize your likelihood of success, you have to take advantage of *every* opportunity at your disposal to strengthen your application.

That's the philosophy behind this Kaplan guide. In our GRE section, we'll tackle that critical element in your application (and in your efforts for financial aid)—your GRE score. We'll give you a quick course in the legendary Kaplan GRE strategies and techniques, and give you tips on how to relax and stay in top form as the day of the test approaches. Then, we'll give you the Kaplan Practice Test to prepare for the real thing, complete with full strategic explanations for every question.

Getting the highest possible GRE score is only part of the battle; all other elements of your application have to be maximized as well, so that they add up to a coherent statement of purpose. In our admissions section, Kaplan's expert in grad-school admissions will take you step-by-step through the application process. You'll find out how to orchestrate everything from recommen-

dations to personal statements, so that your application becomes a powerful marketing tool for yourself. Remember, while every element of your application is an opportunity to present yourself in a favorable light, it's also an opportunity to screw up. We'll show you how to avoid the common pitfalls and make your application stand out from the crowd.

So don't worry about all those other people who are also applying to graduate school. By following the advice in this book, you'll find the perfect graduate school for you.

HOW TO USE THIS BOOK

GRE 1999-2000 is more than just your average test-prep guide. True, because your GRE score is the most important factor that you can still do something about (it's too late to change that D in Geology 101, unfortunately), the bulk of this book is devoted to test prep. But grad schools base their admissions decisions on far more than just the GRE. In fact, there are a host of other parts of your application that can make or break your candidacy. So, to give you the very best odds, we've enlisted the help of an admissions expert to lead you through the application process beyond the GRE.

Step One: Read the GRE Section

Kaplan's live GRE course has been famous for decades. In this section, we've distilled the main techniques and approaches from our course in a clear, easy-to-grasp format. The most important points are summarized in sidebars in the outer margins. We'll introduce you to the mysteries of the GRE and show you how to take control of the test-taking experience on all levels.

Level One: Test Content
Here's where you'll learn specific methods and strategies for every kind of question you're likely to see on the test.

Level Two: Test Mechanics
In addition to the item-specific techniques, you also have to learn how to pace yourself over the entire section, choosing which questions to answer and which to guess on. You should also know how the peculiarities of a standardized test can sometimes be used to your advantage. This is where you'll learn these skills.

Level Three: Test Mentality
Finally, you'll learn how to execute all of what you've learned with the proper test mentality, so you know exactly what you should be doing at every moment on the day of the test.

Step Two: Take Kaplan's GRE Practice Test

After studying the Kaplan methods, you should then take the Practice Test—a timed, simulated GRE—as a test run for the real thing.

Step Three: Use the Printed Strategic Explanations

Explanations for every question on the test will enable you to understand your mistakes, so that you don't make them again on the day of the test. Try not to confine yourself to the explanations for the questions you get wrong. Instead, read all of the explanations—to reinforce good habits and to sharpen your skills so that you can get the right answer even faster and more reliably next time.

Step Four: Review to Shore up Weak Points

Go back to the GRE section and review the topics in which your performance was weak. Read the Tips for the Final Week and the Kaplan Advantage™ Stress Management System to make sure you're in top shape on the day of the test.

Follow these four steps, and you can be confident that your application to grad school will be as strong as it can be.

A Special Note for International Students

About a quarter million international students pursued advanced academic degrees at the master's or Ph.D. level at U.S. universities each year. This trend of pursuing higher education in the United States, particularly at the graduate level, is expected to continue. Business, management, engineering, and the physical and life sciences are particularly popular majors for students coming to the United States from other countries. Along with these academic options, international students are also taking advantage of opportunities for research grants, teaching assistantships, and practical training or work experience in U.S. graduate departments.

If you are not from the United States, but are considering attending a graduate program at a U.S. university, here's what you'll need to get started.

- If English is not your first language, start there. You'll probably need to take the Test of English as a Foreign Language (TOEFL) or show some other evidence that you are proficient in English. Graduate programs will vary on what is an acceptable TOEFL score. For degrees in business, journalism, management or the humanities, a minimum TOEFL score of 600 or better is expected. For the hard sciences and computer technology, a TOEFL score between 500 and 550 may be acceptable.

- You may also need to take the Graduate Record Exam (GRE).

- Since admission to many graduate programs is quite competitive, you may also want to select three or four programs and complete applications for each school..

- Selecting the correct graduate school is very different from selecting a suitable undergraduate institution. You should especially look at the qualifications and interests of the faculty teaching and/or doing research in your chosen field. Look for professors who share your specialty.

- You need to begin the application process at least a year in advance. Be aware that many programs will have September start dates only. Find out application deadlines and plan accordingly.

- Finally, you will need to obtain an I-20 Certificate of Eligibility in order to obtain an F-1 Student Visa to study in the United States.

For details about the admissions requirements, curriculum, and other vital information on top graduate schools in a variety of popular fields, see Kaplan's guide to United States graduate programs, *Graduate School Admissions Adviser*

Access America

If you need more help with the complex process of graduate school admissions and information about the variety of programs available, you may be interested in Kaplan's Access America™ program.

Kaplan created Access America to assist students and professionals from outside the United States who want to enter the U.S. university system. The program was designed for students who have received the bulk of their primary and secondary education outside the United States in a language other than English. Access America also has programs for obtaining professional certification in the United States. Here's a brief description of some of the help available through Access America.

The TOEFL Plus Program

At the heart of the Access America program is the intensive TOEFL Plus Academic English program. This comprehensive English course prepares students to achieve a high level of proficiency in English in order to successfully complete an academic degree. The TOEFL Plus course combines personalized instruction with guided self-study to help students gain this proficiency in a short time. Certificates of Achievement in English are awarded to certify each student's level of proficiency.

Graduate School/GRE Preparation

If your goal is to enter a master's or Ph.D. program in the United States, Kaplan will help you prepare for the GRE, while helping you understand how to choose a graduate degree program in your field.

Preparation for Other Entrance Exams

If you are interested in attending business school, medical school, or law school in the United States, you will probably have to take a standardized entrance exam. Admission to these programs is very competitive, and exam scores are an important criteria.

Graduate Management Admissions Test (GMAT) Preparation. If you are interested in attending business school, you will probably need to take the GMAT. Kaplan can help you prepare for the GMAT, while helping you understand how to choose a graduate management program that's right for you.

Law School Admissions Test (LSAT) Preparation. If you plan to enter a law school in the United States, Kaplan will help you determine whether you need to take the LSAT while helping you to choose an appropriate law program.

Medical College Admissions Test (MCAT) Preparation. If you plan to enter a medical school in the United States, Kaplan can help you prepare for the MCAT. Kaplan also offers professional counseling and advice to help you gain a greater understanding of the American education system. We can help you with every step in the admissions process, from choosing the right medical school, to writing your application, to preparing for the interview.

United States Medical Licensing Exam (USMLE) and Other Medical Licensing. If you are a medical graduate who would like to be FCMFMG certified and obtain a residency in a U.S. hospital, Kaplan can help you prepare for all three steps of the USMLE.

If you are a nurse who wishes to practice in the United States, Kaplan can help you prepare for the Nursing Certification and Licensing Exam (NCLEX) or Commission on Graduates of Foreign Nursing Schools (CGFNS) exam. Kaplan will also prepare you with the English and cross-cultural knowledge that will help you become an effective nurse.

Business Accounting/CPA (Certified Public Accounting). If you are an accountant who would like to be certified to do business in the United States, Kaplan can help you prepare for the CPA exam and assist you in understanding the differences in accounting procedures in the United States.

Applying to Access America

To get more information, or to apply for admission to any of Kaplan's programs for international students or professionals, you can write to us at:

Kaplan Educational Centers, International Admissions Department
888 Seventh Avenue, New York, NY 10106

Or call us at 1-800-522-7700 from within the United States, or at 01-212-262-4980 outside the United States. Our fax number is 01-212-957-1654. Our E-mail address is world@kaplan.com. You can also get more information or even apply through the Internet at http://www.kaplan.com/intl.

GETTING A HIGHER SCORE ON THE GRE

CHAPTER 1

AN INTRODUCTION TO THE GRE

This test preparation section will explain more than just a few basic strategies. It will cover practically everything that's ever on the GRE. No kidding.

We can do this because we don't explain questions in isolation or focus on particular problems. Instead, we explain the underlying principles behind all of the questions on the GRE. What a particular question is *really* testing. We give you the big picture.

One of the keys to getting the big picture is knowing how the test is constructed. Why should you care how the GRE is constructed? Because if you understand the difficulties that the people at ETS have when they make this test, you'll understand what it is you have to do to overcome it. As someone famous once said, "Know thine enemy." And you need to know firsthand the way this test is put together if you want to take it apart.

Before you begin, though, remember that the test makers sometimes change the content, administration, and scheduling of the GRE too quickly for an annual guide to keep up with. For the latest, up-to-the-minute news about the GRE, contact Kaplan's web site at www.kaplan.com.

BE PREPARED

You can't cram for the GRE, but you can prepare for it by learning to think the GRE way.

GET THE EDGE

About half a million people take the GRE each year. By reading the following chapters, you'll learn the underlying principles of GRE questions and acquire test strategies that will help increase your score.

GET TO KNOW THE TEST

By learning how the GRE is created, you can better understand how to beat it.

The Secret Code

There is a sort of unwritten formula at the heart of the GRE. First, there's psychometrics, a peculiar kind of science used to write standardized tests. Also, ETS bases its questions on a certain body of knowledge, which doesn't change. ETS tests the same concepts in every GRE. The useful thinking skills and shortcuts that succeed on one exam—the exam that you're signing up to take, for instance—have already succeeded and will continue to succeed, time and time again.

The Game

If you're like the authors of this book, you weren't too crazy about the idea of taking the SAT back in high school. It seemed unfair that our entire future—where we went to college and where that took us after—would be based on our performance on an unfeeling exam one dreary Saturday morning. Some of us weren't too crazy about our GPAs, which we could no longer do much about.

There are a great many people who think of these exams as cruel exercises in futility, as the oppressive instruments of a faceless societal machine. People who think this way usually don't do very well on these tests.

The key discovery that people who ace standardized tests have made, though, is that fighting the machine doesn't hurt it. If that's what you choose to do, you will just waste your energy. So, instead, they choose to think of the test as a game. Not an instrument of punishment, but an opportunity for reward. And like any game, if you play it enough times, you get really good at it.

GAME THEORY

Think of the GRE as a game—one that you can improve at the more times you play.

Play the Game

You may think that the GRE isn't fair or decent, but that attitude won't help you get into graduate school.

None of the GRE experts who work at Kaplan were *born* acing the GRE. No one is. That's because these tests do not measure innate skills; they measure *acquired* skills. People who are good at standardized tests are, quite simply, people who've already acquired the necessary skills. Maybe they acquired them in math class, or by reading a lot, or by studying logic in college, or perhaps the easiest way—in one of Kaplan's GRE

KAPLAN

classes. But they have, perhaps without realizing it, acquired the skills that bring success on tests like the GRE. And if *you* haven't, you have nothing whatsoever to feel bad about. You simply must acquire them now.

Same Problems—but Different

We know it sounds incredible, but it's true: the test makers use the same problems on every GRE. Only the words and the numbers change. They test the same principles over and over.

Here's an example: This is a type of math problem known as a Quantitative Comparison. Look familiar? These are also on the SAT. Your job is to pick (A) if the term in Column A is bigger, (B) if the term in Column B is bigger, (C) if they're equal, or (D) if there is not enough information given to solve the problem.

Column A	Column B
$2x^2 = 32$	
x	4

Most people answer (C), that they're equal. They divide both sides of the equation by 2 and then take the square root of both sides.

Wrong. The answer isn't (C), because x doesn't have to be 4. It could be 4 *or* –4. Both work. If you just solve for 4 you'll get this problem— and every one like it—wrong. ETS figures that if you get burned here, you'll get burned again next time. Only next time it won't be $2x^2 = 32$; it will be $y^2 = 36$ or $s^4 = 81$.

The concepts that are tested on any particular GRE—Pythagorean triangles, simple logic, word relationships, and so forth—are the underlying concepts at the heart of *every* GRE.

Basically, every GRE is the same as every other one administered that year. In fact, the GREs being given today are extremely similar to those given a decade ago. For instance, most of the math problems you are going to get on the test that you've signed up for are just superficially different from the math problems that have been on every other GRE. To guarantee scores that are almost perfectly consistent, ETS writes tests that are almost perfectly consistent.

WHAT DO STANDARDIZED TESTS MEASURE, ANYWAY?

Standardized tests measure acquired skills. The people who succeed on them are those who have acquired the skills that the test measures.

OLD FAITHFUL

The GRE tests the same principles over and over. Every GRE is virtually the same as every other one because the tests must be consistent from year to year to yield dependable results.

Why Don't They Just Start Testing Something New?

If ETS started testing different principles, it would have to compromise score consistency. Even when it makes very minor changes in test structure or content, it does so between school years, introducing the revisions in the October administration, so that everyone who takes the exam that school year has the same kind of test. That's important, because it means that Kaplan knows what every GRE is going to look like, before it's administered.

ETS makes these minor changes only after testing them exhaustively. This process is called *norming*, which means taking a normal test and a changed test and administering them to a random group of students. As long as the group is large enough for the purposes of statistical validity and the students get consistent scores from one test to the next, then the revised test is just as valid and consistent as any other GRE.

That may sound technical, but norming is actually quite an easy process. We do it at Kaplan all the time—for the tests that we write for our students. The test at the back of this book, for instance, is a normed exam.

They Like Their Test

Another major reason they don't rework most or all of the GRE is that they think it's really a pretty good test. (We know what you're thinking.) To be more specific, they feel that the GRE tests what it's designed to test: various fundamental concepts of algebra, geometry, verbal ability, reasoning ability, and so on. So what if people learn all those principles and get better at the test? That doesn't mean the test is rotten. Quite the opposite: To improve your score, you have to learn a lot of important things.

Let Them Think That

If ETS and the Graduate School Admissions Council want you to learn a bunch of simple concepts and improve your vocabulary, why fight it?

We don't think any of Kaplan's students, after they took the GRE, ever said to themselves, "Now that it's all over, I just wish I hadn't learned all that vocabulary!" Let's face it, none of us would mind being able to read dense, confusing material better and faster. Or being more logical and analytical thinkers. Or improving our vocabulary.

RECOGNIZE WHAT YOU CAN'T CHANGE

Your opinion of the test doesn't matter. Your score on it does.

All those things are good things, and they translate into success on the day of the test. The first step is to take a close look at the setup of this test that you'll be taking.

The Sections

The GRE computer-adaptive test (CAT) consists of three scored sections, with different amounts of time allotted for you to complete each section:

- Verbal: 30 questions, 30 minutes
- Quantitative: 28 questions, 45 minutes
- Analytical: 35 questions, 60 minutes

You'll get a minute break after each section, and an optional 10-minute break in the middle of the test. There are also up to two nonscored sections: an Experimental section and a "Research" section.

The Experimental section is unscored. That means that if you could identify the Experimental section, you could doodle for half an hour, guess in a random pattern, or daydream and still get exactly the same score on the GRE. However, the Experimental section is disguised to look like a real section—there is no way to identify it. All you will really know on the day of the test is that one of the subject areas will have two sections instead of one.

Naturally, many people try to figure out which section is Experimental. But because ETS really wants you to try hard on it, they do their best to keep you guessing. If you guess wrong you could blow the whole test, so we urge you to treat all sections as scored unless you are told otherwise. (Besides, a nap in the middle of the test is pretty unlikely to help you one way or the other.)

The Research section on the CAT is also unscored, and is not always included in the GRE. If you see a Research section on Test Day, ETS will be kind enough to tell you when it appears. So there is no reason whatsoever for you to complete it, unless you feel like doing ETS a favor, or unless they offer you some reward (which they have been known to do).

Scoring

Each of the three sections described above yields a scaled score within a range of 200 to 800. You cannot score higher than 800 on any one section, no matter how hard you try! Similarly, it's impossible (again, no matter how

RUSSIAN ROULETTE

Don't try to figure out which section of your test is experimental. Even if you guess right, it can hurt your score. If you guess wrong . . .

THE NUMBERS GAME

You can't score higher than 800 or lower than 200 on any of the three sections (Verbal, Quantitative, and Analytical).

WHAT'S A PERCENTILE?

The percentile figure tells you how many other test takers scored at or below your level. In other words, a percentile figure of 80 means than 80 percent did as well or worse than you did and that only 20 percent did better.

MEASURE FOR MEASURE

Your percentile rank is the most important result from your GRE. It tells graduate schools how you stack up against other test takers.

hard you try) to have a score lower than 200 on any of the three sections. Scaled scores are much like the old scores that you received if you took the SAT, the major difference being the addition of a score for the Analytical measure, which isn't tested on the SAT.

But you don't receive *only* scaled scores. You will also receive a percentile rank, which will place your performance relative to those of a large sample population of other GRE takers. Percentile scores tell graduate schools just what your scaled scores are worth. For instance, even if everyone got very high scaled scores, universities would still be able to differentiate candidates by their percentile score.

Percentile ranks match with scaled scores differently, depending on the measure. Let's imagine that our founder, Stanley H. Kaplan, were to take the GRE this year. He would (no doubt) get a perfect 800 on each measure type, but that would translate into different percentile ranks. In Verbal, he'd be scoring above 99 percent of the population, so that would be his percentile rank. But in the Quantitative and Analytical sections, many other people will score very high as well. Difficult as these sections may seem, so many people score so well on them that high scaled scores are no big deal. Mr. Kaplan's percentile rank for Quantitative, even if he doesn't miss a single question, would be only in the 96th percentile. So many other people are scoring that high in Quantitative that no one can score above the 96th percentile! Similarly, his Analytical percentile would be 98th.

What this means is that it's pretty easy to get good scaled scores on the GRE and much harder to get good percentile ranks. A Quantitative score of 600, for example, is actually not all that good; if you are applying to science or engineering programs, it would be a handicap at most schools. Even a score of 700 in Quantitative is relatively low for many very selective programs in the sciences or engineering—after all, it's only the 79th or 80th percentile.

The relative frequency of high scaled scores means that universities pay great attention to percentile rank. What you need to realize is that scores that seemed good to you when you took the SAT might not be all that good on the GRE. It's important that you do some real research into the programs you're thinking about. Many schools have cut-off scores below which they don't even consider applicants. But be careful! If a school tells you they look for applicants scoring 600 average per section, that doesn't mean they think those are good scores. That 600 may be the baseline. You owe it to yourself to find out what kinds of scores *impress* the schools you're interested in and work hard until you get those scores. You can definitely get there if you want to and if you work hard enough. We see it every day.

A final note about percentile rank: the sample population that you are compared against in order to determine your percentile is not everyone else who takes the test the same day as you do. ETS doesn't want to penalize an unlucky candidate who takes the GRE on a date when everyone else happens to be a rocket scientist. So they compare your performance with those of a random three-year population of recent GRE test takers. Your score will not in any way be affected by the other people who take the exam on the same day as you. We often tell our students, "Your only competition in this classroom is yourself."

Canceling

When you finish the GRE you will be given the opportunity to cancel your scores, but the only time you can cancel is immediately after the test. That's the only chance you'll have, because if you don't cancel your scores they will be recorded immediately as they are given to you. If you do cancel, you won't be able to undo it.

Pointing and Clicking

The computer on which you will take the GRE has a keyboard and a mouse. You won't use the keyboard to answer questions; instead, you will use the mouse to point at your answer choices and "click" to select them. Also, the test makes use of computer functions, such as HELP and QUIT, which may take some getting used to. For these reasons, make sure you go through the information in the Test Mechanics chapter carefully.

The Kaplan Three-Level Master Plan

To give your best performance on the GRE, you'll need to have the right kind of approach for the entire test as a whole. We've developed a plan to help you, which we call (cleverly enough) "The Kaplan Three-Level Master Plan for the GRE." You should use this plan as your guide to preparing for and taking the GRE. The three levels of the plan are: test content, test mechanics, and test mentality.

MAKING THE CUT

Research the graduate schools that you're interested in to find out what level of scores they're looking for. You'll have to aim higher than their minimum scores to impress them.

Level 1: Test Content

In the first part of the test prep section, we'll talk about how to deal with individual short verbal questions, reading passages, math problems, logic games, and logical reasoning questions. For success on the GRE, you'll need to understand how to work through each of these question types. What's the difference between antonym and analogy questions? What are the best ways of handling each? What's a sentence completion and how do I approach it? How should I read a reading comprehension passage and what should I focus on? What's the best way to approach the Math section? Is there a secret to logic games? How do I solve logical reasoning questions? Our instruction in Level 1 will provide you with all of the information, strategies, and techniques you'll need to answer these questions and more.

Level 2: Test Mechanics

Next, we'll move up the ladder from individual question types to a discussion of how to complete each section within the specified time limit. We'll reveal the test mechanics that will help you to use the strategies you learned in Level 1 to maximum effect.

Level 3: Test Mentality

On this final level, we'll help you pull everything you've learned together. By combining the question strategies and test mechanics, you'll be in control of the entire test experience. With a good test mentality, you can have everything at your fingertips—from building good bridges to gridding techniques, from sequencing game strategies to pacing methods. We'll also outline all of the subtle attitudinal factors that will help you perform your absolute best on the day of the test.

Understanding the three levels, and how they interrelate, is the first step in taking control of the GRE. We'll start, in the next chapter, with the first level, test content.

YOU HAVE TO HAVE A PLAN

The three levels of the Kaplan Master Plan are:

1. Test content
2. Test mechanics
3. Test mentality

KAPLAN

CHAPTER 2

TEST CONTENT: VERBAL

In this chapter and the two chapters that follow, we'll give you the nuts and bolts of GRE preparation—the strategies and techniques for each of the individual question types on the test. For each of the multiple choice sections—Verbal, Quantitative, and Analytical—we'll present you with the following:

- **Directions and General Information**
  The specific directions for each section will introduce you to the question types. We'll also give you some ground rules for each question type.

- **Basic Principles**
  These are the general rules-of-thumb that you need to follow to succeed on this section.

- **Common Question Types**
  Certain types of questions appear repeatedly on each section. We'll show you what these question types are and how best to deal with each one.

- **The Kaplan Method**
  This is a step-by-step way of organizing your work on every question in the section. The Kaplan Method will allow you to orchestrate all of the individual strategies and techniques into a flexible, powerful modus operandi.

VERBAL QUESTION TYPES AT A GLANCE

There are four types of verbal questions:

• Sentence completions
• Analogies
• Antonyms
• Reading comprehension

Now let's begin with an important part of the GRE, the Verbal Section. You'll have 30 minutes to complete 30 questions, which are broken down into four types: sentence completion, analogies, reading comprehension, and antonyms. The chart below shows roughly how many questions correspond to each question type and how much time you should spend on each question type.

	SENTENCE COMPLETION	ANALOGIES	READING COMPREHENSION	ANTONYMS
Number of Questions	about 6	about 7	about 8	about 9
Time per Question	20–45 seconds	30–45 seconds	> 1 minute	30 seconds

The computer determines the sequence and difficulty of the questions, so you will not be able to tell what sort of question you will get next, or how hard it will be.

There are two basic things that the Verbal section tests: your vocabulary and your ability to read a particular kind of passage quickly and efficiently. You may have wondered how the material we covered earlier about test construction is going to help you in the GRE Verbal sections. Well, just like the math questions, which are the same from test to test (just with different numbers), the verbal questions are the same (just with different words). Have you ever heard the expression, "That's an SAT word"? It's a commonly used phrase among high school students, and it refers to any member of a very particular class of prefixed and suffixed words derived from Latin or Greek. For instance, *profligate* is a great SAT word. It's also a great GRE word.

WHAT DOES THE VERBAL SECTION TEST?

The Verbal section tests your vocabulary and your ability to read passages quickly and efficiently.

Vocabulary—the Most Basic Principle for Verbal Success

Many of the same kinds of words that would commonly show up on the SAT are likely candidates for the GRE as well, though GRE words tend to be harder.

The GRE tests the same kinds of words over and over again. (Remember, for ETS, consistency is key.) We'll call these words "GRE words," and we're going to make a point of including them in the rest of

this chapter. That way, you can get a feel for what they look and sound like, and you can see them used in context. So if you see a word in this book that's unfamiliar, take a moment to look it up in the dictionary and reread the sentence with the word's definition in mind. Learning words in context is one of the best ways for the brain to retain their meanings.

The GRE words used in context in this vocabulary section will appear in **boldface.** Look them up while you read. We'll give you an example of what we mean by "the same kinds of words over and over again." The words in the list below all mean nearly the same thing. They all have something to do with the concept of criticism, a concept often tested on the GRE. The GRE that you take could well test you on one of these words or one of the other synonyms for *criticize*. A great way to prepare for GRE Vocabulary, then, is to learn which word concepts are tested most frequently and learn all those words.

CRITICIZE/CRITICISM

calumny
castigate
chastise
deride/derisive
derogate
diatribe
harangue
lambaste
oppugn
pillory
rebuke
remonstrate

On the test, for instance, you might see an antonym question like this:

REMONSTRATE:
(A) show
(B) atone
(C) vouchsafe
(D) laud
(E) undo

Or an analogy question that looks something like this:

VITUPERATE : DISPARAGE ::
(A) profligate : bilk
(B) equivocate : reduce

IT'S DEJA VU ALL OVER AGAIN

The same kinds of vocabulary words that you saw on the SAT may well appear on your GRE.

CONTEXT IS KEY

Learning words in context is a good way to retain their meanings.

WORD FREQUENCY

The best way to prepare for Vocabulary questions is to learn the word concepts that are tested most frequently on the GRE.

A LITTLE KNOWLEDGE IS NOT A DANGEROUS THING

You don't have to know the exact meaning of a word to get the right answer on a GRE vocabulary question.

(C) parody : excuse
(D) lie : prevaricate
(E) brave : succeed

There are many such families of word synonyms whose members appear frequently on the GRE. We'll run across more as we proceed.

What You Need to Know

The GRE does not test whether you know exactly what a particular word means. If you have only an idea what the word means, you will get just as many points for that question as you will if you know the precise dictionary definition of the word. That's because ETS isn't interested in finding out whether you're a walking dictionary. They want to see if you have a broad and diverse (but, of course, classically based) vocabulary.

20% x 500 > 100% x 50

This means, simply, that it's better to know 20 percent of the definition of 500 words than it is to know the exact definition of 50 words. Or, more generally put, it's better to know a little bit about a lot of words than to know a lot about just a few. In fact, it's a lot better. And it's a lot easier.

Thesaurus > Dictionary

The *criticize* family is not the only family of synonyms whose members appear frequently on the GRE. There are plenty of others. And lists of synonyms are much easier to learn than many words in isolation. So don't learn words with a dictionary; learn them with a thesaurus. Make synonym index cards based on the common families of GRE words (listed at the end of this chapter) and **peruse** those lists periodically. It's like weight-lifting for vocabulary. Pretty soon you will start to see results.

If you think this might be **fallacious,** then check this out. The words in the box below all have something to do with the concept of falsehood. Their precise meanings vary: ***erroneous*** means "incorrect," whereas ***mendacious*** means "lying." But unless you're shooting for a very high verbal score (720 or higher) you don't need to know the exact meanings of these words. You will most likely get the question right if you simply know that these words have something to do with the concept of falsehood.

FALSE

apocryphal	guile
dissemble	mendacious
duplicity	mendacity
equivocate	prevaricate
equivocation	prevarication
erroneous	specious
ersatz	spurious
fallacious	

The way that you should use a list like this is to look it over once or twice a week for 30 seconds every week until the test. If you don't have much time until the exam date, look over your lists more frequently. Then, by the day of the test, you should have a rough idea of what most of the words on your lists mean. If you get an antonym question such as:

HONESTY: (A) displeasure (B) mendacity
(C) disrepute (D) resolution (E) failure

you might not know exactly what **mendacity** means, but you'll know that it's "one of those *false* words," which will be enough to get the question right. Your subconscious mind has done most of the work for you!

It might be **vexatious** to learn word meanings the slow way, but you'll be amazed how easy and **facile** vocabulary building can be when you do it this way. Here are some more word families:

ANNOY	BEGINNER	FOUL
aggravate	acolyte	festering
irk	neophyte	fetid
irritate	novice	fulsome
perturb	proselyte	invidious
vex	tyro	noisome

You may not know exactly what **invidious** means, but if you study the last list, pretty soon you will know that it refers to something foul.

We're now going to give you a lot of common GRE words grouped together by meaning. This isn't high-stress learning. All you have to do is make flash cards from these lists and look over your cards a few times a week from now until the day of the test. You'll find that your subconscious mind does much of the work for you.

Note: The categories in which these words are listed are general and should not be understood as the exact definitions of the words.

DIFFICULT TO UNDERSTAND
abstruse
arcane
enigmatic
esoteric
inscrutable
obscure
opaque
rarefied
recondite
turbid

DEBAUCHED/ DEBAUCHERY
bacchanalian
depraved
dissipated
iniquity
libertine
libidinous
licentious
reprobate
ribald
salacious
sordid
turpitude

CRITICIZE/CRITICISM
aspersion
belittle
berate
calumny
castigate
decry
defamation
deride/derisive
diatribe
disparage
excoriate
gainsay
harangue
impugn
inveigh
lambaste
obloquy
objurgate
opprobrium
pillory
rebuke
remonstrate
reprehend
reprove
revile
vituperate

PRAISE
accolade
aggrandize
encomium
eulogize
extol
laud/laudatory
venerate/veneration

FALSEHOOD
apocryphal
dissemble
duplicity
equivocate
equivocation
erroneous
ersatz
fallacious
guile
mendacious/mendacity
prevaricate
prevarication
specious
spurious

BITING (as in wit or
  temperament)
acerbic
acidulous
acrimonious
asperity
caustic
mordant
mordacious
trenchant

RENDER USELESS/
WEAKEN
enervate
obviate
stultify
undermine
vitiate

HARMFUL
baleful
baneful
deleterious
inimical
injurious
insidious
minatory
perfidious
pernicious

TIMID/TIMIDITY
craven
diffident
pusillanimous
recreant
timorous
trepidation

STUBBORN
froward
implacable
inexorable
intractable
intransigent
obdurate
obstinate
pertinaceous
recalcitrant
refractory
renitent
untoward

BEGINNING/YOUNG
burgeoning
callow
inchoate
incipient
nascent

BUILD YOUR VOCABULARY

Make flash cards from
these lists and look over
your cards a few times a
week from now until the
day of the test.

OVERBLOWN/WORDY
bombastic
circumlocution
garrulous
grandiloquent
loquacious
periphrastic
prolix
turgid

HOSTILE/ONE WHO IS HOSTILE
antithetic
churlish
curmudgeon
irascible
malevolent
misanthropic
truculent
vindictive

CLICHÉD/BORING
banal
fatuous
hackneyed
insipid
mundane
pedestrian
platitude
prosaic
quotidian
trite

ALL IN THE FAMILY

Lists of synonyms are easier to learn than long lists of unrelated words.

The Top 100

While we're at it, here, gathered together for easy reference, are the 100 difficult words that appear most frequently on the GRE. You will notice that many of these words are also in the preceding lists of word families.

1. Equivocal/Equivocate/Equivocation—ambiguous, open to two interpretations
2. Tractable (Intractable)—obedient, yielding
3. Placate (Implacable)—to soothe or pacify
4. Miser—person who is extremely stingy
5. Engender—to produce, cause, bring out
6. Dogma(tic)(tism)(tist)—rigidly fixed in opinion, opinionated
7. Garrulous (Garrulity)—very talkative
8. Homogeneous (Homogenize)—composed of identical parts
9. Laconic—using few words
10. Quiescence (Quiescent)—inactivity, stillness
11. Anomalous—irregular or deviating from the norm

12. Venerate (-ion)—to respect

13. Assuage—to make less severe, ease, relieve

14. Misanthrope (-ic)—person who hates human beings

15. Digress(ive)—to turn aside; to stray from the main point

16. Corroborate (-ion)—to confirm, verify

17. Buttress—to reinforce or support

18. Antipathy—dislike, hostility, extreme opposition or aversion

19. Disabuse—to free from error or misconception

20. Feigned (Unfeigned)—pretended

21. Banal(ity)—trite and overly common

22. Desiccate (-ion)—to dry completely, dehydrate

23. Diatribe—bitter verbal attack

24. Pedant(ic)(ry)—uninspired, boring academic

25. Guile(less)—trickery, deception

26. Eulogy (-ize)—high praise, often in public

27. Fawn(ing)—to flatter excessively, seek the favor of

28. Aberrant/Aberration—different from the usual or normal

29. Heresy/Heretic(al)—an act opposed to established religious orthodoxy

30. Obdurate—stubborn

31. Prevaricate(-ion)—to lie, evade the truth

32. Embellish(ment)—to ornament; make attractive with decoration or details; add details to a statement

33. Pragmatic/Pragmatism—practical; moved by facts rather than abstract ideals

34. Precipitate—to cause to happen; to throw down from a height

35. Proximity—nearness

36. Profundity—depth (usually depth of thought)

37. Adulterate—to corrupt or make impure

38. Sanction—permission, support; law; penalty

39. Ameliorate (-ion)—to make better, improve

40. Anachronism/Anachronistic—something chronologically inappropriate

41. Vindictive—spiteful, vengeful, unforgiving

42. Propitiate—to win over, appease

43. Aver—to declare to be true, affirm

44. Burgeon(-ing)—to sprout or flourish

45. Commensurate—proportional

46. Mitigate/Mitigation—to soften, or make milder

47. Culpability—guilt, responsibility for wrong

THE TOP 100

Pencil a check mark by each word you don't know. Quiz yourself on them, erasing check marks as you learn words.

48. Specious(ness)—having a false appearance of truth; showy
49. Turpitude—inherent baseness, depravity
50. Diffident/Diffidence—shy, lacking confidence
51. Repudiate—to reject as having no authority
52. Discrete—individually distinct; consisting of unconnected elements
53. Obviate—to make unnecessary; to anticipate and prevent
54. Dissemble—to pretend, disguise one's motives
55. Implacable—inflexible, incapable of being pleased
56. Emulate—to copy, imitate
57. Complaisance/Complaisant—disposition to please or comply
58. Enervate—to weaken, sap strength from
59. Latency—the condition of being present but hidden
60. Erudite (ition)—learned, scholarly
61. Espouse—to support or advocate; to marry
62. Florid(ness)—gaudy, extremely ornate; ruddy, flushed
63. Occlude—to shut, block
64. Harangue—a ranting writing or speech; lecture
65. Hieroglyph(ic)—pictorial character
66. Iconoclast—one who attacks traditional beliefs
67. Impervious—impossible to penetrate; incapable of being affected
68. Efficacy/Efficacious—effectiveness, efficiency
69. Inchoate—imperfectly formed or formulated
70. Loquacity/Loquacious—talkative
71. Irascible (-ility)—easily angered
72. Ephemeral—momentary, transient, fleeting
73. Laudable (-tory)—deserving of praise
74. Insipid—bland, lacking flavor; lacking excitement
75. Magnanimity/Magnanimous—generosity
76. Precarious(ly)—uncertain
77. Endemic—belonging to a particular area; inherent
78. Mollify—to calm or make less severe
79. Rarefy/Rarefaction—to make thinner, purer, or more refined
80. Disinterest(ed)(edness)—unbiased; not interested
81. Foster—to nourish, cultivate, promote
82. Perennial—present throughout the years; persistent
83. Malevolent—ill-willed; causing evil or harm to others
84. Defer(ence)—to show respect or politeness in a submissive way
85. Precursor(y)—forerunner, predecessor
86. Lucid—clear and easily understood

87. Probity—honesty, high-mindedness
88. Abscond—to depart secretly
89. Propensity—inclination, tendency
90. Audacious—bold, daring, fearless
91. Wheedle—to influence or entice by flattery
92. Prudent—careful, cautious
93. Mundane—worldly; commonplace
94. Diffuse—widely spread out
95. Aggrandize—to make larger or greater in power
96. Decimate—to reduce drastically; to destroy a large part of
97. Succinct—terse, brief, concise
98. Enigma(tic)—a puzzle, something inexplicable
99. Unfettered—free, unrestrained
100. Ascetic—self-denying, abstinent, austere

Roots

You knew that this dreaded word from grade school was going to come up sooner or later. Because GRE words are so heavily drawn from Latin and Greek, roots can be extremely useful, both in deciphering words with obscure meanings and in guessing intelligently.

Use the Kaplan Root List in the back of this book to pick up the most valuable GRE roots. Target these words in your vocabulary prep. Learn a few new roots a day, familiarizing yourself with the meaning.

Learning Vocabulary

In review, the three best ways (in no particular order) to improve your GRE vocabulary are:

- Learning words in context
- Learning families of words
- Deciphering words by their roots

A broader vocabulary will serve you well on all four Verbal question types on the GRE. Now let's look at the Verbal question type that you should tackle first.

GOTTA DIG YOUR ROOTS

The more roots you know, the better you'll be at deciphering perplexing words on the GRE and at coming up with smart guesses.

Sentence Completions

The directions for this section look like this:

> **Directions:** Each of the following questions begins with a sentence that has either one or two blanks. The blanks indicate that a piece of the sentence is missing. Each sentence is followed by five answer choices that consist of words or phrases. Select the answer choice that completes the sentence best.

The difficulty of the sentence completions you will see on the CAT depends on how many questions you get right.

The Four Basic Principles of Sentence Completion

1. Every Clue Is Right in Front of You

Each sentence contains a few crucial clues that determine the answer. In order for a sentence to be used on the GRE, the answer must already be in the sentence. Clues *in the sentence* limit the possible answers, and finding these clues will guide you to the correct answer.

For example, could the following sentence be on the GRE?

> The student thought the test was quite _____.
> (A) long   (B) unpleasant (C) predictable
> (D) ridiculous (E) indelible

No. Because nothing in the sentence hints at which word to choose, it would be a terrible test question. You would *never* see a question like this on the GRE.

Now let's change the sentence to get a question that *could* be answered:

> Since the student knew the form and content of the questions in advance, the test was quite _____ for her.
> (A) long   (B) unpleasant (C) predictable
> (D) ridiculous (E) indelible

What are the important clues in this question? Well, the word *since* is a great structural clue. It indicates that the missing word follows logically from part of the sentence. Specifically, the missing word must follow from "knew the form and content . . . in advance." That means the test was predictable.

2. Look for What's Directly Implied and Expect Clichés

We're not dealing with poetry here. These sentences aren't excerpted from the works of Toni Morrison or William Faulkner. The correct answer is the one most directly implied by the meanings of the words in the sentence.

3. Don't Imagine Strange Scenarios

Read the sentence literally, not imaginatively. Pay attention to the meaning of the words, not associations or feelings that you have.

4. Look for Structural Roadsigns

Structural roadsigns, such as *since,* are keywords that will point you to the right answer. The missing words in sentence completions will usually have a relationship similar or opposite to other words in the sentence. Keywords, such as *and* or *but,* will tell you which it is.

On the GRE, a semicolon by itself always connects two closely related clauses. If a semicolon is followed by another roadsign, then that roadsign determines the direction. Just like on the highway, there are roadsigns on the GRE that tell you to go ahead and that tell you to take a detour.

"Straight ahead" signs are used to make one part of the sentence support or elaborate another part. They continue the sentence in the same direction. The positive or negative charge of what follows is not changed by these clues. Straight-ahead clues include: *and, similarly, in addition, since, also, thus, because, ; (semicolon)*, and *likewise.*

"Detour" signs change the direction of the sentence. They make one part of the sentence contradict or qualify another part. The positive or negative charge of an answer is changed by these clues. Detour signs include: *but, despite, yet, however, unless, rather, although, while, unfortunately,* and *nonetheless.*

In the following examples, test your knowledge of sentence completion roadsigns by finding the right answers (in the parentheses):

1. The winning argument was _____ *and* persuasive. (cogent, flawed)

2. The winning argument was _____ *but* persuasive. (cogent, flawed)

3. The play's script lacked depth and maturity; *likewise,* the acting was altogether _____. (sublime, amateurish)

FILL IN THE BLANK

When working through a sentence completion question:

- Look for clues in the sentence.
- Focus on what's directly implied.
- Pay attention to the meanings of the words.

4. The populace _____ the introduction of the new taxes, *since* they had voted for them overwhelmingly. (applauded, despised)

5. *Despite* your impressive qualifications, I am _____ to offer you a position with our firm. (unable, willing)

6. Scientists have claimed that the dinosaurs became extinct in a single, dramatic event; *yet* new evidence suggests a _____ decline. (headlong, gradual)

7. The first wave of avant-gardists elicited_____from the general population, *while* the second was completely ignored. (indifference, shock)

By concentrating on the roadsigns, wasn't it easy to find your way through the question and arrive at the right answer? (See the "Answers to the Sentence Completion Questions" sidebar when you turn the page.)

The Kaplan Four-Step Method for Sentence Completions
Now that you have the basics, here's how to combine skills.

1. Read the sentence strategically, using your knowledge of scope and structure to see where the sentence is heading.
2. In your own words, anticipate its answer.
3. Look for answers close in meaning to yours and eliminate tempting wrong answers using the clues.
4. Read your choice back into the sentence to make sure it fits.

Using Kaplan's Four-Step Method
Try the following sentence completion questions using the Kaplan Four-Step Method. These are more difficult, but you should be able to do them. Time yourself: you only have 30–45 seconds to do each question.

1. The yearly financial statement of a large corporation may seem _____ at first, but the persistent reader soon finds its pages of facts and figures easy to decipher.
   (A) bewildering
   (B) surprising
   (C) inviting
   (D) misguided
   (E) uncoordinated

   Ⓐ Ⓑ Ⓒ Ⓓ Ⓔ

2. Usually the press secretary's replies are terse, if not downright
    _____, but this afternoon his responses to our questions were
    remarkably comprehensive, almost _____.
    (A) rude . . . concise
    (B) curt . . . verbose
    (C) long-winded . . . effusive
    (D) enigmatic . . . taciturn
    (E) lucid . . . helpful

    Ⓐ Ⓑ Ⓒ Ⓓ Ⓔ

3. Organic farming is more labor intensive and thus initially more
    _____, but its long-term costs may be less than those of con-
    ventional farming.
    (A) uncommon
    (B) stylish
    (C) restrained
    (D) expensive
    (E) difficult

    Ⓐ Ⓑ Ⓒ Ⓓ Ⓔ

4. Unfortunately, there are some among us who equate tolerance
    with immorality; they feel that the _____ of moral values in a
    permissive society is not only likely, but _____.
    (A) decline . . . possible
    (B) upsurge . . . predictable
    (C) disappearance . . . desirable
    (D) improvement . . . commendable
    (E) deterioration . . . inevitable

    Ⓐ Ⓑ Ⓒ Ⓓ Ⓔ

SENTENCE COMPLETION METHOD

- Focus on where the sentence is heading.
- Anticipate the answer in your own words.
- Look for answers that are similar to yours.
- Plug your choice into the sentence to see if it fits.

Think about how you solved these sentence completion questions. You should use the same method when you encounter sentence completion questions on the GRE.

Now let's move to the question type that you should tackle second in the Verbal section: analogies.

Analogies

The directions in this section look like this:

Directions: Each of the following questions consists of a pair of words or phrases that are separated by a colon and followed by five answer choices. Choose the pair of words or phrases in the answer choices that is most similar to the original pair.

On the GRE, the more questions you get right, the harder the analogies you will see.

The Four Basic Principles of Analogies

1. Every Analogy Question Consists of Two Words, Called the Stem Pair, That Are Separated by a Colon

Below the stem pair are five answer choices. That means analogy questions look like this:

MAP : ATLAS ::
(A) key : lock
(B) street : sign
(C) ingredient : cookbook
(D) word : dictionary
(E) theory : hypothesis

2. There Will Always Be a Direct and Necessary Relationship Between the Words in the Stem Pair

You express this relationship by making a short sentence that we call a *bridge*. A bridge is whatever simple sentence you come up with to relate the two words. Your goals when you build your bridge should be to keep it as short and as clear as possible.

A weak bridge expresses a relationship that isn't necessary or direct. For the sample analogy question above, weak bridges include:

* Some maps are put in atlases.
* A map is usually smaller than an atlas.
* Maps and atlases have to do with geography.
* A map is a page in an atlas.

You know you have a weak bridge if it contains such words as *usually, can, might,* or *sometimes.*

A strong bridge expresses a direct and necessary relationship. For the analogy above, strong bridges include:

- Maps are what an atlas contains.
- Maps are the unit of reference in an atlas.
- An atlas collects and organizes maps.

Strong bridges express a definite relationship and can contain an unequivocal word, such as *always, never,* or *must.* The best bridge is a strong bridge that fits exactly one answer choice.

3. Always Try to Make a Bridge Before Looking at the Answer Choices
ETS uses certain kinds of bridges over and over on the GRE. Of these we have identified five classic bridges. Exposing yourself to them now will give you a feel for the sort of bridge that will get you the right answer. Try to answer these questions as you go through them.

1. The definition bridge (*is always* or *is never*)

   PLATITUDE : TRITE ::
   (A) riddle : unsolvable
   (B) axiom : geometric
   (C) omen : portentous
   (D) syllogism : wise
   (E) circumlocution : concise

   Ⓐ Ⓑ Ⓒ Ⓓ Ⓔ

2. The function/purpose bridge

   AIRPLANE : HANGAR ::
   (A) music : orchestra
   (B) money : vault
   (C) finger : hand
   (D) tree : farm
   (E) insect : ecosystem

   Ⓐ Ⓑ Ⓒ Ⓓ Ⓔ

ANSWERS TO THE SENTENCE COMPLETION QUESTIONS

1. A
2. B
3. D
4. E

WHAT'S A STEM PAIR?

Analogy questions consist of two words—the stem pair—that are separated by a colon. Stem pairs look like this:

PREPARATION : SUCCESS

WHAT'S A BRIDGE?

A bridge is a short sentence that connects the two words in the stem pair. You should always make a bridge before you look at the answer choices.

3. The lack bridge

LUCID : OBSCURITY ::
(A) ambiguous : doubt
(B) provident : planning
(C) furtive : legality
(D) economical : extravagance
(E) secure : violence

(A) (B) (C) (D) (E)

4. The characteristic actions/items bridge

PIROUETTE : DANCER ::
(A) sonnet : poet
(B) music : orchestra
(C) building : architect
(D) parry : fencer
(E) dress : seamstress

(A) (B) (C) (D) (E)

5. The degree (often going to an extreme) bridge

ATTENTIVE : RAPT ::
(A) loyal : unscrupulous
(B) critical : derisive
(C) inventive : innovative
(D) jealous : envious
(E) kind : considerate

(A) (B) (C) (D) (E)

WHAT MAKES A STRONG BRIDGE?

You might think that the words *trumpet* and *jazz* have a strong bridge. Don't be fooled. You can play many things on trumpets other than jazz, such as fanfares and rock music. You can also play jazz on things other than trumpets. *Trumpet* and *instrument* do have a strong bridge. A trumpet is a type of instrument. This is always true—it's a strong, definite relationship.

THE FIVE CLASSIC BRIDGES

1. Definition
2. Function/purpose
3. Lack
4. Characteristic action/items
5. Degree

So there you have them, the five classic bridges. Keep them in mind as you practice for the GRE.

4. Don't Fall for Analogies of Type

Analogies of type are pairs of words that are not related to each other but only to a third word.

For instance, it may seem as though there is a strong relationship in RING : NECKLACE; they're both types of jewelry. But this type of relationship will never be a correct answer choice on the GRE. If you see an answer choice like this—where the two words are not directly related to one another but only to a third word (like *jewelry*)—you can always eliminate it.

Now that you have a grasp of the Basic Principles of Analogies, let's take a look at the Kaplan method for solving analogy questions.

KAPLAN

The Kaplan Four-Step Method for Analogies
1. Find a strong bridge between the stem words.
2. Plug the answer choices into the bridge. Be flexible: Sometimes it's easier to use the second word first.
3. Adjust the bridge as necessary. You want your bridge to be simple and somewhat general, but if more than one answer choice fits into your bridge, it was too general. Make it a little more specific and try those answer choices again.
4. If stuck, eliminate all answer choices with weak bridges.

If two choices have the same bridge—for example, (A) TRUMPET : INSTRUMENT or (B) SCREWDRIVER : TOOL—eliminate them both. Try to work backwards from remaining choices to stem pair and make your best guess.

Using the Kaplan Four-Step Method
Let's try an example to learn how to use the four-step method.

> AIMLESS : DIRECTION ::
> (A) enthusiastic : motivation
> (B) wary : trust
> (C) unhealthy : happiness
> (D) lazy : effort
> (E) silly : adventure

For this question, a good bridge is: "Someone *aimless* lacks *direction.*" Now plug that into the answer choices. Only choice (B) fits. If you were stuck, you should have eliminated choices (A), (C), and (E), because their bridges are weak. Remember: If an answer choice has a weak bridge it cannot be correct, because no stem pair that you'll find on the GRE will ever have a weak bridge. To be correct, an answer choice must have a strong, clear relationship.

If you can't build a good bridge because you don't know the definition of one or both stem words, all is not lost. Even when you can't figure out the bridge for the words in the stem pair, you can guess intelligently by eliminating answer choices. In the following questions, there are no stem words. How are you supposed to do them, you ask? Well, do you remember the scene in *Star Wars* when Obi Wan Kenobi is teaching Luke Skywalker about the Force? He put that helmet on Luke's head so that Luke couldn't see when the little robot tried to zap him. This entire scene was actually just a clever (if subtle) metaphor for what it's like to do an analogy when you don't know what the stem words mean.

ANSWERS TO THE FIVE CLASSIC BRIDGES DRILL

1. C
2. B
3. D
4. D
5. B

Take a look at the following sets of answer choices and eliminate all choices that have a weak bridge. Also, if two choices in the same problem have the same bridge, you can eliminate them both (because if one of them were correct the other would have to be also).

1. _ _ _ _ : _ _ _ _ ::
   (A) terrible : appall
   (B) sinister : doubt
   (C) trivial : defend
   (D) irksome : annoy
   (E) noble : admire

   Ⓐ Ⓑ Ⓒ Ⓓ Ⓔ

2. _ _ _ _ : _ _ _ _ ::
   (A) enlist : draft
   (B) hire : promote
   (C) resign : quit
   (D) pacify : mollify
   (E) endanger : enlighten

   Ⓐ Ⓑ Ⓒ Ⓓ Ⓔ

3. _ _ _ _ : _ _ _ _ ::
   (A) congratulate : success
   (B) amputate : crime
   (C) annotate : consultation
   (D) deface : falsehood
   (E) cogitate : habit

   Ⓐ Ⓑ Ⓒ Ⓓ Ⓔ

4. _ _ _ _ : _ _ _ _ ::
   (A) tepid : hot
   (B) lackluster : catatonic
   (C) unusual : rare
   (D) pedantic : didactic
   (E) unique : popular

   Ⓐ Ⓑ Ⓒ Ⓓ Ⓔ

Now let's turn to the third Verbal question type that you'll be dealing with: antonyms.

STRATEGY FOR ANALOGIES

To solve an analogy:

1. Build a bridge between the stem words.
2. Plug in the answer choices.
3. Build a stronger bridge, if necessary.
4. If all else fails, eliminate answer choices with weak bridges.

TICK, TOCK, TICK, TOCK . . .

Don't waste valuable time reading the directions on test day. Learn them now.

Antonyms

The directions for this section will look like this:

> **Directions:** Each of the following questions begins with a single word in capital letters. Five answer choices follow. Select the answer choice that has the meaning most opposite to the word in capital letters.

On the GRE, the more questions you get right, the harder the antonym questions you'll see.

The Seven Basic Principles of Antonyms

1. Think of a Context in Which You've Heard the Word Before
For example, you might be able to figure out the meaning of the italicized words in the following phrases from their context: "*travesty* of justice," "crimes and *misdemeanors*," "*mitigating* circumstances," and "*abject* poverty."

2. Look at Word Roots, Stems, and Suffixes
Even if you don't know what *benediction* means, its prefix (*bene*, which means good) tells you that its opposite is likely to be something bad. Perhaps the answer will begin with *mal*, as in *malefaction*.

3. Use Your Knowledge of a Romance Language
For example, you might guess at the meaning of *credulous* from the Italian, *credere; moratorium* from the French, *morte;* and *mundane* from the Spanish, *mundo.*

4. Use the Positive or Negative Charges of the Words to Help You
Mark up your test booklet with little + signs for words with positive connotations, − signs for those with negative connotations, and = signs for neutral words. This strategy can work wonders. For instance:

$$
\overset{-}{\text{PERDITION}} : \quad \text{(A)} \overset{-}{\text{ deterrent}} \quad \text{(B)} \overset{=}{\text{ rearrangement}}
$$

$$
\text{(C)} \overset{=}{\text{ reflection}} \quad \text{(D)} \overset{+}{\text{ salvation}} \quad \text{(E)} \overset{-}{\text{ rejection}}
$$

5. Eliminate Any Answer Choices That Do Not Have a Clear Opposite
For instance, in the sample problem above, neither choice (B) nor choice (C) has a clear and obvious opposite. They are unlikely to be correct.

STRATEGY FOR ANTONYMS

1. Define the root word.
2. Reverse it.
3. Find a similar opposite in the answer choices.
4. If all else fails, eliminate answer choices and guess.

---

6. On Hard Antonym Questions, Watch out for Trick Choices and Eliminate Them
For instance, if you come across:

CEDE : (A) make sense of   (B) fail
(C) get ahead of   (D) flow out of   (E) retain

you should eliminate B, C, and D. Why? Because *cede* will remind some people of *succeed*, they will pick B. It will remind others of *recede*, as in *receding hairline* or *receding tide*, so they will pick C or D. ETS never rewards people for goofing up. No one ever "lucks" into the right answer on the GRE by making a mistake.

7. Choose Answers Strategically.
When in doubt, try to eliminate incorrect answer choices and then guess.

Now that you have a grasp of the basic principles of antonyms, let's look at the Kaplan method for solving antonym questions.

The Kaplan Four-Step Method for Antonyms
1. Define the root word.
2. Reverse it by thinking about the word's opposite.
3. Now go to the answer choices and find the opposite—that is, the choice that matches your preconceived notion of the choice.
4. If stuck, eliminate any choices you can and guess among those remaining.

Using Kaplan's Four-Step Method
Now let's put this method to the test. Suppose you encounter *loiter* in an antonym question:
Step One: Ask yourself what *loiter* means. Write a definition below:

_____

_____

Step Two: Think about *loiter*'s opposite.

The opposite of *loiter* is _____ .

Step Three: Choose the answers that best matches your reversal of the original word.

LOITER: (A) change direction (B) move purposefully (C) inch forward (D) clean up (E) amble

What's the *opposite* of the choice you picked? Does that match the meaning of the original word?

The opposite of *move purposefully* is *stand around,* or *loiter.*

Step Four: If you get stuck, eliminate choices and guess. (A) change direction, and (D) clean up, seem to be unlikely choices and can be eliminated.

Reading Comprehension

Reading comprehension is the only question type that appears on all major standardized tests, and the reason isn't too surprising. No matter what academic area you pursue, you'll have to make sense of some dense, unfamiliar material. The topics for GRE reading comp passages are taken from three areas: social sciences, natural sciences, and humanities.

These passages tend to be wordy and dull, and you may find yourself wondering where the test makers get them (probably from the same source as computer installation manuals). Well, actually, the test makers go out and collect the most boring and confusing essays available, then chop them up beyond all recognition or coherence. The people behind the GRE know that you'll have to read passages like these in graduate school, so they choose test material accordingly. In a way, reading comp is the most realistic of all the question types on the test. And right now is a good time to start shoring up your critical reading skills, both for the test and for future study in your field.

Format and Directions

The directions in this section look like this:

> **Directions: After each reading passage you will find a series of questions. Select the best choice for each question. Answers are based on the contents of the passage or what the author implies in the passage.**

On the CAT you will see two to four reading comp passages, each with two to four questions. You will have to tackle the passage and questions as they are given to you.

The Seven Basic Principles of Reading Comprehension

To improve your reading comp skills, you'll need a lot of practice—and patience. You may not see dramatic improvement after only one drill. But with ongoing practice, the seven basic principles will help to increase your skill and confidence on this section by the day of the test. After reviewing the following principles, you'll find your first opportunity to apply them by working on a sample passage. And later, on the practice test, you'll have an opportunity to master these skills.

1. Pay Special Attention to the First Third of the Passage

The first third of a reading comp passage usually introduces its topic and scope, the author's main idea or primary purpose, and the author's tone. It almost always hints at the structure that the passage will follow. Let's take a closer look at these important elements of a reading comp passage.

Topic and Scope. *Topic* and *scope* are both objective terms. That means they include no specific reference to the author's point of view. The difference between them is that the topic is broader; the scope narrows the topic. Scope is particularly important because the answer choices (often many) that depart from it will always be wrong. The broad topic of "The Battle of Gettysburg," for example, would be a lot to cover in 450 words. So if you encountered this passage, you should ask yourself, "What aspect of the battle does the author take up?" Because of length limitations, it's likely to be a pretty small chunk. Whatever that chunk is—the prebattle scouting, how the battle was fought—that will be the passage's scope. Answer choices that deal with anything outside of this narrowly defined chunk will be wrong.

Author's Purpose. The distinction between topic and scope ties into another important issue: the author's purpose. In writing the passage, the author has deliberately chosen to narrow the scope by including certain aspects of the broader topic and excluding others. Why the author makes those choices gives us an important clue as to why the passage is being written in the first place. From the objective and broadly stated topic (for instance, a passage's topic might be *solving world hunger*) you zoom in on the objective but narrower scope (*a new technology for solving world hunger*), and the scope quickly leads you to the author's subjective purpose (*the author is writing in order to describe a new technology and its promising uses*). The author's purpose is what turns into the author's main idea, which will be discussed at greater length in the next principle.

So don't just "read" the passage; instead, try to do the following three things:

1. Identify the topic.
2. Narrow it down to the precise scope that the author includes.
3. Make a hypothesis about why the author is writing and where he or she is going with it.

ZOOM IN

As you read the first third of the passage, try to zoom in on the main idea of the passage by first getting a sense of the general topic, then pinning down the scope of the passage, and finally zeroing in on the author's purpose in writing the passage.

BEYOND THE SCOPE

Answer choices that deal with matters outside the passage's scope are always wrong.

Structure and Tone. In their efforts to understand what the author says, test takers often ignore the less glamorous but important structural side of the passage—namely, how the author says it. One of the keys to success on this section is to understand not only the passage's purpose but also the structure of each passage. Why? Simply because the questions at the end of the passage ask both what the author says *and* how he or she says it. Here's a list of the classic GRE passage structures:

• Passages arguing a position (often a social sciences passage)
• Passages discussing something specific within a field of study (for instance, a passage about Shakespearean sonnets in literature)
• Passages explaining some significant new findings or research (often a science passage)

Most passages that you'll encounter will feature one of these classic structures, or a variation thereof. You've most likely seen these structures at work in passages before, even if unconsciously. Your job is to actively seek them out as you begin to read a passage. Usually, the structure is announced within the first third of the passage. Let these classic structures act as a jump start in your search for the passage's "big picture" and purpose.

As for how the author makes his or her point, try to note the author's position within these structures, usually indicated by the author's tone. For example, in passages that explain some significant new findings or research structure, the author is likely to be clinical in description. In passages that argue a position, the opinion could be the author's, in which case the author's tone may be opinionated or argumentative. On the other hand, the author could simply be describing the strongly held opinions of someone else. In the latter case the author's writing style would be more descriptive, factual, even-handed. His or her method may involve mere storytelling or the simple relaying of information, which is altogether different from the former case.

Notice the difference in tone between the two types of authors (argumentative versus descriptive). Correct answer choices for a question about the main idea would, in the former case, use such verbs as *argue for, propose,* or *demonstrate,* whereas correct choices for the same type of question in the latter case would use such verbs as *describe* or *discuss.* Correct answers are always consistent with the author's tone, so noting the author's tone is a good way to understand the passage.

2. Focus on the Main Idea

Every passage boils down to one big idea. Your job is to cut through the fancy wording and focus on this big idea. Very often, the main idea will be presented in the first third of the passage, but occasionally the author will build up to it gradually, in which case you may not have a firm idea of it.

In any case, the main idea always appears somewhere in the passage, and when it does, you must take note of it. For one thing, the purpose of everything else in the passage will be to support this idea. Furthermore, many of the questions—not only "main idea" questions but all kinds of questions—are easier to handle when you have the main idea in the forefront of your mind. Always look for choices that sound consistent with the main idea. Wrong choices often sound inconsistent with it.

3. Get the Gist of Each Paragraph

It will come as no surprise to you that the paragraph is the main structural unit of any passage. After you've read the first third of the passage carefully, you need to find the gist, or general purpose, of each paragraph and then try to relate each paragraph back to the passage as a whole. To find the gist of each paragraph, ask yourself:

- Why did the author include this paragraph?
- What shift did the author have in mind when moving on to this paragraph?
- What bearing does this paragraph have on the author's main idea?

4. Don't Obsess over Details

There are differences between the reading skills required in an academic environment and those that are useful on standardized tests. In school, you probably read to memorize information for an exam. But this isn't the type of reading that's good for racking up points on the GRE reading comprehension section. On the test, you'll need to read for short-term retention. When you finish the questions on a certain passage, that passage is over, gone, done with. Go ahead, forget everything about it!

What's more, there's no need to waste your time memorizing details. The passage will always be right there in front of you. You always have the option to find any details if a particular question requires you to do so. If you have a good sense of a passage's structure and paragraph topics, then you should have no problem navigating back through the text.

WHAT'S THE BIG IDEA?

You should always keep the main idea in mind, even when answering questions that don't explicitly ask for it. Correct answers on even the detail questions tend to echo the main idea in one way or another.

DON'T WASTE YOUR TIME

You don't have to memorize or understand every little thing as you read the passage. Remember, you can always refer back to the passage to clarify the meaning of any specific detail.

5. Attack the Passages, Don't Just Read Them

Remember when you took the SAT? Like some of us, did you celebrate when you finally finished the passage and then treat the questions as afterthoughts? If so, we suggest that you readjust your thinking. Remember: you get no points for just getting through the passage.

When we read most materials, a newspaper, for example, we start with the first sentence and read the article straight through.

The words wash over us and are the only things we hear in our minds. This is typical of a passive approach to reading, and this approach won't cut it on the GRE.

To do well on this test you'll need to do more than just read the words on the page. You'll need to read actively. Active reading involves keeping your mind working at all times, while trying to anticipate where the author's points are leading. It means thinking about what you're reading as you read it. It means paraphrasing the complicated-sounding ideas and jargon. It means asking yourself questions as you read:

- What's the author's main point here?
- What's the purpose of this paragraph? Of this sentence?

While reading actively you keep a running commentary in your mind. You may want to jot down notes in the margin or underline. When you read actively, you don't absorb the passage, you attack it!

6. Beware of Classic Wrong Answer Choices

Knowing the most common wrong answer types can help you to eliminate wrong choices quickly, which can save you a lot of time. Of course, ideally, you want to have prephrased an answer choice in your mind before looking at the choices. When that technique doesn't work, you'll have to go to the choices and eliminate the bad ones to find the correct one. If this happens, you should always be on the lookout for choices that:

- Contradict the facts or the main idea
- Distort or twist the facts or the main idea
- Mention true points not relevant to the question (often from the wrong paragraph)
- Raise a topic that's never mentioned in the passage
- Sound off the wall or have the wrong tone

ATTACK THE PASSAGE

You can be an active reader by:

- Thinking about what you're reading
- Paraphrasing the complicated parts
- Asking yourself questions about the passage
- Jotting down notes or underlining important words

Being sensitive to these classic wrong choices will make it that much easier to zero in on the correct choice quickly and efficiently.

7. Use Outside Knowledge Carefully
You can answer all the questions correctly even if you don't know anything about the topics covered in the passages. Everything you'll need to answer every question is included in the passages themselves. However, as always, you have to be able to make basic inferences and extract relevant details from the texts.

Using outside knowledge that you may have about a particular topic can be beneficial to your cause, but watch out! Outside knowledge can also mess up your thinking. If you use your knowledge of a topic to help you understand the author's points, then you're taking advantage of your knowledge in a useful way. However, if you use your own knowledge to answer the questions, then you may run into trouble because the questions test your understanding of the author's points, not your previous understanding or personal point of view on the topic.

So the best approach is to use your own knowledge and experience to help you to comprehend the passages, but be careful not to let it interfere with answering the questions correctly.

Reading Comprehension Test Run
Here's a chance to familiarize yourself with a short reading comp passage and questions. You'll have more opportunities to practice later, under timed conditions. For now, we want you to take the time to read actively, to give the seven principles a test run.

> **Directions: After each reading passage you will find a series of questions. Select the best choice for each question. Answers are based on the contents of the passage or what the author implies in the passage.**

Migration of animal populations from one region to another is called faunal interchange. Concentrations of species across regional boundaries vary, however, prompting zoologists to classify routes along which
(5) penetrations of new regions occur.

A corridor, like the vast stretch of land from Alaska to the southeastern United States, is equivalent to a path of least resistance. Relative ease of migration often results in the

presence of related species along the entire length of
(10) a corridor; bear populations, unknown in South America, occur throughout the North American corridor. A desert or other barrier creates a filter route, allowing only a segment of a faunal group to pass. A sweepstakes route presents so formidable a barrier that penetration is
(15) unlikely. It differs from other routes, which may be crossed by species with sufficient adaptive capability. As the name suggests, negotiation of a sweepstakes route depends almost exclusively on chance, rather than on physical attributes and adaptability.

1. It can be inferred from the passage that studies of faunal interchange would probably
   (A) fail to explain how similar species can inhabit widely separated areas
   (B) be unreliable because of the difficulty of observing long-range migrations
   (C) focus most directly on the seasonal movements of a species within a specific geographic region
   (D) concentrate on correlating the migratory patterns of species that are biologically dissimilar
   (E) help to explain how present-day distributions of animal populations might have arisen
   Ⓐ Ⓑ Ⓒ Ⓓ Ⓔ

2. The author's primary purpose is to show that the classification of migratory routes
   (A) is based on the probability that migration will occur along a given route
   (B) reflects the important role played by chance in the distribution of most species
   (C) is unreliable because further study is needed
   (D) is too arbitrary, because the regional boundaries cited by zoologists frequently change
   (E) is based primarily on geographic and climatic differences between adjoining regions
   Ⓐ Ⓑ Ⓒ Ⓓ Ⓔ

3. The author's description of the distribution of bear populations (lines 10–11) suggests which of the following conclusions?

    I. The distribution patterns of most other North American faunal species populations are probably identical to those of bears.

    II. There are relatively few barriers to faunal interchange in North America.

    III. The geographic area that links North America to South America would probably be classified as either a filter or a sweepstakes route.

    (A) I only

    (B) II only

    (C) III only

    (D) I and II only

    (E) II and III only

Ⓐ Ⓑ Ⓒ Ⓓ Ⓔ

4. According to the passage, in order to negotiate a sweepstakes route an animal species

    (A) has to spend at least part of the year in a desert environment

    (B) is obliged to move long distances in short periods of time

    (C) must sacrifice many of its young to wandering pastures

    (D) must have the capacity to adapt to a very wide variety of climates

    (E) does not need to possess any special physical capabilities

Ⓐ Ⓑ Ⓒ Ⓓ Ⓔ

How Did You Do?

Were you able to zoom in from the broad topic (migration) to the scope (classification of migration routes)? Did the author's tone and purpose become clear, then, as explanatory rather than argumentative? And were you able to focus on the correct answers and not get distracted by outside knowledge or misleading details? You'll be able to assess your performance and skills as you review the next section, where we'll explore the strategies for dealing with the three question types and how these strategies apply to the above questions.

The Three Common Reading Comp Question Types

We find it useful to break the reading comp section down into the three main question types that accompany each passage: global, explicit detail, and inference. Most test takers find explicit detail questions to

ANSWERS TO READING COMP QUESTIONS

1. E
2. A
3. E
4. E

GLOBAL QUESTIONS AT A GLANCE

Global questions:

- Represent 25 to 30 percent of reading comp questions
- Sum up author's overall intentions or passage structure
- Nouns and verbs must be consistent with the author's tone and the passage's scope
- Main idea and primary purpose, title, structure, and tone questions are related

be the easiest type in the reading comp section, because they're the most concrete. Unlike inferences, which hide somewhere between the lines, explicit details sit out in the open—in the lines themselves. That's good news for you, because when you see an explicit detail question you'll know that the correct answer requires only recall, and not analysis. Let's look at each of these question types more closely, using the sample questions you just dealt with for illustration.

1. Global Questions

Description. A global question will ask you to sum up the author's overall intentions, ideas, or passage structure. It's basically a question whose scope is the entire passage. Global questions account for 25 to 30 percent of all reading comp questions. Question 2 in the preceding sample is a global question because it asks you to identify the author's primary purpose.

Strategy. In general, any global question choice that grabs onto a small detail—or zeroes in on the content of only one paragraph—will be wrong. Often, scanning the verbs in the global question choices is a good way to take a first cut at the question. The verbs must agree with the author's tone and the way in which he or she structures the passage, so scanning the verbs and adjectives can narrow down the options quickly. The correct answer must be consistent with the overall tone and structure of the passage, whereas common wrong-answer choices associated with this type of question are those that are too broad or narrow in scope and those that are inconsistent with the author's tone. You'll often find global questions at the beginning of question sets, and often one of the wrong choices will play on some side issue discussed at the tail end of the passage.

Strategy Applied. Take a closer look at the global question, number 2 in the sample. You've already articulated the passage's topic (migration), scope (the classification of migration routes), and tone (explanatory). The author mentions three different classifications of migration routes—corridors, filter routes, and sweepstakes routes. And what distinguishes one kind of route from another? The likelihood of migration, from the most likely (corridors, with no barriers to migration) to least likely (sweepstakes routes, with barriers that species can cross only by chance). So the author's primary purpose here is to show how the classifications are defined according to how

KAPLAN

likely migration is along each type of route. That should have led directly to choice (A).

A scan of verbs and adjectives is enough to eliminate choices (C) and (D); both imply that the author is making judgments about the classification, but the tone of the passage is objective and explanatory. Meanwhile, (B) focuses too much on one part of the passage—the explanation of sweepstakes routes—where the role of chance is mentioned only in relation to that one classification. And (E) is a distortion, because the author nowhere mentions climatic and geographic differences between adjoining regions.

Main Idea and Primary Purpose Questions. The two main types of global questions are main idea and primary purpose questions. We discussed these types a little earlier, noting that main idea and purpose are inextricably linked, because the author's purpose is to convey his or her main idea. The formats for these question types are pretty self-evident:

> Which one of the following best expresses the main idea of the passage?

> *or*

> The author's primary purpose is to . . .

Title Questions. A very similar form of global question is one that's looking for a title that best fits the passage. A title, in effect, is the main idea summed up in a brief, catchy way. This question may look like this:

> Which of the following titles best describes the content of the passage as a whole?

Be sure not to go with a choice that aptly describes only the latter half of the passage. A valid title, much like a main idea and primary purpose, must cover the entire passage.

Structure Questions. Another type of global question is one that asks you to recognize a passage's overall structure. Here's what this type of question might sound like:

> Which of the following best describes the organization of the passage?

GLOBAL QUESTIONS ALERT

Key phrases in the global question stem include:

- Which of the following best expresses the main idea . . .
- The author's primary purpose is . . .
- Which of the following best describes the content as a whole . . .
- Which of the following best describes the organization . . .
- The author's tone can best be described as . . .

Answer choices to this kind of global question are usually worded very generally; they force you to recognize the broad layout of the passage as opposed to the specific content. For example, here are a few possible ways that a passage could be organized:

A hypothesis is stated and then analyzed.
A proposal is evaluated and alternatives are explored.
A viewpoint is set forth and then subsequently defended.

When picking among these choices, literally ask yourself, "Was there a hypothesis here? Was there an evaluation of a proposal or a defense of a viewpoint?" These terms may all sound similar but, in fact, they're very different things. Learn to recognize the difference between a proposal, a viewpoint, and so on. Try to keep a constant eye on what the author is doing as well as what the author is saying, and you'll have an easier time with this type of question.

Tone Questions. Finally, one last type of global question is the tone question, which asks you to evaluate the style of the writing or how the author sounds. Is the author passionate, fiery, neutral, angry, hostile, opinionated, low-key? Here's an example:

The author's tone in the passage can best be characterized as . . .

Make sure not to confuse the nature of the content with the tone in which the author presents the ideas: a social science passage based on trends in this century's grisliest murders may be presented in a cool, detached, strictly informative way. Once again, it's up to you to separate what the author says from how he or she says it.

2. Explicit Detail Questions

Description. The second major category of reading comprehension questions is the explicit detail question. As the name implies, an explicit detail question is one whose answer can be directly pinpointed and found in the text. This type makes up roughly 20 to 30 percent of the reading comp questions. Question 4 in the sample above is an explicit detail question because it asks you to go back to the passage and examine the description of a sweepstakes route.

Strategy. Often, these questions provide very direct clues as to where an answer may be found, such as a line reference or some text that links up with the passage structure. (Just be careful with line references; they'll bring you to the right area, but usually the actual answer will be found in the lines immediately before or after the referenced line.) Detail questions are usually related to the main idea, and correct choices tend to be related to major points.

Now, you may recall that we advised you to skim over details in reading comp passages in favor of focusing on the big idea, topic, and scope. But now here's a question type that's specifically concerned with details, so what's the deal? The fact is, most of the details that appear in a typical passage aren't tested in the questions. Of the few that are, you'll either:

- Remember them from your reading
- Be given a line reference to bring you right to them
- Simply have to find them on your own in order to track down the answer

In the third case, if your understanding of the purpose of each paragraph is in the forefront of your mind, it shouldn't take long at all to locate the details in question and then choose an answer. And if even that fails, as a last resort you have the option of putting that question aside and returning to it if and when you have the time later to search through the passage. The point is, even with the existence of this question type, the winning strategy is still to note the purpose of details in each paragraph's argument but not to attempt to memorize the details themselves.

Strategy Applied. Take a closer look at the explicit detail question in the sample, Question 4. When an explicit detail question directs you to a specific place in the passage, as Question 4 does to the discussion of sweepstakes routes, your first job is to go right to that spot in the passage and reread it. And if you do that here, you read that negotiation of a sweepstakes route depends "almost entirely on chance, rather than on physical attributes and adaptability." The discounting of physical attributes here should have led you directly to choice (E).

Choice (A)'s mention of a desert environment sinks that choice, because the desert was mentioned by the author in the discussion of filter routes. As for the other choices, "short periods of time," "wandering

EXPLICIT DETAIL QUESTIONS AT A GLANCE

Explicit detail questions:

- Represent 20 to 30 percent of reading comp questions
- Answers can be found in the text
- Sometimes includes line references to help you locate the relevant material
- Are concrete and, therefore, the easiest reading comp question type for most people

EXPLICIT DETAIL ALERT

Key phrases in the explicit detail question stem include:

- According to the passage/author . . .
- The author states that . . .
- The author mentions which one of the following as . . .

pastures," and "a wide variety of climates" aren't mentioned in the passage with regard to sweepstakes routes.

A GOOD INFERENCE

A good inference:

- Stays in line with the gist of the passage
- Stays in line with the author's tone
- Stays in line with the author's point of view
- Stays within the scope of the passage and its main idea
- Is neither denied by, nor irrelevant to, the ideas stated in the passage
- Always makes more sense than its opposite

INFERENCE QUESTIONS AT A GLANCE

Inference questions:

- Involve ability to read between the lines
- Often boil down to paraphrasing the author's points
- Question stem usually provides a clue as to where in the passage the answer will be found
- Will either ask for something that can be inferred from a specific part of the passage, or else for something the author would agree with

3. Inference Questions

Description. An inference is something that is almost certainly true, based on the passage. Inferences require you to "read between the lines." Questions 1 and 3 in the preceding sample are Inference questions. Question 1 specifically asks you what can be "inferred from the passage" and Question 3 asks you to glean possible conclusions based on what is presented.

Strategy. The answer to an inference question is something that the author strongly implies or hints at but does not state directly. Furthermore, the right answer, if denied, will contradict or significantly weaken the passage.

Extracting valid inferences from reading comp passages requires the ability to recognize that information in the passage can be expressed in different ways. The ability to bridge the gap between the way information is presented in the passage and the way it's presented in the correct answer choice is vital. In fact, inference questions often boil down to an exercise in translation.

Strategy Applied. Take a closer look at the sample inference questions, numbers 1 and 3. Question 1 asks you to select a possible application of the migration study, based only on what you know from the passage. Because the different concentrations of animals prompted the zoologists to classify the migration routes in the first place (line 5), it would make sense that the migration study would help explain how these different concentrations, or distributions, would have arisen. So choice (E) is correct.

Choice (A) contradicts the purpose of the passage, which we discussed for question 2, and unless we're told in the passage that the study was a failure (which we are not), we can't guess that it would be one. Choice (B) is outside of the passage's scope because the passage never touches on the reliability of the study or on any difficulties in observing long-range migrations. The answer must be based on the passage. Choices (C) and (D) are misleading distortions. The study does focus on movements of species, but there's no mention of a seasonal influence, and the study does focus on route comparisons but not on species comparisons.

Question 3 asks you to look back to the passage for the example of the bears and decide what conclusion(s) could be drawn based on this example. We read that bear populations occur throughout North America because North America is a "path of least resistance," meaning there are relatively few barriers. The bears did not continue to migrate further south, however, because they're "unknown in South America." This suggests that South America is either a filter or a sweepstakes route. Nowhere are the bears compared with other species, because the focus isn't the bears but the routes. So option I can be eliminated, and options II and III are accurate. This should have led you, then, to choice (E).

The Kaplan Three-Step Method for Reading Comprehension

Now that you've got the basics of GRE reading comp under your belt, you'll want to learn our three-step method that allows you to orchestrate them all into a single modus operandi for the questions.

1. Attack the first third of the passage.
2. Read the rest of the passage.
3. Answer the questions.

1. Attack the First Third of the Passage

As outlined in the basic principles section, read the first part of the passage with care, in order to determine the main idea and purpose (via the zooming-in process we talked about earlier). Two caveats, however. First, in some passages, the author's main idea won't become clear until the end of a passage. Second, occasionally a passage won't include a main idea, which itself is a strong hint that the passage is more of a descriptive, storytelling type of passage, with an even-handed tone and no strong opinions. Bottom line: don't panic if you can't immediately pin down the author's main idea and purpose. Read on.

2. Read the Rest of the Passage

Do so as we described in the basic principles section above, making sure to take note of paragraph topics, location of details, etcetera.

3. Answer the Questions

Look at each question stem carefully. If the question asks something about the passage that isn't clear to you, reread any relevant paragraphs for clarification. Search aggressively for the specific details you need.

Using the Kaplan Three-Step Method

Now let's try the three-step method on another actual GRE-strength reading comp passage. For the time being, we've just included the question stems of the questions attached to this passage, since you don't want to get into individual choices until later.

> **Directions: After each reading passage you will find a series of questions. Select the best choice for each question. Answers are based on the contents of the passage or what the author implies in the passage.**

Although there have been many third-party movements in U.S. history, no third-party candidate has ever been elected president. And except for the Republican Party, which gained prominence when the Whigs were
(5) declining in the 1850s, no third party has ever achieved national major-party status.

The basic factors that have shaped the U.S. political system account for both the frequency and the weakness of peripheral party movements. Chief among these are
(10) the size and widely varying social and economic features of the country. Different interests and voting blocs predominate in various regions, resulting in an electorate that is fragmented geographically. This heterogeneity is heightened by a federal structure that requires major parties
(15) to find support at the state and local levels in different regions. To take one example, the Democratic Party in the mid-twentieth century drew support simultaneously from blacks in the North and white segregationists in the South.
(20) The U.S. electoral system intensifies the difficulties of smaller groups. This system, in which the candidate with the highest vote is the "winner who takes it all"—with no provision for proportional representation, as in many countries—rewards broad-based political strategies that
(25) avoid alienating the mainstream voting population, and, conversely, sharply penalizes parties with more restricted support, whose voters may be left unrepresented.

The nondoctrinal character of U.S. politics has meant that new issues tend to be ignored initially by major parties.

(30) Rather, issues such as opposition to immigration, the abolition of slavery, and the rights of workers and farmers frequently gain entry to the political arena through the creation of a third party. Thus, while mainstream voters have usually viewed certain issues as divisive or threatening,

(35) a dedicated minority has often been instrumental in placing them on the national agenda. Indeed, nearly every major national dilemma has sparked some sort of third-party movement. More ephemeral questions, fringe issues such as vegetarianism and prohibitionism, and highly ideological

(40) programs such as socialism and populism, have also frequently served as seeds for third-party movements.

   Ironically, certain elements that help to give birth to third-party movements also contribute to their failure to thrive. Parties based on narrow or short-term appeals

(45) remain isolated or fade rather rapidly. Parties that raise more salient issues and attract more widespread support face limits of a different kind. Long before a third party might begin to emerge as a truly major political force, major parties will attempt to capture the significant

(50) minority of voters that are represented. The Democratic Party thus pirated much of the platform of the Populists in 1896, and, in subsequent decades, in the eras of Wilson and Roosevelt, sponsored progressive and social welfare programs that relentlessly undercut the influence

(55) of the Socialists.

Question Stems

1. According to the passage, a major factor responsible for the rise of third parties in the United States is the

2. The author cites all of the following as contributing to the weakness of third parties EXCEPT:

3. It can be inferred that which of the following have contributed to the "nondoctrinal character of U.S. politics" (line 28)?

4. It can be inferred from the passage that the probable attitudes of many voters in the general population to the ideas initially put forth by a third party could best be described as

5. It can be inferred that the Republican Party was successful in establishing itself as a major party because

6. The author's description of the U.S. electoral system suggests that it

7. The author of this passage is concerned primarily with

1. Attack the First Third of the Passage
The first few sentences introduce the topic: third-party movements. The scope, as you recall, is the specific angle the author takes on the topic, and this seems to be the factors hindering the success of third-party movements. The author points out that historically the same factors that have shaped the U.S. political system limit the success of the third-party movement, then the author goes on to list those factors. So just from reading the first third of the passage, you have a sense of the overall structure and purpose.

2. Read the Rest of the Passage
Make note of the author's point-by-point examination of the effectiveness of the third-party movement, but don't try to memorize details. You can always refer to the passage to answer questions.

3. Answer the Question Stems
Let's look again at the seven question stems attached to this passage:

1. According to the passage, a major factor responsible for the rise of third parties in the United States is the
2. The author cites all of the following as contributing to the weakness of third parties EXCEPT:
3. It can be inferred that which of the following have contributed to the "nondoctrinal character of U.S. politics" (line 28)?
4. It can be inferred from the passage that the probable attitudes of many voters in the general population to the ideas initially put forth by a third party could best be described as
5. It can be inferred that the Republican Party was successful in establishing itself as a major party because
6. The author's description of the U.S. electoral system suggests that it
7. The author of this passage is concerned primarily with

Global Questions
Question 7 is a global question, since it asks for the main idea or primary purpose of the passage.

7. The author of this passage is concerned primarily with
   (A) suggesting an appropriate role for third parties in U.S. politics
   (B) discussing the decline of third-party movements in U.S. history
   (C) explaining why third-party movements in the U.S. have failed to gain national power
   (D) describing the traditionally nonideological character of U.S. political parties
   (E) suggesting ways in which peripheral parties may increase their influence

The author's primary purpose, as introduced early in the passage, is choice (C), explaining why third-party movements in the U.S. have failed to gain national power. Choices (A) and (E) suggest the author has an advocative tone, but the tone is explanatory. Choice (B), by using the word *decline*, distorts the primary concern, and choice (D) is too narrow in scope, because it doesn't even mention third-party movements.

Questions 1 and 2 are clearly explicit detail questions, while questions 3–6 are inference questions. We've already discussed one of the global questions, Question 7, so let's now conclude this discussion with a brief look at the explicit detail questions and the Inference questions.

Explicit Detail Questions
Here's the complete form (with answer choices) of Question 1:

1. According to the passage, a major factor responsible for the rise of third parties in the U.S. is the
   (A) domination of major parties by powerful economic interests
   (B) ability of third parties to transcend regional interests
   (C) ready acceptance by mainstream voters of issues with strong minority support
   (D) appeal of fringe issues to the average American voter
   (E) slowness of major parties to respond to new issues

By looking back to the passage, specifically to paragraph 4, you can see that one factor listed as responsible for the rise of third parties is choice (E), the slowness of major parties to respond to new issues. As discussed, these issues often act as seeds of third-party movements. Choice (A) is a

factor responsible for the lack of success of the third-party movement, not for its rise, and choices (B), (C), and (D) are distortions.

Question 2 is also an explicit detail question:

2. The author cites all of the following as contributing to the weakness of third parties EXCEPT:
    (A) their tendency to avoid sharply defined political programs
    (B) an electoral system that denies them proportional representation
    (C) their tendency to adopt programs that fail to attract mainstream voter support
    (D) the ability of major parties to undercut their appeal
    (E) the fact that many are based on issues of only temporary relevance

The passage cites all choices as contributing to the weakening of the third-party movement EXCEPT choice (A), the correct choice.

Inference Questions
Finally, let's take a look at a complete inference question, Question 4:

4. It can be inferred from the passage that the probable attitudes of many voters in the general population to the ideas initially put forth by a third party could best be described as
    (A) shocked and disbelieving
    (B) confused and indecisive
    (C) curious and open-minded
    (D) suspicious and disapproving
    (E) apathetic and cynical

You can infer from the passage that the probable attitudes of many voters are listed in choice (D), suspicious and disapproving, because the passage tells us in line 34 that mainstream voters usually view third-party issues as divisive and threatening. None of the other choices are as applicable.

Questions 3, 5, and 6 are also inference questions.

KAPLAN

3. It can be inferred that which of the following have contributed to the "nondoctrinal character of U.S. politics" (line 28)?

    I. The social and economic diversity of the country

    II. The national political structure

    III. The avoidance by major parties of sharply defined ideological programs

(A) I only

(B) III only

(C) I and II only

(D) II and III only

(E) I, II, and III

5. It can be inferred that the Republican Party was successful in establishing itself as a major party because

(A) its political program became essentially nondoctrinal in character

(B) the Whigs were unsuccessful in their attempts to steal from the Republican platform

(C) it was able to abandon its traditional opposition to slavery without alienating its regular supporters

(D) a more established party was simultaneously in decline

(E) it benefited from the experience of previous third parties that had undergone similar transformations

6. The author's description of the U.S. electoral system suggests that it

(A) allows less flexibility than more centralized systems

(B) makes the federal government less important politically than state and local governments

(C) often results in the dominance of third parties in distinct or isolated geographic areas

(D) frequently polarizes the electorate around divisive social and economic issues

(E) fails to represent voters who would be represented in some other electoral systems

For Question 3 the correct inference is (E), all of the given choices contributed to the nondoctrinal character of U.S. politics. For Question 5 the correct choice is (D), a more established party was simultaneously in decline; the other choices can not be inferred solely from the passage. For Question 6 the answer is (E), in that the author suggests the electoral sys-

tem fails to represent voters who would be represented in some other electoral system.

Review Exercises

Because reading comp primarily tests your ability to read actively, paraphrasing as you go and paying attention to purpose and structure, let's spend some time practicing those critical reading skills.

Big Ideas and Details

It's important that you're able to pull out the main idea and supporting details quickly and easily on the day of the test. Covering up the sidebars, read the following three minipassages, jotting down their big ideas in the margin and underlining details. Then compare your answers with ours.

Big Idea: Gutenberg revolutionized book publishing.

Details: late 1400s, moveable type, limitless quantities, typesetting

Few historians would contest the idea that Gutenberg's invention of the printing press revolutionized the production of literature. Before the press became widely available in the late 1400s, every book published had to be individually copied by a scribe working from a master manuscript. With Gutenberg's system of moveable type, however, books could be reproduced in almost limitless quantities once the laborious process of typesetting was complete. . . .

Big Idea: Mantle plume eruptions, not plate tectonics, may explain certain phenomena.

Details: ocean island chains, flood basalt provinces

Plate tectonics, the study of the interaction of the earth's plates, is generally accepted as the best framework for understanding how the continents formed. New research suggests, however, that the eruption of mantle plumes from beneath the plate layer may be responsible for the formation of specific phenomena in areas distant from plate boundaries. A model of mantle plumes appears to explain a wide range of observations relating to both ocean island chains and flood basalt provinces, for example. . . .

Big Idea: Economists disagree on the cost of emission reduction.

Details: international measures, carbon emission, greater savings, conservation

Most of the developed countries are now agreed on the need to take international measures to reduce the emission of carbons into the atmosphere. Despite this consensus, a wide disagreement among economists as to how much emission reduction will actually cost continues to forestall policy making. Analysts who believe the energy market is efficient predict that countries that reduce carbon emis-

sion by as little as 20 percent will experience a significant depreciation in their national product. Those that hold that the market is inefficient, however, estimate much greater long-term savings in conservation and arrive at lower costs for reducing emissions. . . .

Paraphrasing: The Key Skill in Reading Comp

Many people have a hard time paraphrasing passages. Taking dense, academic prose and turning it into everyday English isn't easy under the pressure of time constraints. Yet, this is the most important skill in reading comp. If you are having trouble with paraphrasing, spend some time with the following exercise.

> For centuries, the Roman Empire ruled large parts of Europe, Asia, and Africa. Rome had two assets that made continued domination possible. First, its highly trained army was superior to those of its potential adversaries. Second, and more important, Rome built a sophisticated transportation network linking together all of the provinces of its far flung empire. When necessary, it could deploy powerful military forces to any part of the empire with unmatched speed.

Summarize these lines in your own words:

_____

_____

_____

Now find the best paraphrase:

(A) Rome's army defeated its opponents because it could move quickly along the empire's excellent transportation network.

(B) Rome had a big empire, a powerful army, and a good transportation system.

(C) Rome was able to maintain a big empire because it had an excellent transportation system that allowed its efficient army to move quickly from place to place.

(D) Rome ruled large parts of Europe, Asia, and Africa for centuries because its army was always better than those of its adversaries.

(E) Because it built a sophisticated transportation system, Rome was able to build a big empire in parts of Europe, Asia, and Africa.

Despite overwhelming evidence to the contrary, many people think that flying is more dangerous than driving. Different standards of media coverage account for this erroneous belief. Although extremely rare, aircraft accidents receive a lot of media attention because they are very destructive. Hundreds of people have been killed in extreme cases. Automobile accidents, on the other hand, occur with alarming frequency, but attract little media coverage because few, if any, people are killed or seriously injured in any particular mishap.

Summarize these lines in your own words:

_____

_____

_____

Now find the best paraphrase:

(A) Compared to rare but destructive aircraft accidents, car accidents are frequent but relatively minor.

(B) Because aircraft accidents get a lot of media attention, while car accidents get much less, many people wrongly believe that flying is more dangerous than driving.

(C) Driving is more dangerous than flying because different standards of media coverage have forced airlines to improve their safety standards.

(D) Many people believe that flying is more dangerous than driving, even though overwhelming evidence points to the opposite conclusion.

(E) Media coverage is responsible for the belief that flying is more dangerous than driving, even though every year more people are killed on the roads than in the air.

Keywords

Remember these? Kaplan keywords are words in reading comprehension passages that link the text together structurally and thematically. Paying attention to keywords will help you understand the passage better, and will also help you get some easy points. Here are some keywords that you should look for when reading a passage:

CONTRADICTION:

However
But
Yet
On the other hand
Rather
Instead

SUPPORT:

For example
One reason that
In addition
Also
Moreover
Consequently

EMPHASIS:

Of primary importance
Especially important
Of particular interest
Crucial
Critical
Remarkable

Take a moment and circle the structural signal words in this short reading comprehension passage:

Gettysburg is considered by most historians to be a turning point in the American Civil War. Before Gettysburg, Confederate forces under General Robert E. Lee had defeated their Union counterparts sometimes by
(5) considerable margins—in a string of major battles. In this engagement, however, the Confederate army was defeated and driven back. Even more important than

their material losses, though, was the Confederacy's loss of momentum. Union forces took the initiative, finally
(10) defeating the Confederacy less than two years later. By invading Union territory, the Confederate leadership had sought to shatter the Union's will to continue the war and to convince European nations to recognize the Confederacy as an independent nation. Instead,
(15) the Union's willingness to fight was strengthened and the Confederacy squandered its last chance for foreign support.

Words of emphasis, though rare, are the most important category of Kaplan keywords. Why? Because if you see an emphasis keyword in a sentence, you will always get a question about that sentence. Think about it—if the author thinks something is "of primary importance," it would be pretty silly for the test maker not to ask you about it.

In the Gettysburg passage, did you circle *even more important* (line 7)? You should have. You were bound to get a question about it.

By using all of the techniques discussed above, you will be able to tackle the most difficult reading comprehension questions. And now that you have the tools to handle the Verbal sections of the GRE, take a swing at the set of practice questions that follow. Then we'll move on and take a look at the Quantitative sections of the test.

PARAPHRASE ANSWERS

Roman Empire: (C)
Flying: (B)

Verbal Practice Set

Directions: Each of the following questions begins with a sentence that has either one or two blanks. The blanks indicate that a piece of the sentence is missing. Each sentence is followed by five answer choices that consist of words or phrases. Select the answer choice that completes the sentence best. (Answers and explanations can be found at the end of the set of questions.)

Question 1

Victorien Sardou's play *La Tosca* was originally written as a _____ for Sarah Bernhardt and later _____ into the famous Puccini opera.

(A) role . . . reincarnated
(B) biography . . . changed
(C) metaphor . . . edited
(D) present . . . fictionalized
(E) vehicle . . . adapted

Ⓐ Ⓑ Ⓒ Ⓓ Ⓔ

Question 2

Because the law and custom require that a definite determination be made, the judge is forced to behave as if the verdict is _____, when in fact the evidence may not be _____.

(A) negotiable . . . persuasive
(B) justified . . . accessible
(C) unassailable . . . insubstantial
(D) incontrovertible . . . admissible
(E) self-evident . . . conclusive

Ⓐ Ⓑ Ⓒ Ⓓ Ⓔ

Question 3

The author presumably believes that all businessmen are _____, for her main characters, whatever qualities they may lack, are virtual paragons of _____.

(A) clever . . . ingenuity
(B) covetous . . . greed
(C) virtuous . . . deceit
(D) successful . . . ambition
(E) cautious . . . achievement

Ⓐ Ⓑ Ⓒ Ⓓ Ⓔ

Question 4

Satire is a marvelous reflection of the spirit of an age; the subtle _____ of Swift's epistles mirrored the eighteenth century delight in elegant _____.

(A) profundity . . . ditties
(B) poignancy . . . pejoratives
(C) contempt . . . anachronisms
(D) provinciality . . . rusticity
(E) vitriol . . . disparagement

Ⓐ Ⓑ Ⓒ Ⓓ Ⓔ

Questions 5

Ginnie expects her every submission to be published or selected for performance, and this time her _____ is likely to be _____.

(A) candor . . . dispelled
(B) anticipation . . . piqued
(C) enthusiasm . . . dampened
(D) optimism . . . vindicated
(E) awareness . . . clouded

Ⓐ Ⓑ Ⓒ Ⓓ Ⓔ

Question 6

His opponent found it extremely frustrating that the governor's solid support from the voting public was not eroded by his _____ of significant issues.

(A) exaggeration
(B) misapprehension
(C) discussion
(D) selection
(E) acknowledgment

Ⓐ Ⓑ Ⓒ Ⓓ Ⓔ

Question 7

Our spokesperson seems to be uncertain of our eventual victory but _____ facing the alternative, as if merely admitting the possibility of defeat would lead to the dread thing itself.

(A) unsure of
(B) deterred from
(C) fearful of
(D) certain of
(E) helped by

Ⓐ Ⓑ Ⓒ Ⓓ Ⓔ

Directions: Each of the following questions consist of a pair of words or phrases that are separated by a colon and followed by five answer choices. Choose the pair of words or phrases in the answer choices that is most similar to the original pair.

Question 8

VEGETATE : ACTIVE ::

(A) resist : beaten
(B) mope : gloomy

(C) grow : small
(D) hassle : obnoxious
(E) accept : questioning

Ⓐ Ⓑ Ⓒ Ⓓ Ⓔ

Question 9

PROPONENT : THEORY ::

(A) nonbeliever : sin
(B) traitor : country
(C) adherent : belief
(D) attorney : law
(E) scientist : hypothesis

Ⓐ Ⓑ Ⓒ Ⓓ Ⓔ

Question 10

SPECIES : ORGANISM ::

(A) specialty : physician
(B) origin : idea
(C) language : foreigner
(D) genre : literature
(E) family : ancestry

Ⓐ Ⓑ Ⓒ Ⓓ Ⓔ

Question 11

DISCHARGED : SOLDIER ::

(A) fired : cannon
(B) graduated : student
(C) appointed : judge
(D) transferred : employee
(E) docked : salary

Ⓐ Ⓑ Ⓒ Ⓓ Ⓔ

Question 12

CUT : LACERATION ::

(A) park : place
(B) slit : gap

(C) knife : separation
(D) hole : puncture
(E) boil : blister

Ⓐ Ⓑ Ⓒ Ⓓ Ⓔ

Question 13

OUTFOX : STRATEGY ::

(A) outdo : trickery
(B) defeat : stamina
(C) outlast : force
(D) victimize : terror
(E) outrun : speed

Ⓐ Ⓑ Ⓒ Ⓓ Ⓔ

Question 14

COAX : BLANDISHMENT ::

(A) amuse : platitudes
(B) compel : threats
(C) deter : tidings
(D) batter : insults
(E) exercise : antics

Ⓐ Ⓑ Ⓒ Ⓓ Ⓔ

Question 15

TITLED : NOBLE ::

(A) elected : president
(B) acclaimed : artist
(C) commissioned : officer
(D) deposed : ruler
(E) initiated : argument

Ⓐ Ⓑ Ⓒ Ⓓ Ⓔ

Question 16

NOD : ASSENT ::

(A) glance : beneficence
(B) shudder : rudeness

(C) wink : mystification
(D) shrug : indifference
(E) frown : capriciousness

Ⓐ Ⓑ Ⓒ Ⓓ Ⓔ

Questions 17–19: The four Galilean satellites of Jupiter probably experienced early, intense bombardment. Thus, the very ancient surface of Callisto remains scarred by impact craters. The younger, more varied surface of Ganymede reveals distinct light and dark areas, the light areas featuring networks of intersecting grooves and ridges, probably resulting from later iceflows. The impact sites of Europa have been almost completely erased, apparently by water outflowing from the interior and instantly forming vast, low, frozen seas. Satellite photographs of Io, the closest of the four to Jupiter, were revelatory. They showed a landscape dominated by volcanos, many erupting, making Io the most tectonically active object in the solar system. Since a body as small as Io cannot supply the energy for such activity, the accepted explanation has been that, forced into a highly eccentric orbit, Io is engulfed by tides stemming from a titanic contest between the other three Galilean moons and Jupiter.

Question 17

According to the passage, which of the following is probably NOT true of the surface of Io?

(A) It is characterized by intense tectonic activity.
(B) Its volcanos have resulted from powerful tides.
(C) It is younger than the surface of Callisto.
(D) It is distinguished by many impact craters.
(E) It has apparently not been shaped by internal force.

Ⓐ Ⓑ Ⓒ Ⓓ Ⓔ

Question 18

It can be inferred that the geologic features found in the light areas of Ganymede were probably formed

(A) subsequent to the features found in the dark areas
(B) in an earlier period than those in the dark areas
(C) at roughly the same time as the features found in the dark areas
(D) primarily by early bombardment
(E) by the satellite's volcanic activity

Ⓐ Ⓑ Ⓒ Ⓓ Ⓔ

Question 19

It can be inferred that the author regards current knowledge about the satellites of Jupiter as

(A) insignificant and disappointing
(B) grossly outdated
(C) completed satisfactory
(D) ambiguous and contradictory
(E) persuasive though incomplete

Ⓐ Ⓑ Ⓒ Ⓓ Ⓔ

Answer Key

1. E	8. E	15. C
2. E	9. C	16. D
3. A	10. D	17. D
4. E	11. B	18. A
5. D	12. B	19. E
6. B	13. E	
7. C	14. B	

Explanations

Question 1

Victorien Sardou's play *La Tosca* was originally written as a _____ for Sarah Bernhardt and later _____ into the famous Puccini opera.

Taking a creative work like a play and moving it into another medium is an act of adaptation, so having seen *adapted* among the choices you might have been drawn right away to correct choice (E). And *vehicle* should not have been problematic for you—the sentence refers to *vehicle* not as a means of conveyance but as a means of display or expression. We can infer that *La Tosca* was Bernhardt's vehicle in the sense that it was created for her, to display her particular talents. (The purpose of any "star vehicle" is to showcase that star.)

It is incorrect to say that a play is written as a "role" (A)—written "to provide a role" would be more acceptable grammatically—and a work is not "reincarnated" from one medium into another, that verb being best reserved for the reembodiment or rebirth of living entities. The idea of *La Tosca*'s being a "biography" (B) for Bernhardt doesn't make sense (if the play were about her life, *biography of* would work), so this choice is out even though *changed* isn't bad in the second blank. A metaphor (C) is a poetic or figurative representation of something, and though we might call a play a metaphor for some event or idea we would not be likely to do so for a human being; *edited* provides a further complication, in that the process of editing requires pruning and revision, whereas changing a play into a musical drama requires a great deal more firsthand creativity. And while M. Sardou might well have offered *La Tosca* as a "present" to Mme. Bernhardt (D), *fictionalized* won't do; a real-life event can be fictionalized—into a play or an opera—but that verb cannot apply to something that is already fiction.

Question 2

Because the law and custom require that a definite determination be made, the judge is forced to behave as if the verdict is _____, when in fact the evidence may not be _____.

If the requirement is that the verdict be a "definite determination," then a judge is pressured to consider a verdict to be definitely determined even when there is some room for doubt. (This analysis is supported by the author's use of the phrases *as if,* meaning something hypothetical, and *when in fact,* meaning that which is actually true.) Thus, if the evidence in a case is not "conclusive" (E), if there is room for doubt as to the guilt of the accused, a verdict based upon it probably will not be "self-evident" but will have to be treated as such by a judge (in the face of law and custom, that is).

Certainly if the evidence in a case is not "persuasive" (A), if the conclusion stemming from the evidence is debatable, it surely does suggest there's room for doubt. But pressure for a definite determination would hardly force a judge to view a verdict as "negotiable," that is, open for debate among the interested parties and possibly subject to revision. On the contrary, the more "negotiable" the verdict, the less "definitely determined" it's likely to be. *Justified* (B) works well—the judge might have to consider this verdict warranted even if the evidence didn't support it—but *accessible* in none of its meanings (easily approached; obtainable; open to influence) fits the context. Similarly the first words of (C) and (D), *unassailable* and *incontrovertible* respectively, give us what we need—a verdict that must be seen as a "definite determination"—but their respective second words shoot the choices down. Evidence that's not "insubstantial" is substantial, and there's no contradiction between an "unassailable" verdict and one based on substantial evidence. *Admissible* plays on your associations with real-life law, but the issue of whether or not something may properly be brought into evidence is far removed from the author's central point.

Question 3

The author presumably believes that all businessmen are _____, for her main characters, whatever qualities they may lack, are virtual paragons of _____.

The author mentioned in this sentence believes that businessmen are models of some quality; *whatever qualities they may lack* implies that whatever bad points they possess, there's this one particular good thing about them. All of this should lead you to (A)—if an author's main characters are businessmen, and if they're all paragons of "ingenuity" (meaning inventively talented), one could easily be led to the presumption that the author thinks all businessmen are "clever."

Several of the wrong answers play off your possible biases about people in the business world, (B) being the most blatant in that regard. That choice is tempting only because an author's use of many "greedy" businessman characters might suggest that that author thinks all businessmen are "covetous." But labeling businessmen as greedy contradicts the sense of "whatever qualities they may lack"—as we noted, we need a positive quality. (Also, *paragons of greed* is awkward.) One who is morally upright or "virtuous" (C) would hardly be a paragon of "deceit" (lying, falseness). Characters possessing great "ambition" (D) wouldn't necessarily make one presume that the author believes all such people are "successful," since ambition and success in a field don't always go hand in hand; and there's even less connection between businessman characters who demonstrate great "achievement" (E) and a conclusion that, in the creator's opinion, all businessmen are "cautious."

Question 4

Satire is a marvelous reflection of the spirit of an age; the subtle _____ of Swift's epistles mirrored the eighteenth century delight in elegant _____.

The idea being communicated is that satire reflects the spirit of the age in which it's written, and the semicolon suggests that what's coming up is an example. So Swift must be a satirist, and something "subtle" about his work parallels something "elegant" about his work, which in turn parallels something "elegant" about his era. No outside knowledge of the eighteenth century is needed, because the only choice that works is (E). *Vitriol* is sulfuric acid in chemistry but sarcastic criticism in literature; literary vitriol, especially the "subtle" kind, would certainly mirror the spirit of an age that took delight in suave put-downs, a/k/a elegant "disparagement."

(A) presents a grave contradiction: An age fond of "ditties," brief and insubstantial songs or poems, would hardly be mirrored by "profound" (weight or deep) literary achievement. Two problems with (B): Not only is poignancy—the quality of affecting the emotions in a heartfelt way—not a characteristic usually found in any kind of satire, but poignancy would not "mirror the spirit" of elegant "pejoratives," that is, disparaging remarks. Good, wicked satire might very well display scorn or "contempt" (C), but that would have nothing to do with elegant "anachronisms," obsolete or archaic people or devices. (D) is the only choice providing two words that do, in a sense, mirror each other. Provinciality is rough-hewn unsophistication associated with those living in provinces, or a limited point of view—also called *parochial*. And rusticity, which is a country lifestyle, can also be boorishness and a lack of couth. But the two words aren't associated with the rest of the sentence, in which references to subtlety and (especially) elegance suggest something quite different from rusticity.

Question 5

Ginnie expects her every submission to be published or selected for performance, and this time her _____ is likely to be _____.

The "submissions" described must be manuscripts: Apparently Ginnie is an author who believes she'll strike gold every time she sends in a story or play. The structural signal *and* suggests that her expectations are going to be taken a step further. Now her "optimism" will be "vindicated" (D) and she"ll be published. That structural signal, by the way, is what keeps (C) from being correct: If the signal were *but,* then we'd need a contrast, and Ginnie's "dampened enthusiasm" would contrast strongly with her typical expectations of success. Since Ginnie always figures that her stuff will be accepted, there's no reason for the sentence to point to her "anticipation" being " piqued" (B) on this particular occasion: her anticipation is always piqued (aroused, excited). Nothing in the sentence refers to or even hints at Ginnie's habit of speaking frankly, so it would be improper to conclude with a reference to her "candor" (A), "dispelled" or not. Similarly, Ginnie's perennial optimism about her chances at publication really has nothing to do with her "awareness" (E), but even if you justify it as a reference to "awareness of her chances to be published," a "clouded" awareness would suggest she's going to get shot down this time, and would require a contrast signal like *but* rather than *and.*

Question 6

His opponent found it extremely frustrating that the governor's solid support from the voting public was not eroded by his _____ of significant issues.

No candidate would be pleased at his or her opponent's "solid support from the voting public," but any candidate would become mighty frustrated if such support continued despite overwhelming reasons why

it should cease. In this instance, we can infer that the popular governor remains popular despite the fact that he either doesn't understand "significance issues" or has made foolish choices as to what the "significant issues" are. In line with that analysis only (B) works: if the governor "misapprehended," or misunderstood, the issues, how frustrating it would be to his opponent when the public seemed not to care!

You might have been tempted by (C) or (D), and both are wrong for pretty much the same reason— each is too neutral in tone. The other choices are a good deal worse. It's not clear how a candidate would "exaggerate" (A) significant issues, nor why public support would be expected to erode as result of such overstatement; and a candidate's "acknowledgment" (E) of the key issues—recognition of their existence, perhaps even of their significance—would probably have an effect opposite to the erosion of voter support.

Question 7

Our spokesperson seems to be uncertain of our eventual victory but _____ facing the alternative, as if merely admitting the possibility of defeat would lead to the dread thing itself.

The logic of this sentence is somewhat complex, so the question is tough even though there's only one blank to fill. Actually, it's often the case that hypotheticals—here signalled by the phrase *as if*—are harder to understand and follow than direct statements of fact. Piecing the thing out (rather than attacking it all at once) should help. The alternative of victory is defeat, and that must be the "dread thing" to which the author refers. The spokesperson isn't sure whether our side will win but doesn't want to admit that we will lose, as if (here comes the hypothetical) to say, "We're going to lose!" would cause the defeat to happen. So the spokesperson doesn't want to think about our losing, is "fearful of" it (C).

One might think that either (A) or (D) must be the answer, since they're direct opposites and therefore can't both be incorrect; but they are both wrong and to understand why, we must keep in mind the overall sentence sense. The spokesperson is neither "certain" nor "unsure" of facing the alternative that we might be defeated—rather, he or she dreads facing it. (That's a certainty of sorts, though it doesn't make *certain of* correct.) To be deterred (B) means that someone or something is stopping the spokesperson from facing up to our possible loss. But the sentence seems to suggest that it's the spokesperson, or rather the spokesperson's fear, that's doing the deterring, not some outside influence at all. Finally, (E) makes no sense: how would one be "helped by" being so pessimistic about the outcome?

Question 8

One who vegetates, is inert or inactive, is of course not active. If this gave you trouble, think of vegetables, which are markedly inactive living things, or perhaps the expression "veg out" meaning to stagnate, not think or do anything. Similarly, one who accepts (E) takes things as they are without doubting, doesn't call things into question; he or she is therefore not questioning.

Notice that the stimulus conveys a sense of definition—a vegetator by definition is not active; such a sense is missing from (A), since one who resists may or may not be beaten, now or at some future time. When one mopes (B), it's usually because one feels gloomy— a gloomy person is characterized by moping, not a lack thereof. If a plant grows (C) from a height of 2 feet and 9 inches to a whopping 3 feet, it has indeed grown but remains small. As for (D), someone who hassles you is probably obnoxious, or at least can't be said to show a lack of obnoxiousness.

Question 9

A proponent is someone who argues in favor of or supports some idea or practice. Someone who supports the civil-rights movement can be said to be a proponent of that cause. So we could say that one who supports a particular theory is a proponent of that theory, just as one who support or argues in favor of a particular belief (C) can be called an adherent of that belief. If *adherent* gave you trouble, think of *adhere* (to sick to) or *adhesive* (something that makes things stick together).

While a nonbeliever (A) might be likely to commit a sin in the eyes of others, the term *nonbeliever* is rarely applied to a person who supports a particular sin. A traitor (B) doesn't support his particular country, he betrays it; in betraying one country he may be supporting another, but in that sense he becomes a patriot, not a traitor, so this still doesn't work out. Now an attorney (D) does deal with the law, and must (or should, anyway) support the law. But we really can't say that an attorney is someone who supports a particular law—attorneys deal with all sorts of laws, and are honor-bound to respect them all, whereas adherents or proponents are so-called because of their devotion to one particular idea. And a scientist (E) is just as likely to condemn a given hypothesis as support it—and at any rate, his position on hypotheses is not what defines him as a scientist.

Question 10

Species is a category of living things, and living things are known as *organisms*. *Species* refers to the particular type of organism in question, just as *genre* (D) is a term used to classify or categorize types of literature. Human beings are the species known as *Homo sapiens,* so that's how we're classified among all organisms. The novels of Agatha Christie are classified as mysteries: *mystery* is the classification of them among books.

Physicians (A) are, to an extent, categorized by their specialty—the obstetricians are different from the pediatricians, etcetera. But species and genre are both formal means of classification of their respective worlds, and they specifically break down the various groups for purposes of study. The specialty of a physician, on the other hand, isn't so definitive—it's just the branch of medicine the doctor happens to be most involved with. And no one breaks down the group *physicists* into specialties for the purposes of study. The origin (B) of an idea is not the idea's category—it's where the idea comes from. And the language (C) of a foreigner is not a category. Finally, *family* (E) is used in several figurative senses to classify things (for example, the tiger is a member of the cat family), but as it relates to ancestry it really carries no meaning other than that of a synonym (your ancestry, or the line of your ancestors, is your family).

Question 11

Discharged is what a soldier is said to be when his or her tour of duty is up and he or she is released from commitment to the armed forces, just as graduated (B) is what a student is said to be when he or she has completed a particular stage of schooling, and is released. *Fired* (A) can be applied to a cannon that has made a shot, but it's the cannonball in this case that's released, not the cannon. In states in which judges (C) are not elected they are appointed to the bench—that is, an important official such as the governor or the president appoints the judge to a job. But that begins his or her commitment rather than ends it. An employee (D) who has been transferred has not been released, but had his responsibilities shifted or been relocated. And *docked* (E), as it pertains to *salary,* refers to money withheld from an employee, usually as punishment for poor work. *Docked* does not refer to a salary that has finished its responsibility.

Question 12

A laceration is a large cut, especially as it pertains to a wound. Or you could say a cut is an especially small or minor laceration. (A laceration might require stitches—a cut generally requires a Band-Aid.) Similarly, a slit (B) is a tiny crack, cut, or separation in something, and can be called a small gap.

The precise relationship between *park* (A) and *place* isn't entirely clear—perhaps "the park is a place" or "I can't find a place to park"? (And in Monopoly, there's always Park Place.) At any rate, this isn't what we're after. By cutting, a knife (C) can separate, but a knife isn't a small separation itself. A hole (D) and a puncture are really the same thing, and you can't say that the former is a smaller version of the latter. And as for (E), not only are boils and blisters very different from each other, medically speaking, but even if they were more alike, blisters are often smaller sores than boils.

Question 13

To outfox someone is to defeat or win out over him by means of superior strategy, just as to outrun (E) someone is to beat him by means of superior speed. This is one of those items in which the vocabulary is simple, and you may therefore be prone to overconfidence and careless mistakes. The choices have to be examined carefully here. The most tempting wrong answer was perhaps (A), since it is true that one can outdo another by means of trickery. But outfoxing, outrunning, and outlasting are all specific kinds of outdoing that have to do with specific method, whereas outdoing is very general, and could apply to strategy and speed just as easily as it applies to trickery—there are unlimited ways to outdo someone. The same is true of *defeat* (B)—it's simply too general, and there are many means of defeating someone other than having greater stamina. To outlast someone (C) is, of course, to defeat him by means of superior endurance, or superior patience, not specifically by means of superior force. And victimizing someone (D)

may indeed require terror, but victimizing has nothing to do with winning, and there are means other than terror by which people are victimized.

Question 14

The stimulus and correct answer are really flip sides of each other, but the actions they describe share a common purpose—to get someone to do what you want him to do. In fact, if the stimulus failed—if you couldn't coax (or persuade) someone through a series of flattering and cajoling actions and speeches known as blandishment, you might take the next step and attempt to compel (B) him by making threats.

Platitudes (A) are trite, dull remarks, usually trotted out as if they were fresh and new but not usually trotted out for the purpose of amusing someone. If your aim was to deter (C) someone from a particular course, there are many more effective tools than tidings, which are news, information, or data. You might batter someone (D) with insults—in a figurative sense, at least—but you would not use them to batter someone into doing something. And to exercise (E) often amounts to engaging in antics, but you don't use antics to exercise someone into doing something—not even in the alternative meaning of the verb *exercise,* that is, to annoy or make uneasy.

Question 15

Here's one that's easier to work with if we phrase the relationship from right to left. A person is called a noble (noun) because he or she is titled—that is, he or she has been bestowed with a title, such as Sir, Lord, or Squire. Similarly, an officer (C), especially one of the military, is an officer because he or she's been commissioned.

You may think that a president (A) is the president because he or she has been elected. Not exactly true—a president is chosen by election, but is made president by inauguration. (Besides, presidents have finite terms of office, and can't really be said for the rest of

their lives to be elected, whereas a noble is always titled and an officer is always commissioned unless their title is stripped.) An especially good artist may be acclaimed (B), or highly praised, but that doesn't make him or her an artist. A deposed (D) ruler is one who has been kicked out. What initiates an argument cannot be made analogous to the stimulus.

Question 16

A nod indicates assent, or agreement. A nod doesn't always communicate assent, but it's generally taken that way, just as a shrug (D) is generally taken to be an indication of indifference. If you said to someone, "I make a million dollars a minute!" and that person nodded, he'd be indicating his assent. If he shrugged, he would be indicating his indifference.

A glance (A) has no one particular meaning in terms of what it communicates to the person who's being glanced at, so it can't really be said to commonly indicate beneficence, or kindness. You could shoot someone a nasty glance as easily as a kind one. One might shudder (B) in response to especially noxious rudeness, but one doesn't shudder to indicate rudeness. One winks (C) to say hello, to indicate a secret or in-joke, or to get something out of one's eye—one doesn't wink to show mystification. And one would tend to frown (E) to show sadness or anger, not capriciousness (the state of being changeable or guided by whims).

Question 17

GRE science passages often focus on one big contrast—and the questions will focus on the same contrast, again and again. Thus even detail questions like this one are really "main idea" questions in disguise. In this passage, the big contrast is between moons that have remained unchanged since their "early, intense bombardment" (sentence 1), and those whose surfaces have been altered in more recent epochs. Io is mentioned in the last three sentences. These sentences stress recent, indeed ongoing,

changes in the satellite's surface. By inference, most impact craters from the long-ago bombardments have probably been obliterated (D). Continuing tectonic activity (A) is mentioned explicitly; tides (B) are mentioned in the final sentence as the probable cause of the tectonic activity, and hence the active volcanos. Inferably Io's surface is younger than the "very ancient" surface of Callisto, (C). (E) is the only tricky choice. The phrase *tectonically active* may automatically conjure up the idea of internal forces, since these cause tectonic activity on the earth. But it's explicitly stated (last sentence) that Io is too small to supply its own energy for such activity, so (E) is true.

Question 18

Again, keep your eye on the big contrast. The bombardments, and the craters that record them, were laid down long ago; thus a surface marked by impact craters (the dark areas of Ganymede) is older than one not so marked (the lighter areas). In addition, it's mentioned that some features of the light areas probably result from later iceflows. Thus, (A) is correct, ruling out (B) and (C). The light areas feature grooves and ridges probably resulting from these iceflows, not from early bombardment (D). Volcanic activity, (E), is not mentioned in relation to Ganymede.

Question 19

The passage conveys a great deal of information, which the author implicitly accepts, ruling out (A). (The word used for the photographs of Io—*revelatory*—is enough by itself to eliminate this choice.) The fact that information about Io comes from satellite photographs rules out (B). No contradictions are mentioned, and though areas of uncertainty remain, (D)'s ambiguous and contradictory will not work as a general characterization. On the other hand, (C) is out because of the cautious language used throughout: *probably* (twice), *apparently*, and *the accepted explanation*. Hence the knowledge is persuasive though incomplete, as specified in (E).

CHAPTER 3

TEST CONTENT: QUANTITATIVE

Y ou'll have 45 minutes to complete 28 questions in the Quantitative section. The GRE tests the same sort of mathematical concepts that the SAT does: arithmetic, algebra, and geometry. There is no trigonometry or calculus tested on the GRE. However, the test does contain some mathematical concepts that you didn't see on the SAT: Median, mode, range, standard deviation, and simple probability.

There are three formats of GRE math questions: problem solving, which have five answer choices each; graph questions, which have five answer choices each and which are based on one or more graphs; and quantitative comparisons (QCs), which have only four answer choices each. Because about half of the questions on each math section are QCs, let's look at these first.

Quantitative Comparisons

In QCs, instead of solving for a particular value, you need to compare two quantities. At first, QCs may appear really difficult because of their unfamiliar format. However, once you become used to them, they can be quicker and easier than the other types of math questions.

The Questions

The difficulty of the QCs will depend on how well you are doing in the section. In each question, you'll see two mathematical expressions. They are boxed, one in Column A, the other in Column B. Your job is to compare them. Some questions include additional information about one or both quantities. This information is centered, unboxed, and essential to making the comparison.

LEARN THE ANSWER CHOICES

To score high on QCs, learn what the answer choices stand for, and know these cold.

The Directions

The directions you'll see will look something like this:

Directions: Each of Questions 1–15 consists of two quantities, one in column A and another in Column B. You are to compare the two quantities and answer

(A) if the quantity in Column A is greater

(B) if the quantity in Column B is greater

(C) if the two quantities are equal

(D) if the relationship cannot be determined from the information given

Common information: In a question, information concerning one or both of the quantities to be compared is centered above the two columns. A symbol that appears in both columns represents the same thing in Column A as it does in Column B.

Basic Principles of Quantitative Comparisons

Choices (A), (B), and (C) all represent definite relationships between the quantities in Column A and Column B. But choice (D) represents a relationship that cannot be determined. Here are two things to remember about choice (D) that will help you decide when to pick it:

- Choice (D) is never correct if both columns contain only numbers. The relationship between numbers is unchanging, but choice (D) means more than one relationship is possible.
- Choice (D) is correct if you can demonstrate two different relationships between the columns.

Suppose you ran across the following QC:

Column A	Column B
$2x$	$3x$

If x is a positive number, Column B is greater than Column A. If $x = 0$, the columns are equal. If x equals any negative number, Column B is less than Column A. Because more than one relationship is possible, the answer is (D). In fact, as soon as you find a second possibility, stop work and pick choice (D).

The Kaplan Method for Quantitative Comparisons

Here are six Kaplan strategies that will enable you to make quick comparisons. In the rest of this chapter you'll learn how they work and you'll try them on practice problems.

1. *Compare Piece by Piece.*

   This works on QCs that compare two sums or two products.

2. *Make one column look like the other.*

   This is a great approach when the columns look so different that you can't compare them directly.

3. *Do the same thing to both columns.*

   Change both columns by adding, subtracting, multiplying, or dividing by the same amount on both sides in order to make the comparison more apparent.

4. *Pick numbers.*

   Use this to get a handle on abstract algebra QCs.

5. *Redraw the diagram.*

   Redrawing a diagram can clarify the relationships between measurements.

6. *Avoid QC traps.*

   Stay alert for questions designed to fool you by leading you to the obvious, wrong answer.

Now let's look at how these strategies work.

1. *Compare Piece by Piece*

Column A	Column B
$w > x > 0 > y > z$	
$w + y$	$x + z$

In this problem, there are four variables—w, x, y, and z. Compare the value of each "piece" in each column. If every "piece" in one column is greater than a corresponding "piece" in the other column and the only operation involved is addition, the column with the greater individual values will have the greater total value.

From the given information we know that $w > x$ and $y > z$. Therefore, the first term in Column A, w, is greater than the first term in Column B, x. Similarly, the second term in Column A, y, is greater than the second term in Column B, z. Because each piece in Column A is greater than the corresponding piece in Column B, Column A must be greater; the answer is (A).

2. Make One Column Look Like the Other
When the quantities in Columns A and B are expressed differently, you can often make the comparison easier by changing one column to look like the other. For example, if one column is a percent, and the other a fraction, try converting the percent to a fraction.

Column A	Column B
$x(x-1)$	$x^2 - x$

Here Column A has parentheses, and Column B doesn't. So make Column A look more like Column B: get rid of those parentheses. You then end up with $x^2 - x$ in both columns, which means they are equal and the answer is (C).

Try another example, this time involving geometry.

Column A	Column B

The diameter of circle O is d and the area is a.

Column A	Column B
$\dfrac{\pi d^2}{2}$	a

Make Column B look more like Column A by rewriting a, the area of the circle, in terms of the diameter, d. The area of any circle equals πr^2, where r is the radius.

Since the radius is half the diameter, we can plug in $\frac{d}{2}$ for r in the area formula to get $\pi(\frac{d}{2})^2$ in Column B. Simplifying we get $\frac{\pi d^2}{4}$. Since both columns contain π, we can simply compare $\frac{d^2}{2}$ with $\frac{d^2}{4}$. $\frac{d^2}{4}$ is half as much as $\frac{d^2}{2}$, and since d^2 must be positive, Column A is greater and choice (A) is correct.

3. Do the Same Thing to Both Columns

Some QC questions become much clearer if you change not just the appearances but also the values of both columns. Treat them like two sides of an inequality, with the sign temporarily hidden.

You can add or subtract the same amount from both columns, and multiply or divide by the same positive amount without altering the relationship. You can also square both columns if you're sure they're both positive. But watch out. Multiplying or dividing an inequality by a negative number reverses the direction of the inequality sign. Since it alters the relationship between the columns, avoid multiplying or dividing by a negative number.

In the QC below, what could you do to both columns?

RULE OF THUMB

Don't multiply or divide both QC columns by a negative number.

Column A	Column B
	$4a + 3 = 7b$	
$20a + 10$	$35b - 5$

All the terms in the two columns are multiples of 5, so divide both columns by 5 to simplify. You're left with $4a + 2$ in Column A and $7b - 1$ in Column B. This resembles the equation given in the centered information. In fact, if you add 1 to both columns, you have $4a + 3$ in Column A and $7b$ in Column B. The centered equation tells us they are equal. Thus choice (C) is correct.

In the next QC, what could you do to both columns?

Column A	Column B
	$y > 0$	
$1 + \dfrac{y}{1+y}$	$1 + \dfrac{1}{1+y}$

Solution: First subtract 1 from both sides. That gives you $\dfrac{y}{1+y}$ in Column A, and $\dfrac{1}{1+y}$ in Column B. Then multiply both sides by $1 + y$, which must be positive since y is positive. You're left comparing y with 1.

You know y is greater than 0, but it could be a fraction less than 1, so it could be greater or less than 1. Since you can't say for sure which column is greater, the answer is (D).

4. Pick Numbers

If a QC involves variables, try picking numbers to make the relationship clearer. Here's what you do:

- Pick numbers that are easy to work with.
- Plug in the numbers and calculate the values. Note the relationship between the columns.
- Pick another number for each variable and calculate the values again.

Column A	Column B
$r > s > t > w > 0$	
$\frac{r}{t}$	$\frac{s}{w}$

Try $r = 4$, $s = 3$, $t = 2$, and $w = 1$. Then Column A = $\frac{r}{t}$ = $\frac{4}{2}$ = 2. And Column B = $\frac{s}{w}$ = $\frac{3}{1}$ = 3. So in this case Column B is greater than Column A.

Always Pick More Than One Number and Calculate Again. In the example above, we first found Column B was bigger. But that doesn't mean Column B is always bigger and that the answer is (B). It does mean the answer is not (A) or (C). But the answer could still be (D)—not enough information to decide.

If time is short, guess between (B) and (D). But whenever you can, pick another set of numbers and calculate again.

As best you can, make a special effort to find a second set of numbers that will alter the relationship. Here for example, try making r a lot larger. Pick $r = 30$ and keep the other variables as they were. Now Column A = $\frac{30}{2}$ = 15. This time, Column A is greater than Column B, so answer choice (D) is correct.

Pick Different Kinds of Numbers. Don't assume all variables represent positive integers. Unless you're told otherwise, variables can represent zero, negative numbers, or fractions. Since different kinds of numbers behave differently, always pick a different kind of number the second time around. In the example above, we plugged in a small positive number the first time and a larger number the second.

In the next three examples, we pick different numbers and get different results. Since we can't find constant relationships between Columns A and B, in all these cases the answer is (D).

RULE OF THUMB

If the relationship between Columns A and B changes when you pick other numbers, (D) must be the answer.

<u>Column A</u>   <u>Column B</u>

w $-w$

If $w = 5$, Column A = 5 and Column B = -5, so Column A is greater.
If $w = -5$, Column A = -5 and Column B = 5, so Column B is greater.

<u>Column A</u>   <u>Column B</u>

$$x \neq 0$$

x $\frac{1}{x}$

If $x = 3$, Column A = 3 and Column B = $\frac{1}{3}$, so Column A is greater.

If $x = \frac{1}{3}$, Column A = $\frac{1}{3}$ and Column B = $\dfrac{1}{\frac{1}{3}} = 3$, so Column B is greater.

<u>Column A</u>   <u>Column B</u>
x x^2

If $x = \frac{1}{2}$, Column A = $\frac{1}{2}$ and Column B = $\frac{1}{4}$, so Column A is greater.

If $x = 2$, Column A = 2 and Column B = 4, so Column B is greater.

5. Redraw the Diagram

- Redraw a diagram if the one that's given confuses you.
- Redraw scale diagrams to exaggerate crucial differences.

Some geometry diagrams may be misleading. Two angles or lengths may look equal as drawn in the diagram, but the given information tells you that there is a slight difference in their measures. The best strategy is to redraw the diagram so that their relationship can be clearly seen.

<u>Column A</u>   <u>Column B</u>

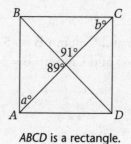

ABCD is a rectangle.

a   *b*

Redraw this diagram to exaggerate the difference between the 89° degree angle and the 91° angle. In other words, make the larger angle much larger, and the smaller angle much smaller. The new rectangle that results is much wider than it is tall.

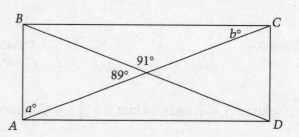

In the new diagram, where the crucial difference jumps out, *a* is clearly greater than *b*.

KAPLAN

6. Avoid QC Traps

To avoid QC traps, always be alert. Don't assume anything. Be especially cautious near the end of the question set.

Don't Be Tricked by Misleading Information

Column A	Column B
	John is taller than Bob.	

Column A	Column B
John's weight in pounds	Bob's weight in pounds

The test makers hope you think, "If John is taller, he must weigh more." But there's no guaranteed relationship between height and weight, so you don't have enough information. The answer is (D). Fortunately, problems like this are easy to spot if you stay alert.

Don't Assume. A common QC mistake is to assume that variables represent positive integers. As we saw in using the picking numbers strategy, fractions or negative numbers often show another relationship between the columns.

Column A	Column B
	When 1 is added to the square of *x* the result is 37.	
x	6

It is easy to assume that *x* must be 6, since the square of *x* is 36. That would make choice (C) correct. However, it is possible that $x = -6$. Since *x* could be either 6 or -6, the answer is (D).

Don't Forget to Consider Other Possibilities

Column A	Column B

$$\begin{array}{r} R \\ S \\ T \\ \hline 1W \end{array}$$

In the addition problem above, *R*, *S*, and *T* are different digits that are multiples of 3, and *W* is a digit.

Column A	Column B
W	8

RULE OF THUMB

Be aware of negative numbers!

Because you're told that R, S, and T are digits and different multiples of 3, most people will think of 3, 6, and 9, which add up to 18. That makes W equal to 8, and Columns A and B equal. But that's too obvious for a QC at the end of the section.

There's another possibility. 0 is also a multiple of 3. So the three digits could be 0, 3, and 9, or 0, 6, and 9, which give totals of 12 and 15, respectively. That means W could be 8, 2, or 5. Since the columns could be equal, or Column B could be greater, answer choice (D) must be correct.

Don't Fall for Look-Alikes

Column A	Column B
$\sqrt{5} + \sqrt{5}$	$\sqrt{10}$

At first glance, forgetting the rules of radicals, you might think these quantities are equal and that the answer is (C). But use some common sense to see this isn't the case. Each $\sqrt{5}$ in Column A is bigger than $\sqrt{4}$, so Column A is more than 4. The $\sqrt{10}$ in Column B is less than another familiar number, $\sqrt{16}$, so Column B is less than 4. The answer is (A).

Now use Kaplan's six strategies to solve nine typical QC questions. Then check your work against our solutions.

Practice Problems

	Column A	Column B
1.	$x^2 + 2x - 2$	$x^2 + 2x - 1$

Ⓐ Ⓑ Ⓒ Ⓓ Ⓔ

$$x = 2y$$
$$y > 0$$

	Column A	Column B
2.	4^{2y}	2^x

Ⓐ Ⓑ Ⓒ Ⓓ Ⓔ

Column A	Column B

$$\frac{x}{y} = \frac{z}{4}$$

x, *y*, and *z* are positive.

3.   6*x*   2*yz*

Ⓐ Ⓑ Ⓒ Ⓓ Ⓔ

---

q, *r*, and *s* are positive integers.

$qrs > 12$

4. $\dfrac{qr}{5}$ $\dfrac{3}{s}$

Ⓐ Ⓑ Ⓒ Ⓓ Ⓔ

$x > 1$

$y > 0$

5. y^x $y^{(x+1)}$

Ⓐ Ⓑ Ⓒ Ⓓ Ⓔ

$7p + 3 = r$

$3p + 7 = s$

6.   *r*   *s*

Ⓐ Ⓑ Ⓒ Ⓓ Ⓔ

In triangle *XYZ*, the measure of angle *X* equals the measure of angle *Y*.

7.   The degree measure   The degree measure
      of angle *Z*   of angle *X* plus the
                                                   degree measure of
                                                   angle *Y*

Ⓐ Ⓑ Ⓒ Ⓓ Ⓔ

$h > 1$

8.   The number of $\dfrac{60}{h}$
      minutes in *h*
      hours

Ⓐ Ⓑ Ⓒ Ⓓ Ⓔ

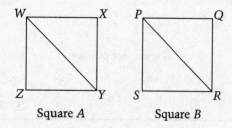

Square A   Square B

9. $\dfrac{\text{Perimeter of square } A}{\text{Perimeter of square } B}$ $\dfrac{\text{Length of } WY}{\text{Length of } PR}$

Ⓐ Ⓑ Ⓒ Ⓓ Ⓔ

Answer Key

1. (B) Comparing the respective pieces of the two columns, the only difference is the third piece: −2 in Column A and −1 in Column B. We don't know the value of x, but whatever it is, x^2 in Column A must have the same value as x^2 in Column B, and $2x$ in Column A must have the same value as $2x$ in Column B. Since any quantity minus 2 must be less than that quantity minus 1, Column B is greater than Column A.

2. (A) Replacing the x exponent in Column B with the equivalent value given in the problem, we're comparing 4^{2y} to 2^{2y}. Since y is greater than zero, raising 4 to the $2y$ power will result in a greater value than raising 2 to the $2y$ power.

3. (B) Do the same thing to both columns until they resemble the centered information. When we divide both columns by $6y$ we get $\dfrac{6x}{6y}$ or $\dfrac{x}{y}$ in Column A, and $\dfrac{2yz}{6y}$, or $\dfrac{z}{3}$ in Column B. Since $\dfrac{x}{y} = \dfrac{z}{4}$, and $\dfrac{z}{3} > \dfrac{z}{4}$ (because z is positive), $\dfrac{z}{3} > \dfrac{x}{y}$.

4. (D) Do the same thing to both columns to make them look like the centered information. When we multiply both columns by $5s$ we get qrs in Column A and 15 in Column B. Since qrs could be any integer greater than 12, it could be greater than, equal to, or less than 15.

KAPLAN

5. (D) Try $x = y = 2$. Then Column A = $y^x = 2^2 = 4$. Column B = y^{x+1} = $2^3 = 8$, making Column B greater. But if $x = 2$ and $y = \frac{1}{2}$, Column A = $(\frac{1}{2})^2 = \frac{1}{4}$ and Column B = $(\frac{1}{2})^3 = \frac{1}{8}$. In this case, Column A is greater than Column B, so the answer is (D).

6. (D) Pick a value for p, and see what effect this has on r and s. If $p = 1$, $r = (7 \times 1) + 3 = 10$, and $s = (3 \times 1) + 7 = 10$, and the two columns are equal. But if $p = 0$, $r = (7 \times 0) + 3 = 3$, and $s = (3 \times 0) + 7 = 7$, and Column A is smaller than Column B. Since there are at least two different possible relationships, the answer is choice (D).

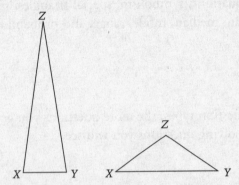

7. (D) Since angle X = angle Y, this is an isosceles triangle. We can draw two diagrams with X and Y as the base angles of an isosceles triangle. In one diagram, make the triangle tall and narrow, so that angle X and angle Y are very large, and angle Z is very small. In this case, column B is greater. In the second diagram, make the triangle short and wide, so that angle Z is much larger than angle X and angle Y. In this case, Column A is greater. Since more than one relationship between the columns is possible, the correct answer is choice (D).

8. (A) The "obvious" answer here is choice (C), because there are 60 minutes in an hour, and 60 appears in Column B. But the number of minutes in h hours would equal 60 times h, not 60 divided by h. Since h is greater than 1, the number in Column B will be less than the actual number of minutes in h hours, so Column A is greater. (A) is correct.

9. (C) We don't know the exact relationship between Square A and Square B, but it doesn't matter. The problem is actually just comparing the ratios of corresponding parts of two squares. Whatever the relationship between them is for one specific length in both squares, the same

relationship will exist between them for any other corresponding length. If a side of one square is twice the length of a side of the second square, the diagonal will also be twice as long. The ratio of the perimeters of the two squares is the same as the ratio of the diagonals. Therefore, the columns are equal. (C) is correct.

Problem Solving

In problem solving, you will have to solve problems that test a variety of mathematical concepts. problem solving questions will cover percentages, simultaneous equations, symbolism, special triangles, multiple and odd-ball figures, mean, median, mode, range, and probability.

The Format

The Questions
As with other question types, the more questions you get right, the harder the problem-solving questions you will see.

The Directions
The directions that you'll see will look something like this:

> **Directions: Each of Questions 16–20 has five answer choices. For each of these questions, select the best answer choices given.**

The Kaplan Approach to Percentages

> Last year Julie's annual salary was $20,000. This year's raise brings her to an annual salary of $25,000. If she gets the same percent raise every year, what will her salary be next year?
>
> (A) $27,500
>
> (B) $30,000
>
> (C) $31,250
>
> (D) $32,500
>
> (E) $35,000

In percent problems, you're usually given two pieces of information and asked to find the third. When you see a percent problem, remember:

- If you are solving for a percent:

$$\text{Percent} = \frac{\text{Part}}{\text{Whole}}$$

- If you need to solve for a part:

$$\text{Percent} \times \text{Whole} = \text{Part}$$

This problem asks for Julie's projected salary for next year—that is, her current salary plus her next raise.

You know last year's salary ($20,000), and you know this year's salary ($25,000), so you can find the difference between the two salaries:

$$\$25{,}000 - \$20{,}000 = \$5{,}000 = \text{her raise}$$

Now find the percent of her raise, by using the formula

$$\text{Percent} = \frac{\text{Part}}{\text{Whole}}$$

Since Julie's raise was calculated on last year's salary, divide by $20,000.

$$\text{Percent raise} = \frac{\$5{,}000}{\$20{,}000} = \frac{1}{4} = 25\%$$

You know she will get the same percent raise next year, so solve for the part. Use the formula: Percent × Whole = Part. Her raise next year will be $25\% \times \$25{,}000 = \frac{1}{4} \times 25{,}000 = \$6{,}250$. Add that amount to this year's salary and you have her projected salary:

$$\$25{,}000 + \$6{,}250 = \$31{,}250 \text{ or answer choice (C).}$$

Make sure that you change the percent to either a fraction or a decimal before beginning calculations.

Practice Problems

Column A	Column B
1.5% of 3% of 45	6.75

Ⓐ Ⓑ Ⓒ Ⓓ Ⓔ

RULE OF THUMB

Be sure you know which whole to plug in. Here you're looking for a percent of $20,000, not $25,000.

MIRROR IMAGE

x percent of y = y percent of x

20% of 50 = 50% of 20

$$\frac{1}{5} \times 50 = \frac{1}{2} \times 20$$

$$10 = 10$$

2. If a sweater sells for $48 after a 25 percent markdown, what was its original price?
(A) $56
(B) $60
(C) $64
(D) $68
(E) $72

Ⓐ Ⓑ Ⓒ Ⓓ Ⓔ

Solutions

1. (B) Percent × Whole = Part. 5% of (3% of 45) = .05 × (.03 × 45) = .05 × 1.35= .0675, which is less than 6.75 in Column B.

2. (C) We want to solve for the original price, the Whole. The percent markdown is 25%, so $48 is 75% of the whole: Percent × Whole = Part.

$$75\% \times \text{Original Price} = \$48$$
$$\text{Original Price} = \frac{\$48}{0.75} = \$64$$

The Kaplan Approach to Simultaneous Equations

If $p + 2q = 14$ and $3p + q = 12$, then $p =$
(A) −2
(B) −1
(C) 1
(D) 2
(E) 3

RULE OF THUMB

Combine the equations—by adding or subtracting them—to cancel out all but one of the variables.

In order to get a numerical value for each variable, you need as many different equations as there are variables to solve for. So, if you have two variables, you need two independent equations.

You could tackle this problem by solving for one variable in terms of the other, and then plugging this expression into the other equation. But the simultaneous equations that appear on the GRE can usually be handled in an easier way.

You can't eliminate p or q by adding or subtracting the equations in their present form. But look what happens if you multiply both sides of the second equation by 2:

$$2(3p + q) = 2(12)$$
$$6p + 2q = 24$$

Now when you subtract the first equation from the second, the qs will cancel out so you can solve for p:

$$6p + 2q = 24$$
$$-[p + 2q = 14]$$

$$5p + 0 = 10$$

If $5p = 10$, $p = 2$.

Practice Problems

1. If $x + y = 8$ and $y - x = -2$, then $y =$
   (A) -2
   (B) 3
   (C) 5
   (D) 8
   (E) 10

   Ⓐ Ⓑ Ⓒ Ⓓ Ⓔ

2. If $m - n = 5$ and $2m + 3n = 15$, then $m + n =$
   (A) 1
   (B) 6
   (C) 7
   (D) 10
   (E) 15

   Ⓐ Ⓑ Ⓒ Ⓓ Ⓔ

Solutions

1. (B) When you add the two equations, the xs cancel out and you find that $2y = 6$, so $y = 3$.

2. (C) Multiply the first equation by 2, then subtract the first equation from the second to eliminate the ms and find that $5n = 5$, or $n = 1$. Plugging this value for n into the first equation shows that $m = 6$, so $m + n = 7$, choice (C).

$$2m + 3n = 15 \qquad m - n = 5 \qquad m + n = 6 + 1 = 7$$
$$-2m + 2n = -10 \qquad m - 1 = 5$$
$$5n = 5 \qquad m = 6$$
$$n = 1$$

The Kaplan Approach to Symbolism

If $a \star b = \sqrt{a + b}$ for all nonnegative numbers, what is the value of $10 \star 6$?

(A) 0

(B) 2

(C) 4

(D) 8

(E) 16

You should be quite familiar with the arithmetic symbols $+$, $-$, $\times$, $\div$, and %. Finding the value of $10 + 2$, $18 - 4$, 4×9, or $96 \div 16$ is easy.

However, on the GRE, you may come across bizarre symbols. You may even be asked to find the value of $10 \star 2$, $5 \ast 7$, $10 \ast 6$, or $65 \heartsuit 2$.

The GRE test makers put strange symbols in questions to confuse or unnerve you. Don't let them. The question stem always tells you what the strange symbol means. Although this type of question may look difficult, it is really an exercise in plugging in.

To solve, just plug in 10 for a and 6 for b into the expression $\sqrt{a + b}$. That equals $\sqrt{10 + 6}$ or $\sqrt{16}$ or 4, choice (C).

How about a more involved symbolism question?

If $a \blacktriangle$ means to multiply a by 3 and $a \ast$ means to divide a by -2, what is the value of $((8 \ast) \blacktriangle) \ast$?

(A) −6

(B) 0

(C) 2

(D) 3

(E) 6

First find $8 \ast$. This means to divide 8 by -2, which is -4. Working out to the next set of parentheses, we have $(-4) \blacktriangle$, which means to multiply -4 by 3, which is -12. Lastly, we find $(-12) \ast$, which means to divide -12 by -2, which is 6, choice (E).

RULE OF THUMB

When a symbolism problem includes parentheses, do the operations inside the parentheses first.

RULE OF THUMB

When two questions include the same symbol, expect the second question to be more difficult and be extra careful.

Practice Problems

Column A	Column B

If $x \neq 0$, let $\spadesuit\, x$ be defined by $\spadesuit\, x = x - \dfrac{1}{x}$

1.   −3 $\spadesuit\,(-3)$

Ⓐ Ⓑ Ⓒ Ⓓ Ⓔ

2. If $r \,\heartsuit\, s = r(r - s)$ for all integers r and s, then $4 \,\heartsuit\, (3 \,\heartsuit\, 5)$ equals

(A) −8

(B) −2

(C) 2

(D) 20

(E) 40

ⒶⒷⒸⒹⒺ

Questions 3–4 refer to the following definition:

$$c \star d = \frac{c - d}{c}, \text{ where } c \neq 0$$

3. $12 \star 3 =$

(A) −3

(B) $\dfrac{1}{4}$

(C) $\dfrac{2}{3}$

(D) $\dfrac{3}{4}$

(E) 3

Ⓐ Ⓑ Ⓒ Ⓓ Ⓔ

4. If $9 \star 4 = 15 \star k$, then $k =$

(A) 3

(B) 6

(C) $\frac{20}{3}$

(D) $\frac{25}{3}$

(E) 9

ⓐ ⓑ ⓒ ⓓ ⓔ

Solutions

1. (B) Plug in –3 for x: $\spadesuit\, x = -3 - \dfrac{1}{-3} = -3 + \dfrac{1}{3} = -2\,\dfrac{2}{3}$, which is greater than –3 in Column A.

2. (E) Start in the parentheses and work out: $(3 \heartsuit 5) = 3(3{-}5) = 3(-2) = -6$; $4 \heartsuit (-6) = 4[4 - (-6)] = 4(10) = 40$.

3. (D) Plug in 12 for c and 3 for d: $12\star 3 = \dfrac{12{-}3}{12} = \dfrac{9}{12} = \dfrac{3}{4}.$

4. (C) Plug in on both sides of the equation:

$$\frac{9{-}4}{9} = \frac{15{-}k}{15}$$

$$\frac{5}{9} = \frac{15{-}k}{15}$$

Cross-multiply and solve for k:

$$75 = 135 - 9k$$

$$-60 = -9k$$

$$\frac{-60}{-9} = k$$

$$\frac{20}{3} = k$$

The Kaplan Approach to Special Triangles

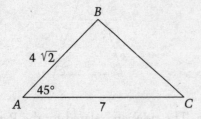

In the triangle above, what is the length of side *BC*?

(A) 4

(B) 5

(C) 4 √2

(D) 6

(E) 5 √2

Special triangles contain a lot of information. For instance, if you know the length of one side of a 30-60-90 triangle, you can easily work out the lengths of the others. Special triangles allow you to transfer one piece of information around the whole figure.

The following are the special triangles you should look for on the GRE.

Equilateral Triangles
All interior angles are 60° and all sides are of the same length.

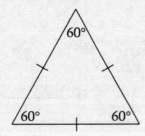

Isosceles Triangles
Two sides are of the same length and the angles facing these sides are equal.

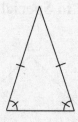

Right Triangles

Contain a 90° angle. The sides are related by the Pythagorean theorem. $a^2 + b^2 = c^2$ where a and b are the legs and c is the hypotenuse.

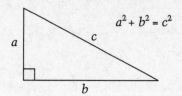

The "Special" Right Triangles

Many triangle problems contain "special" right triangles, whose side lengths always come in predictable ratios. If you recognize them, you won't have to use the Pythagorean theorem to find the value of a missing side length.

The 3-4-5 Right Triangle

(Be on the lookout for multiples of 3-4-5 as well.)

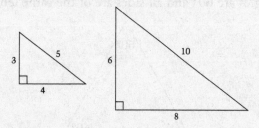

The Isosceles Right Triangle

(Note the side ratio: 1 to 1 to $\sqrt{2}$.)

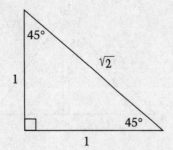

The 30-60-90 Right Triangle
(Note the side ratio: 1 to $\sqrt{3}$ to 2, and which side is opposite which angle.)

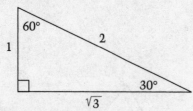

Getting back to our example, you can drop a vertical line from *B* to line *AC*. This divides the triangle into two right triangles.

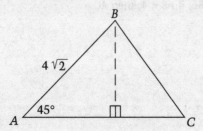

That means you know two of the angles in the triangle on the left: 90° and 45°. So this is an isosceles right triangle, with sides in the ratio of 1 to 1 to $\sqrt{2}$. The hypotenuse here is $4\sqrt{2}$, so both legs have length 4. Filling this in, you have:

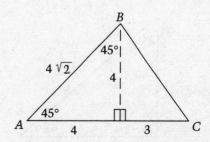

Now you can see that the legs of the smaller triangle on the right must be 4 and 3, making this a 3-4-5 right triangle, and the length of hypotenuse *BC* is 5.

Practice Problems

<u>Column A</u>   <u>Column B</u>

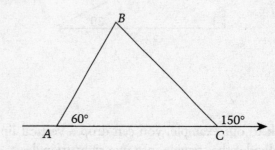

1. In triangle *ABC*, if *AB* = 4, then *AC* =

(A)   10
(B)   9
(C)   8
(D)   7
(E)   6

                                                                    Ⓐ Ⓑ Ⓒ Ⓓ Ⓔ

Column A 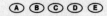 Column B

In the coordinate plane, point *R* has coordinates (0,0) and point *S* has coordinates (9,12).

2. The distance from *R* to *S*   16

Ⓐ Ⓑ Ⓒ Ⓓ Ⓔ

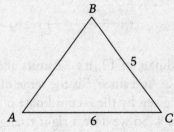

3. If the perimeter of triangle *ABC* above is 16, what is its area?

(A)   8
(B)   9
(C)   10
(D)   12
(E)   15

Ⓐ Ⓑ Ⓒ Ⓓ Ⓔ

Solutions

1. (C) Angle *BCA* is supplementary to the angle marked 150°, so angle *BCA* = 180° − 150° = 30°. Since the interior angles of a triangle sum to 180°, angle *A* + angle *B* + angle *BCA* = 180°, so angle *B* = 180° − 60° − 30° = 90°. So triangle *ABC* is a 30-60-90 right triangle, and its sides are in the ratio 1 : $\sqrt{3}$: 2. The side opposite the 30°, *AB*, which we know has length 4, must be half the length of the hypotenuse, *AC*. Therefore *AC* = 8, and that's answer choice (C).

2. (B) Draw a diagram. Since *RS* isn't parallel to either axis, the way to compute its length is to create a right triangle with legs that are parallel to the axes, so their lengths are easy to find. We can then use the Pythagorean theorem to find the length of *RS*.

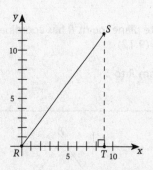

Since S has a y-coordinate of 12, it's 12 units above the x-axis, so the length of ST must be 12. And since T is the same number of units to the right of the y-axis as S, given by the x-coordinate of 9, the distance from the origin to T must be 9. So we have a right triangle with legs of 9 and 12. You should recognize this as a multiple of the 3-4-5 triangle. $9 = 3 \times 3$; $12 = 3 \times 4$; so the hypotenuse RS must be 3×5, or 15. That's the value of Column A, so Column B is greater.

3. (D) To find the area you need to know the base and height. If the perimeter is 16, then $AB + BC + AC = 16$; that is, $AB = 16 - 5 - 6 = 5$. Since $AB = BC$, this is an isosceles triangle. If you drop a line from vertex B perpendicular to AC, it will divide the base in half. This divides the triangle up into two smaller right triangles:

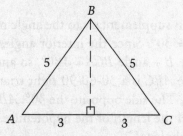

These right triangles each have one leg of 3 and a hypotenuse of 5; therefore they are 3-4-5 right triangles. So the missing leg (which is also the height of triangle ABC) must have length 4. We now know that the base of ABC is 6 and the height is 4, so the area is $\frac{1}{2} \times 6 \times 4$, or 12, answer choice (D).

The Kaplan Approach to Multiple and Oddball Figures

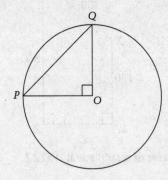

In the figure above, if the area of the circle with center O is 9π, what is the area of triangle POQ?

(A) 4.5

(B) 6

(C) 3.5π

(D) 4.5π

(E) 9

In a problem that combines figures, you have to look for the relationship between the figures. For instance, if two figures share a side, information about that side will probably be the key.

In this case the figures don't share a side, but the triangle's legs are important features of the circle — they are radii. You can see that $OP = OQ$ = the radius of circle O.

The area of the circle is 9π. The area of a circle is πr^2, where r is the radius. So $9\pi = \pi r^2$, $9 = r^2$, and the radius r is 3. The area of a triangle is $\frac{1}{2}$ base times height. Therefore, the area of ΔPOQ is $\frac{1}{2}$ (leg$_1$ × leg$_2$) = $\frac{1}{2}$ (3 × 3) = $\frac{9}{2}$ = 4.5, answer choice (A).

But what if, instead of a number of familiar shapes, you are given something like this?

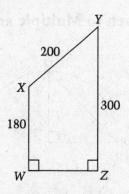

What is the perimeter of quadrilateral *WXYZ* ?

(A) 680

(B) 760

(C) 840

(D) 920

(E) 1,000

Try breaking the unfamiliar shape into familiar ones. Once this is done, you can use the same techniques that you would for multiple figures. Perimeter is the sum of the lengths of the sides of a figure, so you need to find the length of *WZ*. Drawing a perpendicular line from point *X* to side *YZ* will divide the figure into a right triangle and a rectangle. Call the point of intersection *A*.

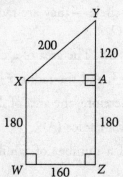

Opposite sides of a rectangle have equal length, so *WZ* = *XA* and *WX* = *ZA*. *WX* is labeled as 180, so *ZA* = 180. Since *YZ* measures 300, *AY* is 300 − 180 = 120. In right triangle *XYA*, hypotenuse *XY* = 200 and leg *AY*

= 120; you should recognize this as a multiple of a 3-4-5 right triangle. The hypotenuse is 5 × 40, one leg is 3 × 40, so *XA* must be 4 × 40 or 160. (If you didn't recognize this special right triangle you could have used the Pythagorean theorem to find the length of *XA*.) Since *WZ = XA* = 160, the perimeter of the figure is 180 + 200 + 300 + 160 = 840, answer choice (C).

Practice Problems

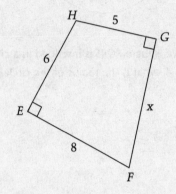

1. What is the value of *x* in the figure above?

   (A) 4
   (B) $3\sqrt{3}$
   (C) $3\sqrt{5}$
   (D) $5\sqrt{3}$
   (E) 9

   Ⓐ Ⓑ Ⓒ Ⓓ Ⓔ

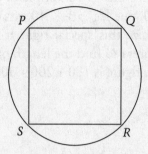

2. In the figure above, square *PQRS* is inscribed in a circle. If the area of square *PQRS* is 4, what is the radius of the circle?

(A) 1

(B) $\sqrt{2}$

(C) 2

(D) $2\sqrt{2}$

(E) $4\sqrt{2}$

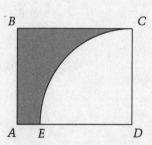

3. In the figure above, the quarter circle with center *D* has a radius of 4 and rectangle *ABCD* has a perimeter of 20. What is the perimeter of the shaded region?

(A) $20 - 8\pi$

(B) $10 + 2\pi$

(C) $12 + 2\pi$

(D) $12 + 4\pi$

(E) $4 + 8\pi$

Solutions

1. (D) Draw a straight line from point *H* to point *F*, to divide the figure into two right triangles.

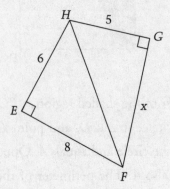

ΔEFH is a 3-4-5 right triangle with a hypotenuse of length 10. Use the Pythagorean theorem in ΔFGH to find *x* :

$$x^2 + 5^2 = 10^2$$
$$x^2 + 5^2 = 100$$
$$x^2 = 75$$
$$x = \sqrt{75}$$
$$x = \sqrt{25}\,\sqrt{3}$$
$$x = 5\,\sqrt{3}$$

2. (B) Draw in diagonal *QS* and you will notice that it is also a diameter of the circle. Since the area of the square is 4 its sides must each be 2. The diagonal of a square is always the length of a side times $\sqrt{2}$.

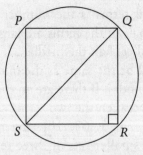

Think of the diagonal as dividing the square into two isosceles right triangles. Therefore, the diagonal = $2\sqrt{2}$ = the diameter; the radius is half this amount or $\sqrt{2}$.

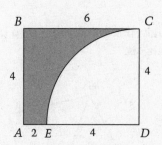

3. (C) The perimeter of the shaded region is $BC + AB + AE +$ arc EC. The quarter circle has its center at D, and point C lies on the circle, so side DC is a radius of the circle and equals 4. Opposite sides of a rectangle are equal so AB is also 4. The perimeter of the rectangle is 20, and since the two short sides account for 8, the two longer sides must account for 12, making BC and AD each 6. To find AE, subtract the length of ED, another radius of length 4, from the length of AD, which is 6; $AE = 2$. Since arc EC is a quarter circle, the length of the arc EC is $\frac{1}{4}$ of the circumference of a whole circle with radius 4: $\frac{1}{4} \times 2\pi r = \frac{1}{4} \times 8\pi = 2\pi$. So the perimeter of the shaded region is $6 + 4 + 2 + 2\pi = 12 + 2\pi$.

Mean, Median, Mode, and Range

The GRE has always tested the concept of a mean, which is also called the arithmetic mean, for no good reason. The mean of several numbers is simply their average. Whenever you see *arithmetic mean* on the GRE, it's not a trick—it just means *average*.

The median of several terms is the number that evenly divides the terms into two groups; half of the terms are larger than the median and half of the terms are smaller than the median. If there is an odd number of terms, the median will be the same as the middle number (not necessarily the average or the mode). If there are an even number of terms, the median will be halfway between the two terms closest to the middle.

For the set {4, 5, 7, 23, 5, 67, 10}, the median is 7, since this divides the set into two smaller sets of three terms each, {4, 5, 5} and {10, 23, 67}.

The mode is even simpler. It's just the term with the most occurrences in a set of numbers. If two or more numbers are tied for the most occurrences, then each is considered a mode.

THE THREE Ms

Mean is the average of a set of numbers.

Median is the term in a set of numbers that evenly divides the terms into two groups.

Mode is the term that occurs the most in a set of numbers.

For the set {4, 5, 7, 23, 5, 67, 10}, the mode is 5, because it occurs the greatest number of times of any of the terms.

The range is the simplest of these four concepts. It's just the difference between the largest term and the smallest term in a set of numbers. Just subtract the smallest from the biggest and you will have the range.

For the set {4, 5, 7, 23, 5, 67, 10}, the range is 63, because the greatest number, 67, minus the smallest, 4, equals 63.

Practice Problems

Column A	Column B

1. The median of the integers from 1 through 31, inclusive. \qquad 16

Ⓐ Ⓑ Ⓒ Ⓓ Ⓔ

2. 2^6 \qquad The range of the series {8, 9, 9, 15, 71}

Ⓐ Ⓑ Ⓒ Ⓓ Ⓔ

3. The only test scores for the students in a certain class are 44, 30, 42, 30, x, 44, and 30. If x equals one of the other scores and is a multiple of 5, what is the mode for the class?
(A) 5
(B) 6
(C) 15
(D) 30
(E) 44

Ⓐ Ⓑ Ⓒ Ⓓ Ⓔ

4. If half the range of the increasing series {11, A, 23, B, C, 68, 73} is equal to its median, what is the median of the series?
(A) 23
(B) 31
(C) 33
(D) 41
(E) 62

Ⓐ Ⓑ Ⓒ Ⓓ Ⓔ

Solutions

1. (C) *Inclusive* just means you should include the numbers on the ends—in this case, 1 and 31. The number right in the middle of this series is 16. There are 15 numbers smaller than it and 15 numbers greater than it.

2. (A) 2^6 equals 64. The range of the series in Column B equals 71 − 8, which equals 63.

3. (D) Since *x* equals one of the other scores, it must equal either 30, 42, or 44. And since it must also be a multiple of 5, we can conclude that *x* equals 30. That means that four of the students—more than earned any other score—earned a score of 30, which makes 30 the mode.

4. (B) Don't get confused by all the variables; just concentrate on what you know. The range must be the difference of the smallest term and the largest term. Since this is an increasing series, the smallest term must be 11 and the largest must be 73. The difference between them is 62, so that's the range. Half of the range, then, is 31, so 31 must equal the median of the series.

Probability

A probability is the fractional likelihood of an event occurring. It can be represented by a fraction ("the probability of it raining today is $\frac{1}{2}$"), a ratio ("the odds of it raining today are 50:50"), or a percent ("the probability of rain today is 50 percent"). You can translate probabilities easily into everyday language: $\frac{1}{100}$ = "one chance in a hundred" or "the odds are one in a hundred."

To find probabilities, count the number of desired outcomes and divide by the number of possible outcomes.

Probability = (Number of Desired Outcomes/Number of Possible Outcomes)

WHAT'S A PROBABILITY?

A probability is the fractional likelihood that a given event will occur. To get a probability, divide the number of desired events by the number of possible events.

For example, what is the probability of throwing a 5 on a six-sided die? There is one desired outcome—throwing a 5. There are six possible outcomes—one for each side of the die. So the probability = $\frac{1}{6}$

All probabilities are between 0 and 1 inclusive. A 0 probability means there is zero chance of an event occurring (i.e., it can't happen). A 1 probability means that an event has a 100 percent chance of occurring (i.e., it must occur). The higher the probability, the greater chance that an event will occur. You can often eliminate answer choices by having some idea where the probability of an event occurring falls within this range.

The odds of throwing a 5 on a die are $\frac{1}{6}$, so the odds of not throwing a 5 are $\frac{5}{6}$. Therefore, you have a much greater probability of not throwing a 5 on a die than of throwing a 5.

Practice Problems

Column A	Column B

The probability of rain on Thursday is 50 percent. The probability that it will not rain on Friday is $\frac{1}{4}$.

1. The probability of rain on Thursday

The probability of rain on Friday

A hat contains an equal number of red, blue, and green marbles.

2. The probability of picking a red marble out of the hat

The probability of picking a green marble out of the hat

Ⓐ Ⓑ Ⓒ Ⓓ Ⓔ

3. If there are 14 women and 10 men employed in a certain office, what is the probability that one employee picked at random will be a woman?

(A) $\frac{1}{6}$

(B) $\frac{1}{14}$

(C) $\frac{7}{12}$

(D) 1

(E) $\frac{7}{5}$

Ⓐ Ⓑ Ⓒ Ⓓ Ⓔ

4. If Tom flips a fair coin twice, what is the probability that at least one head will be thrown?

(A) $\frac{1}{4}$

(B) $\frac{1}{3}$

(C) $\frac{1}{2}$

(D) $\frac{2}{3}$

(E) $\frac{3}{4}$

Ⓐ Ⓑ Ⓒ Ⓓ Ⓔ

Solutions

1. (B) The probability of rain on Thursday is $\frac{1}{2}$ and the probability of rain on Friday is $\frac{3}{4}$.

2. (C) The number of desired outcomes is the same in each case, since there are an equal number of red and green marbles. The number of possible outcomes in each case is also the same, since the marbles are all being pulled from the same hat. Therefore the probabilities are the same.

3. (C) Probability = (Number of Desired Outcomes/Number of Possible Outcomes) = (Number of Women/Number of People) = $\frac{14}{24}$ = $\frac{7}{12}$.

4. (E) Desired outcomes = HH or HT or TH. Possible outcomes = HH or HT or TH or TT. Probability = $\frac{3}{4}$.

Graphs

In every math section, some questions will be based on one or more graphs. Exactly what ETS is trying to test with these questions we have never been able to determine, unless you need to sharpen your clerical skills before you pursue that Ph.D. in electrical engineering.

The Basic Principles of Graphs

Some Will Be Fairly Basic
You will have to do something like find a value from the graph(s) or compare values in the graph(s). They will take only a few steps.

Others Will Be Fairly Difficult
They always require more than just a few steps. They may fool you into thinking that they're easy, but there will be a trick involved. Never, ever pick the obvious answer without checking it first.

The Kaplan Three-Step Method for Graphs
We recommend a three-step method for graph problems:

1. Familiarize Yourself with the Graph(s)
Graph sections usually have more than one graph in them.

- Read the title(s).
- Check the scales to see how the information is measured.
- Read any accompanying notes.
- If there is more than one line on the graph, label each line according to the key so you can reference them easily when working on the questions.

2. Answer the Questions That Follow
Graph questions require a strong understanding of fractions and percents and good attention to detail, but little else.

When there are two graphs in one section, for each of Questions 21 and 22 you'll probably need to use just one graph (though perhaps not

LEVELS OF DIFFICULTY

The first two graph questions are simple; the last two graph questions are hard and can be tricky.

THE KAPLAN METHOD FOR GRAPH PROBLEMS

1. Familiarize yourself with the graph(s).
2. Answer the questions that follow.
3. Work on all the graphs questions in one piece.

the same graph). Questions 23 and 24 may use either one or both of the graphs. For 25, you will need to use both graphs, taking data from one and combining it with data from the other. If you haven't used both graphs for these hard questions, you will almost certainly get them wrong.

3. Work on All the Graph Questions in One Piece
Although the first few graph questions are fairly easy, usually involving something straightforward, the average time per graph question for most students is a little greater than the average time for other problem solving questions, so you may want to save the five graph questions within a math section for last.

Practice Problems
Questions 1–5 are based on the following graphs.

MEGACORP, INC.
REVENUE AND PROFIT DISTRIBUTION FOR FOOD- AND NONFOOD-RELATED OPERATIONS, 1984–1989

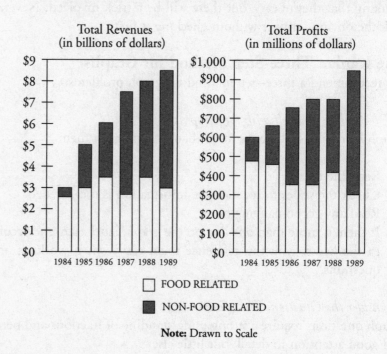

Note: Drawn to Scale

PERCENT OF REVENUES FROM FOOD-RELATED OPERATIONS IN 1989 BY CATEGORY

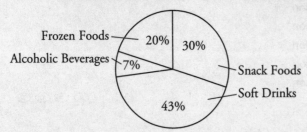

Frozen Foods — 20% 30% Snack Foods

Alcoholic Beverages — 7% Soft Drinks

43%

IT'S ALL RELATIVE

The most common trick in graph questions is to confuse you with the difference between an actual value (a fixed number) and a relative value (a fraction or percent). Make sure you always know which one you're working with.

1. Approximately how much did total revenues increase from 1984 to 1987?
   (A) $0.5 billion
   (B) $1.5 billion
   (C) $4.0 billion
   (D $4.5 billion
   (E) $5.0 billion

2. For the year in which profits from food-related operations increased over the previous year, *total revenues* were approximately
   (A) $3.5 billion
   (B) $4.5 billion
   (C) $5.7 billion
   (D) $6.0 billion
   (E) $8.0 billion

   Ⓐ Ⓑ Ⓒ Ⓓ Ⓔ

3. In 1988, total profits represented approximately what percent of Megacorp's total revenues?
   (A) 50%
   (B) 20%
   (C) 10%
   (D) 5%
   (E) 1%

   Ⓐ Ⓑ Ⓒ Ⓓ Ⓔ

4. For the first year in which revenues from nonfood-related operations surpassed $4.5 billion, total profits were approximately
    (A) $250 million
    (B) $450 million
    (C) $550 million
    (D) $650 million
    (E) $800 million

    Ⓐ Ⓑ Ⓒ Ⓓ Ⓔ

5. In 1989, approximately how many millions of dollars were revenues from frozen food operations?
    (A) 1,700
    (B) 1,100
    (C)   900
    (D)   600
    (E)   450

    Ⓐ Ⓑ Ⓒ Ⓓ Ⓔ

By using all of the techniques discussed above, you will be able to tackle the most difficult Quantitative questions. (You can brush up on all of your math by referring to the Math Reference Appendix in the back of this book.) And now that you have the tools to handle the Quantitative section of the GRE, try the following set of practice questions. After that, we'll move on and take a look at the Analytical section of the test.

ANSWERS TO GRAPH PROBLEMS

1. (D)
2. (E)
3. (C)
4. (E)
5. (D)

Quantitative Practice Set

Numbers
All numbers are real numbers.

Figures
The position of points, lines, angles, etcetera, may be assumed to be in the order shown; all lengths and angle measures may be assumed to be positive.

Lines shown as straight may be assumed to be straight.

Figures lie in the plane of the paper unless otherwise stated.

Figures that accompany questions are intended to provide useful information. However, unless a note states that a figure has been drawn to scale, you should solve the problems by using your knowledge of mathematics, and not by estimation or measurement.

Directions
Each of the Questions 1–15 below consists of two quantities, one in Column A and another in Column B. You are to compare the two quantities and answer

(A) if the quantity in Column A is greater
(B) if the quantity in Column B is greater
(C) if the two quantities are equal
(D) if the relationship cannot be determined from the information given

Common Information
In a question, information concerning one or both of the quantities to be compared is centered above the two columns. A symbol that appears in both columns represents the same thing in Column A as it does in Column B. (Answers and explanations can be found at the end of the set of questions.)

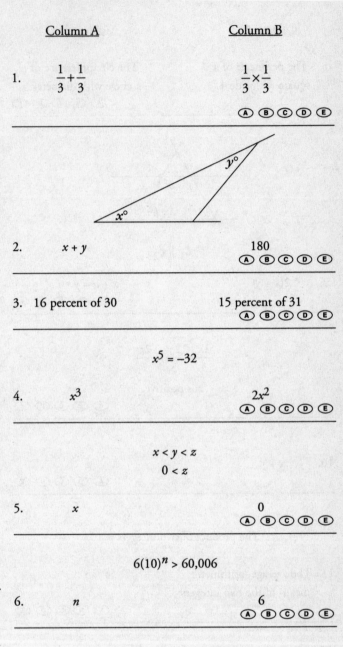

Column A	Column B

1. $\dfrac{1}{3}+\dfrac{1}{3}$ $\dfrac{1}{3}\times\dfrac{1}{3}$

Ⓐ Ⓑ Ⓒ Ⓓ Ⓔ

2. $x + y$ 180

Ⓐ Ⓑ Ⓒ Ⓓ Ⓔ

3. 16 percent of 30   15 percent of 31

Ⓐ Ⓑ Ⓒ Ⓓ Ⓔ

$$x^5 = -32$$

4. x^3 $2x^2$

Ⓐ Ⓑ Ⓒ Ⓓ Ⓔ

$$x < y < z$$
$$0 < z$$

5. x 0

Ⓐ Ⓑ Ⓒ Ⓓ Ⓔ

$$6(10)^n > 60,006$$

6. n 6

Ⓐ Ⓑ Ⓒ Ⓓ Ⓔ

In a three-digit positive integer y, the hundreds' digit is three times the units' digit.

7. The units' digit of y 4

Ⓐ Ⓑ Ⓒ Ⓓ Ⓔ

Column A	Column B

8. The perimeter of a square with side 4 — The circumference of a circle with diameter 5

Ⓐ Ⓑ Ⓒ Ⓓ Ⓔ

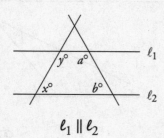

$\ell_1 \parallel \ell_2$

9. $2(x + y)$ — $x + a + y + b$

Ⓐ Ⓑ Ⓒ Ⓓ Ⓔ

$$\frac{2x}{3} = \frac{2y}{5} = \frac{2z}{7}$$

z is positive.

Ⓐ Ⓑ Ⓒ Ⓓ Ⓔ

10. $x + y$ — z

Ⓐ Ⓑ Ⓒ Ⓓ Ⓔ

The product of two integers is 10.

11. The average (arithmetic mean) of the two integers — 3

Ⓐ Ⓑ Ⓒ Ⓓ Ⓔ

The remainder when n is divided by 3 is 1, and the remainder when $n + 1$ is divided by 2 is 1.

12. The remainder when $n - 1$ is divided by 6 — 3

Ⓐ Ⓑ Ⓒ Ⓓ Ⓔ

Column A	Column B

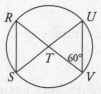

The circle has center T. The measure of angle TVU is 60°.

13. RT — RS

Ⓐ Ⓑ Ⓒ Ⓓ Ⓔ

After five adults leave a party, there are three times as many children as adults. After 25 children leave the party, there are twice as many adults as children.

14. The original number of adults — 14

Ⓐ Ⓑ Ⓒ Ⓓ Ⓔ

There are at least 200 apples in a grocery store. The ratio of the number of oranges to the number of apples is 9 to 10.

15. The number of oranges in the store — 200

Ⓐ Ⓑ Ⓒ Ⓓ Ⓔ

16. How many odd integers are between $\dfrac{10}{3}$ and $\dfrac{62}{3}$?

(A) 19
(B) 18
(C) 10
(D) 9
(E) 8

Ⓐ Ⓑ Ⓒ Ⓓ Ⓔ

Column A   Column B

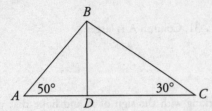

17. In the figure above, if *BD* bisects ∠*ABC*, then the measure of ∠*BDC* is

(A)   50°
(B)   90°
(C)   100°
(D)   110°
(E)   120°

Ⓐ Ⓑ Ⓒ Ⓓ Ⓔ

18. If $d = \dfrac{c-b}{a-b}$, then $b =$

(A) $\dfrac{c-d}{a-d}$

(B) $\dfrac{c+d}{a+d}$

(C) $\dfrac{ca-d}{ca+d}$

(D) $\dfrac{c-ad}{1-d}$

(E) $\dfrac{c+ad}{d-1}$

Ⓐ Ⓑ Ⓒ Ⓓ Ⓔ

Column A   Column B

19. If $x > 0$, then $(4x)(8x) =$

(A) $2^9 x$
(B) $2^8 x$
(C) $2^6 x$
(D) $2^5 x$
(E) $2^4 x$

Ⓐ Ⓑ Ⓒ Ⓓ Ⓔ

20. In one class in a school, 30 percent of the students are boys. In a second class that is half the size of the first, 40 percent of the students are boys. What percent of both classes are boys?

(A)   20%
(B)   25%
(C)   28%
(D)   30%
(E) $33\dfrac{1}{3}\%$

Ⓐ Ⓑ Ⓒ Ⓓ Ⓔ

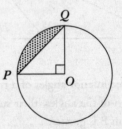

21. In a circle *O* above, if Δ*POQ* is a right triangle and radius *OP* is 2, what is the area of the shaded region?

(A) $4\pi - 2$
(B) $4\pi - 4$
(C) $2\pi - 2$
(D) $2\pi - 4$
(E) $\pi - 2$

Ⓐ Ⓑ Ⓒ Ⓓ Ⓔ

Answer Key

1. A	8. A	15. D
2. B	9. C	16. E
3. A	10. A	17. C
4. B	11. D	18. D
5. D	12. C	19. D
6. D	13. C	20. E
7. B	14. A	21. E

Explanations

Question 1

$\frac{1}{3}+\frac{1}{3}$ is $\frac{2}{3}$. $\frac{1}{3}\times\frac{1}{3}=\frac{1}{9}$. Since $\frac{2}{3}>\frac{1}{9}$, column A is larger.

Question 2

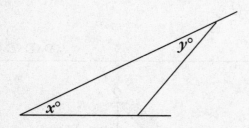

The sum of the three interior angles of a triangle is 180°. Since x and y are only two of the angles, their sum must be less than 180 degrees. Column B is greater.

Question 3

Sixteen percent of 30 is $\frac{16}{100}(30)$ or $\frac{(16)(30)}{100}$. Similarly, 15 percent of 31 is $\frac{15}{100}(31)$ or $\frac{(15)(31)}{100}$. We can ignore the denominator of 100 in both columns, and just compare (16)(30) in Column A to (15)(31) in column B. Divide both columns by

15; we're left with 31 in Column B and (16)(2) or 32 in column A. Since 32 > 31, Column A is larger.

Question 4

Start by working with the sign of x, and hope that you won't have to go any further than that. If x^5 is negative, then what is the sign of x? It must be negative—if x were positive, then any power of x would also be positive. Since x is negative, Column A, x^3, which is a negative number raised to an odd exponent, must also be negative. But what about column B? Whatever x is, x^2 must be positive (or zero, but we know that x can't be zero); therefore, the quantity in Column B must be positive. We have a positive number in column B and a negative number is column A; Column B must be greater.

Question 5

We could pick numbers here, or else just use logic. We know that z is positive, and that x and y are less than z. But does that mean that x or y must be negative? Not at all—they could be, but they could also be positive. For instance, suppose $x = 1$, $y = 2$, and $z = 3$. Then Column A would be larger. However, if $x = -1$, $y = 0$, and $z = 1$, then Column B would be larger. We need more information to determine the relationship between the columns.

Question 6

Divide both sides of the inequality by 6. We're left with $(10)n > 10,001$. 10,001 can also be written at $10^4 + 1$, so we know that $(10)n > 10^4 + 1$. Therefore, the quantity in Column A, n, must be 5 or greater. Column B is 6; since n could be less than, equal to, or greater than 6, we need more information.

Question 7

Try to set the columns equal. Could be units' digit of y be 4? If it is, and the hundreds' digit is three times the units' digit, then the hundreds' digit must be . . . 12? That can't be right. A digit must be one of the integers 0 through 9; 12 isn't a digit. Therefore, 4 is too big to be the units' digit of y. We don't know what the units' digit of y is (and we don't care either), but we know that it must be less than 4. Column B is greater than Column A.

Question 8

The perimeter of a square with side 4 is 4(4) or 16. The circumference of a circle is the product of π and the diameter, so the circumference in Column B is 5π. Since π is approximately 3.14, $5(\pi)$ is approximately 5(3.14) or 15.70, which is less than 16. Column A is greater.

Question 9

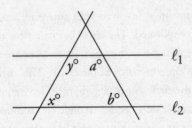

Column B is the sum of all the angles in the quadrilateral. The sum of the angles in any quadrilateral is 360 degrees. Column B is 360. In Column A, angle x and angle y are angles made when a transversal cuts a pair of parallel lines, in this case, ℓ_1 and ℓ_2. Such angles are either equal or supplementary. Angles x and y obviously aren't equal, so they must be supplementary, and their sum is 180. Then Column A is $2(x + y) = 2 \times 180$ or 360. The columns are equal.

Question 10

One way to work here is to pick numbers. Just make sure that anything you pick satisfies the requirements of the problem. How about picking $x = 3$, $y = 5$, and $z = 7$, since in the equation these numbers would cancel with their denominators, thus leaving us with the equation $2 = 2 = 2$. Therefore, we know that these values satisfy the equation. In addition, if $z = 7$, then z is positive, so we have satisfied the other requirement as well. Then the sum of x and y, in Column A, is $3 + 5$ is 8. This is larger than z, so in this case, Column A is larger. That's just one example though; we should really try another one. In fact, any other example we pick that fits the initial information will have Column A larger. To see why, we have to do a little messy work with the initial equations; on the test, you should just pick a couple of sample values, then go on to the next questions.

Start by dividing all of the equations through by 2, and multiply all of the terms through by $3 \times 5 \times 7$, to eliminate all the fractions. This leaves us with:

$$35x = 21y = 15z$$

Now let's put everything in terms of x.

$$x = x \qquad y = \frac{35}{21}x = \frac{5}{3}x \qquad z = \frac{35}{15}x = \frac{7}{3}x$$

Then in Column A, the sum of x and y is $x + \frac{5}{3}x = \frac{8}{3}x$. In column B, the value of z is $\frac{7}{3}x$. Now since z is positive, x and y must also be positive. (If one of them is negative, that would make all of them negative.) Since x is positive, $\frac{8}{3}x > \frac{7}{3}x$.

Column A is greater.

The moral here is that proving that one column must be bigger can involve an awful lot of time on some GRE QC questions—more time than you can afford on the test. Try to come up with a good answer, but don't spend a lot of time proving it. Even if you end up showing that your original suspicion was wrong, it's not worth it if it took five minutes away from the rest of the problems.

Question 11

The best place to start here is with some pairs of integers that have a product of 10. The numbers 5 and 2 have a product of 10, as do 10 and 1, and the average of each of these pairs is greater than 3, so you may have thought that (A) was the correct answer. If so, you should have stopped yourself, saying, "That seems a little too easy for such a late QC question. They're usually trickier than that." In fact, this one was. There's nothing in the problem that limits the integers to positive numbers: they can just as easily be negative. The numbers −10 and −1 also have a product of 10, but their average is a negative number—in other words, less than Column B. We need more information here; the answer is (D).

Question 12

The best way to do this question is to pick numbers. First we have to figure out what kind of number we want. Since $n - 1$ leaves a remainder of 1 when it's divided by 2, we know that $n + 1$ must be an odd number. Then n itself is an even number. We're told that n leaves a remainder of 1 when it's divided by 3. Therefore, n must be 1 more than a multiple of 3, or $n - 1$ is a multiple of 3. So what are we looking for? We've figured out that n should be an even number, that's one more than a multiple of 3. So let's pick a number now. How about 10 ? That's even, and it's one more than a multiple of 3. Then what's the remainder when we divide $n - 1$, or $10 - 1 = 9$, by 6? We're left with a remainder of 3: 6 divides into 9 one time, with 3 left over. In this case, the columns are equal.

Now since this a QC question and there's always a possibility that we'll get a different result if we pick a different number, we should either pick another case, or else use logic to convince ourselves that the columns will always be equal. Let's do the latter here. Since n is even, $n - 1$ must be odd. We saw before that $n - 1$ is a multiple of 3, so we now know that it is an odd multiple of 3. Does this tell us anything about $n - 1$'s relation to 6? Yes, it does: $n - 1$ is 3 multiplied by an odd number m, which can be written as $2p + 1$ where p is an integer. So $n - 1 = 3(2p + 1) = 6p + 3$. $6p$ is a multiple of 6, so the remainder when $n - 1$ is divided by 6 must be 3. The answer is (C).

Question 13

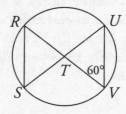

There are many steps involved with this problem, but none of them is too complicated. The circle has its center at point T. To start with the triangle at the right, its vertices are at T and two points on the circumference of the circle. This makes two of its sides radii of the circle. Since all radii must have equal length, this makes the triangle an isosceles triangle. In addition, we're told one of the base angles of this triangle has measure 60°. Then the other base angle must also have measure 60° (since the base angles in an isosceles triangle have equal measure). Then the sum of the two base angles is 120°, leaving 180 − 120 or 60° for the other angle, the one at point T.

Now, $\angle RTS$ is opposite this 60° angle; therefore, its measure must also be 60°. $\triangle RST$ is another isosceles triangle; since $\angle RTS$ has measure 60°, the other two angles in the triangle must also measure 60°. So what we have in the diagram is two equilateral triangles. RT and RS are two sides in one of these triangles; therefore, they must be of equal length, and the two columns are equal.

Question 14

Start by setting the columns equal. Suppose there were originally 14 adults at the party. Then after five of them leave, there are 14 − 5 or nine adults left. There are three times as many children as adults, so there are 3 × 9 or 27 children. Then 25 children leave the party, so there are 27 − 25 or two children left. So nine adults and two children remain at this party. Is that twice as many adults as children? No, it is more than four times as many, So this clearly indicates that the columns can't be equal—but does it mean that Column A is bigger or Column B is bigger? Probably the simplest way to decide is to pick another number for the original number of adults, and see whether the ratio gets better or worse. Suppose we start with 13 adults. After five adults leave, there are 13 − 5 or eight adults. Multiplying 3 times 8 gives 24 children. Now if 25 children leave, we're left with 24 − 25 or −1 children. But that's no good; how can you have a negative number of children? This means we've gone the wrong way; our ratio has gotten worse instead of better. So 14 isn't right for the number of adults, and 13 is even worse, so the correct number must be something more than 14, and Column A is larger.

Question 15

We know that the ratio of oranges to apples is 9 to 10, and that there are "at least" 200 apples. The ratio tells us that there are more apples than oranges. How does that help us? Good question. It helps us because it tells us that there could be fewer than 200 oranges in the store. Could there be more than 200? Sure. If there were a lot more than 200 apples, say 600 apples, then there would be a lot more than 200 oranges. So we have one situation in which Column A is larger, and another case in which Column B is larger. We need more information to decide.

Question 16

Here we're asked for the odd integers between $\frac{10}{3}$ and $\frac{62}{3}$.

First let's be clearer about this range. $\frac{10}{3}$ is the same as $3\frac{1}{3}$, and $\frac{62}{3}$ is the same as $20\frac{2}{3}$. So we need to count the odd integers between $3\frac{1}{3}$ and $20\frac{2}{3}$. We can't include 3, since 3 is less than $3\frac{1}{3}$. Similarly, we can't include 21, since it's larger than $20\frac{2}{3}$.

So the odd integers in the appropriate range are 5, 7, 9, 11, 13, 15, 17, and 19. That's a total of 8.

Question 17

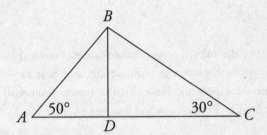

Notice that we're given the measures of two interior angles in $\triangle ABC$: $\angle BAC$ measures 50° and $\angle BCA$ measures 30°. Therefore, $\angle ABC$, the third interior angle in $\triangle ABC$, measures 180 − (50 + 30), or 180 − 80, or 100°. Since BD bisects $\angle ABC$, BD splits up $\angle ABC$ into two smaller angles equal in measure, $\angle ABD$ and $\angle DBC$. Therefore, the measure of $\angle DBC$ is half the measure of $\angle ABC$, so $\angle DBC$ measures $\frac{1}{2}(100)$, or 50°. Now we can use this information along with the fact that $\angle BCD$ measures 30° to find $\angle BDC$. Since these three angles are interior angles of $\triangle BDC$, their measures sum to 180°. So $\angle BDC$ measures 180 − (50 + 30), or 100°.

Question 18

Solve for b in terms of a, c, and d.

$$d = \frac{c - b}{a - b}$$

Clear the denominator by multiplying both sides by $a - b$.　　$d(a - b) = c - b$

Multiply out parentheses.　　$da - db = c - b$

Gather all bs on one side.　　$b - db = c - da$

Factor out the bs on the left hand side.　　$b(1 - d) = c - da$

Divide both sides by $1 - d$ to isolate b.　　$b = \frac{c - ad}{1 - d}$

Question 19

Remember the rules for operations with exponents. First you have to get both powers in terms of the same base so you can combine the exponents. Note that the answer choices all have base 2. Start by expressing 4 and 8 as powers of 2.

$$(4x)(8x) = (2^2)x \times (2^3)x$$

To raise a power to an exponent, multiply the exponents:

$$(2^2)x = 2^2 x$$
$$(2^3)x = 2^3 x$$

To multiply powers with the same base, add the exponents:

$$2^2 x \times 2^3 x = 2^{(2x + 3x)}$$
$$= 2^5 x$$

Question 20

Pick a sample value for the size of one of the classes. The first class might have 100 students. That means there are 30 percent of 100 or 30 boys in the class. The second class is half the size of the first, so it has 50 students, of which 40 percent of 50 = 20 are boys. This gives us 100 + 50 = 150 students total, of whom 30 + 20 = 50 are boys. So $\frac{50}{150} = \frac{1}{3}$ of both classes are boys. Now convert $\frac{1}{3}$ to a percent. $\frac{1}{3} = \frac{1}{3} \times 100\% = 33\frac{1}{3}\%$.

Question 21

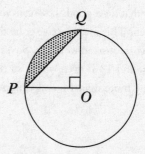

The area of the shaded region is the area of the quarter-circle (sector *OPQ*) minus the area of right triangle *OPQ*. The radius of circle *O* is 2, so the area of the quarter-circle is

$$\frac{1}{4}\pi r^2 = \frac{1}{4} \times \pi(2)^2 = \frac{1}{4} \times 4\pi = \pi$$

Each leg of the triangle is a radius of circle *O*, so the area of the triangle is

$$\frac{1}{2}bh = \frac{1}{2} \times 2 \times 2 = 2$$

Therefore, the area of the shaded region is $\pi - 2$.

CHAPTER 4

TEST CONTENT: ANALYTICAL

Y ou'll have 60 minutes to complete 35 questions in the Analytical section, which has two question types in the Analytical section: logic games and logical reasoning. (ETS once planned to introduce a third type: pattern identification. However, they withdrew that question type after Kaplan developed a strategy so devastating that a test taker could get any pattern identification question correct right away!) On the test, you should do the logical reasoning questions before the logic games.

We're going to tackle logic games first, because they are the most difficult Analytical question type for many students.

Logic Games

> **Directions:** Each group of questions is based on a passage or a set of conditions. You may wish to draw a diagram to answer some of the questions. Choose the best answer for each question.

Nothing inspires more fear in the hearts of GRE test takers than logic games. Why? Partly because the skills tested on the section seem so unfamiliar. You need to turn a game's information to your advantage by organizing your thinking and spotting key deductions, and that's not easy to do.

Games tend to give the most trouble to students who don't have a clearly defined method of attack. And that's where Kaplan's basic princi-

ples, game-specific strategies, and five-step method for logic games will help most, streamlining your work so you can rack up points quickly and confidently.

The Four Basic Principles of Logic Games

The rallying cry of the logic games-impaired is, "I could do these, if only I had more time!" Well, that's true of everybody. You can spend as much time on a game as you like when you're sitting in your own kitchen, but when your proctor says, "You have 30 minutes . . . begin," he or she is not kidding around.

Logic games are perhaps the most speed-sensitive questions on the test. The test makers know that if you could spend hours methodically trying out every choice in every question, you'd probably get the right one. But it's all about efficiency, both on the test and in your future studies.

And that brings us to the first (and somewhat paradoxical-sounding) logic games principle.

1. To Go Faster, You Need to Slow Down

To gain time in logic games, you must spend more time thinking through and analyzing the setup and the rules. This is not only the most important principle for logic games success, it's also the one that's most often ignored, probably because it just doesn't seem intuitively right; people who have timing difficulties tend to speed up, not slow down. But take our word for it: by spending a little extra time up front thinking through the setup, the "action," and the rules, you'll be able to recognize the game's key issues and make important deductions that will actually save you time in the long run.

Games are structured so that, in order to answer the questions quickly and correctly, you need to search out relevant pieces of information that combine to form valid new statements, called deductions. Now, you can either do this once, up front, and then utilize the same deductions throughout the game, OR you can choose to piece together the same basic deductions—essentially repeating the same work—for every single question.

For instance, let's say that two of the rules for a logic game go as follows:

> If Bob is chosen for the team, then Eric is also chosen.
> If Eric is chosen for the team, then Pat will not be chosen.

You can, as you read through the rules of the game, just treat those rules as two separate pieces of independent information. But there's a deduction to be made from them. Do you see it? If Bob is chosen, Eric is

too. If Eric is chosen, Pat is not. That means that, if Bob is chosen, Pat is not chosen. That's an important deduction—one that will undoubtedly be required from question to question. If you don't take the time to make it up front, when you're first considering the game, you'll have to make it over and over again, every time it's necessary to answer a question. But if you do take the time to make it up front and build it into your entire conception of the game, you'll save that time later.

So, always try to take the game scenario and the rules as far as you can before moving on to the questions. Look for common elements among the rules (like Eric in the previous rules); this will help you combine them and weed out major deductions. The stimulus creates a situation, and the rules place restrictions on what can and cannot happen within that situation. If you investigate the possible scenarios and look for and find major deductions up front, you'll then be able to rack up points quickly and confidently.

2. Understand What a Rule Means, Not Just What It Says

If you're interested in demonstrating how well you can read a statement and then spit it back verbatim, you'd be better off training to be a clerk instead of a scholar. That's why you'll never see this on the GRE:

> Rule: Arlene is not fifth in line.
> Question: Which one of the following people is not fifth in line?
> Answer: Arlene.

Some LG questions are easy, but not that easy. The GRE, after all, measures critical thinking, and virtually every sentence in logic games has to be filtered through some sort of analytical process before it will be of any use. You may have to use the information about Arlene to help you eliminate a choice or lead you to the right answer, but even in the simplest of cases, this will involve the application, as opposed to the mere parroting, of the rule.

So, getting back to the principle, it's not enough to just copy a rule off the page (or shorthand it, as we'll discuss momentarily); it's imperative that you think through its exact meaning, including any implications that it might have. And *don't* limit this behavior to the indented rules; statements in the games' introductions are very often rules in themselves and warrant the same meticulous consideration.

For instance, let's say a game's introduction sets up a scenario in which you have three boxes, each containing at least two of the following three types of candy—chocolates, gumdrops, and mints. Then you get the following rule:

No Parrots, Please

To fully grasp a rule in logic games, you must know more than just what it says. You've got to know what the rule means in the context of the game and in combination with other rules.

Kaplan Rules

Always try to turn negative rules—Box 2 does not contain any gumdrops—into a positive statement —Box 2 must contain chocolates and mints.

Game Wisdom

You must know the rules of a logic game cold— what they mean, how they impact on other rules, what implications they have in the context of the game scenario.

Box 2 does not contain any gumdrops.

What does that rule say? That there aren't any gumdrops in box 2. But what does that rule mean, when you think about it in the context of the game? That Box 2 does contain chocolates and mints. Each box contains at least two of three things, remember. So, once you eliminate one of the three things for any particular box, you know that the other two things *must* be in that box.

Part of understanding what a rule means, moreover, is grasping what the rule doesn't mean. For example, take the rule we mentioned earlier:

RULE: If Bob is chosen for the team, then Eric is also chosen.
MEANS: Whenever Bob is chosen, Eric is, too.
DOESN'T MEAN: Whenever Eric is chosen, Bob is, too.

3. Use Scratchwork and Shorthand

The proper use of scratchwork can help you do your best on logic games. As you may recall, the directions state, "You may wish to draw a rough sketch to help answer some of the questions." Notice that they use the wording *rough sketch,* not *masterpiece, work of art,* or *classic portrait for the ages.* The GRE is not a drawing contest; you get no points for creating beautiful visual imagery on the page.

Although some recent games aren't even amenable to scratchwork, for most games you'll find that it is helpful to create a master sketch, one that encapsulates all of the game's information in one easy-to-reference picture. Doing so will not only give your eyes a place to gravitate toward when you need information, but it will also help to solidify in your mind the action of the game, the rules, and whatever deductions you come to up front.

Remember to keep your scratchwork simple; the less time you spend drawing, the more time you'll have for thinking and answering questions. Pay careful attention to the scratchwork suggestions in the explanations to the four games on the Practice Test in this book.

The part of your scratchwork where you jot down on your page a quick and shortened form of each rule is called shorthand. Shorthand is a visual representation of a mental thought process and is useful only if it reminds you at a glance of the rule's meaning. Whether you shorthand a rule or commit it to memory, you should never have to look back at the game itself once you get to the questions. The goal of the entire scratchwork process is to condense a lot of information into manageable, user-friendly visual cues.

IT'S NOT ART SCHOOL

You're applying to grad school, not art school. Don't worry about making elaborate diagrams in logic games. There is no "right diagram" for any game. But there is good scratchwork that will help you get points quicker and more accurately.

WHAT IS SHORTHAND?

Shorthand is a visual representation of a mental thought process. Use shorthand to remind yourself of the meaning of a rule in a logic games problem.

It's much easier to remember rules written like so:

B → E
No G in 3

than rules written like this:

> If Bob is chosen for the team, then Eric is also chosen.
> Box 3 does not contain any gumdrops.

Just remember what your shorthand means— for instance, what the arrow from B to E means—and be consistent in using it. If you can develop a personal shorthand that is instantly understandable to you, you will have a decided advantage come the day of the test.

4. Try to Set off Chains of Deduction

When hypothetical information is offered in a question stem, try to use it to set off a chain of deductions. Consider the following question. (Because this question is excerpted without the accompanying introduction and rules, ignore the specific logic of the discussion; it's just presented to make a point.)

> If the speedboat is yellow, which one of the following must be true?
>
> (A) The car is green.
> (B) The airplane is red.
> (C) The train is black.
> (D) The car is yellow.
> (E) The train is red.

The question stem contains a hypothetical, which is an if-clause offering information pertaining only to that particular question. The wrong approach is acknowledging that the speedboat is yellow and then proceeding to test out all of the choices. The muddled mental thought process accompanying this tragic approach might sound something like this:

"All right, the speedboat's yellow, does the car have to be green? Well, let's see, if the speedboat's yellow, *and* the car is green, then the train would have to be yellow, but I can't tell what color the airplane is, and I guess this is okay, I don't know, I better try the next choice. Let's see what happens if the speedboat's yellow and the airplane is red. . . ."

A LAST RESORT

Trial and error with the answer choices should be your last resort, not your first. It's much quicker to follow a chain of deduction until it leads you to the answer. In some cases, trial and error is necessary, but don't turn to it unless you're really stuck.

Don't do this kind of dithering! Notice that the question doesn't ask, "What happens, if in addition to this, the car is green?" or "What happens if this is true and the airplane is red?" So why is the confused test taker above intent on answering all of these irrelevant questions? Our second point is that you should never begin a question by trying out answer choices; that's going about it backwards. Only if you're entirely stuck or are faced with a question stem that leaves you no choice, should you resort to trial and error.

Most logic games questions are amenable to another, more efficient and systematic methodology. The correct approach is to incorporate the new piece of information into your view of the game, creating one quick sketch if you wish. How do you do this? Apply the rules and any previous deductions to the new information in order to set off a new chain of deductions. Then follow through until you've taken the new information as far as it can go. Just as you must take the game and rules as far as you can before moving on to the questions, you must carry the information in a question stem out as far as you can before moving on to the choices.

So be sure to stay out of answer-choice land until you have sufficiently mined the hypothetical. If the question stem contains a hypothetical, then your job is to get as much out of that piece of information as you can before even looking at the choices. This way, *you* dictate to the test, not the other way around. You'll then be able to determine the answer and simply pick it off the page.

You'll have the chance to see these major logic games principles in action when you review the explanations to the games in the Practice Test in the back of this book.

Common Logic Game Types

Although the logic games section can contain a wide variety of situations and scenarios, certain game types appear again and again. These are the most common:

1. Sequencing Games

Logic games that require sequencing skills have long been a favorite of the test makers. No matter what the scenario in games of this type, the common denominator is that in some way, shape, or form, they all involve putting entities in order. In a typical sequence game, you may be asked to arrange the cast of characters numerically from left to right, from top to bottom, during days of the week, in a circle, and so on. The sequence may be a sequence of degree; ranking the eight smartest test takers from 1 to 8

SEQUENCING GAMES AT A GLANCE

Sequencing games:

- Are historically the most common game type
- Involve putting entities in order
- Involve orderings, which can be in time (runners finishing a race), space (people standing next to one another in line), or degree (shortest to tallest, worst to best, etcetera)

falls into this category. On the other hand, the sequence may be based on time, such as one that involves the finishing times of runners in a race. In some cases, there are two or even three orderings to keep track of in a single game.

Typical Issues. The following is a list of the key issues that underlie sequencing games. Each key issue is followed by a corresponding rule— in some cases, with several alternative ways of expressing the same rule. At the end, we'll use these rules to build a miniature logic game, so that you can see how rules work together to define and limit a game's "action." These rules all refer to a scenario in which eight events are to be sequenced from first to eighth.

- Which entities are concretely placed in the ordering?

   X is third.

- Which entities are excluded from a specific position in the ordering?

   Y is not fourth.

- Which entities are next to, adjacent to, or are immediately preceding or following one another?

   X and Y are consecutive.
   X is next to Y.
   No event comes between X and Y.
   X and Y are consecutive in the ordering.

- Which entities CANNOT BE next to, adjacent to, or immediately preceding or following one another?

   X does not immediately precede or follow Z.
   X is not immediately before or after Z.
   At least one event comes between X and Z.
   X and Z are not consecutive in the sequence.

- How far apart in the ordering are two particular entities?

   Exactly two events come between X and Q.
   At least two events come between X and Q.

- What is the relative position of two entities in the ordering?

> Q comes before T in the sequence.
> T comes after Q in the sequence.

How a Sequence Game Works. Let's see how rules like those above might combine to create a simple logic game.

Eight events—Q, R, S, T, W, X, Y, and Z—are being sequenced from first to eighth.

> X is third.
> Y is not fourth.
> X and Y are consecutive in the sequence.
> Exactly two events come between X and Q.
> Q comes before T in the sequence.

How would you approach this simplified game? Remember our fourth basic principle: Use scratchwork and shorthand. With eight events to sequence from first to eighth, you'd probably want to draw eight dashes in the margin of your test booklet, maybe in two groups of four (so you can easily determine which dash is which). Then take the rules in order of concreteness, starting with the most concrete of all—Rule 1—which tells you that X is third. Fill that into your sketch:

$$\_ \ \_ \ \underset{}{X} \ \_ \quad \_ \ \_ \ \_ \ \_$$

Jump to the next most concrete rule—Rule 4, which tells you that exactly two events come between X and Q. Well, Q must be sixth, then:

$$\_ \ \_ \ \underset{}{X} \ \_ \quad \_ \ \underset{}{Q} \ \_ \ \_$$

Rule 5 says that Q comes before T. Because Q is sixth, T must be either seventh or eighth. To indicate this, under the sketch, write T with two arrows pointing to the seventh and eighth dashes.

Rule 3 says that X and Y are consecutive. X is third, so Y will be either second or fourth. Rule 2 clears up that matter. Y can't be fourth, says Rule 2, so it will have to be second:

$$\_ \ \underset{}{Y} \ \underset{}{X} \ \_ \quad \_ \ \underset{}{Q} \ \_ \ \_$$
$$\nwarrow_{T}\nearrow$$

This is how the rules work together to build a sequence game.

The questions might then present hypothetical information that would set off the chain of deduction we mentioned in the basic principles section.

2. Grouping Games

A grouping game, much like every other type of game, begins with a set of entities. What sets grouping apart is the "action" of the game, or specifically, what you're asked to do with the entities. In a pure grouping game, unlike sequencing, there's no call for putting the entities in order. Instead, you'll usually be required to select a smaller group from the initial group or distribute the entities in some fashion into more than one subgroup. As a distinct skill, grouping differs from sequencing in that you're not really concerned with what order the entities are in, but rather how they're grouped—who's in, who's out, and who can and cannot be with whom in various subgroups.

Grouping Games of Selection and of Distribution. The two varieties of grouping games are really very similar under the skin. In "distribution" types of grouping games, we are told every entity has to go somewhere—as in a game where we have to distribute eight marbles into two jars with four marbles each.

In "selection" types of grouping games, we're primarily concerned with which entities go into one subgroup—as in a game where we have eight marbles and we have to select five marbles to go into a jar (and nothing is said about what happens to the other three marbles). What makes these games just two variants on a theme is the fact that you can think of selection games as distribution games. For example, in the second game you can think of the task as that of dividing the marbles into two groups, the group in the jar and the group outside. Grouping games can get quite complicated, involving different sorts of entities (marbles and rocks) and many different groups (three jars and two pockets).

Like sequencing games, grouping games have a language all their own, and it's up to you to speak that language fluently when you come across games that require this particular skill on your test.

Typical Issues—Grouping Games of Selection. The following is a list of the key issues that underlie grouping games. Each key issue is followed by a corresponding rule—in some cases, with several alternative ways of expressing the same rule. At the end, again, we'll use these rules to build a miniature logic game.

First, grouping games of selection. These rules all refer to a scenario in which you are to select a subgroup of four from a group of eight entities—Q, R, S, T, W, X, Y, and Z:

- Which entities are definitely chosen?

    Q is selected.

- Which entities rely on a different entity's selection in order to be chosen?

    If X is selected, then Y is selected.
    X will be selected only if Y is selected.
    X will not be selected unless Y is selected.

Note: A common misconception surrounds the rule "If X is selected, then Y is selected." The rule works only in one direction; if X is chosen, Y must be, but if Y is chosen, X *may or may not be.* Remember the discussion of Basic Principle Number 2 above—understand not only what a rule means, but also what it doesn't mean!

- Which entities must be chosen together, or not at all?

    If Y is selected, Z is selected, and vice versa.
    Y will not be selected unless Z is selected, and vice versa.

- Which entities cannot both be chosen?

    If R is selected, Z is not selected.
    If Z is selected, R is not selected.
    R and Z can't both be selected.

How Grouping Games of Selection Work. We can combine these rules to create a rudimentary grouping game of selection:

    A professor must choose a group of four books for her next seminar. She must choose from a pool of eight books—Q, R, S, T, W, X, Y, and Z.
    Q is selected.
    If X is selected then Y is selected.
    If Y is selected, Z is selected, and vice versa.
    If R is selected, Z is not selected.

A good way of dealing with this kind of game might be to write out the eight letters—four on top, four on the bottom—and then circle the ones that are definitely selected while crossing out the ones that are definitely not selected. Thus, Rule 1 would allow you to circle the Q:

$$Q\hspace{-0.3em}\text{R S T}$$
$$\text{W X Y Z}$$

The other rules can't be built into the sketch just yet, because they describe eventualities (what happens if something else happens). Here's where you would want to use shorthand:

Rule 2 translates as: "If X, then Y" or "X —> Y"
Rule 3 might be rendered as: "Y <—> Z" (because the requirement is vice versa).
Rule 4 could be written in shorthand as: "R ≠ Z" (because R and Z are mutually exclusive).

The rules would then be poised to take effect whenever a question would add new hypothetical information, setting off a chain of deduction. For instance, let's say a question reads like so:

If R is selected, which of the following must be true?

This new information would put the rules into motion. R's inclusion would set off Rule 4, "R ≠ Z," so we'd have to circle R and cross out Z:

$$Q\hspace{-0.2em}R\hspace{-0.2em}\text{S T}$$
$$\text{W X Y } \cancel{Z}$$

This would in turn set off Rule 3, "Y <—> Z." Because Z is out, Y is out, since they are chosen together or not at all:

$$Q\hspace{-0.2em}R\text{S T}$$
$$\text{W X } \cancel{Y}\cancel{Z}$$

Now Rule 2 comes into play. "X —> Y" means that if Y is not chosen, X can't be either (since X's inclusion would require Y's). So we can take the chain of deduction one step further:

Ⓠ⟨R⟩S T
W X̶ X̶ X̶

A correct answer to this question, then, might be "X is not included." And that, in a nutshell, is how a (simplified) grouping game of selection works.

Typical Issues—Grouping Games of Distribution. Here are the issues involved in the other kind of grouping games—grouping games of distribution—along with the rules that govern them. These rules, by the way, refer to a scenario in which our old favorite group of eight entities—Q, R, S, T, W, X, Y, Z—have to be distributed into three different classes:

- Which entities are concretely placed in a particular subgroup?

    X is placed in class 3.

- Which entities are barred from a particular subgroup?

    Y is not placed in class 2.

- Which entities must be placed in the same subgroup?

    X is placed in the same class as Z.
    Z is placed in the same class as X.
    X and Z are placed in the same class.

- Which entities cannot be placed in the same subgroup?

    X is not placed in the same class as Y.
    Y is not placed in the same class as X.
    X and Y are not placed in the same class.

- Which entity's placement depends on the placement of a different entity?

    If Y is placed in class 1, then Q is placed in class 2.

How Grouping Games of Distribution Work. The above rules, neatly enough, also can combine to form a miniature grouping game of distribution:

> Eight students—Q, R, S, T, W, X, Y, and Z—must be subdivided into three different classes—Classes 1, 2, and 3.
>
> > X is placed in Class 3.
> > Y is not placed in Class 2.
> > X is placed in the same class as Z.
> > X is not placed in the same class as Y.
> > If Y is placed in Class 1, then Q is placed in Class 2.

A good scratchwork scheme for games of this type would be to draw three circles in your booklet, one for each of the three classes. Then put the eight entities in the appropriate circles as that information becomes known.

Here again, start with the most concrete rule first, which is Rule 1, which definitively places X in Class 3. Rule 2 just as definitively precludes Y from Class 2, so build that into the scratchwork, too:

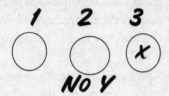

Rule 3 requires Z to join X in Class 3:

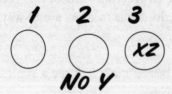

Rule 4, prohibiting Y from being in the same class as X, means that Y can't be in Class 3. But we already know that Y can't be in Class 2. We can deduce, therefore, that Y must go in Class 1. That in turn puts Rule 5 into play: if Y is in Class 1 (as it is here), Q is in Class 2:

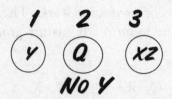

And that is the dynamic of most grouping games of distribution (though, again, in much simplified form).

3. Mapping Games

Mapping games are significantly less common than either sequencing games or grouping games, but you may well see one on the test. We hope you do, because once you know how to set up a mapping game, you will probably find them to be the easiest type on the test.

Typical Issues. The key to identifying a mapping game is to see whether the game asks you to make connections between entities in which each entity connects to one or more others. For example, the rules may tell you how rooms in a house are connected to other rooms, or the rules may tell you how bus or rail stops are connected to each other in a transit system.

We call them mapping games because the best way to handle them is to draw a quick-and-dirty map of the connections between entities. Once you've drawn the map, all the answers to the game's questions should fall into your lap.

There are two things to pay special attention to when you've drawn your map. Note the entities that are especially well or poorly connected to others (some will serve as "hubs" while others are "dead ends"). Also note if the connections from one entity to another don't go both ways (for example, you may be able to get from A to B, but you can't go back from B to A). The best way to do this is to draw a one-way arrow in the direction of such connections.

How Mapping Games Work. Now let's see a typical mapping game in action:

Exactly three public transportation routes serve a municipal airport. The vehicles on each route go in both directions and only pick up or deposit passengers at the designated stops.

Bus Line A stops at the main terminal, the airport hotel, the convention center, the downtown bus depot, and the train station.

Bus Line B stops at the main terminal, Dykman Avenue, Charles Street, 32nd Street, 25th Street, 16th Street, the downtown bus depot, and 4th Street.

The express bus line stops at the main terminal and the train station.

The scratchwork for this game should incorporate all three rules into one easy-to-read diagram. Make it as simple as possible. Start by treating Line A in Rule 1 as a straight line, and symbolize each stop with a letter.

$$M — H — C — D — T$$

Now add Rule 2. Make the stops that are the same for Lines A and B the same on the map (if it helps, also distinguish Lines A and B by using a dotted or a squiggly line for the connections on Line B, and using lowercase letters for the stops).

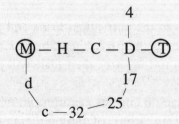

For Rule 3, we could either add a new line from M to T, or simply circle these stops to indicate that the express bus stops there.

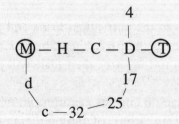

Now we have all we need to answer any question you're likely to face about the game. Note that M and D are the only places where a transfer is possible between the lines, and there will certainly be questions about making transfers. Likewise, notice that the 4th Street stop and the train station stop are dead ends. There will almost certainly be one or more questions testing your awareness of this in some way.

For example, a question would probably ask where any bus on this transit system must stop before arriving at the 4th Street stop. Hopefully, by now the answer will jump out at you: the bus must stop at the down-town bus depot first.

General Logic Games Tips

Hybrid Games

Some games are what you might call hybrid games, requiring you to combine sequencing and grouping. Keep in mind that while we try to recognize games as a particular type, it's not necessary to attach a strict name to every game you encounter. For example, it really doesn't matter if you categorize a game as a sequencing game with a grouping element or as a grouping game with a sequencing element, as long as you're comfortable with both sets of skills.

No "Best" Choice

Unlike the answer choices in logical reasoning, in which the correct answer is the "best" choice, the answers in logic games are objectively correct or incorrect. Therefore, when you find an answer that's definitely right, have the confidence to circle it and move on, without wasting time to check the other choices. This is one way to improve your timing on the section.

Common Elements and Deductions

Rules that contain common elements are often the ones that lead to deductions. Consider the following three rules:

> If Sybil goes to the party, then Edna will go to the party.
> If Jacqui goes to the party, then Sherry will not go to the party.
> If Edna goes to the party, then Dale will go to the party.

Rules 1 and 2 have no entities in common, which is a good sign that we can't deduce anything from combining them. The same goes for Rules

IT'S LIKE MATH

Like math questions, logic game questions have definite right and wrong answers. Once you find the answer that works, pick it and move on. There's no need to check out the other choices.

LOOK FOR THE COMMON ELEMENT

Rules that deal with one or more of the same entities can often be combined to make important deductions.

2 and 3. But because Rules 1 and 3 have Edna in common, a deduction is likely (although not guaranteed). In this case, combining Rules 1 and 3 would allow us to deduce another rule: If Sybil goes to the party, then Dale will go also.

Focus on the Important Rules
Not all rules are created equal—some are inherently more important than others. Try to focus first on the ones that have the greatest impact on the situation, specifically the ones that involve the greatest number of the entities. These are also the rules to turn to first whenever you're stuck on a question and don't know how to set off the chain of deductions.

The Kaplan Five-Step Method for Logic Games
Now that you have some logic games background, it's time to see how you can marshal that knowledge into a systematic approach to games.

1. Get an Overview
Read carefully the game's introduction and rules, to establish the "cast of characters," the "action," and the number limits governing the game.

2. Visualize and Map out the Game
Make a mental picture of the situation and let it guide you as you create a sketch, or some other kind of scratchwork, if need be, to help you keep track of the rules and handle new information.

3. Consider the Rules Individually
After you've thought through the meaning and implications of each rule, you have three choices. You can:

- Build it directly into your sketch of the game situation
- Jot down the rule in shorthand form to help you remember it
- Underline or circle rules that don't lend themselves to the first two techniques

4. Combine the Rules
Look for common elements among the rules; that's what will lead you to make deductions. Treat these deductions as additional rules, good for the whole game.

IS NOTHING CLICKING?

If you find that you can't make a single important deduction by combining rules, you're probably missing something. Check the game introduction and rules again to make sure that you're not misinterpreting something.

THE KAPLAN FIVE-STEP METHOD FOR LOGIC GAMES

1. Get an overview.
2. Visualize and map out the game.
3. Consider the rules individually.
4. Combine the rules.
5. Work on the questions systematically.

WHAT'S IN A NAME?

Remember, you get no points for categorizing a game; you get points for answering questions correctly. Don't worry about what to call a game. Just decide what skills it will require.

5. Work on the Questions Systematically
Read the question stem carefully! Take special notice of such words as *must, could, cannot, not, impossible,* and *except.* As always, use the hypothetical information offered in *if*-clauses to set off a chain of deductions.

Using the Kaplan Five-Step Method
Here's how the approach can work with an actual logic game:

Five repair people—Mona, Patrick, Renatta, Saffie, and Will—are assigned shifts to repair appliances on five days of a single week, Monday to Friday. There are exactly three shifts available to each repair person each day—a morning shift, an afternoon shift, and an evening shift. No more than one repair person works on any given shift. Each repair person works exactly two shifts during the week, but no repair person works more than one shift in a single day.

Exactly two repair people work on each day of the week.
Mona and Will work a shift on the same days of the week.
Patrick doesn't work on any afternoon or evening shifts during the week.
Will doesn't work on any morning or afternoon shifts during the week.
Mona works shifts on two consecutive days of the week.
Saffie's second shift of the week is on an earlier day of the week than Mona's first shift.

1. Which one of the following must be true?

   (A) Saffie works a shift on Tuesday afternoon.
   (B) Patrick works a shift on Monday morning.
   (C) Will works a shift on Thursday evening.
   (D) Renatta works a shift on Friday afternoon.
   (E) Mona works a shift on Tuesday morning.

2. If Will does not work a shift on Friday, which one of the following could be false?

   (A) Renatta works a shift on Friday.
   (B) Saffie works a shift on Tuesday.

    (C) Mona works a shift on Wednesday.

    (D) Saffie works a shift on Monday.

    (E) Patrick works a shift on Tuesday.

(Note that there are only two questions accompanying this game; a typical logic game will have three to six questions. This game is as complicated as any you're likely to see on the GRE.)

1. Get an Overview

We need to schedule five repair people, abbreviated M, P, R, S, and W, in a particular order during a five-day calendar week, Monday to Friday. The ordering element tells us we're dealing with a sequencing game, though there is a slight grouping element involved in that a couple of rules deal with grouping issues—namely, which people can or cannot work on the same day of the week as each other.

Be very careful about the numbers governing this game; they go a long way in defining how the game works. There are to be exactly two repair people per day (never working on the same shift). Each repair person must work exactly two shifts, and because repair people are forbidden to take two shifts in the same day, this means that each repair person will work on exactly two days. So, in effect, 10 out of the 15 available shifts will be taken, and five will be left untouched.

2. Visualize and Map out the Game

Go with whatever you feel is the most efficient way to keep track of the situation. Most people would settle on a sketch of the five days, each broken up into three shifts, like so:

Into this sketch—one letter per box—each entity will have to go twice (each repair person does two shifts, remember). So your pool of entities to place would be: MMPPRRSSWW. You might want to include five Xs (or Øs) for the five shifts that won't be taken by any of the repair people.

3. Consider the Rules Individually

We've already dealt with some of the number-related rules hidden in the game's introduction. Now let's consider this statement from the intro:

No more than one repair person works in any given shift.

Make sure you interpret rules like this correctly. You may have to para-phrase, in your own words, its exact meaning. In this case: two repair peo-ple per shift is no good, three is out of the question, etcetera. But it doesn't mean that any given shift *must* have a repair person. If the test makers meant to imply that, they would have written, "Exactly one repair person works on every given shift." Notice the difference in wording. It's subtle, but it has a huge impact on the game.

Let's consider the other rules. We've already handled Rule 1. You may wish to jot down "2 a day," or something like that, to remind you of this important information.

Rule 2: Mona and Will work on the same days, and that holds for both of the days these repair people work. Write this in any shorthand that seems fitting (one suggestion is to draw MW with a circle around it on your page).

Rules 3 and 4: We can handle these two rules together because they're so similar. You can write shorthand for these rules as they are, but you'd be doing yourself a great disservice. Instead, first work out their impli-cations, which is actually a pretty simple matter: if Patrick doesn't work afternoons or evenings, he must work mornings. If Will doesn't work mornings or afternoons, he must work evenings. Always take the rules as far as you can, and then jot their implications down on your page for reference.

Rule 5: This one is pretty self-explanatory; Mona's shifts must be on consecutive days, such as Thursday and Friday. M = M might be a good way to write this in shorthand.

Rule 6: Here's another sequencing rule—you must place both Ss for Saffie on earlier days of the week than the two Ms for Mona. That means that Saffie and Mona can't work on the same day (although we already knew that from Rule 2), and that Mona's shifts can't come before Saffie's.

Try writing this in shorthand as S . . . S . . . M . . . M.

4. Combine the Rules

This is the crucial stage for most games. Here, notice that Mona appears in three of the six indented rules; that's a good indication that combining these rules should lead somewhere useful. Combining Rule 2 and Rule 5 gives us two Mona/Will days in a row:

$$\frac{M}{W} = \frac{M}{W}$$

Will must be scheduled for evening shifts (remember, we turned Rule 4 into this positive statement). That means that Mona would take the morning or afternoon shift on these consecutive days.

Rule 6 concerns Mona as well: two Saffies before two Monas. How is this possible? We need two Ss on different days to come before the two consecutive Ms. If Saffie's shifts are as early in the week as possible, she'll work on Monday and Tuesday. That means that the earliest day that Mona can work (and Will as well, thanks to Rule 2) is Wednesday. There's our first really key deduction:

> Mona and Will cannot work on Monday and Tuesday; they must work Wednesday, Thursday, or Friday.

Do we stop there? No, of course not. The difference between the logic games expert and the logic games novice is that the expert knows how to press on when further useful deductive possibilities exist. If you relate this deduction back to Rule 5, it becomes clear that Mona and Will must work on Wednesday and Thursday *or* on Thursday and Friday. This brings us to another big deduction:

> Either way, Mona and Will must work on Thursday. Thanks to Rule 4, we can slot Will in for Thursday evening. Mona will then take Thursday morning OR afternoon. The other Mona/Will day must be either Wednesday or Friday, to remain consecutive.

Here's what your completed sketch may look like, with as many of the rules built into it as possible:

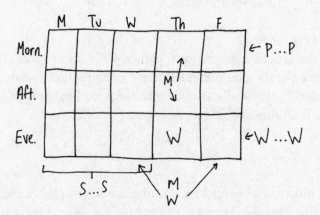

Now that we've combined the rules, and have even uncovered a few big deductions, it's time to move on to the questions.

5. Work on the Questions Systematically

Now you'll see how all the work we did up front pays off. Question 1 offers no hypothetical information; it simply asks what must be true. And because we've already deduced a few things that must be true, we can scan the choices for one that matches any one of our newly discovered pieces of information. It doesn't take long to spot choice (C)—it's our big deduction staring us right in the face. You shouldn't even waste time checking the other choices. Instead, have the confidence that you've done the right work the right way, and circle (C) and move on. [Just for the record, for those of you who are curious, (A), (B), and (D) *could* be true, but don't have to be, whereas (E), as we discovered earlier, is an impossibility.]

Question 2 contains a hypothetical: no Will on Friday. One glance at our sketch tells us that the second Mona/Will cluster must therefore be placed on Wednesday, next to the Thursday Mona/Will group. Saffie must then work on Monday and Tuesday, in order to satisfy Rule 6 (although we don't yet know the exact shifts she takes during those days).

That brings us to the two questions that test takers ask all too infrequently, "Who's left?" and more importantly, "Where can they go?" Two Ps and two Rs are left to place, with one spot on Monday, one spot on Tuesday, and two spots on Friday open to place them. How can this be done? Friday can't get both Ps or both Rs (from the last sentence in the introduction), so it will have to get one of each, with P in the morning and R in either the afternoon or evening. The other P and the other R

will join S on Monday or Tuesday, in either order. Of course, whichever day P is on, he must be in the morning, whereas the exact shifts for R and S are ambiguous.

Look at how far the chain of deductions takes us, beginning with the simple statement in the question stem:

If Will doesn't work a shift on Friday, then . . .

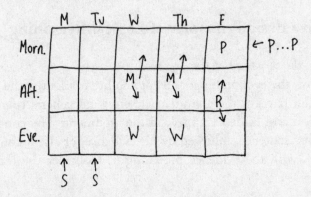

With all of this information at our disposal, there's not a question in the world that we can't answer correctly. This one asks for a statement that could be false—which means that the four wrong choices will all be things that must be true. And in fact, choices (A) through (D) match the situation in this question perfectly, while (E) merely *could* be true: Patrick's first shift of the week could be on Tuesday, but it just as easily could be on Monday as well. (His second shift must be on Friday, of course.) (E) is therefore the only choice that could be false.

Logical Reasoning

Directions: Each group of questions is based on a passage or a set of conditions. You may wish to draw a diagram to answer some of the questions. Choose the best answer for each question.

Most people find GRE logical reasoning to be much easier than logic games. In fact, many people don't do any preparation for this question type at all. It's ironic, then, that this is also one of the most stubborn item

YOU'VE GOT WHAT IT TAKES

There's nothing bizarre or esoteric about the skills you need for logical reasoning. You've just got to learn how to adapt those skills to the peculiar requirements of a timed, standardized test.

EVERYTHING'S AN ARGUMENT

Virtually every logical reasoning stimulus is an argument that consists of two major parts—evidence and conclusion. Every GRE logical reasoning stimulus—that is, every argument—is made up of two basic parts:

- The conclusion (the point that the author is trying to make)
- The evidence (the support that the author offers for the conclusion).

STRUCTURAL SIGNALS

Certain keywords can help you isolate the conclusion and the evidence in a stimulus. Clues that signal evidence include: *because, since, for, as a result of,* and *due to.* Clues that signal the conclusion include: *consequently, hence, therefore, thus, clearly, so,* and *accordingly.*

types on the test—that is, it's one of the toughest to improve at. That's because there aren't really any quick fixes in logical reasoning. You can only get better by understanding arguments better.

On the GRE CAT, you will probably see one to three logical reasoning questions at the start of the section, followed by a logic game, then one to three more logical reasoning questions. After that each logical reasoning question is usually followed by a logic game. That is, you will see about eight logical reasoning questions.

The Five Basic Principles of Logical Reasoning

1. Know the Structure of the Argument
Success on this section hinges on your ability to identify the evidence and conclusion parts of the argument. There is no general rule about where conclusion and evidence appear in the argument: The conclusion could be the first sentence, followed by the evidence, or else it could be the last sentence, with the evidence preceding it. Consider the following short stimulus:

> The Brookdale Public Library will require extensive physical rehabilitation to meet the new building codes just passed by the town council. For one thing, the electrical system is inadequate, causing the lights to flicker sporadically. Furthermore, there are too few emergency exits, and even those are poorly marked and sometimes locked.

Let's suppose that the author of the argument above were only allowed one sentence to convey her point. Do you think that she would waste her lone opportunity on the statement, "The electrical system at the Brookdale Public Library is inadequate, causing the lights to flicker sporadically"? Would she walk away satisfied that she got her main point across? Probably not. Given a single opportunity, she would have to state the first sentence: "The Brookdale Public Library will require extensive physical rehabilitation, etcetera." This is her conclusion. If you pressed her for her *reasons* for making this statement, she would then cite the electrical and structural problems with the building. This is the evidence for her conclusion.

2. Preview the Question Stem
Looking over the question stem before you read the stimulus will alert you in advance of what to focus on in your initial reading of the stimu-

lus. In effect, it gives you a jump on the questions. For example, let's say the question attached to the original library argument above asked the following:

> The author supports her point about the need for rehabilitation at the Brookdale Library by citing which of the following?

If you were to preview this question stem before reading the stimulus, you would know what to look for in advance—namely, evidence, the "support" provided for the conclusion. Similarly, if the question asked you to find an assumption that the author is relying on, this would tell you in advance that there was a crucial piece of the argument missing, and you could begin to think about it right off the bat.

3. Paraphrase the Author's Point

After you read the stimulus, you'll want to paraphrase the author's main argument, that is, restate the author's ideas in your own words. Frequently, the authors in logical reasoning say pretty simple things in complex ways. But if you mentally translate the verbiage into a simpler form, you'll find the whole thing more manageable.

In the library argument, for instance, you probably don't want to deal with the full complexity of the author's stated conclusion:

> The Brookdale Public Library will require extensive physical rehabilitation to meet the new building codes just passed by the town council.

Instead, you probably want to carry a much simpler form of the point in your mind, something like:

> The library will need fixing up to meet new codes.

Often, by the time you begin reading through answer choices, you run the risk of losing sight of the gist of the stimulus. After all, you can only concentrate on a certain amount of information at one time. Restating the argument in your own words will not only help you get the author's point in the first place, but it'll also help you hold on to it until you've found the correct answer.

KNOW WHAT YOU'RE LOOKING FOR

Previewing the question stem before reading the stimulus makes you a better, more directed reader. You'll know what you're looking for in advance.

IN YOUR OWN WORDS

It's much easier to understand and remember an argument if you restate it simply, in your own words.

4. Try to Prephrase an Answer

You must try to approach the answer choices with at least a faint idea of what the answer should look like. This is not to say that you should ponder the question for minutes until you're able to write out your own answer; it's still a multiple-choice test, so the right answer is on the page. Just try to get in the habit of instinctively thinking through the question and framing an answer in your own mind.

For instance, let's say a question for the library argument went like this:

> The author's argument depends on which of the following assumptions about the new building codes?

Having thought about the stimulus argument, an answer to this question may have sprung immediately to mind: namely, the assumption that the new codes are tougher than the old codes. After all, the library will have to be rehabilitated to meet the new codes, according to the author. Clearly, the assumption is that the new codes are more stringent than the old. And that's the kind of statement you would look for among the choices.

5. Keep in Mind the Scope of the Argument

One of the most important logical reasoning skills, particularly when you're at the point of actually selecting one of the five choices, is the ability to focus in on the scope of the argument. The majority of wrong choices on this section are wrong because they are "outside the scope." In everyday language, that simply means that these choices contain elements that don't match the author's ideas or that simply go beyond the context of the stimulus.

Some common examples of scope problems are choices that are too narrow, or too broad, or literally have nothing to do with the author's points. Also, watch for and eliminate choices that are too extreme to match the argument's scope; they're usually signaled by words such as *all, always, never, none,* and so on. Choices that are more qualified are often correct for arguments that are moderate in tone and contain such words as *usually, sometimes, probably,* etcetera. To illustrate the scope principle, let's look again at the question mentioned above:

> The author's argument depends on which of the following assumptions about the new building codes?

SCOPE IT OUT

A remarkable number of wrong answers in Logical Reasoning have scope problems. Always be on the lookout for choices that are too extreme, that contain value judgments that are not relevant to the argument, or that don't match the stimulus in tone or subject matter.

Let's say one of the choices read as follows:

(A) The new building codes are far too stringent.

Knowing the scope of the argument would help you to eliminate this choice very quickly. You know that this argument is just a claim about what the new codes will require—that the library be rehabilitated. It's not an argument about whether the requirements of the new codes are good, or justifiable, or ridiculously strict. That kind of value judgment is outside the scope of this argument. Recognizing scope problems is a great way of eliminating dozens of wrong answers quickly.

The Five Common Logical Reasoning Question Types

Now that you're familiar with the basic principles of logical reasoning, let's look at the most common types of questions that you'll be asked. As we said earlier, certain question types crop up again and again on the GRE, and it pays to be familiar with them. Of the types discussed below, the first three predominate, but try to become familiar with the others as well.

1. Assumption Questions

An assumption bridges the gap between an argument's evidence and conclusion. It's a piece of support that isn't explicitly stated but that is required for the conclusion to remain valid. When a question asks you to find an author's assumption, it's asking you to find the statement without which the argument falls apart.

In order to test whether a statement is necessarily assumed by an author, therefore, we can employ the denial test. Here's how it works: simply deny or negate the statement and see if the argument falls apart. If it does, that choice is the correct assumption. If, on the other hand, the argument is unaffected, the choice is wrong. Consider, as an example, this simple stimulus:

Allyson plays volleyball for Central High School.
Therefore, Allyson must be over 6 feet tall.

You should recognize the second sentence as the conclusion, and the first sentence as the support, or evidence, for it. But is the argument complete? Obviously not. The piece that's missing—the unstated link between the evidence and conclusion—is the assumption, and you could probably prephrase this one pretty easily:

THE MISSING LINK

Some arguments lack an important bridge between their evidence and their conclusion. That bridge is the necessary assumption—a key part of many arguments that remains unspoken.

ASSUMPTION QUESTIONS AT A GLANCE

- Represent one of the most popular LR question types
- Assumptions unstated in the stimulus
- Bridge the gap between evidence and conclusion
- Must be true in order for the conclusion to remain valid
- Can be checked by applying the denial test

STRENGTHEN/ WEAKEN QUESTIONS AT A GLANCE

Weaken questions are very popular; strengthen questions less so. Strengthen/ weaken questions are:

- Related to assumption: strengthener often shores up central assumption; weakener often shows central assumption to be unreasonable
- You must evaluate each choice as to the effect it would have on the argument if true
- Correct choices don't prove or disprove argument, but simply tip the scale the most in the desired direction

All volleyball players for Central High School are over 6 feet tall.

To test whether this really is a necessary assumption of the argument, let's apply the "denial test" to it, by negating it. What if it's not true that all volleyball players for Central High School are over 6 feet tall? Can we still logically conclude that Allyson must be taller than 6 feet? No, we can't. Sure, it's possible that she is, but just as possible that she's not. By denying the statement, then, the argument falls to pieces; it's simply no longer valid. And that's our conclusive proof that the statement above is a necessary assumption that the author of this stimulus is relying on.

As we've just seen, you can often prephrase the answer to an assumption question. By previewing the question stem, you'll know what to look for. And stimuli for assumption questions just "feel" like they're missing something. Often, the answer will jump right out at you, as in this case. In more difficult assumption questions, the answers may not be as obvious. But in either case, you can use the denial test to check whichever choice seems correct.

Here are some of the ways in which assumption questions are worded:

- Which one of the following is assumed by the author?
- The argument depends on the assumption that . . .
- The validity of the argument above depends on which one of the following?

2. Strengthen/Weaken the Argument
Determining an argument's necessary assumption, as we've just seen, is required to answer assumption questions. But it also is required for another common question type, strengthen-and-weaken-the-argument questions.

One way to weaken an argument is to break down a central piece of evidence. Another way is to attack the validity of any assumptions that the author may be making. The answer to many weaken-the-argument questions is the one that reveals an author's assumption to be unreasonable; conversely, the answer to many strengthen-the-argument questions provides additional support by affirming the truth of an assumption or by presenting more persuasive evidence.

Weakening questions tend to be more common on the GRE than strengthening questions. But here are a few concepts that apply to both question types:

- Weakening an argument is not the same thing as disproving it, whereas strengthening is not the same as proving the conclusion to be true. A strengthener tips the scale toward believing in the validity of the conclusion, whereas a weakener tips the scale in the other direction, toward doubting the conclusion.
- Don't be careless. Wrong answer choices in these question types often have exactly the opposite of the desired effect. That is, if you're asked to strengthen a stimulus argument, it's quite likely that one or more of the wrong choices will contain information that actually weakens the argument. By the same token, weaken questions may contain a choice that strengthens the argument. So once again, pay close attention to what the question stem asks.

The stems associated with these two question types are usually self-explanatory. Here's a list of what you can expect to see on the test:

Weaken:

- Which one of the following, if true, would most weaken the argument above?
- Which one of the following, if true, would most seriously undermine the argument above?

Strengthen:

- Which one of the following, if true, would most strengthen the argument?
- Which one of the following, if true, would provide the most support for the conclusion in the argument above?
- The argument above would be more persuasive if which one of the following were found to be true?

3. Inference Questions
Another of the most common question types you'll encounter on the logical reasoning section is the inference question. The process of inferring is a matter of considering one or more statements as evidence and then drawing a conclusion from them.

INFERENCE QUESTIONS AT A GLANCE

- Are one of the most popular LR question types
- Answer must be true if statements in the stimulus are true
- Often stick close to the author's main point
- Question stems vary considerably in appearance
- Can be checked by applying the denial test

Sometimes the inference is very close to the author's overall main point. Other times, it deals with a less central point. In logical reasoning, the difference between an inference and an assumption is that the conclusion's validity doesn't logically depend on an inference, as it does on a necessary assumption. A valid inference is merely something that must be true if the statements in the passage are true—an extension of the argument rather than a necessary part of it.

Be careful. Unlike an assumption, an inference need not have anything to do with the author's conclusion; it may simply be a piece of information derived from one or more pieces of evidence. However, the denial test works for inferences as well as for assumptions: a valid inference always makes more sense than its opposite. If you deny or negate an answer choice, and it has little or no effect on the argument, then chances are that the choice is not inferable from the passage.

Here are some tips for making proper inferences (useful for reading comprehension, as well!). A good inference:

- Stays in line with the gist of the passage
- Stays in line with the author's tone
- Stays in line with the author's point of view
- Stays within the scope of the argument or the main idea
- Is neither denied by, nor irrelevant to, the argument or discussion
- Always makes more sense than its opposite

Here's a quick rundown of the various forms that inference questions are likely to take on your test:

- Which one of the following is implied by the argument above?
- The author suggests that . . .
- If all the statements above are true, which one of the following must also be true?
- The author of the passage would most likely agree with which one of the following?

PARALLEL REASONING AT A GLANCE

- Must mimic structure or form, not content, of stimulus
- Sometimes are amenable to algebraic symbolization
- Key is to summarize argument's overall form and match it to that of the correct choice

4. Parallel Reasoning Questions

Parallel Reasoning questions require you to identify the answer choice that contains the argument most similar, or parallel, to that in the stimulus in terms of the reasoning employed. Your task is to abstract the stimulus argument's form, with as little content as possible, and then

KAPLAN

locate the answer choice that has the form most similar to that of the stimulus. Do not let yourself be drawn to a choice based on its subject matter. A stimulus about music may have an answer choice that also involves music, but that doesn't mean that the reasoning in the two arguments are similar.

5. Paradox Questions

A paradox exists when an argument contains two or more seemingly inconsistent statements. You'll know you're dealing with a paradoxical situation if the argument ends with what seems to be a bizarre contradiction. Another sure sign of a paradox is when the argument builds to a certain point, and then the exact opposite of what you would expect to happen happens.

The Kaplan Four-Step Method for Logical Reasoning

Now that you've learned the basic logical reasoning principles and have been exposed to the full range of question types, it's time to learn how to use all of that knowledge to formulate a systematic approach to logical reasoning. We've developed a four-step method that you can use to attack every question on the section.

1. Preview the Question Stem

As we mentioned in the discussion of basic principles, previewing the stem is a great way to focus your reading of the stimulus, so that you know exactly what you're looking for.

2. Read the Stimulus for Structure

With the question stem in mind, read the stimulus, paraphrasing as you go. Remember to read actively and critically, pinpointing evidence and conclusion. Also get a sense for how strong or weak the argument is.

3. Try to Prephrase an Answer

Sometimes, if you've read the stimulus critically enough, you'll know the answer without even looking at the choices. It will be much easier to find it if you have a sense of what you're looking for among the choices.

4. Choose an Answer

Yes, this is obvious, but it matters how you do it. If you were able to prephrase an answer, skim the choices looking for something that sounds like what you have in mind. If you couldn't think of anything, read and

---

PARADOX QUESTIONS AT A GLANCE

- correct choice will resolve apparent discrepancy or contradiction
- correct choice should have an intuitive click
- correct choice will often involve realizing that two groups presented as identical are actually not

THE KAPLAN FOUR-STEP METHOD FOR LOGICAL REASONING

1. Preview the question stem.
2. Read the stimulus.
3. Try to prephrase an answer.
4. Choose an answer.

evaluate each choice, throwing out the ones that are outside the scope of the argument. After settling on an answer, you may wish to briefly double-check the question stem to make sure that you're indeed answering the question that was asked.

Using the Kaplan Four-Step Method

Now let's try this method on a genuine logical reasoning item:

> A study of 20 overweight men revealed that each man experienced significant weight loss after adding SlimDown, an artificial food supplement, to his daily diet. For three months, each man consumed one SlimDown portion every morning after exercising, and then followed his normal diet for the rest of the day. Clearly, anyone who consumes one portion of SlimDown every day for at least three months will lose weight and will look and feel their best.
>
> Which one of the following is an assumption on which the argument depends?
>
> (A) The men in the study will gain back the weight if they discontinue the SlimDown program.
> (B) No other dietary supplement will have the same effect on overweight men.
> (C) The daily exercise regimen was not responsible for the effects noted in the study.
> (D) Women won't experience similar weight reductions if they adhere to the SlimDown program for three months.
> (E) Overweight men will achieve only partial weight loss if they don't remain on the SlimDown program for a full three months.

1. Preview the Question Stem

We see, quite clearly, that we're dealing with an assumption question. Good. We can immediately adopt an "assumption mindset," which basically means that, before even reading the first word of the stimulus, we know that the conclusion will be lacking an important piece of supporting evidence. We now turn to the stimulus, already on the lookout for this missing link.

2. Read the Stimulus for Structure

The first sentence introduces a study of 20 men using a food supplement product, resulting in weight loss for all 20 of them. The second sentence

STEP BY STEP

The Kaplan four-step method is designed to give structure to your work on the logic reasoning section. But be flexible in using it. These are guidelines, not commandments.

KAPLAN

describes how they used it: once a day, for three months, after morning exercise. So far so good; it feels as if we're building up to something. The structural signal usually indicates that some sort of conclusion follows, and in fact it does: The author concludes in the third sentence that anyone who has one portion of the product daily for three months will lose weight, too.

You must read critically! Notice that the conclusion doesn't say that anyone who follows the same routine as the 20 men will have the same results; it says that anyone who simply consumes the product in the same way will have the same results. You should have begun to sense the inevitable lack of crucial information at this point. The evidence in the second sentence describes a routine that includes taking the supplement after daily exercise, whereas the conclusion focuses primarily on the supplement and entirely ignores the part about the exercise. The conclusion, therefore, doesn't stem logically from the evidence in the first two sentences. This blends seamlessly into the third step.

3. Prephrase an Answer

As expected, the argument is beginning to look as if it has a serious shortcoming. Of course, we expected this because we previewed the question stem before reading the stimulus.

In really simplistic terms, the argument proceeds like so: "A bunch of guys did A and B for three months and had X result. If anyone does A for three months, that person will experience X result, too." Sound a little fishy? You bet. The author must be assuming that A (the product), not B (exercise), must be the crucial thing that leads to the result. If not (the denial test), the conclusion makes no sense.

So, you might prephrase the answer like this, "Something about the exercise thing needs to be cleared up." That's it. Did you think your prephrasing had to be something fancy and glamorous? Well, it doesn't. All you need is an inkling of what the question is looking for, and in this case, it just seems that if we don't shore up the exercise issue, the argument will remain invalid and incomplete. So, with our vague idea of a possible assumption, we can turn to the fourth step.

4. Choose an Answer

Because we were able to prephrase something, it's best to skim the choices looking for it. And, lo and behold, there's our idea, stated in choice (C). (C) clears up the exercise issue. Yes, this author must assume (C) to make the conclusion that eating SlimDown alone will cause the men to lose weight.

THE ART OF PREPHRASING

Your prephrasing of an answer need not be elaborate or terribly specific. Your goal is just to get an idea of what you're looking for, so the correct answer will jump out at you.

At this point, if you're stuck for time, you simply choose (C) and move on. If you have more time, you may as well quickly check the remaining choices, to find (we hope) that none of them fits the bill.

Of course, once you grasp the structure of the argument and have located the author's central assumption, you should be able to answer any question they throw at you. This one takes the form of an assumption question. But it could have just as easily been phrased as a weaken-the-argument-question.

What's Next?

Now that you have the skills to handle the Analytical section, try them out on the following practice set.

After that, we'll move on to Level 2 in the Kaplan Three-Level Master Plan for the GRE: test mechanics.

KAPLAN

Analytical Practice Set

Directions: Each question or group of questions is based on a passage or a set of conditions. You may wish to draw a diagram to answer some of the questions. Choose the *best* answer for each question and fill in the corresponding space on your answer sheet. (Answers and explanations can be found at the end of the set of questions.)

Questions 1–5: The creative director of an ad agency wants to select four employees to work on different aspects of a new campaign. The seven employees available are Alan, Beatrice, Cindy, Dalen, Enid, Felicity, and Godfrey.

Alan and Beatrice will not work together.
Enid and Beatrice will not work together.
Felicity will not work on the new campaign unless Dalen does, and vice versa.

Question 1

Which of the following is an acceptable group of four employees?

(A) Alan, Cindy, Enid, Dalen
(B) Beatrice, Dalen, Cindy, Godfrey
(C) Beatrice, Dalen, Enid, Felicity
(D) Beatrice, Dalen, Alan, Felicity
(E) Beatrice, Dalen, Felicity, Godfrey

Ⓐ Ⓑ Ⓒ Ⓓ Ⓔ

Question 2

If Beatrice is selected, which of the following must also be selected?

(A) Godfrey
(B) Dalen
(C) Enid
(D) Cindy
(E) Alan

Ⓐ Ⓑ Ⓒ Ⓓ Ⓔ

Question 3

Which of the following must be true?

(A) If Alan is not selected, Felicity must be selected.
(B) If Godfrey is not selected, Cindy must be selected.
(C) If Beatrice is not selected, Cindy must be selected.
(D) If Enid is selected, Cindy must not be selected.
(E) If Enid is not selected, Beatrice is not selected.

Ⓐ Ⓑ Ⓒ Ⓓ Ⓔ

Question 4

If Cindy and Enid are to be selected as two of the employees, which of the following pairs could be the other two employees?

(A) Alan and Dalen
(B) Felicity and Alan
(C) Alan and Beatrice
(D) Alan and Godfrey
(E) Felicity and Godfrey

Ⓐ Ⓑ Ⓒ Ⓓ Ⓔ

Question 5

If Dalen is not chosen, all of the following must be chosen EXCEPT

(A) Enid
(B) Cindy
(C) Alan
(D) Beatrice
(E) Godfrey

ⒶⒷⒸⒹⒺ

Questions 6–9: Individual members from seven animal species are to be chosen for a special exhibit habitat. The seven species are P, Q, R, S, T, U, V, and W. Because of the way these animals interact, certain guidelines must be followed.

Animals that will fight cannot be placed in the habitat together.

Members of species V will fight with members of species S, T, and U.

A member of species R will fight with a member of species Q, but only if a member of species V is present.

If a member of species W is present, a member of species P will not fight with any animal.

If a member of species W is not present, a member of species P will fight with members of species Q and R.

No fights other than those described above will occur.

Question 6

If V is chosen for the habitat, which of the following CANNOT also be chosen?

(A) P
(B) Q
(C) T
(D) R
(E) W

ⒶⒷⒸⒹⒺ

Question 7

If two other animals are to be added to a habitat containing a member of species Q and a member of species V in the habitat, which of the following could be those two animals?

(A) members of species W and P
(B) members of species R and P
(C) members of species S and W
(D) members of species W and R
(E) members of species U and R

ⒶⒷⒸⒹⒺ

Question 8

If two habitats are set up, one containing members of species P, Q, W, and V, and the other containing members of species S, U, R, and T, which animals could be switched one for the other without provoking any fights?

(A) species W and U
(B) species Q and R
(C) species P and R
(D) species V and S
(E) species W and T

ⒶⒷⒸⒹⒺ

Question 9

If S, P, and R are chosen for the habitat, which of the following must also be chosen?

(A)   W
(B)   V
(C)   U
(D)   T
(E)   Q

Question 10

Those who believe that the stars are influential in shaping our personalities fail to see that astrologers, alleged experts on the heavens, often present widely disparate interpretations of the same individual's astrological chart. If astrology were indeed a science, individuals would be able to present the same information—their birth date, place, and time—to many astrologers and receive the same evaluation and description of their personalities from each one. Since this is never the case, astrology cannot be considered a true science.

A believer in astrology can most effectively counter the author's objection by contending that

(A) most astrologers are not expert enough to recognize the one correct interpretation of a chart.
(B) the definition of astrology implies that it is a true science.
(C) even if astrologers' interpretations aren't exactly the same, they usually contain similarities.
(D) believers in astrology actually like the element of variability inherent in astrological interpretations.
(E) some people do not know exactly what time they were born, and therefore present inaccurate and misleading information to the astrologist.

Ⓐ Ⓑ Ⓒ Ⓓ Ⓔ

Question 11

Plant Y thrives in environments of great sunlight and very little moisture. Desert X is an environment with constant, powerful sunlight, and next-to-no moisture. Although Plant Y thrives in the areas surrounding Desert X, it does not exist naturally in the desert, nor does it survive long when introduced there.

Which of the following would be most useful in explaining the apparent discrepancy above?

(A) Desert X's climate is far too harsh for the animals that normally feed on Plant Y.
(B) For one week in the fall, Desert X gets consistent rainfall.
(C) The environment around Desert X is ideally suited to the needs of Plant Y.
(D) Due to the lack of sufficient moisture, Desert X can support almost no plant life.
(E) Plant Y cannot survive in temperatures as high as those normally found in Desert X.

Ⓐ Ⓑ Ⓒ Ⓓ Ⓔ

Question 12

Although air pollution was previously thought to exist almost exclusively in our nation's cities, the recent increase in the number of persons suffering from illnesses attributed to excessive air pollution leaves us no choice but to conclude that other, nonurban areas are now affected.

Which of the following, if true, would most seriously weaken the conclusion of the argument above?

(A) The nation's cities have seen a marked decrease in their levels of air pollution.
(B) The nation has experienced a sharp decrease in the number of people moving out of its cities.

(C) Illnesses due to air pollution are among the least common causes of death to urban dwellers.

(D) Many illnesses previously thought unrelated to air pollution are now considered to be caused by it.

(E) As a result of the problems in urban areas, non-urban areas have passed strict pollution control measures.

Ⓐ Ⓑ Ⓒ Ⓓ Ⓔ

Questions 13–18: An assistant principal must assign seven classes—K, L, M, N, O, P, and Q—to the seven classrooms lining the north corridor of her high school. The rooms are numbered 1 through 7, from west to east. There is a broom closet between Rooms 4 and 5. Only Rooms 1 and 7 have taping facilities.

L and M may not be in adjacent rooms.
N and O must be in adjacent rooms.
L must be in Room 3.
Language classes must be in a room with taping facilities.

Question 13

If M and N are both language classes, and if O and L are in adjacent rooms, it must be true that M is in

(A) Room 1
(B) Room 7
(C) a room adjacent to Q's room
(D) a room adjacent to O's room
(E) a room adjacent to K's room

Ⓐ Ⓑ Ⓒ Ⓓ Ⓔ

Question 14

If K and O are both language classes, and if M is not in a room adjacent to the broom closet, all of the following must be true EXCEPT:

(A) O is in room 1.
(B) N is in room 2.
(C) M is in room 6.
(D) P and L are in adjacent rooms.
(E) Q is in a room adjacent to the broom closet.

Ⓐ Ⓑ Ⓒ Ⓓ Ⓔ

Question 15

If O is a language class, and if K and Q are in the two rooms adjacent to the broom closet, in which room could M be?

(A) Room 1
(B) Room 2
(C) Room 3
(D) Room 4
(E) Room 5

Ⓐ Ⓑ Ⓒ Ⓓ Ⓔ

Question 16

If K and M are in the two rooms adjacent to the broom closet, and if O is in Room 2, which of the following must be true?

(A) N is in Room 7.
(B) P and M are in adjacent rooms.
(C) Q and P are in adjacent rooms.
(D) N and K are in adjacent rooms.
(E) P is in a room with taping facilities.

Ⓐ Ⓑ Ⓒ Ⓓ Ⓔ

Question 17

If O is a language class, and if N and L are not in adjacent rooms, all of the following are possible EXCEPT:

(A) M is in Room 1.
(B) M and N are in adjacent rooms.
(C) M and O are in adjacent rooms.
(D) M and K are in adjacent rooms.
(E) M is in a room adjacent to the broom closet.

Ⓐ Ⓑ Ⓒ Ⓓ Ⓔ

Question 18

If O and P are language classes, and M is in a room adjacent to N, which of the following could be true?

(A) P is in Room 7.
(B) N is in Room 5.
(C) K is in Room 5.
(D) Q is in Room 2.
(E) O is in Room 1.

Ⓐ Ⓑ Ⓒ Ⓓ Ⓔ

Questions 19–22: Four teams of tennis players, A, B, C, and D, have to be redistributed into three new teams. The players on team A are L, M, and N. The players on team B are O, P, Q, and R. The players on team C are S and T. And the players on team D are U, V, and W.

Each of the three new teams must contain at least one member from three of the original four teams (teams A, B, C, and D).

No new team can contain all the members of any of the original teams (teams A, B, C, and D).

V and W cannot be on the same new team.

Question 19

If one new team is made up of L, N, O, and V only, which of the following groups CANNOT completely represent one of the other new teams?

(A) M, P, T
(B) P, S, U, W
(C) M, R, W
(D) Q, R, M, S
(E) R, T, W

Ⓐ Ⓑ Ⓒ Ⓓ Ⓔ

Question 20

If one new team is made up of N, O, P, Q, and U only, which of the following groups CANNOT completely represent one of the other new teams?

(A) R, T, W
(B) L, M, T
(C) M, S, V
(D) M, R, T, W
(E) L, R, S, W

Ⓐ Ⓑ Ⓒ Ⓓ Ⓔ

Question 21

What is the maximum number of people who can be on the same new team?

(A) five
(B) six
(C) seven
(D) eight
(E) nine

Ⓐ Ⓑ Ⓒ Ⓓ Ⓔ

Question 22

If one group is made up of L, O, and U only, which of the following could be the complete roster of another new team?

(A) M, N, S, P
(B) M, N, P, Q
(C) S, V, W, R
(D) T, S, V, M
(E) T, W, M, R

Ⓐ Ⓑ Ⓒ Ⓓ Ⓔ

Answer Key

1. E	6. C	11. E	16. C	21. B
2. B	7. A	12. D	17. C	22. E
3. A	8. B	13. B	18. D	
4. D	9. A	14. D	19. C	
5. D	10. A	15. A	20. B	

Explanations

Questions 1–5: In this game we're to select groups of exactly four employees from a field of seven, now known as A, B, C, D, E, F, and G. The rules are only three in number, and amount to the facts that one cannot include both A and B, nor can once include both E and B. Furthermore, if either F or D is included in the group, then so is the other, and if either F or D isn't included, then neither is the other.

Question 1

For Question 1 we can just eliminate the four incorrect groups. Rule 1 says A and B can't both be selected. That eliminates choice (D). Rule 2 says that E and B can't both be selected. That eliminates choice

(C). And rule 3 says that F and D must both be in or both be out; that eliminates choices (A) and (B), leaving the only correct choice, (E).

Question 2

Question 2 asks who must be selected if B is selected. It can't be A; the first rule prohibits that. And it can't be E; the second rule prohibits that. So choices (C) and (E) are out. That leaves, C, D, F, and G, from which we must pull three others. If we were to leave out the F and D pair, we'd have only three employees selected. Clearly, then, F and D must both be selected. D is listed as choice (B), which is correct. G and C, choices (A) and (D) could go, either one or the other, but neither must go.

Question 3

We're asked to find something that must be true. The best approach is to deny the answer choices, and see what happens. The one that can't be denied must be true. (A) turns out to be correct here. If A is not selected F must be selected. To not select F would require also not selecting D. With A, D, and F out, we'd have only four employees left. That would require that they all be selected. E and B, however, would be two of those employees, and the second rule says that they can't both go. (B) need not be true because we could have a group of A, E, D, and F. (C) need not be true because E, D, F, and G make an acceptable group. (D) can be eliminated with E, C, D, and F. And (E) can be denied with B, C, D, and F.

Question 4

Question 4 says that C and E are selected, and asks which of the answer choices lists an acceptable pair to be selected with them. Well, with E chosen, we know B can't be chosen. So choice (C) is immediately out.

Furthermore, any pair that lists either D or F must also include the other one. That eliminates (A), (B), and (E). Thus choice (D) is correct. And indeed, there were only two possible pairs to go with C and E: A and G, as (D) says, or D and F, which wasn't listed.

Question 5

Question 5 says D is not selected, and asks which of the listed answer choices need not be chosen. If D isn't chosen, then F isn't, either. That leaves five employees (A, B, C, E, and G) from which we can get four. Now suppose B were one of those selected. That would mean neither A nor E could be selected, leaving only three employees, which isn't allowed. So B cannot be selected and A, C, E, and G must be selected. That means that choice (D) is correct for question 5.

Questions 6–9: This is a rather odd grouping game of selection, a type that has, however, appeared on the GRE. The key to this game is realizing that "Members of species V will fight with members of species S, T, and U" is another way of saying: "V and S cannot be included in the same habitat, V and T cannot be included in the same habitat . . ." and so forth—because the first rule of the game tells you that fighting species cannot share a habitat. Although there's no "right" way of shorthanding all this, we've chosen an old familiar method for "cannot be together" rules:

$$V \neq S \quad V \neq T \quad V \neq U$$
$$RV \neq Q$$
$$P \neq Q \quad P \neq R$$
$$PW = P \text{ no fighting}$$

Question 6

Question 6 is a gift to you, as long as you understand that each rule's references to fighting are a backhand way of saying, "This species cannot share the habitat with that species." A quick check of the rules gives you the answer: V cannot share a habitat with S, T, or U, and one of those three—T—is correct choice (C). The remaining choices are all possible roommates for V; their inclusion or exclusion in this habitat depends on their interactions with other species.

Question 7

Question 7 is a standard grouping-game question—two members of the group have been chosen for you, and your job is to choose two more acceptable members. You're given species V and Q. The first step is to check your rules to eliminate known quarrelsome species. With V in, S, T, and U are definitely out, leaving you with P, R, and W. (You can eliminate answer choices (C) and (E) at this point.) Now you must find an acceptable pair from the remaining three contenders. Can P and R be chosen? No, because the pairing of R and V will cause fighting with Q, and that eliminates choice (B). Can R and W be chosen? No, for the same reason, and now you've eliminated choice (D). That leaves (A), which must be correct—and it is, since W's presence prevents P from fighting with Q and makes a happy habitat.

Question 8

Question 8 is the odd duck in this game. You're asked about switching between two habitats without causing fighting. The correct choice is (B). Switching species Q and R creates no problems, since V is still separated from the species it fights with and the P-W pair keeps P under control. (A) doesn't work because without the calming influence of W, P is going to

fight with Q. (C) unwisely puts both R and V in the habitat with Q—guaranteeing that R and Q will fight. (D) places V in a group with two of its archenemies, T and U, so fighting is certain to result, and (E) creates chaos—V is grouped with U and S, and without W's calming presence, P will certainly fight with Q.

Question 9

Another must-be situation appears in Question 9. This time the question stem supplies three species and asks you for a fourth, so go to the rules for guidance. You begin with species S, P, and R—wait a minute! P and R should not be in a habitat together, since the fifth rule specifically stated that P fights with R. But there's another rule about P. The fourth rule says that P will not fight if W is present—and there's your must be. W, correct choice (A), must be included in this habitat to make the given group of S, P, and R acceptable.

Question 10

The argument in Question 10 is an attack on astrology. The author says that since different astrologers, using the same data, arrive at different evaluations of a person, astrology cannot be considered a hard science. This is because if astrology were a hard science (the author implies), identical data would lead to identical interpretations, and therefore someone's astrological chart would have only one scientific interpretation.

We're asked to weaken the argument. Choice (A) does so by pointing out that not every astrologer is an expert. It could be that expert astrologers do agree on the one correct interpretation, and only the inexpert ones differ. (B) is no good because, quite simply, the definition of astrology may be completely wrong. A cardinal rule of thinking says don't confuse the name

of a thing with the thing itself. You can call a rose a fish, but it's still not a fish. (C) is too weak, in that being "similar" isn't enough. If a science must provide a single answer, then multiple similar answers still don't cut it. (D) claims that the variability is appealing. So what? Astrology may be appealing even though it's not a science, so (D) is out. And finally, (E) says that inaccurate information is used in constructing the charts. Yet if all the astrologers are given the same inaccurate information then they should still give the same interpretation. The author said astrology wasn't a science because the interpretations differ, not because the data are inaccurate.

Question 11

Question 11 is a type of question that appears increasingly on the GRE. We have to explain an apparent discrepancy; the key is to find the explanation that is consonant with all of the stimulus information. We're told enough to expect that Plant Y would thrive in Desert X. Yet it doesn't. Why doesn't it? For this we go to what we're told. X is a desert, which means it's dry and sunny. Y likes dry and sunny regions—but we're told nothing about its temperature preferences. So the best explanation is, as choice (E) says, that it's the high temperatures in Desert X that Y can't handle. If the information in (E) were true, you would actually expect Y not to grow in Desert X. (A) won't work because a plant usually doesn't need animals to feed on it in order to survive. One would expect quite the opposite. (B)'s out because one week of consistent rainfall hardly explains why Y isn't in Desert X at other times. The fact that it can easily grow elsewhere isn't relevant to the question of why it doesn't grow in Desert X, so (C) is out. And the ability of other plants to survive in Desert X by itself has nothing to do with Plant Y, so (D) also fails.

Question 12

In Question 12 the author gives you an in-your-face signal of what the conclusion is when she says, "leaves us no choice but to conclude that. . . ." Her conclusion then is that nonurban areas are now affected by air pollution. She makes this conclusion based on evidence that there has recently been an increase in the number of persons suffering from illnesses attributable to air pollution. Now we hope you see that this argument has a few holes in it. We're asked to weaken it, and since there are several holes, there are many ways to weaken it. So our best bet may be to go through the answer choices and see what effect they each have.

(A) says that air pollution in the cities is decreasing. Well, that certainly won't weaken an argument that air pollution is rising elsewhere, especially when the number of people suffering from it is on the increase. (B) says that fewer people are moving out of the cities. Well, that won't have effect. The argument draws a link between those suffering from air pollution–related illnesses and the movement of air pollution out to nonurban areas. The fact that fewer people are leaving the cities doesn't make a difference one way or the other. (C) says that air pollution doesn't kill very many city dwellers. Happy thought, but it would seem that the new sufferers from air pollution (not necessarily death victims) that the author speaks of would-be noncity dwellers. So the fact that not many urban dwellers die from air pollution has no effect, or a very weakly strengthening effect.

As for (D), ah ha! Here we go. (D) is going to weaken the argument by providing another explanation for the increase in air pollution sufferers. The number of sufferers from air pollution–related illnesses has risen, not because pollution is spreading out to nonurban areas, but because a lot more illnesses are now considered to be caused by air pollution. If that's true, then nothing at all need have changed. It could be the same pollution in the same place, with

the same victims, only now, more of them have been identified. So (D) is our answer. (E) is incorrect. Whether or not antipollution measures have been passed is irrelevant. There could still be plenty of pollution in nonurban areas. For example, they could have passed the measures only yesterday.

Questions 13–18

The scenario and rules to this sequencing game don't start you off with much in the way of concrete information, and they leave one big mystery: Which are the language classes? A quick overview of the question stems clears this up. The classes designated as "language" change from question to question, so there's no point in worrying about them unduly during your work with the stimulus. As for the rest of it, this is about as concrete as we can get:

$$— — L —/— — —$$

tape   tape

$$L \neq M$$

$$N = O$$

Note that half the challenge of this game is keeping track of special conditions (broom closet, taping rooms) as well as characters. Most of the characters are still in the realm of "could be true" at this point, so let's go on to the questions.

Question 13

Since Room 1 and 7 are taping rooms, the first if-clause in the stem of Question 13 means that either M is first and N seventh or vice versa. However, you should see that "vice versa" is in fact what it must be: N seventh would require that O be sixth, yet doing so does not allow O to be adjacent to L, which the

question stem specifies. The only possible combination is

$$N \quad O \quad L \quad — \quad / \quad — \quad — \quad M$$

—and notice that these are the only placements that must be true, thus making (B) the correct answer. Notice that (C) and (E) are only possible, while (A) and (D) are impossible.

Question 14

Question 14 starts off in much the same way as Question 13: the stem tells us that K and O are either first and seventh or seventh and first, respectively. Wherever O is, N must be next to it, so we can set up two options:

$$K \quad — \quad L \quad — \quad / \quad — \quad N \quad O$$
$$O \quad N \quad L \quad — \quad / \quad — \quad — \quad K$$

But the first option must be discarded: if M is prohibited from being next to the broom closet, M can't be fifth; and M can't be second or fourth, either, since L ≠ M. Option 2 is the only prospect, and in it M must be sixth for the reasons we just discussed. P and Q, then, will occupy the positions next to the broom closet, though we don't know in which order. The answer choice that need not be true is (D): P could very well be in Room 5, far from L in Room 3.

Question 15

The stem of Question 15 doesn't exactly start you off with truly concrete information, but that doesn't matter; to get this one, you need not know precisely where each class is as long as you know which rooms are in use and which are not. The quickest approach to this question is to first note that the answers do not include Rooms 6 and 7; you're to worry about the first five rooms only. Second, you can eliminate answer choices before dealing with the information in the

question stem. L always occupies Room 3, so choice (C) is out; since L ≠ M, M cannot be in Rooms 2 or 4, and choices (B) and (D) are out. This leaves only Rooms 1 and 5 as possibilities; what does the question stem tell you about them? It tells you that Room 5 is definitely occupied, by either K or Q—Room 5 is to the left of the broom closet, so either K or Q will be there. This leaves only one alternative, choice (A).

Question 16

The stem of Question 16 coyly hints that K and M can be interchangeable between the fourth and fifth spots, but we know better: M can't be fourth, because it would be right next to L in Room 3, which is forbidden. So M is fifth and K fourth. O is second, which means N must be first (again, O = N), leaving only the last two spots to be filled by P and Q in either order. The only answer choice, then, that must be true is (C). Note that (B) and (E) could be true, while (A) and (D) cannot be true.

Question 17

If, as in Question 17, O is a language course but N is not adjacent to L, we should see that O must be seventh: since O = N, O first would mean N second and that would violate the question stem. So we're left with:

$$— \quad — \quad L \quad — \quad / \quad — \quad N \quad O$$

—and a lot of possible locations for the other four classes, though the five answer choices are mainly concerned with where M can possibly go. In point of fact, M can't be next to O, since N is there, so (C) is the correct answer.

Question 18

Answering Question 18 is a matter of putting the two pieces of information in the question stem together.

O and P will be first and seventh, or vice versa, so we know that N will be either second (if O is 1st) or sixth (if O is seventh). And M will be adjacent to N, so M will be either third (if N is second) or fifth (if N is sixth). But putting O first, N second, and M third won't work, since 3 is permanently occupied by L. So this is the only option:

$$ P \; — \; L \; —\! / \; M \quad N \quad o $$

That's what must be true; this question asks for what could be true. The only options left open are for K and Q; they could be either second and fourth or vice versa. Look among the choices for mentions of K and Q: choice (C) mentions K but places K in Room 5, which is already occupied by M. Choice (D), however, states that Q could be in Room 2, which is what we just deduced, and so is the correct answer.

Questions 19–22: The last game is a grouping game of distribution. Four teams are being broken up to form three new ones. The first two rules are the type it's best to just remember, or circle and refer to. One of them says that each of the new teams must contain a member from three of the old teams. That tells us each of the three new teams will contain at least three players. The other says that all the members of one of the old teams can't be on the same new team. So, in the case of team C, players S and T cannot be on the same new team. As if that weren't enough to juggle, we're also told in the last rule V and W can't be on the same team.

Question 19

Question 19 asks which answer choice can't be one of the new teams, when another of the new teams is L, N, O, and V. Now the key here and throughout the game is to realize that there must always be three new teams, each of which must conform to the rules. So

here it's not merely a matter of checking the listed teams for discrepancies; you must also check the third, unmentioned team, which is composed of the remaining players.

Well, we know the first team is L, N, O, and V. That teams M, P, Q, R, S, T, U, and W are left to be distributed into the other two teams. Choice (A) would have M, P, and T on one of the other teams. That would leave Q, R, S, U, and W for the third new team. We have to check both because if the remaining players cannot from the third team, then the second, listed new team is illegitimate. In looking, we see that S and T are split; V and W are split; and the other requirements are met. (B) lists P, S, U, and W. That means the third team would be composed of M, Q, R, T. Again, no problem with this distribution. (C), however, cannot be the new distribution. If M, R, and W are on the second new team, then S and T would be among the members of the third team. That's a no-no because S and T are all the members of one of the original teams, and rule 2 prohibits this kind of distribution.

Question 20

Question 20 is exactly the same type of question that we just saw. We use the same procedure and find that the new teams listed in choices (A), (C), (D), and (E) are all acceptable, as the remaining players also form an acceptable team. Choice (B), on the other hand, is unacceptable. With N, O, P, Q, and U on one of the teams, and L, M, and T on another, we would have S, R, V, and W on the third. Since V and W cannot be on the same team, this arrangement is unacceptable.

Question 21

Question 21 asks for the maximum number of players that can be put on one of the new teams. The thing to see here is that we don't just want to divvy them up in any old order, so that at the end we have rules telling

us where we can and can't put the last players. (If you did this, you may have wound up with incorrect choice (A).) We want to satisfy the rules with as few distributions as possible, so that we're free to throw the remaining players all on on team.

Let's start with V and W, and S and T. We know these pairs must be split. We don't want to put S and T on the same two teams as V and W. That would be maximizing the number of two teams and minimizing the number on one team. We want just the opposite. So, let's say we've got S on Team 1, T and V on Team 2, and W on Team 3. That takes care of original team C, and leaves only U from team D. Since V and W are already split, U will go anywhere we like. We need to get one member from three of the original teams on each of the new teams. Team A has three members— L, M, and N— and we can just put one on each of the new teams: L and S on Team 1, M, T, and V on Team 2, and N and W on Team 3. Original Team A is now exhausted. That leaves all of Team B, and U from Team D. If we look at our new teams, Team 2 has three members from three different original teams. Teams 1 and 3, however, have only two members from two different original teams. Let's fill them out: Put O on Team 1 and P on Team 3. We've now satisfied all the rules, and can place Q, R, and U anywhere we like. Since we're to maximize the players on one team, we'll add all three players to one of the three teams. That'll mean, regardless of which team we add them to, six players maximum—choice (B). If you had trouble with this one, you could eliminate choices by noting that Rule 1 alone requires at least three players per new team, and that in itself makes seven-, eight-, or nine-player teams impossible.

Question 22

Question 22 tells us that L, O, and U comprise one new team, and then asks which answer choice could comprise another complete team. That leaves M, N, P, Q, R, S, T, V, and W to distribute into the other teams. We know S and T will have to be on separate teams, and that means that any acceptable team must have either S or T listed, and not both. This criterion tells us that (B) and (D) are out. If (B) were a second new team, S and T would both be on Team 3, breaking Rule 2. And (D) blatantly puts them on the same team. The same applies to V and W. Each acceptable second team must have V or W, but not both. That eliminates (A) and (C). (C) puts them on the same team, which isn't allowed, and (A) omits both from the second team, thus forcing them together on the third. That leaves correct answer (E), which is acceptable, and which puts N, P, Q, S, and V on the third team, which is also acceptable.

CHAPTER 5

TEST MECHANICS

The first year of graduate school is a frenzied experience for many students. It's no surprise, then, that the GRE, the test specifically designed to predict success in the first year of graduate school, is a speed-intensive test that demands good time-management skills.

So when you're comfortable with the content of the test, namely, the type of material discussed in the previous chapters, your next challenge will be to take it to the next level—test mechanics—which will enable you to manage an entire section at a time.

On most of the tests you take in school, you wouldn't dream of not making at least a try at every single one of the questions. If a question seems particularly difficult, you spend significantly more time on it, because you'll probably be given more points for answering a hard question correctly. Not so on the GRE.

You've got to develop a way of handling the test sections to make sure you get as many points as you can as quickly and easily as you can. The following principles will help you do just that.

IT'S NOT JUST ABOUT CORRECT ANSWERS

For complete GRE success, you've got to get as many correct answers as possible in the time you're given. Knowing the strategies is not enough. You've got to perfect your time management skills so that you get a chance to use those strategies on as many questions as possible.

Mechanics of the GRE CAT

The CAT is in some ways quite different from the traditional paper-and-pencil tests you've probably taken in the past. In fact, it's pretty weird at first. Here's how it works. You will only see one question at a time. Instead of having a predetermined mixture of basic, medium, and hard questions, the computer will select questions for you based on how well you are doing. The first question will be of medium difficulty. If you get it right, the second question will be a little harder; if you get the first question wrong, the second will be a little more basic.

If you keep getting questions right, the test will get harder and harder; if you slip and make some mistakes, the test will adjust and start giving you easier problems, but if you answer them correctly, it will go back to the hard ones. Ideally, the test gives you enough questions to ensure that scores are not based on luck. If you get one hard question right you might just have been lucky, but if you get 10 hard questions right, then luck has little to do with it. So the test is self-adjusting and self-correcting.

Because of this format, the CAT is very different structurally from a paper-and-pencil test. After the first problem, every problem that you see is based on how you answered the prior problem. That means you cannot return to a question once you've answered it, because that would throw off the sequence. Once you answer a question, it's part of your score, for better or worse. That means you can't skip around within a section and do questions in the order that you like.

Another major consequence is that hard problems count more than easy ones. It has to be this way, because the very purpose of this adaptive format is to find out at what level you reliably get about *half* the questions right; that's your scoring level. It actually makes a lot of sense. Imagine two students—one who does 10 basic questions, half of which she gets right and half of which she gets wrong, and one who does 10 very difficult questions, half of which she gets right and half of which she gets wrong. The same number of questions have been answered correctly in each case, but this does not reflect an equal ability on the part of the two students.

In fact, the student who answered five out of ten very difficult questions incorrectly could still get a very high score on the CAT GRE. But in order to get to these hard questions, she first had to get medium-difficulty questions right. So, no matter how much more comfortable you might be sticking to the basic questions, you definitely want to get to the hard questions if you can, because that's where the points are.

First Impressions Count

One of the most important things to know is that the early questions are vital for a good score on the CAT. As in life, first impressions make a big difference.

Why? Because the computer doesn't have information about you at the start of the test, and its goal is to get an accurate estimate of your score as quickly as possible. In order to do that, the computer has to make large jumps in the estimation of your score for each of the first few questions.

It's a lot like how you would act if you were trying to guess which number a person had picked from one to 10, and the only thing you could be told was whether the number was higher, lower, or the same as what you guessed. To do this most efficiently, you'd guess five first, since if the right number were higher or lower you could eliminate about half the choices. If you were told the actual number was lower than five, you'd guess three next, since that cuts the possibilities down the most. If the number were higher than three, it would have to be four. If it were lower, it would have to be one or two. Using this method, at most you would have to take three guesses before you knew the answer, whereas if you just started guessing randomly, or started from one and worked your way up, you could guess as many as 10 times before getting the right answer.

Like the efficient guesser, the computer doesn't use intuition to find the right answer, but uses the most effective method. Instead of using numbers to "guess" your score, though, the computer gives you questions that have a precise difficulty level assigned to them. In effect, you tell it whether your score is higher or lower than this difficulty level by getting the question right or wrong.

What's the upshot of all this? Simple: *Pay extra attention to the first few questions, and do all that you can to get them right!* Feel free to spend a little extra time double-checking the first five problems or so, and make sure you try every elimination technique you know before guessing on one of these problems if you don't know the answer.

Three More Section Management Techniques

First, if you get a lot of mileage from the strategy of eliminating answer choices based on difficulty level, you can apply it on the CAT, though in a different and limited way. It won't be spelled out for you as it is on the paper-and-pencil test, but as you progress through the questions, you should have a good idea of how you're doing. If you've practiced a lot on real questions, it's fairly easy to maintain a pretty clear sense of the difficulty level of your questions and to eliminate answer choices accordingly.

DON'T BE STUBBORN . . .

It's difficult for some of us to give up on a tough, time-consuming question, but it must be done occasionally. Remember, there's no point of honor at stake here, but there are GRE points at stake.

For instance, if you're confident that you've been answering most of the questions correctly, then you should be seeing harder and harder questions. If that seems to be the case, you can safely eliminate answer choices that look too obvious or basic for a difficult question.

Secondly, if crossing off answer choices on paper tests really helps to clarify your thinking (using a process of elimination), you may want to consider making a grid on your scratch paper before you begin the CAT. Use it to mark off answer choices that you have eliminated, as shown below. That way you can tell at a glance which answer choices are still in the running. If you end up using it often, it will be worth the 10 seconds it takes to draw a simple grid, like this one:

A	×	×		×		×			×		×		
B		×	×	×			×	×	×	×		×	
C					×				×				×
D	×		×		×			×		×		×	
E	×	×		×			×			×			

Finally, the timer in the corner can work to your advantage, but if you find yourself looking at it so frequently that it becomes a distraction, you should turn it off for 10 or 15 minutes and try to refocus your attention on the test, even if you lose track of time somewhat. *The CAT rewards focus and accuracy more than it does raw speed.*

Navigating the CAT: Computer Basics

Let's preview the primary computer functions that you will use to move around on the CAT. ETS calls them "testing tools," but they're basically just boxes that you can click with your mouse. The following screen is typical for an adaptive test.

Here's what the various buttons do.

The Time Button
Clicking on this button turns the time display at the top of the screen on and off. When you have five minutes left in a section, the clock flashes and the display changes from Hours/Minutes to Hours/Minutes/Seconds.

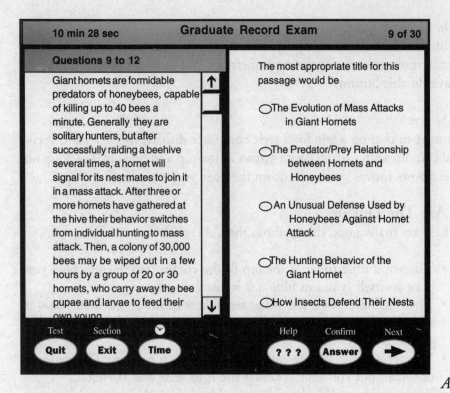

Questions 9 to 12

Giant hornets are formidable predators of honeybees, capable of killing up to 40 bees a minute. Generally they are solitary hunters, but after successfully raiding a beehive several times, a hornet will signal for its nest mates to join it in a mass attack. After three or more hornets have gathered at the hive their behavior switches from individual hunting to mass attack. Then, a colony of 30,000 bees may be wiped out in a few hours by a group of 20 or 30 hornets, who carry away the bee pupae and larvae to feed their own young

The most appropriate title for this passage would be

○ The Evolution of Mass Attacks in Giant Hornets

○ The Predator/Prey Relationship between Hornets and Honeybees

○ An Unusual Defense Used by Honeybees Against Hornet Attack

○ The Hunting Behavior of the Giant Hornet

○ How Insects Defend Their Nests

| Test | Section | | | Help | Confirm | Next |
| Quit | Exit | Time | | ??? | Answer | → |

A typical screen display in the CAT.

The Exit Button
This allows you to exit the section before the time is up. If you budget your time wisely you should never have to use this button—time will run out just as you are finishing the section.

The Help Button
This one leads to directions and other stuff from the tutorial. You should know all this already, and besides, the test clock won't pause just because you click on Help.

The Quit Button
Hitting this button ends the test.

The Next Button
Hit this when you want to move on to the next question. After you press Next, you must hit Confirm.

The Confirm Button

This button tells the computer you are happy with your answer and are really ready to move to the next question. You cannot proceed until you have hit this button.

The Scroll Bar

Similar to that on a windows-style computer display, this is a thin, vertical column with up and down arrows at the top and bottom. Clicking on the arrows moves you up or down the page you're reading.

CAT: The Upside

There are many good things about the CAT, such as:

- There is a little timer at the top of the computer screen to help you pace yourself (you can hide it if it distracts you).
- There will be only a few other test takers in the room with you—it won't be like taking it in one of those massive lecture halls with distractions everywhere.
- You get a pause of one-minute between each section. The pause is optional, but you should always use it to relax and stretch.
- You can sign up for the CAT two days before the test, and registration is very easy.
- The CAT is convenient to schedule. It's offered at more than 175 centers three to five days a week (depending on the center) all year long.
- You don't have to take it on the same day as a subject test, which can greatly reduce fatigue.
- You can see your scores before you decide which schools you want to send them to.
- Perhaps the CAT's best feature is that it gives you your scores immediately and will send them to schools just 10 to 15 days later.

CAT: The Downside

There are also not-so-good things about the CAT.

- You cannot skip around on this test; you must answer the questions one at a time in the order the computer gives them to you.
- If you realize later that you answered a question incorrectly, you can't go back and change your answer.
- You can't cross off an answer choice and never look at it again, so you have to be more disciplined about not reconsidering choices you've already eliminated.
- You have to scroll through reading comprehension passages, graphs, and some games, which means you won't be able to see the whole thing on the screen at once.
- You can't write on your computer screen the way you can on a paper test (though some have tried), so you have to use scratch paper they give you, which will be inconveniently located away from the computer screen.
- Lastly, many people find that computer screens tire them and cause eyestrain—especially after three hours.

CHAPTER 6

TEST MENTALITY

In this test prep section, we first looked at the content that makes up each specific section of the GRE, focusing on the strategies and techniques that you'll need to tackle individual questions, games, and passages. Then we discussed the mechanics involved in moving from individual items to working through full-length sections. Now we're ready to turn our attention to the often overlooked attitudinal aspects of the test. We'll then combine these factors with what we learned in the chapters on test content and test mechanics to put the finishing touches on your comprehensive GRE approach.

We've already armed you with the weapons that you need to do well on the GRE. But you must wield those weapons with the right frame of mind and in the right spirit. This involves taking a certain stance toward the entire test. Here's what's involved.

Test Awareness

To do your best on the GRE, you must always keep in mind that the test is like no other test you've taken before, both in terms of its content and in terms of its scoring system. If you took a test in high school or college and got a quarter of the questions wrong, you'd probably receive a pretty lousy grade. Not so with the GRE CAT. The test is geared so that only the very best test takers are able to finish every section. But even these people rarely get every question right.

THE BASIC PRINCIPLES OF GOOD TEST MENTALITY

- Test awareness
- Stamina
- Confidence
- The right attitude

NOBODY'S PERFECT

Remember that the GRE isn't like most tests you've taken. You can get a lot of questions wrong and still get a great score. So don't get rattled if you miss a few questions.

BE COOL

Losing a few extra points here and there won't do serious damage to your score but losing your head will. Keeping your composure is an important test-taking skill.

What does this mean for you? Well, just as you shouldn't let one tough reading comp passage ruin an entire section, you shouldn't let what you consider to be a subpar performance on one section ruin your performance on the entire test. A lousy performance on one single section will not by itself spoil your score—unless you literally miss almost every question. If you allow that subpar section to rattle you, however, it can have a cumulative negative effect that sets in motion a downward spiral. It's that kind of thing that could potentially do serious damage to your score. Losing a few extra points won't do you in, but losing your head will.

Remember, if you feel that you've done poorly on a section, don't sweat it. It could be the experimental one. And even if it's not, you must remain calm and collected. Simply do your best on each section, and once a section is over, forget about it and move on.

Stamina

You must work on your test taking stamina. Overall, the GRE is a fairly grueling experience, and some test takers simply run out of gas on the final few sections. To avoid this, you must prepare by taking as many full-length practice tests as possible in the week or two before the test, so that on the test, five sections plus a writing sample will seem like a breeze (well, maybe not a breeze, but at least not a hurricane).

GET IN SHAPE

You wouldn't run a marathon without working on your stamina well in advance of the race, would you? The same goes for taking the GRE.

One option is to buy *Practicing to Take the GRE General Test*, which is a book of released tests from ETS. There are six real GREs in this book, and they can be great practice for all the strategies that you're learning in this book. Since that book was published, ETS has revised the Analytical and Quantitative sections, adding more logical reasoning questions and questions on probability, median, mode, and range in Quantitative.

Another option, if you have some time and need a really great score, would be to take the full Kaplan course. We'll give you access to every released test plus loads of additional material, so you can really build up your GRE stamina. But most importantly, you'll also have the benefit of our expert live instruction. In fact, you could even set up special one-on-one tutoring sessions with Kaplan experts. If you decide to go this route, call 1-800-KAP-TEST for a Kaplan center location near you.

Confidence

Confidence feeds on itself, and unfortunately, so does the opposite of confidence—self-doubt. Confidence in your ability leads to quick, sure answers and an ease of concentration that translates into more points. If you lack concentration, you end up reading sentences and answer choices two, three, or four times. This leads to timing difficulties, which only continue the downward spiral, causing anxiety and a tendency to rush.

If you subscribe to the GRE mindset that we've described, however, you'll be ready and able to take control of the test. Learn our techniques and then practice them on real test material, such as that found in *Practicing to Take the GRE*. That's the way to score your best on the test.

Attitude

Those who approach the GRE as an obstacle and who rail against the necessity of taking it usually don't fare as well as those who see the GRE as an opportunity to show off the reading and reasoning skills that graduate schools are looking for. Those who look forward to doing battle with the GRE—or, at least, who enjoy the opportunity to distinguish themselves from the rest of the applicant pack—tend to score better than do those who resent or dread it.

It may sound a little dubious, but take our word for it: attitude adjustment is proven to raise points. Here are a few steps you can take to make sure you develop the right GRE attitude:

- Look at the GRE as a challenge, but try not to obsess over it; you certainly don't want to psyche yourself out of the game.
- Remember that, yes, the GRE is obviously important, but, contrary to what some people think, this one test will not single-handedly determine the outcome of your life. In many cases, it's not even the most important piece of your graduate application.
- Try to have fun with the test. Learning how to match your wits against the test makers can be a very satisfying experience, and the reading and thinking skills you'll acquire will benefit you in graduate school as well as in your future career.
- Remember that you're more prepared than most people. You've trained with Kaplan. You have the tools you need, plus the know-how to use those tools.

ATTITUDE ADJUSTMENT

Your attitude towards the test does affect your performance. You don't have to "think nice thoughts" about the GRE, but you should take a good mental stance toward the test.

CHAPTER 7

THE SUBJECT TESTS

Subject Tests are designed to test the fundamental knowledge most important for successful graduate study in a particular subject area. In order to do well on a Subject Test, you basically need to have an extensive background in the particular subject area—the sort of background you would be expected to have if you majored in the subject. In this section, we'll answer the most common questions about Subject Tests.

Do You Have to Take a Subject Test?

Not every graduate school or program requires Subject Tests, so check admissions requirements at those schools in which you're interested.

What's the Purpose of Subject Tests?

Unlike the GRE General Test, which assesses skills that have been developed over a long period of time and are not related to a particular subject area, Subject Tests assess knowledge of a particular field of study. They enable admissions officers to compare students from different colleges with different standards and curricula.

What Are Subject Tests Like?

All Subject Tests are administered in paper-and-pencil format. Except for the Revised Music test, Subject Tests consist exclusively of multiple-choice questions that are designed to assess knowledge of the areas of the subject that are included in the typical undergraduate curriculum.

On Subject Tests, you'll earn one point for each multiple choice question that you answer correctly but lose one-quarter point for each incorrectly answered question. Unanswered questions aren't counted in the scoring. Your raw score is the number of correctly answered questions minus one-quarter of the incorrectly answered questions. This raw score is then converted into a scaled score, which can range from 200 to 900. The range varies from test to test.

Some Subject Tests also contain subtests, which provide more specific information about your strengths and weaknesses. The same questions that contribute to your subtest scores also contribute to your overall score. Subtest scores, which range from 20 to 99, are reported along with the overall score. For further information on scoring, you should consult the relevant Subject Test Descriptive Booklet, available from ETS.

Are There Any Different Test-Taking Strategies for the Subject Tests?

Because the multiple-choice questions on Subject Tests have a wrong-answer penalty of one-quarter point, you should adopt a different test-taking strategy for your Subject Test than the one you're going to use for the GRE General Test. On the Subject Tests, you shouldn't attempt to fill in an answer for every question on the test, nor do you have to guess at every question you see if you want to get another question (like on the CAT). On Subject Tests, you should guess only if you can eliminate one or more of the answer choices.

When Should You Take the Subject Test?

Subject Tests are offered on the same days as the paper-and-pencil versions of the GRE General Test, but they are not offered in June. On these days, the General Test is given in the morning, and the Subject Tests are

DON'T GUESS

On a Subject Test, you gain one point for each correct answer—but lose one-quarter point for each wrong answer.

KAPLAN

given in the afternoon. Although you can take the General and Subject Test on the same day, we don't recommend it. Your testing will last six hours or longer, and by the afternoon you may find your concentration waning, which will probably result in a lower score on the Subject Test.

A good alternative is to take the CAT at least a couple days before or after you take the Subject Test.

GIVE YOURSELF A BREAK

Don't take the General Test and the Subject Test on the same day. You'll end up testing for six hours or longer.

How Many Subject Tests Are There and What Fields Do They Cover?

There are 14 Subject Tests. A list of them follows, along with brief descriptions.

Biochemistry, Cell, and Molecular Biology

This test consists of 180 questions and is divided among three subscore areas: biochemistry, cell biology, and molecular biology and genetics.

Biology

This test consists of about 200 questions divided among three subscore areas: cellular and molecular biology, organismal biology, and ecology and evolution.

Chemistry

This test consists of about 136 questions. There are no subscores, and the questions cover the following topics: analytical chemistry, inorganic chemistry, organic chemistry, and physical chemistry.

Computer Science

This test consists of approximately 70 questions. There are no subscores, and the questions cover the following topics: software systems and methodology, computer organization and architecture, theory, mathematical background, and other, more advanced topics, such as modeling, simulation, and artificial intelligence.

Economics

This test consists of 130 questions. A majority of the questions cover micro- and macroeconomics, but about 7 percent of the questions cover basic statistics. The rest of the test covers other areas of economics.

SUBJECT TESTS ONLINE

Visit the Kaplan Web site (http:/www.kaplan.com/gre) for a more detailed description of each subject test and some free sample questions.

Engineering

This test consists of 140 questions divided among two subscore areas: engineering and mathematics. The engineering subscore is taken from the 105 questions based on material learned by most engineers during their first two years of college, including basic physics and chemistry. The mathematics subscores is taken from the 35 questions based on the mathematical facts needed to work efficiently in engineering and on the application of calculus.

Geology

This test consists of about 185 questions. Most of the test is divided among three major areas: stratigraphy and sedimentology, structural geology and tectonics, and mineralogy and petrology. The remainder of the test questions cover general geology, hydrogeology, paleontology, surficial processes and geomorphology, and geophysics. These questions contribute to the overall score but not to the subscores.

History

This test consists of 195 questions divided between two subscore areas: European history and U.S. history. There are also a few questions that deal with African, Asian, or Latin American history. These questions contribute to the overall score but not to the subscores.

Literature in English

This test consists of 230 questions on literature in the English language. There are two basic types of questions: factual questions that test the student's knowledge of writers typically covered in the undergraduate curriculum, and interpretive questions that test the student's ability to read various types of literature critically.

Mathematics

This test consists of 66 questions on the content of various undergraduate courses in mathematics. Most of the test assesses the student's knowledge of calculus, abstract algebra, linear algebra, and real analysis. About a quarter of the test, however, requires knowledge in other areas of math.

Music

This test consists of about 111 multiple choice questions and 23 free-response questions. The test has two sections. In Section One, the multiple choice questions test knowledge of the history and theory of music

from the middle ages to the modern era. The free-response questions in this section cover the fundamentals of music theory. In Section Two, the multiple choice questions require the student to analyze taped excerpts of music. The free-response questions consist of dictation, part-writing exercises, and counterpoint exercises. This test has three subscores—history and theory, listening and literature, and aural skills.

Physics

This test consists of 100 questions covering mostly material covered in the first three years of undergraduate physics. Topics include classical mechanics, electromagnetism, atomic physics, optics and wave phenomena, quantum mechanics, thermodynamics and statistical mechanics, special relativity, and laboratory methods. About 9 percent of the test covers advanced topics, such as nuclear and particle physics, condensed matter physics, and astrophysics.

Psychology

This test consists of 220 questions drawn from courses most commonly included in the undergraduate curriculum. Questions fall into three categories. The experimental or natural science-oriented category includes questions in learning, cognitive psychology, sensation and perception, ethology and comparative psychology, and physiological psychology. The social or social science-oriented category includes questions in abnormal psychology, developmental psychology, social psychology, and personality. Together, these make up about 85 percent of the test, and each of the two categories provides its own subscore. The other 15 percent or so of the questions fall under the "general" category, which includes the history of psychology, tests and measurements, research design and statistics, and applied psychology.

Sociology

This test consists of 190 questions drawn from the major subfields in the undergraduate sociology curriculum: general theory, methodology and statistics, criminology and deviance, demography, family and gender roles, organizations, race and ethnic relations, social change, social institutions, social psychology, social stratification, and urban, rural, and community sociology.

For more information, you should consult the relevant Subject Test Descriptive Booklet, available from ETS. You can also visit the Kaplan Web site (http:/www.kaplan.com/gre) for a more detailed description of each subject test and some free sample questions.

A Final Word

Before setting you off on the Practice Test, let's conclude with a recap of some of the most important general principles for success on the GRE. (Remember, this is a pencil-and-paper test, so not all of these principles apply to the CAT.)

- In many sections of the test, the test questions are presented in order of difficulty. Because you get the same number of points for getting a basic or a hard question correct, try to answer the questions that are easiest for you first.
- There is no penalty for wrong answers on the GRE General Test. Always guess if you can't answer a question or don't get to it. Never, ever, leave a question blank.
- Never spend an excessive amount of time on any one question. If you're stumped, skip it and return to it when you've finished the other questions in the section.

Use these points—and all of the content-specific strategies that you've learned—when you work on the Practice Test. Good luck!

CHAPTER 8

THE GRE WRITING ASSESSMENT

The ability to write clearly about complex subjects and fine distinctions is an important part of graduate school, and until now there was no standardized way for schools to evaluate a student's academic writing ability. Enter the GRE Writing Assessment, a new test being introduced in October 1999 by ETS. The Writing Assessment, which is a separate test from the GRE general exam, will remedy that problem for the schools you're applying to, *if* they require you to take it. The admissions staff at the schools you're applying to can tell you whether you need a Writing Assessment score or not.

At the time this book went to press, the Writing Assessment was being finalized by the test makers and the details of the actual test may differ slightly from what you'll see on the following pages. For the most up-to-date information on the test, visit Kaplan's Web site (www.kaplan.com).

If you need to take the Writing Assessment, don't consider it another hurdle to potentially trip over on your way to grad school. This is just another way for you to show off your skills. . . which, we might add, will be finely polished with a little practice on the strategies in this chapter.

TYPING IS OPTIONAL

You'll have the choice of typing or handwriting your essays.

An Essay Overview: Issues and Arguments

The Writing Assessment consists of two timed essay sections. The first is what ETS calls an "Issue" essay: You'll be shown two essay topics—each a sentence or paragraph that expresses an opinion on an issue of general interest. You'll be asked to choose one of the two topics, and then you'll be given 45 minutes to plan and write an essay that communicates your own view on the issue. Whether your agree or disagree with the opinion on the screen is irrelevant: What matters is that you back up your view with relevant examples and statements.

The second of the two writing tasks is the "Argument" essay, which is somewhat different. This time, you will be shown a paragraph that argues a certain point. You will then be given 30 minutes to assess that argument's logic. As with the "Issue" essay, it won't matter whether you agree with what you see on the screen. The testmakers want you to critique the *reasoning behind the argument*, and not the argument itself.

How the Essays Are Administered

At the start of the Writing Assessment, you will be asked whether you would like to handwrite your essays or type them using a simplified word processor. If you choose to type, you will be given a brief tutorial on how to use the word processor. If you aren't comfortable with complex word processing programs, don't worry. The program you'll use on the GRE is quite simple, and you'll be well acquainted with it by the time you start writing.

The Four Basic Principles of Analytical Writing

Writing analytically can be boiled down to a simple, two-stage process: First you decide what you want to say about a topic, and then you figure out how to say it. If your writing style isn't clear, your ideas won't come across, no matter how brilliant they are. Good GRE English is not only grammatical but also clear and concise, and by using some basic principles, you'll be able to express your ideas clearly and effectively in both of your essays.

1. Your control of language is important.

Writing that is grammatical, concise, direct, and persuasive displays the "superior control of language" (as the test makers term it) that earns top

GRE Analytical Writing scores. To achieve what we call "effective GRE style," you should pay attention to the following points.

Grammar
Your writing must follow the rules of standard written English. If you're not confident of your mastery of grammar, brush up before the test.

Diction
Diction means word choice. For example, do you use the words affect and effect correctly? Be careful with such commonly confused words as *precede/proceed*, *principal/principle*, *whose/who's*, and *stationary/stationery*.

Syntax
Syntax refers to sentence structure. Do you construct your sentences so that your ideas are clear and understandable? Do you vary your sentence structure, sometimes using simple sentences and other times using sentences with clauses and phrases?

2. It's better to keep things simple.
Perhaps the single most important thing to bear in mind when writing GRE essays is to keep everything simple. This rule applies to word choice, sentence structure, and organization. If you obsess about how to spell an unusual word, you can lose your flow of thought. The more complicated your sentences are, the more likely it is that they will be plagued by errors. The more complex your organization gets, the more likely it is that your argument will get bogged down in convoluted sentences that obscure your point. But keep in mind that simple does not mean simplistic. A clear, straightforward approach can still be sophisticated and convey perceptive insights.

3. Minor grammatical flaws won't torpedo your score.
Many test takers mistakenly believe that they'll lose a few points because of a few mechanical errors such as misplaced commas, misspellings, or other minor glitches. Occasional mistakes of this type will not dramatically affect your GRE essay score. In fact, the test makers' description of a top-scoring essay acknowledges that there may be minor grammatical flaws. The essay graders understand that you are writing first-draft essays. They will not be looking to take points off for minor errors, provided you don't make them consistently. However, if your essays are littered with

> ### THE BASIC PRINCIPLES OF GRE WRITING
>
> 1. Use language effectively.
> 2. Keep it simple.
> 3. Don't worry excessively about making minor errors.
> 4. Keep sight of your goal.

misspellings and grammar mistakes, then the graders may conclude that you have a serious communication problem.

To write an effective essay, you must be concise, forceful, and correct. An effective essay wastes no words, makes its point in a clear, direct way, and conforms to the generally accepted rules of grammar and form.

4. Keep sight of your goal.

Remember, your goal isn't to become a prize-winning stylist. It's to write two solid essays that will convince admissions officers you can write well enough to clearly communicate your ideas to a reader—or business associate. GRE essay graders don't expect rhetorical flourishes, but they do expect effective expression.

This chapter will give you the chance to sharpen your GRE writing skills.

The Two Essay Types

As explained before, the two types of essay you'll meet on the GRE, the Issue and the Argument, require generally similar tasks. You must analyze a subject, take an informed position, and explain that position in writing. The two essay types, however, require different specific tasks.

The Issue Essay

The directions for the issue essay will probably be something similar to this:

> Directions: You will have 45 minutes to plan and write an essay that communicates your perspective on a given topic. No other topics are admissible for this essay.
>
> You will see the topic as a short quotation that expresses an issue of general interest. Write an essay that agrees with, refutes, or qualifies the quotation, and support your opinion with relevant information drawn from your academic studies, reading, observation, or other experiences.
>
> Feel free to consider the issue for a few minutes before planning your response and beginning your writing. Be

WHAT'S YOUR JOB?

For both essay types, you must:

- Analyze a topic
- Take a position
- Explain your position

KAPLAN

certain that your ideas are fully developed and organized logically, and make sure you have enough time left to review and revise what you've written.

Some topics may be a single-sentence quotation while others will be in paragraph form, but they will all state an argument for which one or more counter-arguments could be constructed. In short, the topic will present a point of view: Your job is to choose an opinion on that point of view and make a case for that opinion.

The topics you see in the Issues section may be similar to these:

> "The invention of gunpowder is the single most destructive achievement in history."

> "If extraterrestrial beings whose intelligence was comparable to humans' visited Earth, they would judge humans by their potential and achievements rather than by their weaknesses and mistakes."

> "The drawbacks to the use of nuclear power mean that it is not a long-term solution to the problem of meeting ever-increasing energy needs."

The Kaplan Five-Step Method for Issue Essays

Here's the deal: You have half an hour to show the graduate school admissions people that you can think logically and express yourself in clearly written English. They don't care how many syllables you can cram into a sentence or how fancy your phrases are. They care that you're making sense. Whatever you do, don't try to hide beneath a lot of hefty words and abstractions. Just make sure that everything you say is clearly written and relevant to the topic. Get in there, state your main points, back them up, and get out.

1. Interpret the issue.
Spend about a minute giving the question a thorough read. Check out the opposing viewpoints and get a feel for the assignment.

2. Select a position.

Spend another minute figuring out which side you're going to take. Remember, there isn't one right answer, so don't waste time agonizing over your choice. Just figure out which of the two sides you can argue more easily, and then stick with it.

3. Flesh out your argument and outline.

Take about six minutes to write down points that support your argument and points that weaken the opposing argument. Get down whatever comes to mind, so you have a pool from which to select your strongest points.

4. Compose your essay.

Take about 20 minutes to type or write. Remember to keep things simple; think in terms of around five paragraphs. Open up with a general statement that recaps the issue at hand, and then state your position. Spend your next few paragraphs stating your main points, generally opening each paragraph with a point. Fill out the rest of the paragraph by backing up your point—giving examples, refuting the opposing point of view, etcetera. Make sure your essay doesn't just trail away—spend at least a sentence at the end reiterating your general point of view.

5. Proofread your work.

You should have about two minutes left—take a quick look back and fix any glaring errors.

THE KAPLAN METHOD FOR ISSUE ESSAYS

1. Interpret the issue
2. Select a position
3. Flesh out your argument and outline.
4. Compose your essay
5. Proofread your work

Using the Kaplan Five-Step Method

Let's use the Kaplan Five-Step method on one of the sample issue topics we saw before:

> "The drawbacks to the use of nuclear power mean that it is not a long-term solution to the problem of meeting ever-increasing energy needs."

1. Interpret the Issue.

It's simple enough. The person who wrote this believes that nuclear power is not a suitable replacement for other forms of energy.

2. Select an Argument

Your job, as stated in the directions, is to decide whether or not you agree and explain your decision. Some would argue that the use of nuclear power is too dangerous, while others would say that we can't afford not to use it. So which side do you take? Remember, this isn't about showing the admissions people what your deep-seated beliefs about the environment are—it's about showing that you can formulate an argument and write it down. Quickly think through the pros and cons of each side, and choose the side for which you have the most relevant things to say. For this topic, that process might go something like this:

Arguments for the use of nuclear power:
- *Inexpensive compared to other forms of energy*
- *Fossil fuels will eventually be depleted*
- *Solar power still too problematic and expensive*

Arguments against the use of nuclear power:
- *Radioactive byproducts are deadly.*
- *Safer alternatives like nuclear fusion may be viable in the future*
- *Solar power already in use*

Again, it doesn't matter which side you take. Let's say that in this case you decide to argue against nuclear power. Remember, the question is asking you to argue *why* the cons of nuclear power outweigh the pros—the inadequacy of this power source is the end you're arguing toward, so don't list it as a supporting argument.

3. Flesh out your argument and outline.

You've already begun to think out your arguments—that's why you picked the side you did in the first place. Now's the time to write them all out, including ones that weaken the opposing side.

Nuclear power is not a viable alternative to other sources of energy because:

- *Radioactive, spent fuel has leaked from storage sites (too dangerous)*
- *Reactor accidents can be catastrophic—Three Mile Island, Chernobyl (too dangerous)*

- *More research into solar power will bring down its cost (weakens opposing argument)*
- *Solar-powered homes and cars already exist (alternatives proven viable)*
- *No serious effort to research other alternatives like nuclear fusion (better alternatives lie undiscovered)*
- *Energy companies don't spend money on alternatives; no vested interest (better alternatives lie undiscovered)*

4. Compose your essay.

Remember, open up with a general statement and then assert your position. From there, get down your main points. Your essay for this assignment might look like the following.

Sample Essay 1

At first glance, nuclear energy may seem to be the power source for the future. It's relatively inexpensive, it doesn't produce smoke or harmful chemical byproducts, and its fuel supply is virtually inexhaustible. But a close examination of the issue reveals that nuclear energy is more problematic and dangerous than other forms of energy production.

A main reason that nuclear energy is undesirable is the problem of radioactive waste storage. Highly toxic fuel left over from nuclear fission remains toxic for thousands of years, and the spills and leaks from existing storage sites are hazardous and costly to clean up. Even more appalling is the prospect of accidents at the reactor itself: Incidents at the Three Mile Island and Chernobyl power plants have proven that the consequences of a nuclear meltdown can be catastrophic and have consequences that are felt worldwide.

Environmental and health problems aside, the bottom line for the production of energy is profit. Nuclear power is a business just like any other, and the large companies that produce this country's electricity and gas claim they can't make alternatives like solar power affordable. Yet—largely due to incentives from the federal government—

there exist today homes that are heated by solar power, and cars that are fueled by the sun have already hit the streets. If the limited resources that have been devoted to energy alternatives have already produced working models, a more intensive effort is likely to make those alternatives less expensive and problematic.

Options like solar power, hydroelectric power, and nuclear fusion are far better in the long run in terms of cost and safety. The only money required for these alternatives is for the materials required to harvest them: Sunlight, water, and the power of the atom are free. They also don't produce any toxic byproducts for which long-term storage—a hidden cost of nuclear power—must be found. And, with the temporary exception of nuclear fusion, these sources of energy are already being harnessed today.

While there are arguments to be made for both sides, it is clear that the drawbacks to the use of nuclear power are too great. If other alternatives are explored more seriously than they have been in the past, safer and less expensive sources of power will undoubtedly prove to be better alternatives.

5. Proofread your work.

Take that last couple of minutes to catch any glaring errors.

The Argument Essay

Now let's look at how you might approach the other type of essay topic, the Argument. The directions for this section will probably look something like this:

> Directions: You will have 30 minutes to plan and write a critique of an argument presented in the form of a short passage. You will be asked to consider the logical soundness of the argument by:
>
> - Identifying and analyzing the argument's important points

- Organizing, developing, and expressing your ideas
- Supporting your ideas with relevant reasons and/or examples
- Demonstrating a knowledge of standard written English

Feel free to consider the issue for a few minutes before planning your response and beginning your writing. Be certain that your ideas are fully developed and organized logically, and make sure you have enough time left to review and revise what you've written.

THERE'S NO "RIGHT" ANSWER

You can take either side in an Argument essay. Take the position that you prefer and argue it well.

The author of an argument topic is trying to persuade you of something—his conclusion—by citing some evidence. So look for these two basic components of an argument: a conclusion and supporting evidence. You should be on the lookout for assumptions—the ways the writer makes the leap from evidence to conclusion.

The topics you see in the Argument section may be similar to these:

"The problem of poorly trained teachers that has plagued the state public school system is bound to become a good deal less serious in the future. The state has initiated comprehensive guidelines that oblige state teachers to complete a number of required credits in education and educational psychology at the graduate level before being certified."

"The commercial airline industry in the country of Freedonia has experienced impressive growth in the past three years. This trend will surely continue in the years to come, since the airline industry will benefit from recent changes in Freedonian society: Incomes are rising, most employees now receive more vacation time, and interest in travel is rising, as shown by an increase in media attention devoted to foreign cultures and tourist attractions."

"Insurance policies guaranteeing the policyholder's income if he or she becomes permanently disabled will surely provide the insurance industry with a popular and

profitable product. The cost to a worker who is at average risk is very little, and the benefits paid to the disabled far outweigh this cost."

The Kaplan Six-Step Method for Argument Essays

1. Take the argument apart.
Take about two minutes to identify the argument's conclusion and evidence; then sum up the argument in your own words.

2. Evaluate the argument's persuasiveness.
Take about two minutes to determine how the argument uses its evidence to reach a conclusion. Are there any gaps in the logic?

3. Manipulate the argument.
Take about two minutes to determine what evidence would make an argument stronger or more valid (or possibly weaker and less valid).

4. Write your opening sentence and create an outline.
Sum up the evidence and conclusion in your opening sentence. Use your notes from Steps 1–3 to make a rough outline. Take about two minutes to do both.

5. Compose your essay.
Take about 20 minutes to type.

6. Proofread.
Spend about two minutes reviewing your essay to catch any obvious errors.

Using the Kaplan Six-Step Method for Analysis of an Argument Essays

Let's use the Kaplan Six-Step method on the Analysis of an Argument topic we saw before:

THE KAPLAN METHOD FOR ARGUMENT ESSAYS

1. Take the argument apart.
2. Evaluate the argument's persuasiveness.
3. Manipulate the argument.
4. Write your opening sentence and create an outline.
5. Compose your essay.
6. Proofread.

The problem of poorly trained teachers that has plagued the state public school system is bound to become a good deal less serious in the future. The state has initiated comprehensive guidelines that oblige state teachers to complete a number of required credits in education and educational psychology at the graduate level before being certified.

Explain how logically persuasive you find this argument. In discussing your viewpoint, analyze the argument's line of reasoning and its use of evidence. Also explain what, if anything, would make the argument more valid and convincing or help you to better evaluate its conclusion.

1. Take the argument apart.
First, identify the conclusion—the point the argument's trying to make. Here, the conclusion is:

> The problem of poorly trained teachers that has plagued the state public school system is bound to become a good deal less serious in the future.

Next, identify the evidence—the basis for the conclusion. Here, the evidence is:

> The state has initiated comprehensive guidelines that oblige state teachers to complete a number of required credits in education and educational psychology at the graduate level before being certified.

Finally, sum up the argument in your own words:

The problem of badly trained teachers will become less serious because of the better training they'll be getting.

2. Evaluate the argument's persuasiveness.
It's your job to decide how well the argument uses its evidence—does it do so in a convincing manner? This particular essay topic concludes future improvements in teaching will come from better training regulations. Then identify any assumptions, or gaps, between premise and conclusion. These gaps are going to lay the groundwork for your essay. In this case, there are several assumptions.

- *Credits in education will improve teachers' classroom performance*
- *Present bad teachers haven't already met this standard of training*
- *Current poor teachers will not still be teaching in the future, or will have to be trained, too*

3. Manipulate the argument.

Determine whether there's anything relevant that's not discussed.

- *Whether the training will actually address the cause of the problems*
- *How to either improve or remove the poor teachers now teaching*

Also determine what types of evidence would make the argument stronger or more valid. In this case, we need some new evidence supporting the assumptions.

- *Evidence verifying that this training will make better teachers*
- *Evidence making it clear that present bad teachers haven't already had this training*
- *Evidence suggesting why all or many bad teachers won't still be teaching in the future (or why they'll be better trained)*

4. Write your opening sentence and create an outline.

For an essay on this topic, your opening sentence might look like this:

The writer concludes that the present problem of poorly trained teachers will become less severe in the future because of required coursework in education and psychology.

Then use your notes for steps 1 through 3 as a working outline. If you have a lot of points, take a moment to number or renumber them in a way that will clearly organize your ideas.

The argument says that:
    The problem of poorly trained teachers will become less serious with better training.
It assumes that:
- *Coursework in education will improve teachers' classroom performance*

WRITE ON!

In writing your essay, cover each point in your outline and employ the four basic principles of analytical writing.

- *Present bad teachers haven't already met this standard of class-room training*
- *Current poor teachers will not be teaching in the future or will get training, too*

5. Compose your essay.

Now's the time to take apart the argument in a clear, logical way. Keep in mind the basic principles of writing that we discussed earlier. Your essay might look something like this:

Sample Essay 2

The writer concludes that the present problem of poorly trained teachers will become less severe in the future because of required credits in education and psychology. However, the conclusion relies on assumptions for which there is no clear evidence.

First, the writer assumes that the required courses will make better teachers. In fact, the courses might be entirely irrelevant to the teachers' failings. If, for example, the prevalent problem is cultural and linguistic gaps between teacher and student, graduate level courses that do not address these specific issues probably won't do much good. The argument that the courses will improve teachers would be strengthened if the writer provided evidence that the training will be relevant to the problems.

In addition, the writer assumes that current poor teachers have not already had this training. In fact, the writer doesn't mention whether or not some or all of the poor teachers have had similar training. The argument would be strengthened considerably if the writer provided evidence that current poor teachers have not had training comparable to the new requirements.

Finally, the writer assumes that poor teachers currently working will either stop teaching in the future or will have received training. The writer provides no evidence, though, to indicate that this is the case. As the argument stands, it's highly possible that only brand-new teachers

will receive the training, and the bright future to which the writer refers is decades away. Only if the writer provides evidence that all teachers in the system will receive training—and will then change their teaching methods accordingly—does the argument hold.

6. Proofread.

Save a few minutes to go back over your essay and catch any obvious errors.

GRE Style Checklist

Cut the fat.

❏ Cut out words, phrases, and sentences that don't add any information or serve a purpose.

❏ Watch out for repetitive phrases such as refer back or serious crisis.

❏ Don't use conjunctions to join sentences that would be more effective as separate sentences.

Be forceful.

❏ Avoid jargon and pompous language; it won't impress anybody.

❏ Avoid clichés and overused terms or phrases. (For example, beyond the shadow of a doubt.)

❏ Don't be vague. Avoid generalizations and abstractions when more specific words would be clearer. (For example, write a waste of time and money instead of pointless temporal and financial expenditure.)

❏ Don't use weak sentence openings. Avoid beginning a sentence with there is or there are.

❏ Don't refer to yourself needlessly. Avoid pointless phrases like in my personal opinion.

❏ Don't be monotonous: Vary sentence length and style.

❏ Use transitions to connect sentences and make your essay easy to follow. Paragraphs should clarify the different parts of your essay.

Be correct.

❏ Stick to the rules of standard written English.

CHAPTER 9

TIPS FOR THE FINAL WEEK

Is it starting to feel as if your whole life is a buildup to the GRE? You've known about it for years, worried about it for months, and now spent at least a few weeks in solid preparation for it. As the test gets closer, you may find that your anxiety is on the rise. You shouldn't worry. Armed with the preparation strategies that you've learned from this book, you're in good shape for the day of the test.

To calm any pretest jitters that you may have, though, let's go over a few strategies for the couple of days before and after the test.

The Week Before the Test

In the week or so leading up to the test, you should do the following:

- Recheck your admission ticket for accuracy; call ETS if corrections are necessary.
- Visit the testing center, if you can. Sometimes seeing the actual room where your test will be administered and taking notice of little things—such as the kind of computer you'll be working on, whether the room is likely to be hot or cold, and so forth—may help to calm your nerves. And if you've never been to the campus or building where your test will take place, this is a good way to ensure that you don't get lost.

THE FINAL DAYS FOR YOUR PREPARATION ARE KEY

The tendency among students is to study too hard during the last few days before the test and then to forget the important, practical matters until the last minute. Part of taking control means avoiding this last-minute crush.

PREPARATION, NOT DESPERATION

Don't try to cram a lot of studying into the last day before the test. It probably won't do you much good, and it could bring on a case of test burnout.

YOUR SURVIVAL KIT

On the night before the test, get together a "GRE survival kit" containing the following items:

- A watch
- Bottle of water (but don't drink too much)
- Pain-killer, in case you get a headache
- Photo ID card
- Your admission ticket
- A snack (You must eat a high-energy snack during the break or you'll run out of gas during later parts of the test. Fruit or energy bars are good snacks. Candy bars aren't.)

- Practice getting up early and working on test material, preferably a full-length test, as if it were the day of the test.
- Time yourself accurately, so that you'll know how to pace yourself on the day of the test.
- Evaluate thoroughly where you stand. Use the time remaining before the test to shore up your weak points, rereading the appropriate sections of this book. But don't neglect your strong areas—after all, this is where you'll rack up most of your points.

D-Day Minus One

Try to avoid doing intensive studying the day before the test. There's little you can do to help yourself at this late date, and you may just wind up exhausting yourself. Our advice is to review a few key concepts, get together everything that you'll need for the test, and then take the night off entirely. Go to see an early movie or watch some TV. Try not to think too much about the test.

Test Day!!!

Let's now discuss what you can expect on the day of the test itself. The day should start with a moderate, high energy breakfast. Cereal, fruit, bagels, or eggs are good. Avoid donuts, danishes, or anything else with a lot of sugar in it. Also, unless you are utterly catatonic without it, it's a good idea to stay away from coffee. Yeah, yeah, you drink two cups every morning and don't even notice it. But it's different during the test. Coffee won't make you alert (your adrenaline will do that much more effectively); it will just give you the jitters. Kaplan has done experiments in which test takers go into one exam having drunk various amounts of coffee and another exam without having drunk coffee. The results indicate that even the most caffeine-addicted test takers will lose their focus midway through the second section if they've had coffee, but they report no alertness problems without it.

When you get to the test center, you will be seated at a computer station. Some administrative questions will be asked before the test begins, and once you're done with those . . . it's showtime. While you're taking the test, a small clock will count down the time you have left in each section. The computer will tell you when you're done with each section, and when you're completed the test itself.

Finally, here are some last-minute reminders to help guide your work on the test:

- Give all five answer choices a fair shot in Verbal (especially reading comp) and in logical reasoning, time permitting. For Quantitative and logic games, go with the objectively correct answer as soon as you find it and blow off the rest.
- Don't bother trying to figure out which section is unscored. It can't help you, and you might very well be wrong. Instead, just determine to do your best on every section.
- Pay no attention to people who are chattering on their break. Just concentrate on how well prepared you are.
- Dress in layers for maximum comfort. This way, you can adjust to the room's temperature accordingly.
- Take a few minutes now to look back over your preparation and give yourself credit for how far you've come. Confidence is key. *Accentuate the positives and don't dwell on the negatives!* Your attitude and outlook are crucial to your performance on the test.
- During the exam, try not to think about how you're scoring. It's like a baseball player who's thinking about the crowd's cheers, the sportswriters, and his contract as he steps up to the plate. It's a great way to strike out. Instead, focus on the question-by-question task of picking the correct answer choice. After all, the correct answer is *there*. You don't have to come up with it; it's sitting right there in front of you! Concentrate on each question, each passage, each game—on the mechanics, in other words—and you'll be much more likely to hit a home run.

Cancellation and Multiple Scores Policy

Unlike many things in life, the GRE allows you a second chance. If, at the end of the test, you feel that you've definitely not done as well as you can, you have the option to cancel your score. The trick is, you must decide whether you want to keep your scores *before* the computer shows them to you. If you cancel, your scores will be disregarded. (You also won't get to see them.) Canceling a test means that it won't be scored. It will just appear on your score report as a canceled test. No one will know how well or poorly you really did—not even you.

ABOUT CANCELING

Students sometimes underestimate their performance immediately after the test and think that they should cancel their scores. Doing poorly on one part of the test does not necessarily mean that they've bombed. Scores should be canceled only after you've given the issue careful thought and decided that your overall performance was poorer than usual.

Two legitimate reasons to cancel your test are illness and personal circumstances that cause you to perform unusually poorly on that particular day. Also, if you feel that you didn't prepare sufficiently, then it may be acceptable to cancel your score and approach your test preparation a little more seriously the next time.

But keep in mind that test takers historically underestimate their performance, especially immediately following the test. This underestimation is especially true on the CAT, which is designed to give you questions at the limits of your abilities. They tend to forget about all of the things that went right and focus on everything that went wrong. So unless your performance is terribly marred by unforeseen circumstances, don't cancel your test

If you do cancel, your future score reports will indicate that you've canceled a previous score. But since the canceled test was never scored, you don't have to worry about bad numbers showing up on any subsequent score report. If you take more than one test without canceling, then all the scores will show up on each score report, so the graduate schools will see them all. Most grad schools average GRE scores, although there are a few exceptions. Check with individual schools for their policies on multiple scores.

Post-GRE Festivities

After all the hard work that you've put in preparing for and taking the GRE, make sure you take time to celebrate afterwards. Plan to get together with friends the evening after the test. Relax, have fun, let loose. After all, you've got a lot to celebrate. You prepared for the test ahead of time. You did your best. You're going to get a good score.

DO YOU REALLY WANT TO CANCEL?

The key question to ask yourself when deciding whether to cancel is this: "Will I really do significantly better next time?"

THE KAPLAN ADVANTAGE™ STRESS MANAGEMENT SYSTEM

The countdown has begun. Your date with THE TEST is looming on the horizon. Anxiety is on the rise. The butterflies in your stomach have gone ballistic. Perhaps you feel as if the last thing you ate has turned into a lead ball in your stomach. Your thinking is getting cloudy. Maybe you think you won't be ready. Maybe you already know your stuff, but you're going into panic mode anyway. Worst of all, you're not sure of what to do about it.

Don't freak! It is possible to tame that anxiety and stress—before and during the test. We'll show you how. You won't believe how quickly and easily you can deal with that killer anxiety.

Making the Most of Your Prep Time

Lack of control is one of the prime causes of stress. A ton of research shows that if you don't have a sense of control over what's happening in your life you can easily end up feeling helpless and hopeless. So, just having concrete things to do and to think about—taking control—will help reduce your stress. This section shows you how to take control during the days leading up to taking the GRE—or any other test.

STRESS TIP

Don't forget that your school probably has counseling available. If you can't conquer test stress on your own, make an appointment at the counseling center. That's what counselors are there for.

Identify the Sources of Stress

The first step in gaining control is identifying the sources of your test-related stress. The idea is to pin down that free-floating anxiety so that you can take control of it. Here are some examples:

- I always freeze up on tests.
- I'm nervous about the reading comprehension questions (or the quantitative comparison questions or the logic games).
- I need a good/great score to go to Acme Graduate School.
- My older brother/sister/best friend/girl- or boyfriend did really well. I must match their scores or do better.
- My parents, who are paying for school, will be really disappointed if I don't test well.
- I'm afraid of losing my focus and concentration.
- I'm afraid I'm not spending enough time preparing.
- I study like crazy but nothing seems to stick in my mind.
- I always run out of time and get panicky.
- I feel as though thinking is becoming like wading through thick mud.

Take a few minutes to think about your own particular sources of test-related stress. Then write them down in some sort of order. List the statements you most associate with your stress and anxiety first, and put the least disturbing items last. As you write the list, you're forming a hierarchy of items so you can deal first with the anxiety provokers that bug you most. Very often, taking care of the major items from the top of the list goes a long way toward relieving overall testing anxiety. You probably won't have to bother with the stuff you placed last.

Take Stock of Your Strengths and Weaknesses

Take one minute to list the areas of the test that you are good at. They can be general (reading comprehension questions) or specific (inference questions). Put down as many as you can think of, and if possible, time yourself. Write for the entire time; don't stop writing until you've reached the one-minute stopping point.

VERY SUPERSTITIOUS

Stress expert Stephen Sideroff, Ph.D., tells of a client who always stressed out before, during, and even after taking tests. Yet she always got outstanding scores. It became obvious that she was thinking superstitiously—subconsciously believing that the great scores were a result of her worrying. She also didn't trust herself, and believed that if she didn't worry she wouldn't study hard enough. Sideroff convinced her to take a risk and work on relaxing before her next test. She did, and her test results were still as good as ever—which broke her cycle of superstitious thinking.

Next, take one minute to list areas of the test you're not so good at, just plain bad at, have failed at, or keep failing at. Again, keep it to one minute, and continue writing until you reach the cutoff. Don't be afraid to identify and write down your weak spots! In all probability, as you do both lists you'll find you are strong in some areas and not so strong in others. Taking stock of your assets *and* liabilities lets you know the areas you don't have to worry about, and the ones that will demand extra attention and effort.

Now, go back to the "good" list, and expand it for two minutes. Take the general items on that first list and make them more specific; take the specific items and expand them into more general conclusions. Naturally, if anything new comes to mind, jot it down. Focus all of your attention and effort on your strengths. Don't underestimate yourself or your abilities. Give yourself full credit. At the same time, don't list strengths you don't really have; you'll only be fooling yourself.

Every area of strength and confidence you can identify is much like having a reserve of solid gold at Fort Knox. You'll be able to draw on your reserves as you need them. You can use your reserves to solve difficult questions, maintain confidence, and keep test stress and anxiety at a distance. The encouraging thing is that every time you recognize another area of strength, succeed at coming up with a solution, or get a good score on a test, you increase your reserves. And, with a plan to strengthen a weak area or get a good score on a practice test, there is absolutely no limit to how much self-confidence you can have or how good you can feel about yourself.

What Do You Want to Accomplish in the Time Remaining?

The whole point of this next exercise is sort of like checking out a used car you might want to buy. You'd want to know up front what the car's weak points are, right? Knowing that influences your whole shopping-for-a-used-car campaign. So it is with your conquering-test-stress campaign: Knowing your weak points ahead of time helps you prepare.

So let's get back to the list of your weak points. Take two minutes to expand it just as you did with your "good" list. Be honest with yourself without going overboard. It's an accurate appraisal of the test areas that give you troubles.

Facing your weak spots gives you some distinct advantages. It helps a lot to find out where you need to spend extra effort. Increased exposure

THE "NEW AGE" OF RELAXATION

Here are some more tips for beating stress:

- Find out if massage, especially shiatsu, is offered through your school's phys ed department, or at the local "Y."
- Check out a book on acupressure, and find those points on your body where you can press a "relax button."
- If you're especially sensitive to smells, you might want to try some aromatherapy. Lavender oil, for example, is said to have relaxing properties. Health food stores, drug stores, and New Age bookstores may carry aromatherapy oils.
- Many health food stores carry herbs and supplies that have relaxing properties, and they often have a specialist on staff who can tell you about them.

STRESS TIP

If you want to play music, keep it low and in the background. Music with a regular, mathematical rhythm—reggae, for example—aids the learning process. A recording of ocean waves is also soothing.

to tough material makes it more familiar and less intimidating. (After all, we mostly fear what we don't know and are probably afraid to face.) You'll feel better about yourself because you're dealing directly with areas of the test that bring on your anxiety. You can't help feeling more confident when you know you're actively strengthening your chances of earning a higher overall test score.

Imagine Yourself Succeeding

This next little group of exercises is both physical and mental. It's a natural followup to what you've just accomplished with your lists.

First, get yourself into a comfortable sitting position in a quiet setting. Wear loose clothes. If you wear glasses, take them off. Then, close your eyes and breathe in a deep, satisfying breath of air. Really fill your lungs until your rib cage is fully expanded and you can't take in any more. Then, exhale the air completely. Imagine you're blowing out a candle with your last little puff of air. Do this two or three more times, filling your lungs to their maximum and emptying them totally. Keep your eyes closed, comfortably but not tightly. Let your body sink deeper into the chair as you become even more comfortable.

With your eyes shut you can notice something very interesting. You're no longer dealing with the worrisome stuff going on in the world outside of you. Now you can concentrate on what happens inside you. The more you recognize your own physical reactions to stress and anxiety, the more you can do about them. You may not realize it, but you've begun to regain a sense of being in control.

Let images begin to form on the "viewing screens" on the back of your eyelids. You're experiencing visualizations from the place in your mind that makes pictures. Allow the images to come easily and naturally; don't force them. Imagine yourself in a relaxing situation. It might be in a special place you've visited before or one you've read about. It can be a fictional location that you create in your imagination, but a real-life memory of a place or situation you know is usually better. Make it as detailed as possible and notice as much as you can.

If you don't see this relaxing place sharply or in living color, it doesn't mean the exercise won't work for you. Some people can visualize in great detail, while others get only a sense of an image. What's important is not how sharp the details or colors, but how well you're able to manipulate the images. If you can conjure up finely detailed images, great. If you only

have a faint sense of the images, that's okay—you'll still experience all the benefits of the exercise.

Think about the sights, the sounds, the smells, even the tastes and textures associated with your relaxing situation. See and feel yourself in this special place. Say you're special place is the beach, for example. Feel how warm the sand is. Are you lying on a blanket, or sitting up and looking out at the water? Hear the waves hitting the shore, and the occasional seagull. Feel a comfortable breeze. If your special place is a garden or park, look up and see the way sunlight filters through the trees. Smell your favorite flowers. Hear some chimes gently playing and birds chirping.

Stay focused on the images as you sink farther back into your chair. Breathe easily and naturally. You might have the sensations of any stress or tension draining from your muscles and flowing downward, out your feet and away from you.

Take a moment to check how you're feeling. Notice how comfortable you've become. Imagine how much easier it would be if you could take the test feeling this relaxed. You've coupled the images of your special place with sensations of comfort and relaxation. You've also found a way to become relaxed simply by visualizing your own safe, special place.

Now, close your eyes and start remembering a real-life situation in which you did well on a test. If you can't come up with one, remember a situation in which you did something (academic or otherwise) that you were really proud of—a genuine accomplishment. Make the memory as detailed as possible. Think about the sights, the sounds, the smells, even the tastes associated with this remembered experience. Remember how confident you felt as you accomplished your goal. Now start thinking about the upcoming test. Keep your thoughts and feelings in line with that successful experience. Don't make comparisons between them. Just imagine taking the upcoming test with the same feelings of confidence and relaxed control.

This exercise is a great way to bring the test down to earth. You should practice this exercise often, especially when the prospect of taking the exam starts to bum you out. The more you practice it, the more effective the exercise will be for you.

Exercise Your Frustrations Away

Whether it is jogging, walking, biking, mild aerobics, pushups, or a pick-up basketball game, physical exercise is a very effective way to stimulate

OCEAN DUMPING

Visualize a beautiful beach, with white sand, blue skies, sparkling water, a warm sun, and seagulls. See yourself walking on the beach, carrying a small plastic pail. Stop at a good spot and put your worries and whatever may be bugging you into the pail. Drop it at the water's edge and watch it drift out to sea. When the pail is out of sight, walk on.

TAKE A HIKE, PAL

When you're in the middle of studying and hit a wall, take a short, brisk walk. Breathe deeply and swing your arms as you walk. Clear your mind. (And don't forget to look for flowers that grow in the cracks of the sidewalk.)

CYBERSTRESS

If you spend a lot of time in cyberspace anyway, do a search for the phrase *stress management*. There's a ton of stress advice on the Net, including material specifically for students.

NUTRITION AND STRESS: THE DOS AND DON'TS

Do eat:

- Fruits and vegetables (raw is best, or just lightly steamed or nuked)
- Low-fat protein such as fish, skinless poultry, beans, and legumes (like lentils)
- Whole grains such as brown rice, whole wheat bread, and pastas

Don't eat:

- Refined sugar; sweet, high-fat snacks (simple carbohydrates like sugar make stress worse, and fatty foods lower your immunity)
- Salty foods (they can deplete potassium, which you need for nerve functions)

both your mind and body and to improve your ability to think and concentrate. A surprising number of students get out of the habit of regular exercise, ironically because they're spending so much time prepping for exams. Also, sedentary people—this is medical fact—get less oxygen to the blood and hence to the head than active people. You can live fine with a little less oxygen; you just can't think as well.

Any big test is a bit like a race. Thinking clearly at the end is just as important as having a quick mind early on. If you can't sustain your energy level in the last sections of the exam, there's too good a chance you could blow it. You need a fit body that can weather the demands any big exam puts on you. Along with a good diet and adequate sleep, exercise is an important part of keeping yourself in fighting shape and thinking clearly for the long haul.

There's another thing that happens when students don't make exercise an integral part of their test preparation. Like any organism in nature, you operate best if all your "energy systems" are in balance. Studying uses a lot of energy, but it's all mental. When you take a study break, do something active instead of raiding the fridge or vegging out in front of the TV. Take a five- to ten-minute activity break for every 50 or 60 minutes that you study. The physical exertion gets your body into the act which helps to keep your mind and body in sync. Then, when you finish studying for the night and hit the sack, you won't lie there, tense and unable to sleep, because your head is overtired and your body wants to pump iron or run a marathon.

One warning about exercise, however: It's not a good idea to exercise vigorously right before you go to bed. This could easily cause sleep-onset problems. For the same reason, it's also not a good idea to study right up to bedtime. Make time for a "buffer period" before you go to bed: For 30 to 60 minutes, just take a hot shower, meditate, simply veg out.

Get High . . . Naturally

Exercise can give you a natural high, which is the only kind of high you can afford right now. Using drugs (prescription or recreational) specifically to prepare for and take a big test is definitely self-defeating. Except for the drugs that occur naturally in your brain, every drug has major drawbacks—and a false sense of security is only one of them.

You may have heard that popping uppers helps you study by keeping you alert. You heard wrong: Don't waste your time. Amphetamines make

it hard to retain information. So you'll stay awake, but you probably won't remember much of what you read. And, taking an upper before you take the test could really mess things up. You're already going to be a little anxious and hyper; adding a strong stimulant could easily push you over the edge into panic. Remember, a little anxiety is a good thing. The adrenaline that gets pumped into your bloodstream helps you stay alert and think more clearly. But, too much anxiety and you can't think straight at all.

Mild stimulants, such as coffee, cola, or over-the-counter caffeine pills can sometimes help as you study, since they keep you alert. On the down side, they can also lead to agitation, restlessness, and insomnia. Some people can drink a pot of high-octane coffee and sleep like a baby. Others have one cup and start to vibrate. It all depends on your tolerance for caffeine.

Alcohol and other depressants are out, too. Again, if they're illegal, forget about it. Depressants wouldn't work, anyway, since they lead to the inevitable hangover/crash, the fuzzy thinking, and lousy sense of judgment. These are not going to help you ace the test.

Instead, go for endorphins—the "natural morphine." Endorphins have no side effects and they're free—you've already got them in your brain. It just takes some exercise to release them. Running around on the basketball court, bicycling, swimming, aerobics, power walking—these activities cause endorphins to occupy certain spots in your brain's neural synapses. In addition, exercise develops staying power and increases the oxygen transfer to your brain. Go into the test naturally.

Take a Deep Breath . . .

Here's another natural route to relaxation and invigoration. It's a classic isometric exercise that you can do whenever you get stressed out—just before the test begins, even *during* the test. It's very simple and takes just a few minutes.

Close your eyes. Starting with your eyes and—*without holding your breath*—gradually tighten every muscle in your body (but not to the point of pain) in the following sequence:

1. Close your eyes tightly.
2. Squeeze your nose and mouth together so that your whole face is scrunched up. (If it makes you self-conscious to do this in the test room, skip the face-scrunching part.)

STRESS TIP

Don't study on your bed, especially if you have problems with insomnia. Your mind may start to associate the bed with work, and make it even harder for you to fall asleep.

THE RELAXATION PARADOX

Forcing relaxation is like asking yourself to flap your arms and fly. You can't do it, and every push and prod only gets you more frustrated. Relaxation is something you don't work at. You simply let it happen. Think about it. When was the last time you tried to force yourself to go to sleep, and it worked?

3. Pull your chin into your chest, and pull your shoulders together.
4. Tighten your arms to your body, then clench your hands into tight fists.
5. Pull in your stomach.
6. Squeeze your thighs and buttocks together, and tighten your calves.
7. Stretch your feet, then curl your toes (watch out for cramping in this part).

At this point, every muscle should be tightened. Now, relax your body, one part at a time, in reverse order, starting with your toes. Let the tension drop out of each muscle. The entire process might take five minutes from start to finish (maybe a couple of minutes during the test). This clenching and unclenching exercise should help you to feel very relaxed.

And Keep Breathing

Conscious attention to breathing is an excellent way of managing test stress (or any stress, for that matter). The majority of people who get into trouble during tests take shallow breaths. They breathe using only their upper chests and shoulder muscles, and may even hold their breath for long periods of time. Conversely, the test taker who by accident or design keeps breathing normally and rhythmically is likely to be more relaxed and in better control during the entire test experience.

So, now is the time to get into the habit of relaxed breathing. Do the next exercise to learn to breathe in a natural, easy rhythm. By the way, this is another technique you can use during the test to collect your thoughts and ward off excess stress. The entire exercise should take no more than three to five minutes.

With your eyes still closed, breathe in slowly and deeply through your nose. Hold the breath for a bit, and then release it through your mouth. The key is to breathe slowly and deeply by using your diaphragm (the big band of muscle that spans your body just above your waist) to draw air in and out naturally and effortlessly. Breathing with your diaphragm encourages relaxation and helps minimize tension.

As you breathe, imagine that colored air is flowing into your lungs. Choose any color you like, from a single color to a rainbow. With each breath, the air fills your body from the top of your head to the tips of your toes. Continue inhaling the colored air until it occupies every part of you,

BREATHE LIKE A BABY

A baby or young child is the best model for demonstrating how to breathe most efficiently and comfortably. Only her stomach moves as she inhales and exhales. The action is virtually effortless.

STRESS TIP

A lamp with a 75-watt bulb is optimal for studying. But don't put it so close to your study material that you create a glare.

bones and muscles included. Once you have completely filled yourself with the colored air, picture an opening somewhere on your body, either natural or imagined. Now, with each breath you exhale, some of the colored air will pass out the opening and leave your body. The level of the air (much like the water in a glass as it is emptied) will begin to drop. It will descend progressively lower, from your head down to your feet. As you continue to exhale the colored air, watch the level go lower and lower, farther and farther down your body. As the last of the colored air passes out of the opening, the level will drop down to your toes and disappear. Stay quiet for just a moment. Then notice how relaxed and comfortable you feel.

Thumbs Up for Meditation

Once relegated to the fringes of the medical world, meditation, biofeedback, and hypnosis are increasingly recommended by medical researchers to reduce pain from headaches, back problems—even cancer. Think of what these powerful techniques could do for your test-related stress and anxiety.

Effective meditation is based primarily on two relaxation methods you've already learned: body awareness and breathing. A couple of different meditation techniques follow. Experience them both, and choose the one that works best for you.

Breath Meditation

Make yourself comfortable, either sitting or lying down. For this meditation you can keep your eyes open or close them. You're going to concentrate on your breathing. The goal of the meditation is to notice everything you can about your breath as it enters and leaves your body. Take three to five breaths each time you practice the meditation, which should take about a minute for the entire procedure.

Take a deep breath and hold it for five to ten seconds. When you exhale, let the breath out very slowly. Feel the tension flowing out of you along with the breath that leaves your body. Pay close attention to the air as it flows in and out of your nostrils. Observe how cool it is as you inhale and how warm your breath is when you exhale. As you expel the air, say a cue word such as *calm* or *relax* to yourself. Once you've exhaled all the air from your lungs, start the next long, slow inhale. Notice how relaxed feelings increase as you slowly exhale and again hear your cue words.

THINK GOOD THOUGHTS

Create a set of positive, but brief affirmations and mentally repeat them to yourself just before you fall asleep at night. (That's when your mind is very open to suggestion.) You'll find yourself feeling a lot more positive in the morning. Periodically repeating your affirmations during the day makes them even more effective.

Mantra Meditation

For this type of meditation experience you'll need a mental device (a mantra), a passive attitude (don't try to do anything), and a position in which you can be comfortable. You're going to focus your total attention on a mantra you create. It should be emotionally neutral, repetitive, and monotonous, and your aim is to fully occupy your mind with it. Furthermore, you want to do the meditation passively, with no goal in your head of how relaxed you're supposed to be. This is a great way to prepare for studying or taking the test. It clears your head of extraneous thoughts and gets you focused and at ease.

Sit comfortably and close your eyes. Begin to relax by letting your body go limp. Create a relaxed mental attitude and know there's no need for you to force anything. You're simply going to let something happen. Breathe through your nose. Take calm, easy breaths and as you exhale, say your mantra (*one, ohhm, aah, soup*—whatever is emotionally neutral for you) to yourself. Repeat the mantra each time you breathe out. Let feelings of relaxation grow as you focus on the mantra and your slow breathing. Don't worry if your mind wanders. Simply return to the mantra and continue letting go. Experience this meditation for 10 to 15 minutes.

Quick Tips for the Days Just Before the Exam

- The best test takers do less and less as the test approaches. Taper off your study schedule and take it easy on yourself. You want to be relaxed and ready on the day of the test. Give yourself time off, especially the evening before the exam. By that time, if you've studied well, everything you need to know is firmly stored in your memory banks.

- Positive self-talk can be extremely liberating and invigorating, especially as the test looms closer. Tell yourself things such as, "I choose to take this test" rather than "I have to"; "I will do well" rather than "I hope things go well"; "I can" rather than "I cannot." Be aware of negative, self-defeating thoughts and images and immediately counter any you become aware of. Replace them with affirming statements that encourage your self-esteem and confidence. Create and practice doing visualizations that build on your positive statements.

- Get your act together sooner rather than later. Have everything (including choice of clothing) laid out days in advance. Most important, know where the test will be held and the easiest, quickest way to get there.

DRESS FOR SUCCESS

When you dress on the day of the test, do it in loose layers. That way you'll be prepared no matter what the temperature of the room is. (An uncomfortable temperature will just distract you from the job at hand.)

And, if you have an item of clothing that you tend to feel "lucky" or confident in—a shirt, a pair of jeans, whatever—wear it. A little totem couldn't hurt.

You will gain great peace of mind if you know that all the little details—gas in the car, directions, etcetera—are firmly in your control before the day of the test.

- Experience the test site a few days in advance. This is very helpful if you are especially anxious. If at all possible, find out what room your part of the alphabet is assigned to, and try to sit there (by yourself) for a while. Better yet, bring some practice material and do at least a section or two, if not an entire practice test, in that room. In this case, familiarity doesn't breed contempt; it generates comfort and confidence.

- Forego any practice on the day before the test. It's in your best interest to marshal your physical and psychological resources for 24 hours or so. Even race horses are kept in the paddock and treated like princes the day before a race. Keep the upcoming test out of your consciousness; go to a movie, take a pleasant hike, or just relax. Don't eat junk food or tons of sugar. And—of course—get plenty of rest the night before. Just don't go to bed too early. It's hard to fall asleep earlier than you're used to, and you don't want to lie there thinking about the test.

Handling Stress During the Test

The biggest stress monster will be the day of the test itself. Fear not; there are methods of quelling your stress during the test.

- Keep moving forward instead of getting bogged down in a difficult question or passage. You don't have to get everything right to achieve a fine score. So, don't linger out of desperation on a question that is going nowhere even after you've spent considerable time on it. The best test takers skip (temporarily) difficult material in search of the easier stuff. They mark the ones that require extra time and thought. This strategy buys time and builds confidence so you can handle the tough stuff later.

- Don't be thrown if other test takers seem to be working more busily and furiously than you are. Continue to spend your time patiently but doggedly thinking through your answers; it's going to lead to higher-quality test taking and better results. Don't mistake the other people's sheer activity as signs of progress and higher scores.

WHAT ARE "SIGNS OF A WINNER," ALEX?

Here's some advice from a Kaplan instructor who won big on *Jeopardy!*™ In the green room before the show, he noticed that the contestants who were quiet and "within themselves" were the ones who did great on the show. The contestants who did not perform as well were the ones who were fact-cramming, talking a lot, and generally being manic before the show. Lesson: Spend the final hours leading up to the test getting sleep, meditating, and generally relaxing.

- *Keep breathing!* Weak test takers tend to share one major trait: they forget to breathe properly as the test proceeds. They start holding their breath without realizing it, or they breathe erratically or arrhythmically. Improper breathing hurts confidence and accuracy. Just as importantly, it interferes with clear thinking.

- Some quick isometrics during the test—especially if concentration is wandering or energy is waning—can help. Try this: put your palms together and press intensely for a few seconds. Concentrate on the tension you feel through your palms, wrists, forearms, and up into your biceps and shoulders. Then, quickly release the pressure. Feel the difference as you let go. Focus on the warm relaxation that floods through the muscles. Now you're ready to return to the task.

- Here's another isometric that will relieve tension in both your neck and eye muscles. Slowly rotate your head from side to side, turning your head and eyes to look as far back over each shoulder as you can. Feel the muscles stretch on one side of your neck as they contract on the other. Repeat five times in each direction.

With what you've just learned here, you're armed and ready to do battle with the test. This book and your studies will give you the information you'll need to answer the questions. It's all firmly planted in your mind. You also know how to deal with any excess tension that might come along, both when you're studying for and taking the exam. You've experienced everything you need to tame your test anxiety and stress. You are going to get a great score.

THE PRACTICE TEST
FOR THE GRE

How to Take This Practice Test

Since the interactive test experience of a CAT is impossible to reproduce in a book, this is a paper-and-pencil test. Before taking this Practice Test, find a quiet place where you can work uninterrupted for three hours. Make sure you have a comfortable desk and several pencils. Time yourself according to the time limits shown at the beginning of each section. It's okay to take a short break between sections, but for best results you should go through all three sections in one sitting. Use the answer grid on the following page to record your answers.

You'll find the answer key and score converter following the test. Good luck.

If you purchased the edition of this book that is bundled with a CD-ROM, you will find four realistic CATs to practice with in the software. If you purchased the edition that doesn't come with a CD-ROM, you can take a mini-CAT on Kaplan's Web site (www.kaplan.com) or purchase Kaplan's *Higher Score on the GRE* at your local software store.

How to Take This Practice Test

PRACTICE TEST

Remove or photocopy this answer sheet and use it to complete the Practice Test.
See the answer key at the end of the test to correct your answers when finished.

SECTION 1 — VERBAL

1 Ⓐ Ⓑ Ⓒ Ⓓ Ⓔ	20 Ⓐ Ⓑ Ⓒ Ⓓ Ⓔ	39 Ⓐ Ⓑ Ⓒ Ⓓ Ⓔ	58 Ⓐ Ⓑ Ⓒ Ⓓ Ⓔ
2 Ⓐ Ⓑ Ⓒ Ⓓ Ⓔ	21 Ⓐ Ⓑ Ⓒ Ⓓ Ⓔ	40 Ⓐ Ⓑ Ⓒ Ⓓ Ⓔ	59 Ⓐ Ⓑ Ⓒ Ⓓ Ⓔ
3 Ⓐ Ⓑ Ⓒ Ⓓ Ⓔ	22 Ⓐ Ⓑ Ⓒ Ⓓ Ⓔ	41 Ⓐ Ⓑ Ⓒ Ⓓ Ⓔ	60 Ⓐ Ⓑ Ⓒ Ⓓ Ⓔ
4 Ⓐ Ⓑ Ⓒ Ⓓ Ⓔ	23 Ⓐ Ⓑ Ⓒ Ⓓ Ⓔ	42 Ⓐ Ⓑ Ⓒ Ⓓ Ⓔ	61 Ⓐ Ⓑ Ⓒ Ⓓ Ⓔ
5 Ⓐ Ⓑ Ⓒ Ⓓ Ⓔ	24 Ⓐ Ⓑ Ⓒ Ⓓ Ⓔ	43 Ⓐ Ⓑ Ⓒ Ⓓ Ⓔ	62 Ⓐ Ⓑ Ⓒ Ⓓ Ⓔ
6 Ⓐ Ⓑ Ⓒ Ⓓ Ⓔ	25 Ⓐ Ⓑ Ⓒ Ⓓ Ⓔ	44 Ⓐ Ⓑ Ⓒ Ⓓ Ⓔ	63 Ⓐ Ⓑ Ⓒ Ⓓ Ⓔ
7 Ⓐ Ⓑ Ⓒ Ⓓ Ⓔ	26 Ⓐ Ⓑ Ⓒ Ⓓ Ⓔ	45 Ⓐ Ⓑ Ⓒ Ⓓ Ⓔ	64 Ⓐ Ⓑ Ⓒ Ⓓ Ⓔ
8 Ⓐ Ⓑ Ⓒ Ⓓ Ⓔ	27 Ⓐ Ⓑ Ⓒ Ⓓ Ⓔ	46 Ⓐ Ⓑ Ⓒ Ⓓ Ⓔ	65 Ⓐ Ⓑ Ⓒ Ⓓ Ⓔ
9 Ⓐ Ⓑ Ⓒ Ⓓ Ⓔ	28 Ⓐ Ⓑ Ⓒ Ⓓ Ⓔ	47 Ⓐ Ⓑ Ⓒ Ⓓ Ⓔ	66 Ⓐ Ⓑ Ⓒ Ⓓ Ⓔ
10 Ⓐ Ⓑ Ⓒ Ⓓ Ⓔ	29 Ⓐ Ⓑ Ⓒ Ⓓ Ⓔ	48 Ⓐ Ⓑ Ⓒ Ⓓ Ⓔ	67 Ⓐ Ⓑ Ⓒ Ⓓ Ⓔ
11 Ⓐ Ⓑ Ⓒ Ⓓ Ⓔ	30 Ⓐ Ⓑ Ⓒ Ⓓ Ⓔ	49 Ⓐ Ⓑ Ⓒ Ⓓ Ⓔ	68 Ⓐ Ⓑ Ⓒ Ⓓ Ⓔ
12 Ⓐ Ⓑ Ⓒ Ⓓ Ⓔ	31 Ⓐ Ⓑ Ⓒ Ⓓ Ⓔ	50 Ⓐ Ⓑ Ⓒ Ⓓ Ⓔ	69 Ⓐ Ⓑ Ⓒ Ⓓ Ⓔ
13 Ⓐ Ⓑ Ⓒ Ⓓ Ⓔ	32 Ⓐ Ⓑ Ⓒ Ⓓ Ⓔ	51 Ⓐ Ⓑ Ⓒ Ⓓ Ⓔ	70 Ⓐ Ⓑ Ⓒ Ⓓ Ⓔ
14 Ⓐ Ⓑ Ⓒ Ⓓ Ⓔ	33 Ⓐ Ⓑ Ⓒ Ⓓ Ⓔ	52 Ⓐ Ⓑ Ⓒ Ⓓ Ⓔ	71 Ⓐ Ⓑ Ⓒ Ⓓ Ⓔ
15 Ⓐ Ⓑ Ⓒ Ⓓ Ⓔ	34 Ⓐ Ⓑ Ⓒ Ⓓ Ⓔ	53 Ⓐ Ⓑ Ⓒ Ⓓ Ⓔ	72 Ⓐ Ⓑ Ⓒ Ⓓ Ⓔ
16 Ⓐ Ⓑ Ⓒ Ⓓ Ⓔ	35 Ⓐ Ⓑ Ⓒ Ⓓ Ⓔ	54 Ⓐ Ⓑ Ⓒ Ⓓ Ⓔ	73 Ⓐ Ⓑ Ⓒ Ⓓ Ⓔ
17 Ⓐ Ⓑ Ⓒ Ⓓ Ⓔ	36 Ⓐ Ⓑ Ⓒ Ⓓ Ⓔ	55 Ⓐ Ⓑ Ⓒ Ⓓ Ⓔ	74 Ⓐ Ⓑ Ⓒ Ⓓ Ⓔ
18 Ⓐ Ⓑ Ⓒ Ⓓ Ⓔ	37 Ⓐ Ⓑ Ⓒ Ⓓ Ⓔ	56 Ⓐ Ⓑ Ⓒ Ⓓ Ⓔ	75 Ⓐ Ⓑ Ⓒ Ⓓ Ⓔ
19 Ⓐ Ⓑ Ⓒ Ⓓ Ⓔ	38 Ⓐ Ⓑ Ⓒ Ⓓ Ⓔ	57 Ⓐ Ⓑ Ⓒ Ⓓ Ⓔ	76 Ⓐ Ⓑ Ⓒ Ⓓ Ⓔ

SECTION 2 — QUANTITATIVE

1 Ⓐ Ⓑ Ⓒ Ⓓ Ⓔ	16 Ⓐ Ⓑ Ⓒ Ⓓ Ⓔ	31 Ⓐ Ⓑ Ⓒ Ⓓ Ⓔ	46 Ⓐ Ⓑ Ⓒ Ⓓ Ⓔ
2 Ⓐ Ⓑ Ⓒ Ⓓ Ⓔ	17 Ⓐ Ⓑ Ⓒ Ⓓ Ⓔ	32 Ⓐ Ⓑ Ⓒ Ⓓ Ⓔ	47 Ⓐ Ⓑ Ⓒ Ⓓ Ⓔ
3 Ⓐ Ⓑ Ⓒ Ⓓ Ⓔ	18 Ⓐ Ⓑ Ⓒ Ⓓ Ⓔ	33 Ⓐ Ⓑ Ⓒ Ⓓ Ⓔ	48 Ⓐ Ⓑ Ⓒ Ⓓ Ⓔ
4 Ⓐ Ⓑ Ⓒ Ⓓ Ⓔ	19 Ⓐ Ⓑ Ⓒ Ⓓ Ⓔ	34 Ⓐ Ⓑ Ⓒ Ⓓ Ⓔ	49 Ⓐ Ⓑ Ⓒ Ⓓ Ⓔ
5 Ⓐ Ⓑ Ⓒ Ⓓ Ⓔ	20 Ⓐ Ⓑ Ⓒ Ⓓ Ⓔ	35 Ⓐ Ⓑ Ⓒ Ⓓ Ⓔ	50 Ⓐ Ⓑ Ⓒ Ⓓ Ⓔ
6 Ⓐ Ⓑ Ⓒ Ⓓ Ⓔ	21 Ⓐ Ⓑ Ⓒ Ⓓ Ⓔ	36 Ⓐ Ⓑ Ⓒ Ⓓ Ⓔ	51 Ⓐ Ⓑ Ⓒ Ⓓ Ⓔ
7 Ⓐ Ⓑ Ⓒ Ⓓ Ⓔ	22 Ⓐ Ⓑ Ⓒ Ⓓ Ⓔ	37 Ⓐ Ⓑ Ⓒ Ⓓ Ⓔ	52 Ⓐ Ⓑ Ⓒ Ⓓ Ⓔ
8 Ⓐ Ⓑ Ⓒ Ⓓ Ⓔ	23 Ⓐ Ⓑ Ⓒ Ⓓ Ⓔ	38 Ⓐ Ⓑ Ⓒ Ⓓ Ⓔ	53 Ⓐ Ⓑ Ⓒ Ⓓ Ⓔ
9 Ⓐ Ⓑ Ⓒ Ⓓ Ⓔ	24 Ⓐ Ⓑ Ⓒ Ⓓ Ⓔ	39 Ⓐ Ⓑ Ⓒ Ⓓ Ⓔ	54 Ⓐ Ⓑ Ⓒ Ⓓ Ⓔ
10 Ⓐ Ⓑ Ⓒ Ⓓ Ⓔ	25 Ⓐ Ⓑ Ⓒ Ⓓ Ⓔ	40 Ⓐ Ⓑ Ⓒ Ⓓ Ⓔ	55 Ⓐ Ⓑ Ⓒ Ⓓ Ⓔ
11 Ⓐ Ⓑ Ⓒ Ⓓ Ⓔ	26 Ⓐ Ⓑ Ⓒ Ⓓ Ⓔ	41 Ⓐ Ⓑ Ⓒ Ⓓ Ⓔ	56 Ⓐ Ⓑ Ⓒ Ⓓ Ⓔ
12 Ⓐ Ⓑ Ⓒ Ⓓ Ⓔ	27 Ⓐ Ⓑ Ⓒ Ⓓ Ⓔ	42 Ⓐ Ⓑ Ⓒ Ⓓ Ⓔ	57 Ⓐ Ⓑ Ⓒ Ⓓ Ⓔ
13 Ⓐ Ⓑ Ⓒ Ⓓ Ⓔ	28 Ⓐ Ⓑ Ⓒ Ⓓ Ⓔ	43 Ⓐ Ⓑ Ⓒ Ⓓ Ⓔ	58 Ⓐ Ⓑ Ⓒ Ⓓ Ⓔ
14 Ⓐ Ⓑ Ⓒ Ⓓ Ⓔ	29 Ⓐ Ⓑ Ⓒ Ⓓ Ⓔ	44 Ⓐ Ⓑ Ⓒ Ⓓ Ⓔ	59 Ⓐ Ⓑ Ⓒ Ⓓ Ⓔ
15 Ⓐ Ⓑ Ⓒ Ⓓ Ⓔ	30 Ⓐ Ⓑ Ⓒ Ⓓ Ⓔ	45 Ⓐ Ⓑ Ⓒ Ⓓ Ⓔ	60 Ⓐ Ⓑ Ⓒ Ⓓ Ⓔ

SECTION 3 — ANALYTICAL

1 Ⓐ Ⓑ Ⓒ Ⓓ Ⓔ	14 Ⓐ Ⓑ Ⓒ Ⓓ Ⓔ	27 Ⓐ Ⓑ Ⓒ Ⓓ Ⓔ	40 Ⓐ Ⓑ Ⓒ Ⓓ Ⓔ
2 Ⓐ Ⓑ Ⓒ Ⓓ Ⓔ	15 Ⓐ Ⓑ Ⓒ Ⓓ Ⓔ	28 Ⓐ Ⓑ Ⓒ Ⓓ Ⓔ	41 Ⓐ Ⓑ Ⓒ Ⓓ Ⓔ
3 Ⓐ Ⓑ Ⓒ Ⓓ Ⓔ	16 Ⓐ Ⓑ Ⓒ Ⓓ Ⓔ	29 Ⓐ Ⓑ Ⓒ Ⓓ Ⓔ	42 Ⓐ Ⓑ Ⓒ Ⓓ Ⓔ
4 Ⓐ Ⓑ Ⓒ Ⓓ Ⓔ	17 Ⓐ Ⓑ Ⓒ Ⓓ Ⓔ	30 Ⓐ Ⓑ Ⓒ Ⓓ Ⓔ	43 Ⓐ Ⓑ Ⓒ Ⓓ Ⓔ
5 Ⓐ Ⓑ Ⓒ Ⓓ Ⓔ	18 Ⓐ Ⓑ Ⓒ Ⓓ Ⓔ	31 Ⓐ Ⓑ Ⓒ Ⓓ Ⓔ	44 Ⓐ Ⓑ Ⓒ Ⓓ Ⓔ
6 Ⓐ Ⓑ Ⓒ Ⓓ Ⓔ	19 Ⓐ Ⓑ Ⓒ Ⓓ Ⓔ	32 Ⓐ Ⓑ Ⓒ Ⓓ Ⓔ	45 Ⓐ Ⓑ Ⓒ Ⓓ Ⓔ
7 Ⓐ Ⓑ Ⓒ Ⓓ Ⓔ	20 Ⓐ Ⓑ Ⓒ Ⓓ Ⓔ	33 Ⓐ Ⓑ Ⓒ Ⓓ Ⓔ	46 Ⓐ Ⓑ Ⓒ Ⓓ Ⓔ
8 Ⓐ Ⓑ Ⓒ Ⓓ Ⓔ	21 Ⓐ Ⓑ Ⓒ Ⓓ Ⓔ	34 Ⓐ Ⓑ Ⓒ Ⓓ Ⓔ	47 Ⓐ Ⓑ Ⓒ Ⓓ Ⓔ
9 Ⓐ Ⓑ Ⓒ Ⓓ Ⓔ	22 Ⓐ Ⓑ Ⓒ Ⓓ Ⓔ	35 Ⓐ Ⓑ Ⓒ Ⓓ Ⓔ	48 Ⓐ Ⓑ Ⓒ Ⓓ Ⓔ
10 Ⓐ Ⓑ Ⓒ Ⓓ Ⓔ	23 Ⓐ Ⓑ Ⓒ Ⓓ Ⓔ	36 Ⓐ Ⓑ Ⓒ Ⓓ Ⓔ	49 Ⓐ Ⓑ Ⓒ Ⓓ Ⓔ
11 Ⓐ Ⓑ Ⓒ Ⓓ Ⓔ	24 Ⓐ Ⓑ Ⓒ Ⓓ Ⓔ	37 Ⓐ Ⓑ Ⓒ Ⓓ Ⓔ	50 Ⓐ Ⓑ Ⓒ Ⓓ Ⓔ
12 Ⓐ Ⓑ Ⓒ Ⓓ Ⓔ	25 Ⓐ Ⓑ Ⓒ Ⓓ Ⓔ	38 Ⓐ Ⓑ Ⓒ Ⓓ Ⓔ	
13 Ⓐ Ⓑ Ⓒ Ⓓ Ⓔ	26 Ⓐ Ⓑ Ⓒ Ⓓ Ⓔ	39 Ⓐ Ⓑ Ⓒ Ⓓ Ⓔ	

PRACTICE TEST

SECTION
1
VERBAL

SECTION
2
QUANTITATIVE

SECTION
3
ANALYTICAL

SECTION ONE: VERBAL
Time—60 minutes   76 questions

Directions: Each of the following questions begins with a sentence that has either one or two blanks. The blanks indicate that a piece of the sentence is missing. Each sentence is followed by five answer choices that consist of words or phrases. Select the answer choice that completes the sentence best.

1. The fundamental _____ between dogs and cats is for the most part a myth; members of these species often coexist _____.

   (A) antipathy . . amiably
   (B) disharmony . . uneasily
   (C) compatibility . . together
   (D) relationship . . peacefully
   (E) difference . . placidly

2. His desire to state his case completely was certainly reasonable; however, his lengthy technical explanations were monotonous and tended to _____ rather than _____ the jury.

   (A) enlighten . . inform
   (B) interest . . persuade
   (C) provoke . . influence
   (D) allay . . pacify
   (E) bore . . convince

3. In some countries, government restrictions are so _____ that businesses operate with nearly complete impunity.

   (A) traditional
   (B) judicious
   (C) ambiguous
   (D) exacting
   (E) lax

4. The recent Oxford edition of the works of Shakespeare is _____ because it not only departs frequently from the readings of most other modern editions, but also challenges many of the basic _____ of textual criticism.

   (A) controversial . . conventions
   (B) typical . . innovations
   (C) inadequate . . norms
   (D) curious . . projects
   (E) pretentious . . explanations

5. The early form of writing known as Linear B was _____ in 1952, but no one has yet succeeded in the _____ of the still more ancient Linear A.

   (A) superseded . . explanation
   (B) encoded . . transcription
   (C) obliterated . . analysis
   (D) deciphered . . interpretation
   (E) discovered . . obfuscation

GO ON TO THE NEXT PAGE

6. Considering everything she had been through, her reaction was quite normal and even _____; I was therefore surprised at the number of _____ comments and raised eyebrows that her response elicited.

(A) commendable . . complimentary
(B) odious . . insulting
(C) apologetic . . conciliatory
(D) commonplace . . typical
(E) laudable . . derogatory

7. The purpose of the proposed insurance policy is to _____ the burden of medical costs, thereby removing what is for many people a major _____ medical care.

(A) augment . . problem with
(B) eliminate . . perquisite of
(C) ameliorate . . study of
(D) assuage . . impediment to
(E) clarify . . explanation for

Directions: Each of the following questions consists of a pair of words or phrases that are separated by a colon and followed by five answer choices. Choice the pair of words or phrases in the answer choices that are most similar to the original pair.

8. NOVEL : BOOK ::

(A) epic : poem
(B) house : library
(C) tale : fable
(D) number : page
(E) play : theater

9. HUNGRY : RAVENOUS ::

(A) thirsty : desirous
(B) large : titanic
(C) famous : eminent
(D) dizzy : disoriented
(E) obese : gluttonous

10. BOUQUET : FLOWER ::

(A) humidor : tobacco
(B) mosaic : tile
(C) tapestry : color
(D) pile : block
(E) sacristy : vestment

11. REALIST : QUIXOTIC ::

(A) scholar : pedantic
(B) fool : idiotic
(C) idler : lethargic
(D) tormentor : sympathetic
(E) diner : dyspeptic

12. SHARD : GLASS ::

(A) grain : sand
(B) morsel : meal
(C) strand : rope
(D) scrap : quilt
(E) splinter : wood

13. FILTER : IMPURITY ::

(A) expurgate : obscenity
(B) whitewash : infraction
(C) testify : perjury
(D) perform : penance
(E) vacuum : carpet

GO ON TO THE NEXT PAGE

14. PARAPHRASE : VERBATIM ::

  (A) approximation : precise
  (B) description : vivid
  (C) quotation : apt
  (D) interpretation : valid
  (E) significance : uncertain

15. ONCOLOGY : TUMOR ::

  (A) chronology : time
  (B) theology : tenet
  (C) oral : sound
  (D) philology : religion
  (E) taxonomy : classification

16. INTRANSIGENT : FLEXIBILITY ::

  (A) transient : mobility
  (B) disinterested : partisanship
  (C) dissimilar : variation
  (D) progressive : transition
  (E) ineluctable : modality

Directions: After each reading passage you will find a series of questions. Select the best choice for each question. Answers are based on the contents of the passage or what the author implies in the passage.

There can be nothing simpler than an elementary particle: it is an indivisible shard of matter, without internal structure and without detectable shape or size. One might expect (5) commensurate simplicity in the theories that describe such particles and the forces through which they interact; at the least, one might expect the structure of the world to be explained with a minimum number of (10) particles and forces. Judged by this criterion of parsimony, a description of nature that has evolved in the past several years can be accounted a reasonable success. Matter is built out of just two classes of elementary particles: (15) the leptons, such as the electron, and the quarks, which are constituents of the proton, the neutron, and many related particles. Four basic forces act between the elementary particles. Gravitation and electromagnetism have long (20) been familiar in the macroscopic world; the weak force and the strong force are observed only in subnuclear events. In principle this complement of particles and forces could account for the entire observed hierarchy of material structure, (25) from the nuclei of atoms to stars and galaxies. An understanding of nature at this level of detail is a remarkable achievement; nevertheless, it is possible to imagine what a still simpler theory might be like. The existence of two disparate (30) classes of elementary particles is not fully satisfying; ideally, one class would suffice. Similarly, the existence of four forces seems a needless complication; one force might explain all the interactions of elementary particles. An (35) ambitious new theory now promises at least a partial unification along these lines. The theory does not embrace gravitation, which is by far the feeblest of the forces and may be fundamentally different from the (40) others. If gravitation is excluded, however, the theory unifies all elementary particles and forces. The first step in the construction of the unified theory was the demonstration that the weak, the strong, (45) and the electromagnetic forces could all be described by theories of the same general kind. The three forces remained distinct, but they could be seen to operate through the same mechanism. In the course of this (50) development a deep connection was discovered between the weak force and

GO ON TO THE NEXT PAGE

electromagnetism, a connection that hinted at a still grander synthesis. The new theory is the leading candidate for accomplishing (55) the synthesis. It incorporates the leptons and the quarks into a single family and provides a means of transforming one kind of particle into the other. At the same time the weak, the strong, and the electro- (60) magnetic forces are understood as aspects of a single underlying force. With only one class of particles and one force (plus gravitation), the unified theory is a model of frugality.

17. All of the following are differences between the two theories described by the author EXCEPT

(A) the second theory is simpler than the first
(B) the first theory encompasses gravitation while the second does not
(C) the second theory includes only one class of elementary particles
(D) the first theory accounts for only part of the hierarchy of material structure
(E) the second theory unifies forces that the first theory regards as distinct

18. The primary purpose of the passage is to

(A) correct a misconception in a currently accepted theory of the nature of matter
(B) describe efforts to arrive at a simplified theory of elementary particles and forces
(C) predict the success of a new effort to unify gravitation with other basic forces
(D) explain why scientists prefer simpler explanations over more complex ones
(E) summarize what is known about the basic components of matter

19. According to the passage, which of the following are true of quarks?

I. They are the elementary building blocks for neutrons.
II. Scientists have described them as having no internal structure.
III. Some scientists group them with leptons in a single class of particles.

(A) I only
(B) III only
(C) I and II only
(D) II and III only
(E) I, II, and III

20. The author considers which of the following in judging the usefulness of a theory of elementary particles and forces?

I. The simplicity of the theory
II. The ability of the theory to account for the largest possible number of known phenomena
III. The possibility of proving or disproving the theory by experiment

(A) I only
(B) II only
(C) I and II only
(D) I and III only
(E) II and III only

21. It can be inferred that the author considers the failure to unify gravitation with other forces in the theory he describes to be

(A) a disqualifying defect
(B) an unjustified deviation
(C) a needless oversimplification
(D) an unfortunate oversight
(E) an unavoidable limitation

GO ON TO THE NEXT PAGE

22. The author organizes the passage by

    (A) enumerating distinctions among several different kinds of elementary particles
    (B) stating a criterion for judging theories of nature, and using it to evaluate two theories
    (C) explaining three methods of grouping particles and forces
    (D) criticizing an inaccurate view of elemental nature and proposing an alternative approach
    (E) outlining an assumption about scientific verification, then criticizing the assumption

23. It can be inferred that the author would be likely to consider a new theory of nature superior to present theories if it were to

    (A) account for a larger number of macroscopic structures than present theories
    (B) reduce the four basic forces to two more fundamental, incompatible forces
    (C) propose a smaller number of fundamental particles and forces than current theories
    (D) successfully account for the observable behavior of bodies due to gravity
    (E) hypothesize that protons but not neutrons are formed by combinations of more fundamental particles

The majority of white abolitionists and the majority of suffragists worked hard to convince their compatriots that the changes they advocated were not revolutionary, that far from
(5) undermining the accepted distribution of power they would eliminate deviations from the democratic principle it was supposedly based on. Non-Garrisonian abolitionists repeatedly disavowed miscegenationist or revolutionary
(10) intentions. And as for the suffragists, despite the presence in the movement of socialists, and in the final years of a few blacks, immigrants, and workers, the racism and nativism in the movement's thinking were not an aberration and
(15) did not conflict with the movement's objective of suffrage. Far from saying, as presentist historians do, that the white abolitionists and suffragists compromised the abiding principles of equality and the equal right of all to life, liberty, and the
(20) pursuit of happiness, I suggest just the opposite: the non-Garrisonian majority of white abolitionists and the majority of suffragists showed what those principles meant in their respective generations, because they traced the
(25) farthest acceptable boundaries around them.

GO ON TO THE NEXT PAGE

24. The author's main point is that

(A) the actions of the abolitionist and suffragist movements compromised their stated principles
(B) the underlying beliefs of abolitionists and suffragists were closer than is usually believed
(C) abolitionists' and suffragists' thinking about equality was limited by the assumptions of their time
(D) presentist historians have willfully misrepresented the ideology of abolitionists and suffragists
(E) historians should impose their own value systems when evaluating events of the past

25. Which of the following does the author imply about the principle of equality?
    I. It does not have a fixed meaning.
    II. Suffragists applied it more consistently than abolitionists.
    III. Abolitionists and suffragists compromised it to gain their political objectives.

(A) I only
(B) II only
(C) III only
(D) I and II only
(E) II and III only

26. The author takes exception to the views of presentist historians by

(A) charging that they ignore pertinent evidence
(B) presenting new information that had not been available before
(C) applying a different interpretation to the same set of facts

(D) refuting the accuracy of their historical data
(E) exposing a logical contradiction in their arguments

27. Which of the following is suggested about the abolitionist movement?

(A) Its members disguised their objectives from the public.
(B) It contained different groupings characterized by varied philosophies.
(C) It undermined its principles by accommodating public concerns.
(D) A majority of its members misunderstood its objectives.
(E) Its progress was hindered by the actions of radical factions within it.

Directions: Each of the following questions begins with a single word in capital letters. Five answer choices follow. Select the answer choice that has the most opposite meaning of the word in capital letters.

Since some of the questions require you to distinguish fine shades of meaning, be sure to consider all the choices before deciding which one is best.

28. UNDERMINE:

(A) appreciate   (B) donate
(C) bolster   (D) decay
(E) simplify

GO ON TO THE NEXT PAGE

29. OBSEQUIOUS:

    (A) original   (B) haughty
    (C) casual   (D) virtuous
    (E) informative

30. BLANCH:

    (A) stand   (B) repay
    (C) flush   (D) relax
    (E) cope

31. DISSIPATED:

    (A) temperate   (B) pleased
    (C) inundated   (D) encouraged
    (E) planned

32. FECUNDITY:

    (A) levity   (B) sanity
    (C) cowardice   (D) sterility
    (E) ventilation

33. ENCUMBER:

    (A) animate   (B) inaugurate
    (C) bleach   (D) disburden
    (E) obliterate

34. DISSEMINATE:

    (A) fertilize   (B) ordain
    (C) suppress   (D) explain thoroughly
    (E) make an impression

35. RESTIVE:

    (A) morose   (B) intangible
    (C) fatigued   (D) patient
    (E) curious

36. SYNCOPATED:

    (A) carefully executed
    (B) normally accented
    (C) brightly illuminated
    (D) easily understood
    (E) justly represented

37. VITUPERATIVE:

    (A) lethal   (B) incapacitated
    (C) laudatory   (D) insulated
    (E) prominent

38. SATURNINE:

    (A) magnanimous   (B) ebullient
    (C) finicky   (D) unnatural
    (E) impoverished

GO ON TO THE NEXT PAGE

Directions: Each of the following questions begins with a sentence that has either one or two blanks. The blanks indicate that a piece of the sentence is missing. Each sentence is followed by five answer choices that consist of words or phrases. Select the answer choice that completes the sentence best.

39. Her concern for the earthquake victims _____ her reputation as a callous person.

(A) restored
(B) rescinded
(C) created
(D) proved
(E) belied

40. Due to unforeseen circumstances, the original plans were no longer _____ and were therefore _____.

(A) relevant . . adaptable
(B) applicable . . rejected
(C) expedient . . adopted
(D) acceptable . . appraised
(E) capable . . allayed

41. The microscopic cross section of a sandstone generally shows a _____ surface, each tiny layer representing an _____ of deposition that may have taken centuries or even millennia to accumulate.

(A) ridged . . enlargement
(B) multifaceted . . angle
(C) distinctive . . area
(D) stratified . . interval
(E) coarse . . episode

42. The convict has always insisted upon his own _____ and now at last there is new evidence to _____ him.

(A) defensiveness . . incarcerate
(B) culpability . . exonerate
(C) blamelessness . . anathematize
(D) innocence . . vindicate
(E) contrition . . condemn

43. The theory of plate tectonics was the subject of much _____ when it was first proposed by Alfred Wegener, but now most geophysicists _____ its validity.

(A) opposition . . grant
(B) consideration . . see
(C) acclamation . . boost
(D) prognostication . . learn
(E) contention . . bar

44. Despite her professed _____, the glint in her eyes demonstrated her _____ with the topic.

(A) intelligence . . obsession
(B) interest . . concern
(C) obliviousness . . confusion
(D) indifference . . fascination
(E) expertise . . unfamiliarity

GO ON TO THE NEXT PAGE

45. Lacking sacred scriptures or _____, Shinto is more properly regarded as a legacy of traditional religious practices and basic values than as a formal system of belief.

    (A) followers
    (B) customs
    (C) dogma
    (D) relics
    (E) faith

<u>Directions:</u> Each of the following questions consists of a pair of words or phrases that are separated by a colon and followed by five answer choices. Choice the pair of words or phrases in the answer choices that are most similar to the original pair.

46. IMPECCABLE : FLAW ::

    (A) impeachable : crime
    (B) obstreperous : permission
    (C) impetuous : warning
    (D) moribund : living
    (E) absurd : sense

47. SEISMOGRAPH : EARTHQUAKE ::

    (A) stethoscope : health
    (B) speedometer : truck
    (C) telescope : astronomy
    (D) thermometer : temperature
    (E) abacus : arithmetic

48. GUZZLE : DRINK ::

    (A) elucidate : clarify
    (B) ingest : eat
    (C) boast : describe
    (D) stride : walk
    (E) admonish : condemn

49. ORATOR : ARTICULATE ::

    (A) soldier : merciless
    (B) celebrity : talented
    (C) judge : unbiased
    (D) novice : unfamiliar
    (E) dignitary : respectful

50. BADGE : POLICEMAN ::

    (A) placard : demonstrator
    (B) tattoo : sailor
    (C) dog-tag : soldier
    (D) pedigree : dog
    (E) fingerprint : defendant

51. SCRUTINIZE : OBSERVE ::

    (A) excite : pique
    (B) beseech : request
    (C) search : discover
    (D) smile : grin
    (E) dive : jump

52. INDULGE : EPICUREAN ::

    (A) frighten: ugly
    (B) retract : revocable
    (C) hesitate : unproductive
    (D) revenge : vindictive
    (E) understand : comprehensible

GO ON TO THE NEXT PAGE

53. FLOOD : DILUVIAL ::

(A) punishment : criminal
(B) bacteria : biological
(C) verdict : judicial
(D) light : candescent
(E) heart : cardiac

54. SPHINX : PERPLEX ::

(A) oracle : interpret
(B) prophet : prepare
(C) siren : lure
(D) jester : astound
(E) minotaur : anger

Directions: After each reading passage you will find a series of questions. Select the best choice for each question. Answers are based on the contents of the passage or what the author implies in the passage.

Although the schooling of fish is a familiar form of animal social behavior, how the school is formed and maintained is only beginning to be understood in detail. It had been thought that
(5) each fish maintains its position chiefly by means of vision. Our work has shown that, as each fish maintains its position, the lateral line, an organ sensitive to transitory changes in water displacement, is as important as vision. In each
(10) species a fish has a "preferred" distance and angle from its nearest neighbor. The ideal separation and bearing, however, are not maintained rigidly. The result is a probabilistic arrangement that appears like a random aggregation. The
(15) tendency of the fish to remain at the preferred distance and angle, however, serves to maintain the structure. Each fish, having established its position, uses its eyes and its lateral lines

simultaneously to measure the speed of all the
(20) other fish in the school. It then adjusts its own speed to match a weighted average that emphasizes the contribution of nearby fish.

55. According to the passage, the structure of a fish school is dependent upon which of the following?
  I. rigidly formed random aggregations
  II. the tendency of each fish to remain at a preferred distance from neighboring fish
  III. measurements of a weighted average by individual fish

(A) II only
(B) III only
(C) I and II only
(D) I and III only
(E) II and III only

56. Which of the following best describes the author's attitude toward the theory that the structure of fish schools is maintained primarily through vision?

(A) heated opposition
(B) careful neutrality
(C) considered dissatisfaction
(D) cautious approval
(E) unqualified enthusiasm

GO ON TO THE NEXT PAGE

KAPLAN

57. The passage suggests that, after establishing its position in the school formation, an individual fish will subsequently

(A) maintain its preferred position primarily by visual and auditory means
(B) rigorously avoid changes that would interfere with the overall structure of the school
(C) make continuous sensory readjustments to its position within the school
(D) make unexpected shifts in position only if threatened by external danger
(E) surrender its ability to make quick, instinctive judgments

Whether as a result of some mysterious tendency in the national psyche or as a spontaneous reaction to their turbulent historical experience after the breakup of the Mycenaean
(5) world, the Greeks felt that to live with changing, undefined, unmeasured, seemingly random impressions—to live, in short, with what was expressed by the Greek word *chaos*—was to live in a state of constant anxiety.
(10)   If the apparent mutability of the physical world and of the human condition was a source of pain and bewilderment to the Greeks, the discovery of a permanent pattern or an unchanging substratum by which apparently chaotic experi-
(15) ence could be measured and explained was a source of satisfaction, even joy, which had something of a religious nature. For the recognition of order and measure in phenomena did more than simply satisfy their intellectual
(20) curiosity or gratify a desire for tidiness; it also served as the basis of a spiritual ideal. "Measure and commensurability are everywhere identified with beauty and excellence," was Plato's way of putting it in a dialogue in which measure is
(25) identified as a primary characteristic of the ultimate good. Rational definability and spirituality were never mutually exclusive categories in Greek thought. If the quest for order and clarity was in essence the search for a kind of spiritual
(30) ideal, it was not an ideal to be perceived in rapturous emotional mysticism but rather one to be arrived at by patient analysis.

We see this process at work especially in Greek philosophy, which in various ways was
(35) aimed at alleviating the anxiety that is inherent in the more spontaneous expression of lyric poetry. The Milesian philosophers of the sixth century were interested above all in discovering a primary substance from which all other phe-
(40) nomena could be explained. Neat, clear, and sublimely undisturbed by the social world of humanity, which took shape and dissolved within the natural order of things, it was an austere ideal, an astringent antidote to the appar-
(45) ent senselessness of life. The person who contemplated it deeply could feel a part of a great system that was impersonal but predictable, and, like Lucretius, who revived the Milesian attitude in a later age, he or she could derive a
(50) peculiar peace from it. As time passed and Greek philosophy developed, the urge to find order in experience was shifted from physics to the realm of mathematical abstraction by the Pythagoreans, and to the world of human
(55) behavior by various thinkers of the later fifth century; and, finally, Plato and Aristotle attempted to weave all these foci of interest into comprehensive pictures of the relationship between human life and the world as a whole.
(60) But in all these epochs the basic quest—the search for a "kosmos"—remained the same.

GO ON TO THE NEXT PAGE

58. The author's primary purpose is to

    (A) evaluate conflicting viewpoints
    (B) challenge an accepted opinion
    (C) question philosophical principles
    (D) enumerate historical facts
    (E) describe a cultural phenomenon

59. The author indicates that the discovery of "an unchanging substratum" (lines 13–14) served primarily to

    (A) alter the Greeks' perception of the mutability of existence
    (B) help eradicate severe social problems
    (C) alleviate painful memories of national suffering
    (D) calm a restless intellectual curiosity
    (E) foster a more mystical understanding of the physical world

60. It can be inferred from the passage that rational thought and spiritual ideals were categories of experience that were

    (A) unimportant and unfamiliar to most ordinary Greeks
    (B) advocated by the Milesians and rejected by the Pythagoreans
    (C) neglected by most philosophers before Plato and Aristotle
    (D) seen by the Greeks as essentially compatible
    (E) embraced mainly by Greek poets

61. All of the following can be inferred about the Greeks' anxiety over the possibility of "chaos" EXCEPT that it

    (A) had sources in their national consciousness
    (B) was reflected in specific aspects of their religion
    (C) was related to their sense of change in the physical world
    (D) led to a striving for order in their philosophy
    (E) was expressed in their lyric poetry

62. The author implies that the Milesian philosophers of the sixth century sought relief from worldly anxiety by

    (A) focusing narrowly on inherently human questions
    (B) establishing sharp distinctions between spiritual and rational understanding
    (C) focusing primarily on an impersonal natural order
    (D) attempting to integrate rational and mystical worldviews
    (E) withdrawing from the physical world into the realm of mathematical abstraction

GO ON TO THE NEXT PAGE

63. Which of the following best describes the organization of lines 17–28 of the passage ("For . . . thought")?

(A) The author summarizes two viewpoints, cites historical evidence, and then declines to support either of the viewpoints.
(B) The author makes an observation, admits to evidence that weakens the viewpoint, and then revises his observation.
(C) The author specifies two distinct arguments, examines both in detail, then advances a third argument that reconciles the other two.
(D) The author clarifies a previous statement, offers an example, and then draws a further conclusion based on these ideas.
(E) The author states a thesis, mentions an opposed thesis, and cites evidence supporting it, and then restates his original thesis.

64. According to the passage, the Pythagoreans differed from the Milesians primarily in that the Pythagoreans

(A) focused on mathematical abstractions rather than physical phenomena
(B) placed a renewed emphasis on understanding human behavior
(C) focused primarily on a rational means to understanding truth
(D) attempted to identify a fundamental physical unit of matter
(E) stressed concrete reality over formal theory

65. In the context of the author's overall argument, which of the following best characterizes the Greeks' "search for a 'kosmos'" (line 61)?

(A) a mystical quest for a strong national identity
(B) efforts to replace a sterile philosophical rationalism with revitalized religious values
(C) attempts to end conflict among key philosophical schools
(D) a search for order and measure in an unpredictable world
(E) a search for an alternative to a narrow preoccupation with beauty and excellence

Directions: Each of the following questions begins with a single word in capital letters. Five answer choices follow. Select the answer choice that has the most opposite meaning of the word in capital letters.

Since some of the questions require you to distinguish fine shades of meaning, be sure to consider all the choices before deciding which one is best.

66. ENMITY:

(A) friendship
(B) reverence
(C) boredom
(D) stylishness
(E) awkwardness

GO ON TO THE NEXT PAGE

67. DILATE:

(A) enclose
(B) shrink
(C) hurry
(D) inflate
(E) erase

68. CHARLATAN:

(A) genuine expert
(B) powerful leader
(C) false idol
(D) unknown enemy
(E) hardened villain

69. PERIPHERAL:

(A) civilized
(B) partial
(C) central
(D) unharmed
(E) stable

70. MERITORIOUS:

(A) effulgent
(B) stationary
(C) uneven
(D) narrow-minded
(E) unpraiseworthy

71. DISCHARGE:

(A) heal
(B) advance
(C) enlist
(D) penalize
(E) delay

72. MALEDICTION:

(A) blessing
(B) preparation
(C) good omen
(D) liberation
(E) pursuit

73. MAWKISH:

(A) unsentimental
(B) sophisticated
(C) graceful
(D) tense
(E) descriptive

74. TEMERITY:

(A) blandness
(B) caution
(C) severity
(D) strength
(E) charm

75. JEJUNE:

(A) morose
(B) natural
(C) mature
(D) contrived
(E) accurate

76. VITIATE:

(A) deaden
(B) trust
(C) rectify
(D) drain
(E) amuse

END OF SECTION

SECTION TWO—QUANTITATIVE
Time—60 minutes   60 questions

Numbers: The numbers in this section are real numbers.

Figures: You may assume that the position of points, lines, angles, etcetera are in the order shown and that all lengths and angle measures may be assumed to be positive.

You may assume that lines that look straight are straight.

Figures are in a plane unless otherwise stated.

Figures are not drawn to scale unless otherwise stated.

Directions: Questions 1–15 provide two quantities, one in Column A and another in Column B. Compare the two quantities and answer

(A)   if the quantity in Column A is greater
(B)   if the quantity in Column B is greater
(C)   if the two quantities are equal
(D)   if the relationship cannot be determined from the information given

Common
Information: In each question, information relating to one or both of the quantities in Column A and Column B is centered above the two columns. A symbol that appears in both columns represents the same thing in Column A as it does in Column B.

Column A	Column B	Sample Answers

Example 1: 3×4 $3 + 4$ ● Ⓑ Ⓒ Ⓓ Ⓔ

GO ON TO THE NEXT PAGE

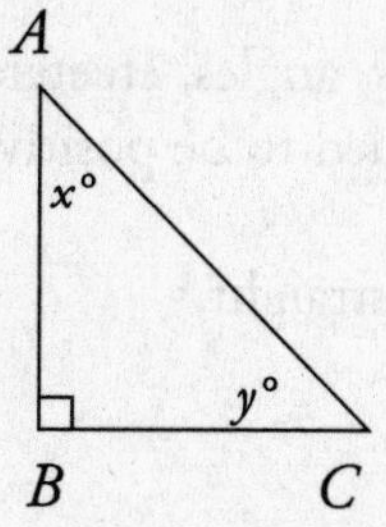

Column A	Column B	Sample Answers

Examples 2–4 refer to the figure below.

A

$x°$

$y°$

B   C

Example 2: x y

Ⓐ Ⓑ Ⓒ ● Ⓔ

(Because we cannot assume
the angles are equal, even
though they appear that way.)

Example 3: $x + y$ 90

Ⓐ Ⓑ ● Ⓓ Ⓔ

(Because the sum of the angles
is 180°.)

Example 4: x 90

Ⓐ ● Ⓒ Ⓓ Ⓔ

(Since $\triangle ABC$ is a right
triangle, x is less than 90°.)

GO ON TO THE NEXT PAGE

KAPLAN

Column A	Column B
1. 0.0260	0.0256

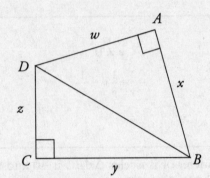

$\triangle ABD$ and $\triangle CDB$ are right triangles.

Column A	Column B
2. $w^2 + x^2$	$y^2 + z^2$

$$x + 4y = 6$$
$$x = 2y$$

Column A	Column B
3. x	y

Column A	Column B
4. $\sqrt{4^2 + 5^2}$	$\sqrt{3^2 + 6^2}$

Column A	Column B

In a certain accounting firm, there are exactly three types of employees: managerial, technical, and clerical. The firm has 120 employees and 25 percent of the employees are managerial.

Column A	Column B
5. The number of managerial employees	Two-thirds of the number of clerical employees

| 6. $\dfrac{12 \times 1}{12 + 1}$ | $\dfrac{12 + 1}{12 \times 1}$ |

| 7. $(a + 1)(b + 1)$ | $ab + 1$ |

In the two-digit number jk, the value of the digit j is twice the value of the digit k.

| 8. k | 6 |

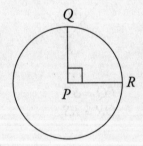

P is the center of the circle and the area of sector PQR is 4.

| 9.   The area of circle P | 4π |

GO ON TO THE NEXT PAGE

Column A | Column B

Henry purchased x apples and Jack purchased 10 apples fewer than one-third of the number of apples Henry purchased.

10. The number of apples Jack purchased | $\dfrac{x-30}{3}$

11. The volume of a rectangular solid with a length of 5 feet, a width of 4 feet, and a height of x feet | The volume of a rectangular solid with a length of 10 feet, a width of 8 feet, and a height of y feet

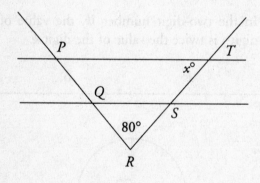

$PQ = ST$
$QR = RS$

12. x | 50

Column A | Column B

$2 \times 16 \times 64 = 2 \times 4n \times 256$

13. n | 2

$y \neq 0$
$$-\frac{2}{y} + \frac{1}{3} = -\frac{1}{2y}$$

14. y | 4

The perimeter of isosceles $\triangle ABC$ is 40 and the length of side BC is 12.

15. The length of side AB | 14

Directions: Each of Questions 16–30 has five answer choices. For each of these questions, select the best of the answer choices given.

16. If $\dfrac{p-q}{p} = \dfrac{2}{7}$, then $\dfrac{q}{p} =$

(A) $\dfrac{2}{5}$

(B) $\dfrac{5}{7}$

(C) 1

(D) $\dfrac{7}{5}$

(E) $\dfrac{7}{2}$

GO ON TO THE NEXT PAGE

17. Which of the following numbers is both a multiple of 8 and a factor of 72?

    (A)   4
    (B)   9
    (C) 16
    (D) 24
    (E) 36

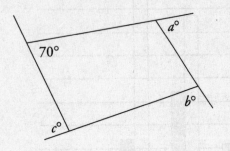

18. In the figure above, what is the value of $a + b + c$?

    (A) 110
    (B) 250
    (C) 290
    (D) 330
    (E) 430

19. John has four ties, 12 shirts, and three belts. If each day he wears exactly one tie, one shirt, and one belt, what is the maximum number of days he can go without repeating a particular combination?

    (A)   12
    (B)   21
    (C)   84
    (D) 108
    (E) 144

20. Which of the following is the greatest?

    (A) $\dfrac{0.00003}{0.0007}$

    (B) $\dfrac{0.0008}{0.0005}$

    (C) $\dfrac{0.007}{0.0008}$

    (D) $\dfrac{0.006}{0.0005}$

    (E) $\dfrac{0.01}{0.008}$

GO ON TO THE NEXT PAGE

Questions 21–25 refer to the charts below.

U.S. PHYSICIANS IN SELECTED SPECIALTIES BY SEX, 1986

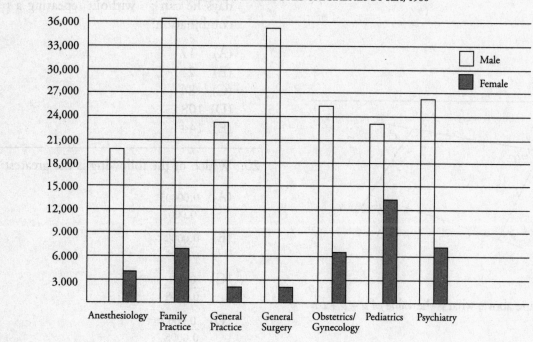

GENERAL SURGERY PHYSICIANS BY AGE, 1986

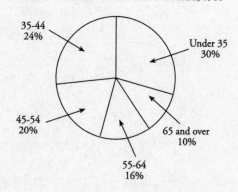

GO ON TO THE NEXT PAGE

KAPLAN

21. Approximately what percent of all general practice physicians in 1986 were male?

    (A) 23%
    (B) 50%
    (C) 75%
    (D) 82%
    (E) 90%

22. Which of the following physician specialties had the lowest ratio of males to females in 1986?

    (A) Family practice
    (B) General surgery
    (C) Obstetrics/gynecology
    (D) Pediatrics
    (E) Psychiatry

23. In 1986, approximately how many general surgery physicians were between the ages of 45 and 54, inclusive?

    (A) 5,440
    (B) 6,300
    (C) 7,350
    (D) 7,800
    (E) 8,900

24. If in 1986 all the family practice physicians represented 7.5 percent of all the physicians in the United States, approximately how many physicians were there total?

    (A) 300,000
    (B) 360,000
    (C) 430,000
    (D) 485,000
    (E) 570,000

25. If the number of female general surgeon physicians in the under-35 category represented 3.5 percent of all the general surgeon physicians, approximately how many male general surgeon physicians were under 35 years?

    (A) 9,200
    (B) 9,800
    (C) 10,750
    (D) 11,260
    (E) 11,980

26. $|3| + |-4| + |3 - 4| =$

    (A) 14
    (B) 8
    (C) 7
    (D) 2
    (E) 0

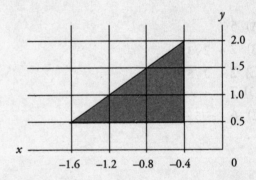

27. What is the area of the shaded region in the figure above?

    (A) 0.5
    (B) 0.7
    (C) 0.9
    (D) 2.7
    (E) 4.5

GO ON TO THE NEXT PAGE

28. A computer can perform 30 identical tasks in six hours. At that rate, what is the minimum number of computers that should be assigned to complete 80 of the tasks within three hours?

(A) 6
(B) 7
(C) 8
(D) 12
(E) 16

29. The volume of the cube in the figure above is 8. If point A is the midpoint of an edge of this cube, what is the perimeter of $\triangle ABC$?

(A) 5
(B) $2 + 2\sqrt{3}$
(C) $2 + 2\sqrt{5}$
(D) 7
(E) $6 + \sqrt{5}$

30. Which of the following is 850 percent greater than 8×10^3 ?

(A) 8.5×10^3
(B) 6.4×10^4
(C) 6.8×10^4
(D) 7.6×10^4
(E) 1.6×10^5

GO ON TO THE NEXT PAGE

Column A	Column B

$$y = (x + 3)^2$$

31. The value of y when $x = 1$ 9

---

32. The number of miles traveled by a car that traveled for four hours at an average speed of 40 miles per hour | The number of miles traveled by a train that traveled for two and a half hours at an average speed of 70 miles per hour

---

33. The number of cookies in a bag that weighs 3 kilograms | The number of grapes in a bag that weighs 2 kilograms

---

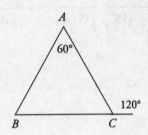

34. AB BC

Column A	Column B

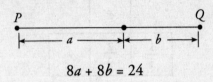

$$8a + 8b = 24$$

35. The length of segment PQ 2

$$x < y$$

36. $y - x$ $x - y$

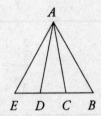

The area of triangular region ABE is 75.

37. The area of $\triangle ABC$ | The area of $\triangle ADE$

GO ON TO THE NEXT PAGE

Column A   Column B

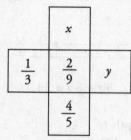

The sum of the numbers in the horizontal row of boxes equals the sum of the numbers in the vertical row of boxes.

38. x y

39. $\dfrac{\frac{1}{3}\times\frac{1}{4}}{\frac{2}{3}\times\frac{1}{2}}$ $\dfrac{\frac{2}{3}\times\frac{1}{2}}{\frac{1}{3}\times\frac{1}{4}}$

Eileen drives due north from town A to town B for a distance of 60 miles, then drives due east from town B to town C for a distance of 80 miles.

40.   The distance from town   120
      A to town C in miles

41. $(\sqrt{7}-2)(\sqrt{7}+2)$ $(2-\sqrt{7})(-\sqrt{7}-2)$

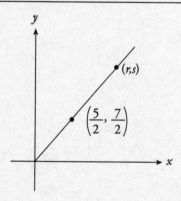

42. r s

Column A   Column B

x is an integer greater than 0.

43. $1-$ 0.95

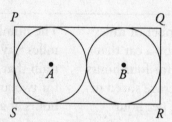

The two circles with centers A and B have the same radius.

44.   The sum of the   The perimeter of
     circumferences of   rectangles $PQRS$
     of the two circles

45. $3^{17}+3^{18}+3^{19}$ 3^{20}

GO ON TO THE NEXT PAGE

Directions: Questions 16–30 each have five answer choices. For each of these questions, select the best of the answer choices given.

46. If $4 + y = 14 - 4y$, then $y =$

(A) -4

(B) 0

(C) $\frac{5}{8}$

(D) $\frac{4}{5}$

(E) 2

47. $\frac{4}{5} + \frac{5}{4} =$

(A) 1

(B) $\frac{9}{8}$

(C) $\frac{6}{5}$

(D) $\frac{41}{20}$

(E) $\frac{23}{10}$

48. If $3m = 81$, then $m^3 =$

(A) 9
(B) 16
(C) 27
(D) 54
(E) 64

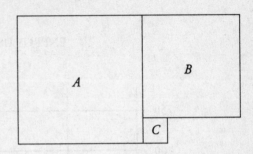

49. In the figure above, there are three square gardening areas. The area of square A is 81 square meters and the area of square B is 49 square meters. What is the area, in square meters, of square C?

(A) 2
(B) 4
(C) 9
(D) 27
(E) 32

50. In a certain history class, all except 23 students scored under 85 on a test. If 18 students scored over 85 on this test, how many students are there in this history class?

(A) 33
(B) 37
(C) 39
(D) 41
(E) It cannot be determined from the information given.

GO ON TO THE NEXT PAGE

Questions 51–55 refer to the following graphs

ENERGY USE BY YEAR, COUNTRY Y, 1950-1980
(in millions of kilowatt-hours)

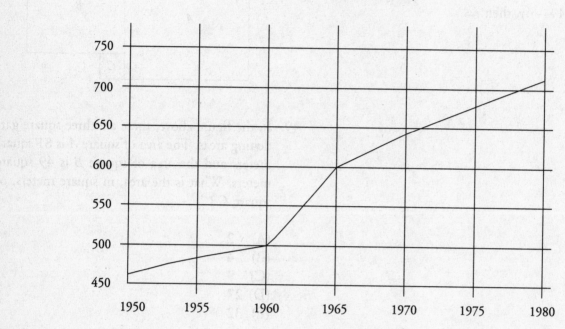

ENERGY USE BY TYPE, COUNTRY Y

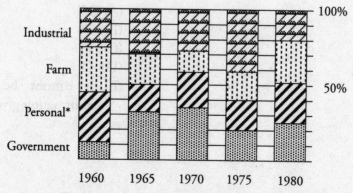

*Total personal use = population ∞ per-capita personal use

GO ON TO THE NEXT PAGE

51. In which of the following years was the energy use in country Y closest to 650 million kilowatt-hours?

    (A) 1960
    (B) 1965
    (C) 1970
    (D) 1975
    (E) 1980

52. In 1965, how many of the categories shown had energy use greater than 150 million kilowatt-hours?

    (A) none
    (B) one
    (C) two
    (D) three
    (E) four

53. In which of the following years was industrial use of energy greatest in country Y?

    (A) 1960
    (B) 1965
    (C) 1970
    (D) 1975
    (E) 1980

54. If the population of country Y increased by 20 percent from 1960 to 1965, approximately what was the percent decrease in the per-capita personal use of energy between those two years?

    (A) 0%
    (B) 17%
    (C) 25%
    (D) 47%
    (E) It cannot be determined from the information given.

55. Which of the following can be inferred from the graphs?

    I. Farm use of energy increased between 1960 and 1980.
    II. In 1980, industrial use of energy was greater than industrial use of energy in 1965.
    III. More people were employed by the government of country Y in 1980 than in 1960.

    (A) I only
    (B) II only
    (C) I and II only
    (D) II and III only
    (E) I, II, and III

56. If the average of two numbers is $3y$ and one of the numbers is $y - z$, what is the other number, in terms of y and z?

    (A) $y + z$
    (B) $3y + z$
    (C) $4y - z$
    (D) $5y - z$
    (E) $5y + z$

GO ON TO THE NEXT PAGE

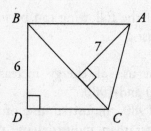

57. In the figure above, the area of $\triangle ABC$ is 35. What is the length of DC?

(A) 6
(B) 8
(C) $6\sqrt{2}$
(D) 10
(E) $6\sqrt{3}$

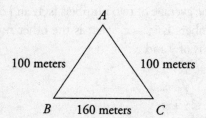

58. In the figure above is a triangular field. What is the minimum distance, in meters, that a person would have to walk to go from point A to a point on side BC?

(A) 60
(B) 80
(C) 100
(D) 140
(E) 180

59. If the ratio of $2a$ to b is 8 times the ratio of b to a, then $\frac{b}{a}$ could be

(A) 4
(B) 2
(C) 1
(D) $\frac{1}{2}$
(E) $\frac{1}{4}$

60. A certain dentist earns n dollars for each filling she puts in, plus x dollars for every 15 minutes she works. If in a certain week she works 14 hours and puts in 21 fillings, how much does she earn for the week, in dollars?

(A) $\frac{7}{2}x + 21n$
(B) $7x + 14n$
(C) $14x + 21n$
(D) $56x + 21n$
(E) $56x + \frac{21}{4}n$

END OF SECTION

SECTION THREE—ANALYTICAL
Time—60 minutes   50 questions

Directions: Each group of questions is based on a passage or a set of conditions. You may wish to draw a diagram to answer some of the questions. Choose the best answer for each question.

Questions 1–3

A spice farmer must harvest exactly five spices grown on her farm. The spices must be harvested consecutively, the harvest of one being completed before the harvest of the next begins. The five spices to be harvested are allspice, cloves, nutmeg, sage, and thyme.

> Nutmeg must be harvested before thyme.
> Cloves must be harvested immediately after allspice.
> Sage must not be harvested first.

1. Which of the following is an acceptable order for the harvesting of the five spices?

   (A) nutmeg, sage, allspice, cloves, thyme
   (B) sage, nutmeg, thyme, allspice, cloves
   (C) allspice, sage, thyme, cloves, nutmeg
   (D) cloves, nutmeg, allspice, sage, thyme
   (E) allspice, cloves, thyme, sage, nutmeg

2. If nutmeg is the fourth spice harvested, which of the following must be false?

   (A) Allspice is the first spice harvested.
   (B) Sage is harvested immediately after cloves.
   (C) Exactly one crop is harvested between sage and thyme.
   (D) Nutmeg is harvested immediately after cloves.
   (E) Thyme is the last spice harvested.

3. If sage is the second spice harvested, allspice must be which of the following?

   (A) the first or the third spice harvested
   (B) the first or the fourth spice harvested
   (C) the third or the fourth spice harvested
   (D) the third or the fifth spice harvested
   (E) the fourth or the fifth spice harvested

4. If a judge is appointed for life, she will make courtroom decisions that reflect the accumulated wisdom inherent in this country's judicial history, relying upon the law and reason rather than upon trends in political thinking. If, on the other hand, the judge is appointed or elected for short terms in office, her decisions will be heavily influenced by the prevailing political climate. In sum, the outcome of many court cases will be determined by the method by which the presiding judge has been installed in her post.

   Which one of the following, if true, does NOT support the argument in the passage above?

   (A) Surveys indicate that judges enjoy their work and want to remain in office as long as possible.
   (B) Judges appointed for life are just as informed about political matters as are judges who are elected or appointed for short terms.

(C) The rulings of judges who must run for re-election are generally approved of by the voters who live in their elective districts.

(D) Most judges appointed for life hand down identical rulings on similar cases throughout their long careers.

(E) Only judges who are elected or appointed for short terms of office employ pollsters to read the mood of the electorate.

5. There are those who claim that reductions in the spending on and deployment of weapons systems would result in a so-called "climate of peace," thereby diminishing the likelihood of armed conflict. The facts show otherwise. These self-proclaimed pacifists are either the victims or the propagators of a false argument.

Which of the following is an assumption underlying the conclusion of the passage above?

(A) Military actions involving our forces can be instigated by any number of different factors.

(B) Our buildup of weapons systems and combat personnel has prevented our adversaries from increasing their own spending on defense.

(C) The increased defense spending of the past 10 years has lessened the need for significant military expenditure in future decades.

(D) At the present time, state-of-the-art weapons systems and the augmentation of combat personnel are equally important to a nation's defense.

(E) An established correlation between greater spending on weapons systems and a decreased incidence of conflict will persist.

6. Should present trends continue, within five years it will be cheaper for audio enthusiasts to build their stereo systems around sets of separate, high quality tuners and amplifiers, rather than around integrated tuners and amplifiers, known as receivers. While receivers have been considered the necessary compromise for those with budget restrictions, recent trends in retail pricing seem destined to change that perception. The average retail price of a high-quality tuner has declined at a rate of 20 percent each of the last two years, and the average retail price of a high-quality amplifier has declined at the rate of 35 percent for each of those years. At the same time, the average retail price of integrated receivers has declined only 12 percent.

In evaluating the claim made in the passage above, information about which of the following would be most useful?

(A) the average life expectancy of stereo tuners as compared to the average life expectancy of stereo amplifiers

(B) the number of integrated receivers sold each year and the number of sets of separate tuners and amplifiers sold each year

(C) the present average retail price of an integrated receiver and the present average retail price of a tuner and amplifier set

(D) the number of separate tuner and amplifier sets expected to be purchased over the next five years and the number of integrated receivers expected to be purchased over the next five years

(E) the percentage of audio enthusiasts who prefer separate tuner and amplifier sets to integrated receivers

GO ON TO THE NEXT PAGE

Questions 7–10

An editor must choose five articles to be published in the upcoming issue of an arts review. The only articles available for publication are theater articles *F*, *G*, *H*, and *J*, and dance articles *K*, *L*, *M*, and *O*.

At least three of the five published articles must be dance articles.

If *J* is chosen, then *M* cannot be.

If *F* is chosen, then *J* must also be chosen.

7. If *M* is not chosen for the issue, which of the following must be chosen?

   (A) *F*
   (B) *G*
   (C) *H*
   (D) *J*
   (E) *K*

8. How many acceptable groupings of articles include *J*?

   (A) one
   (B) two
   (C) three
   (D) four
   (E) five

9. The choice of which article makes only one group of articles acceptable?

   (A) *F*
   (B) *G*
   (C) *J*
   (D) *L*
   (E) *M*

10. If *G* is chosen for the issue, which of the following must be true?

    (A) *J* is not chosen.
    (B) Exactly three dance articles are chosen.
    (C) *H* is not chosen.
    (D) All four of the dance articles are chosen.
    (E) *F* is not chosen.

Questions 11–14

An obedience school is experimenting with a new training system. To test the system, three trainers (Luis, Molly, and Oprah) and three dogs (Lassie, Mugs, and Onyx) are assigned to three different rooms, one trainer, and one dog per room. The initial assignment is as follows:

Room 1: Luis and Lassie
Room 2: Molly and Mugs
Room 3: Oprah and Onyx

The participants have learned five different commands, each of which they will execute as soon as the command is given.

Command *W* requires the trainer in Room 1 to move to Room 2, the trainer in Room 2 to move to Room 3, and the trainer in Room 3 to move to Room 1.

Command *X* requires the dogs in Rooms 1 and 2 to change places.

Command *Y* requires the dogs in Rooms 2 and 3 to change places.

Command *Z* requires the dogs in Rooms 3 and 1 to change places.

Command *A* requires each of the dogs to go to the room containing the trainer it was matched with in the initial assignment.

GO ON TO THE NEXT PAGE

11. If the participants in the initial assignment are given exactly one command, Command W, which of the following will be true in the resulting arrangement?

(A) Oprah and Mugs will be in the same room.
(B) Molly will be in Room 3.
(C) Molly and Lassie will be in the same room.
(D) Luis will be in Room 3.
(E) Luis and Onyx will be in the same room.

12. Which of the following commands or series of commands will yield a final arrangement in which Onyx is in Room 2?

(A) One call of W
(B) Two calls of X
(C) Two calls of W followed by one call of A
(D) Two calls of W followed by one call of Z
(E) Two calls of X followed by one call of Z

13. Which of the following sequences of commands will yield a final arrangement in which Oprah and Lassie are in Room 2?

(A) X, Y, W
(B) X, W, W
(C) Z, W, A
(D) X, Y, A, W
(E) Z, W, W, X

14. Which of the following sequences of commands could result in a final arrangement in which Molly and Onyx are in Room 1, Oprah and Mugs are in Room 2, and Luis and Lassie are in Room 3?

(A) Z, W, X
(B) W, Y, Z
(C) W, A, Y, W
(D) W, Z, W, X
(E) X, Z, W, W

Questions 15–18

There are eight apartments in a two-story building, four on each floor. The top floor is called Level A, the bottom floor is Level B. The rooms on each level are numbered 1 through 4 in order from one end of the building to the other, such that the apartments on Level A are directly above the apartments with the same numbers on Level B. Exactly seven people—P, Q, R, S, T, V, and W—live in the building, one to an apartment. One of the apartments is empty.

W's apartment is directly above S's apartment.
S and Q live on different levels.
P's apartment is adjacent to T's apartment on the same level.
T's apartment is directly between two other apartments on the same level.
W's apartment is adjacent to the empty apartment on the same level.

GO ON TO THE NEXT PAGE

15. Which of the following must be on Level B?

    (A) P's apartment
    (B) Q's apartment
    (C) R's apartment
    (D) V's apartment
    (E) the empty apartment

16. If W lives in Apartment 2 on Level A, which of the following must be true?

    (A) V lives in Apartment 1 on Level B.
    (B) The empty apartment is Apartment 3 on Level A.
    (C) R's apartment is on Level A.
    (D) P lives in Apartment 4 on Level A.
    (E) T lives in Apartment 3 on Level B.

17. If R lives in Apartment 3 on Level A, directly above P's apartment, in which apartment must V live?

    (A) Apartment 1 on Level A
    (B) Apartment 4 on Level A
    (C) Apartment 1 on Level B
    (D) Apartment 2 on Level B
    (E) Apartment 4 on Level B

18. If Q lives in Apartment 2 on Level A, directly above T's apartment, which of the following could possibly be Apartment 1 on Level A?

    (A) P's apartment
    (B) S's apartment
    (C) V's apartment
    (D) W's apartment
    (E) the empty apartment

Questions 19–22

Exactly seven people are present in the game room of a club. Three of those present—F, G, and H—are senior club members, two—K and M—are junior club members, and two—P and R—are club applicants. They decide that two of those present will play backgammon, two will play chess, and three will play dominoes.

> Each person present can play only one of the three games.
> There must be a senior club member playing each game.
> G cannot play the same game that R plays.
> H and P must play the same game.
> M cannot play dominoes.

19. Which of the following is an acceptable grouping of people playing backgammon, chess, and dominoes, respectively?

    (A) G, K; H, P; F, M, R
    (B) G, M; K, R; F, H, P
    (C) F, R; G, P; H, K, M
    (D) H, P; G, M; F, K, R
    (E) F, M; H, P; G, K, R

20. If K and R play the same game, which of the following must be true?

    (A) H plays dominoes.
    (B) P plays chess.
    (C) G plays backgammon.
    (D) F plays dominoes.
    (E) M plays backgammon.

GO ON TO THE NEXT PAGE

21. If R plays backgammon, how many different groupings of people and games are possible?

   (A) one
   (B) two
   (C) three
   (D) four
   (E) six

22. Which of the following pairs CANNOT play the same game?

   (A) H and R
   (B) K and M
   (C) F and M
   (D) G and M
   (E) P and R

23. European nations are starting to decrease the percentage of their foreign aid that is "tied"—that is, given only on the condition that it be spent to obtain goods and materials produced by the country from which the aid originates. By doing so, European nations hope to avoid the ethical criticism that has been recently leveled at some foreign aid donors, notably Japan.

   Which of the following can most reasonably be inferred from the passage?

   (A) Many non-European nations give foreign aid solely for the purpose of benefiting their domestic economies.
   (B) Only ethical considerations, and not those of self-interest, should be considered when foreign aid decisions are made.
   (C) Many of the problems faced by underdeveloped countries could be eliminated if a smaller percentage of the foreign aid they obtain were "tied" to specific purchases and uses.

   (D) Much of Japan's foreign aid returns to Japan in the form of purchase orders for Japanese products and equipment.
   (E) Non-European nations are unwilling to offer foreign aid that is not "tied" to the purchase of their own manufactures.

24. Our environment can stand only so much more "progress." We must take a few steps backward and accept some inconvenience if we want to secure the health and well-being of our planet. This is not merely a matter of using manual mowers instead of power mowers, or foregoing a few outdoor barbecues. Something must be done about the 51.1 percent of total ozone that is contributed by vehicles and fuel. The percentage must be cut regardless of the cost or inconvenience. Such concerns are irrelevant here; what needs to be done must be done.

   The author of the passage above makes which of the following arguments?

   (A) People will have to go back to living as they did a century ago if they want to save the environment.
   (B) If people would be willing to drive their cars less, pollution would be drastically reduced.
   (C) People can continue to use power lawn mowers and have barbecues as long as industry cuts down on its use of fuel.
   (D) People must accept drastic and costly measures as they are necessary to save the environment.
   (E) Lack of concern for the environment leads people to continue their overuse of the automobile.

GO ON TO THE NEXT PAGE

25. If you stop in the movie studio's commissary during lunch time, you may be able to meet the actors. Although the actors always eat elsewhere on workdays when the commissary does not serve fish, they always eat there on workdays when the commissary does serve fish.

If all the statements above are true, and it is true that the actors are eating in the commissary, which of the following must also be true?

(A) It is not a workday, or the commissary is serving fish, or both.
(B) It is a workday, or the commissary is serving fish, or both.
(C) It is not a workday and the commissary is not serving fish.
(D) It is not a workday and the commissary is serving fish.
(E) It is a workday and the commissary is serving fish.

GO ON TO THE NEXT PAGE

Questions 26–29

A new kind of lock is opened by pushing symbols in sequence on a keyboard. The sequence is called a combination. All acceptable combinations must consist of exactly five symbols—four letters and one single-digit number. Acceptable combinations must also conform to the following rules:

> The number must be either the second or third symbol in the combination.
> The fourth and fifth symbols in the combination must not be the same.
> If the third symbol is a number, then the fifth must be either B or D.
> If the third symbol is a letter, then there must be no Fs or Gs in the combination.
> The first symbol must be a letter closer to the beginning of the alphabet than any other symbol in the combination.

26. Which of the following sequences of symbols is an acceptable combination?

(A) E, R, 2, K, B
(B) F, 6, T, T, Y
(C) B, W, 4, G, G
(D) C, 7, M, Q, D
(E) A, X, L, 3, P

27. Which of the following could possibly be the first symbol in an acceptable sequence?

(A) F
(B) 7
(C) Y
(D) 3
(E) E

28. A combination whose first symbol is B and whose fourth symbol is G could have which of the following as its second, third, and fifth symbols, respectively?

(A) J, 6, D
(B) A, 9, T
(C) 9, Z, X
(D) 3, H, G
(E) M, 4, S

29. The combination C, Q, 8, P, F can be made acceptable by doing which of the following?

(A) replacing the F with a B
(B) reversing the C and the P
(C) reversing the Q and the 8
(D) replacing the F with a D
(E) replacing the C with an A

GO ON TO THE NEXT PAGE

30. Some scientists argue that if fish are as common in unfished areas of the oceans as they are in the areas we now fish, current estimates of the amount of protein that our planet supports are far too low. Thus, even if Earth's population continues to grow at its present rate, we can ensure the availability of protein for even the poorest of countries over the next two decades.

Which of the following, if true, would most weaken the argument above?

(A) Some scientists believe that the unfished areas of the ocean support substantially fewer fish per cubic kilometer than do the areas currently fished.

(B) The technology needed to fish new areas of the oceans is more expensive than that now used in ocean fishing.

(C) Increasing the supply of other sources of protein, such as beef and poultry, would be less expensive than fishing new parts of the oceans.

(D) The rate of increase of Earth's population will slowly decline over the next two decades.

(E) It will take at least 30 years to develop the technology necessary for fishing the unfished areas of the ocean.

31. Travelers may enter and remain in the Republic for up to 59 days. If a traveler is to stay for more than seven days, however, a special visa is required.

If the statements above are true, which of the following must also be true?

(A) A traveler who is staying in the Republic for 14 days must have a special visa.

(B) Many travelers who stay in the Republic do not need visas.

(C) Some travelers who stay in the Republic for more than seven days do not have the appropriate visas.

(D) Travelers who stay less than seven days in the Republic do not need visas.

(E) Travelers who merely pass through the Republic while en route to other destinations do not need visas.

32. Despite a steady stream of pessimistic forecasts, our economy continues to grow and prosper. Over the last 15 years the service sector of our economy has greatly expanded. Last year alone, 500,000 Americans found employment in the service sector. In the face of evidence such as this, one cannot argue that our economy is wilting.

Which of the following, if true, would most seriously undermine the conclusion drawn above?

(A) Many Americans who took jobs in the service sector last year were also offered jobs in other sectors of the economy.

(B) Most of the job growth in the service sector can be attributed to people forced out of the declining manufacturing sector.

(C) American society has developed many programs that greatly offset the consequences of a sluggish economy.

(D) Forty years ago the American economy experienced a period of prosperity far greater than that of today.

(E) The importance of the service sector in determining the well-being of the overall American economy has decreased somewhat in the past ten years.

Questions 33–37

There are three bells in a clock tower. One of the bells produces a low-pitched ring, one produces a medium-pitched ring, and one produces a high-pitched ring. The bell ringer must decide on a sequence of eight rings to play on special occasions. He decides that, for the sequence, the low bell must be rung exactly three times, the medium bell must be rung exactly three times, and the high bell must be rung exactly twice. The bell ringer's choice of sequence is further limited by the following rules:

The sixth ring must be that of the medium bell.
The low bell must not be rung twice in succession.
The high bell must be rung twice in succession.

33. Which of the following is an acceptable eight-ring sequence?

(A) medium, low, high, low, high, medium, low, medium

(B) low, high, high, low, medium, medium, low, medium

(C) medium, low, high, high, medium, low, medium, low

(D) medium, high, high, low, low, medium, low, medium

(E) low, medium, low, low, medium, medium, high, high

34. If the high bell is rung fifth in the sequence, all of the following must be true EXCEPT:

(A) The low bell is rung first.
(B) The medium bell is rung second.
(C) The low bell is rung third.
(D) The high bell is rung fourth.
(E) The low bell is rung seventh.

35. If the medium bell is rung fourth, the high bell CANNOT be rung

(A) first
(B) second
(C) third
(D) fifth
(E) eighth

GO ON TO THE NEXT PAGE

36. Which of the following CANNOT be the order of bells rung third, fourth, and fifth, respectively?

    (A) high, medium, low
    (B) low, medium, low
    (C) high, high, low
    (D) high, medium, medium
    (E) high, low, medium

37. Which of the following is IMPOSSIBLE?

    (A) The high bell is rung first.
    (B) The low bell is rung second.
    (C) The medium bell is rung third.
    (D) The high bell is rung fourth.
    (E) The low bell is rung fifth.

Questions 38–42

A large corporation has branches in the following six cities—Atlanta, Beijing, Caracas, Dakar, Edinburgh, and Fresno. Memos of two types, Priority 1 and Priority 2, are sent from the head office to the branches.

Priority 1 memos are sent directly from the head office to either Atlanta or Dakar.

Priority 2 memos are sent directly from the head office to either Atlanta or Beijing.

Any branch that receives a memo directly from the head office must pass it on to at least one other branch. That other branch can pass it on to yet another branch, though it is not required to do so. The passing of memos from branch to branch must conform to the following rules:

Atlanta can send memos of either type to Caracas only.

Beijing can send Priority 1 memos to Edinburgh only and Priority 2 memos to Fresno only.

Caracas can send memos of either type to either Beijing or Dakar.

Dakar can send Priority 1 memos to Caracas only and Priority 2 memos to Edinburgh only.

Edinburgh can send memos of either type to either Fresno or Atlanta.

Fresno cannot send memos to any other branches.

38. A memo that is sent from the home office to Atlanta must be sent on to which of the following?

    (A) Beijing
    (B) Caracas
    (C) Dakar
    (D) Edinburgh
    (E) Fresno

39. A memo that is sent from Edinburgh to Fresno could NOT be which of the following?

    (A) A Priority 1 memo that was initially sent to Atlanta
    (B) A Priority 1 memo that was sent to Edinburgh from Beijing
    (C) A Priority 1 memo that was initially sent to Dakar
    (D) A Priority 2 memo that was sent to Edinburgh from Dakar
    (E) A Priority 2 memo that was initially sent to Beijing

GO ON TO THE NEXT PAGE

40. A Priority 2 memo that was not originally sent to Atlanta could have been seen by a maximum of how many branches?

   (A) two
   (B) three
   (C) four
   (D) five
   (E) six

41. A memo that reaches Edinburgh without having passed through Atlanta must have been seen in a minimum of how many branches besides Edinburgh?

   (A) one
   (B) two
   (C) three
   (D) four
   (E) five

42. Which of the following cannot be the complete progress of a memo from the head office?

   (A) Atlanta to Caracas to Beijing
   (B) Atlanta to Caracas to Beijing to Edinburgh
   (C) Atlanta to Caracas to Dakar to Edinburgh
   (D) Beijing to Edinburgh to Fresno
   (E) Dakar to Caracas to Beijing

Questions 43–47

An athlete has six trophies to place on an empty three-shelf display case. The six trophies are bowling trophies F, G, and H and tennis trophies J, K, and L. The three shelves of the display case are labeled 1 to 3 from top to bottom. Any of the shelves can remain empty. The athlete's placement of trophies must conform to the following conditions:

   J and L cannot be on the same shelf.
   F must be on the shelf immediately above the shelf that L is on.
   No single shelf can hold all three bowling trophies.
   K cannot be on Shelf 2.

43. If G and H are on Shelf 2, which of the following must be true?

   (A) K is on Shelf 1.
   (B) L is on Shelf 2.
   (C) J is on Shelf 3.
   (D) G and J are on the same shelf.
   (E) F and K are on the same shelf.

44. If no tennis trophies are on Shelf 3, which pair of trophies must be on the same shelf?

   (A) F and G
   (B) L and H
   (C) L and G
   (D) K and J
   (E) G and H

GO ON TO THE NEXT PAGE

45. If J is on Shelf 2, which of the following must also be on Shelf 2?

   (A) K
   (B) G
   (C) F
   (D) L
   (E) H

46. If Shelf 1 remains empty, which of the following must be FALSE?

   (A) H and F are on the same shelf.
   (B) There are exactly three trophies on Shelf 2.
   (C) G and H are on the same shelf.
   (D) There are exactly two trophies on Shelf 3.
   (E) G and K are on the same shelf.

47. If L and G are on the same shelf, and if one of the shelves remains empty, which of the following must be true?

   (A) If H is on Shelf 3, then J is on Shelf 2.
   (B) K and L are on the same shelf.
   (C) If H is on Shelf 2, then J is on Shelf 3.
   (D) F and K are on the same shelf.
   (E) If J is on Shelf 2, then H is on Shelf 1.

48. Painting wood furniture requires less time than does finishing the furniture with a stain and polyurethane. On the other hand, a finish of stain and polyurethane lasts much longer than does paint. Yet one further fact in favor of paint is that it costs significantly less than does stain and polyurethane. Therefore, if reducing work time and saving money are more important to people, they will paint their wood furniture rather than finish it with stain and polyurethane.

The argument in the passage above makes which of the following assumptions?

   (A) It is better to paint wood furniture than it is to stain and polyurethane it.
   (B) Most people consider reducing work time and saving money to be more important than the longevity of a finish.
   (C) Most people prefer to paint or to stain and polyurethane wood furniture, rather than to leave the wood unfinished.
   (D) Work time, cost, and longevity are equally important factors in deciding whether to paint wood furniture or stain and polyurethane it.
   (E) Work time, cost, and longevity are the only important differences between painting wood furniture and finishing it with stain and polyurethane.

GO ON TO THE NEXT PAGE

49. Young Cowonga lion cubs in the wild often engage in aggressive play with their siblings. This activity is instigated by the cubs' mother. Cowonga lion cubs born in captivity, however, never engage in this aggressive play. Some zoologists have concluded that this particular form of play teaches the young lions the skills needed for successful hunting in the wild, and that such play is not instigated in captivity because the development of hunting skills is unnecessary in such an environment.

The zoologists' conclusion would be most strengthened if it could be demonstrated that

(A) all Cowonga lion cubs raised in the wild are capable of hunting successfully
(B) other predatory animals also engage in aggressive play at a young age
(C) no Cowonga lion cub that has been raised in captivity is able to hunt successfully in the wild
(D) the skills used in aggressive play are similar to the skills necessary for successful hunting
(E) female lions that were raised in captivity will not instigate aggressive play among their offspring

50. According to a recent school survey, the number of students who regularly attend religious services on campus has increased 50 percent from the figure 10 years ago. It must be an increased religiosity at our college that has massively reduced incidences of cheating on exams during this period.

Which of the following, if true, most significantly weakens the inference above?

(A) Most of the students who now attend campus services do so only for social reasons.
(B) Campus chaplains have time and again spoken about the importance of academic honesty.
(C) Fifteen years ago, the college switched from an honor system to faculty-proctored exams.
(D) Not all students responded to the survey.
(E) Cheating was never a major problem at this school.

END OF SECTION

PRACTICE TEST ANSWER KEY

VERBAL

1. A	14. A	27. B	40. B	53. E	66. A
2. E	15. E	28. C	41. D	54. C	67. B
3. E	16. B	29. B	42. D	55. E	68. A
4. A	17. D	30. C	43. A	56. C	69. C
5. D	18. B	31. A	44. D	57. C	70. E
6. E	19. E	32. D	45. C	58. E	71. C
7. D	20. C	33. D	46. E	59. A	72. A
8. A	21. E	34. C	47. D	60. D	73. A
9. B	22. B	35. D	48. D	61. B	74. B
10. B	23. C	36. B	49. C	62. C	75. C
11. D	24. C	37. C	50. C	63. D	76. C
12. E	25. A	38. B	51. B	64. A	
13. A	26. C	39. E	52. D	65. D	

QUANTITATIVE

1. A	11. D	21. E	31. A	41. C	51. C
2. C	12. C	22. D	32. B	42. B	52. C
3. A	13. B	23. C	33. D	43. D	53. D
4. B	14. A	24. E	34. C	44. A	54. D
5. D	15. D	25. B	35. A	45. B	55. A
6. B	16. B	26. B	36. A	46. E	56. E
7. D	17. D	27. C	37. D	47. D	57. B
8. B	18. B	28. A	38. B	48. E	58. A
9. A	19. E	29. C	39. B	49. B	59. D
10. C	20. D	30. D	40. B	50. E	60. D

ANALYTICAL

1. A	10. E	19. D	28. A	37. C	46. D
2. D	11. B	20. D	29. D	38. B	47. A
3. C	12. C	21. A	30. E	39. E	48. E
4. B	13. B	22. B	31. A	40. A	49. C
5. E	14. C	23. D	32. B	41. C	50. A
6. C	15. A	24. D	33. B	42. D	
7. E	16. E	25. A	34. E	43. B	
8. C	17. E	26. D	35. D	44. D	
9. A	18. C	27. E	36. D	45. C	

CALCULATE YOUR SCORE

Step 1

Add together your total number correct for each section: Analytical, Verbal, and Quantitative. This is your raw score for each measure.

VERBAL

Total Correct [ ] (raw score)

_____

QUANTITATIVE

Total Correct [ ] (raw Score)

_____

ANALYTICAL

Total Correct [ ] (raw score)

KAPLAN

Step 2

Find your raw score on the table below and read across to find your scaled score and your percentile.

Verbal

RAW SCORE	SCALED SCORE	PERCENTILE RANK	RAW SCORE	SCALED SCORE	PERCENTILE RANK
0	200	1	39	420	41
1	200	1	40	420	41
2	200	1	41	430	44
3	200	1	42	440	48
4	200	1	43	450	51
5	200	1	44	460	54
6	200	1	45	470	56
7	200	1	46	470	59
8	200	1	47	480	61
9	200	1	48	490	67
10	200	1	49	510	69
11	200	1	50	520	72
12	200	1	51	530	74
13	220	1	52	530	76
14	230	1	53	540	78
15	240	1	54	550	80
16	250	1	55	560	82
17	260	1	56	570	84
18	270	2	57	580	85
19	270	3	58	590	87
20	280	4	59	590	89
21	290	5	60	600	90
22	300	6	61	610	92
23	310	7	62	620	93
24	320	9	63	630	94
25	320	10	64	640	95
26	330	12	65	650	95
27	330	14	66	660	96
28	340	16	67	670	97
29	350	16	68	680	98
30	360	20	69	690	98
31	360	22	70	700	99
32	360	24	71	710	99
33	370	24	72	720	99
34	380	26	73	740	99
35	390	30	74	760	99
36	400	33	75	780	99
37	410	36	76	800	99
38	420	38			

Quantitative

RAW SCORE	SCALED SCORE	PERCENTILE RANK	RAW SCORE	SCALED SCORE	PERCENTILE RANK
0	200	1	31	430	24
1	200	1	32	440	26
2	200	1	33	460	28
3	200	1	34	470	32
4	200	1	35	480	35
5	200	1	36	490	37
6	200	1	37	500	40
7	200	1	38	510	42
8	200	1	39	520	45
9	200	1	40	530	48
10	210	1	41	540	49
11	220	1	42	540	51
12	240	1	43	560	57
13	260	1	44	570	59
14	270	1	45	580	61
15	280	2	46	600	66
16	290	2	47	610	68
17	300	3	48	630	72
18	310	4	49	640	74
19	310	5	50	660	78
20	320	5	51	670	80
21	330	6	52	690	84
22	340	7	53	690	86
23	360	9	54	700	89
24	370	10	55	720	92
25	380	13	56	730	94
26	390	14	57	750	97
27	400	16	58	770	97
28	410	18	59	780	97
29	420	20	60	800	97
30	420	22			

Analytical

Raw Score	Scaled Score	Percentile Rank	Raw Score	Scaled Score	Percentile Rank
0	200	1	26	530	38
1	210	1	27	550	40
2	230	1	28	560	46
3	230	1	29	580	49
4	250	1	30	580	55
5	260	1	31	590	58
6	280	1	32	600	64
7	300	1	33	620	67
8	310	1	34	640	72
9	330	1	35	650	74
10	330	1	36	660	76
11	340	1	37	670	81
12	360	2	38	680	85
13	370	3	39	690	87
14	390	4	40	700	88
15	400	6	41	710	91
16	420	7	42	720	92
17	420	10	43	730	95
18	430	12	44	740	96
19	450	15	45	750	97
20	460	17	46	760	98
21	470	20	47	770	99
22	490	23	48	780	99
23	500	27	49	790	99
24	500	31	50	800	99
25	520	35			

PRACTICE TEST
EXPLANATIONS

Section One—Verbal

1 **(A)** We're told that the fundamental (blank) between cats and dogs is a myth, that the species actually coexist quite (blankly). We need a contrast, and we find it in (A)—*antipathy* means aversion or dislike, and *amiably* means agreeably.

In (B), if the members of the species coexisted "uneasily," their "disharmony" wouldn't be a myth. In (C), both *compatibility* and *together* imply that dogs and cats are good friends. In (D), it doesn't make sense to say that the "relationship" between dogs and cats is a myth. In (E), no one could claim that there's no "difference" between dogs and cats.

2 **(E)** The clue is the signal *rather than*: We need a contrast between what the speaker intended and what he achieved. The word *monotonous* clues you into boredom, and *bore* in (E), followed by *convince* makes the contrast we need. In (A), *enlighten* and *inform* are similar. *Interest* and *persuade,* (B), don't show contrast. In (C), *provoke* and *influence* don't express a contrast. *Allay* in (D) means to relieve, which is similar to *pacify,* which means to calm or to make peace. No contrast here, and again, it's (E) for this question.

3 **(E)** The blank is part of a cause-and-effect structure as the keyword *that* indicates. Because government restrictions are so something, businesses can operate with nearly complete impunity. There's an absence of restrictions, so we need a word that cancels out restrictions. Would a "traditional" restriction, (A), be canceled out? No. (B), *judicious,* means wise or having sound judgment, but a wise restriction would probably be effective. In (C), *ambiguous* means unclear, but though ambiguity might interfere with the effectiveness of restrictions, it doesn't cancel them out. (D), *exacting,* means very strict, which is the

opposite of what we want. (E), *lax,* means loose, careless, or sloppy. This describes restrictions that aren't very strict, and it's correct for this question.

4 **(A)** The first blank describes a book—the recent Oxford edition of the works of Shakespeare is (blank). The word *because* tells us that what follows is an explanation of why this book is whatever it is. The "not only but also" structure tells us that there are *two* reasons why: it departs from the readings of other editions, and it challenges basic (blanks) of textual criticism. In (A), we could say that challenging conventions could make a book "controversial." Conventions are accepted practices, so challenging conventions would make a book controversial. What else have we got? (B) gives us the book is typical because it challenges innovations. *Typical* doesn't fit in with *departs from other editions.* How about (C)? Challenging norms, which are rules or patterns, wouldn't make something "inadequate." (D)—a book that is different might be called curious, but could you call a book curious for challenging projects? Finally, (E) says the book is pretentious because it challenges explanations—no good. So the best answer is (A).

5 **(D)** We learn that an early form of writing, Linear B, was (blank) in 1952. The keyword *but* tells us that Linear A, an older form, met with a contrasting fate, so we'll look for a pair of contrasting words. The words *no one has yet succeeded in* precede the second blank, so instead of a word that is contrasted with the first blank, we need a word that means about the same thing. That leads us to pick (D)—the words *deciphered* and *interpretation* are similar since both imply understanding.

The word *superseded* in (A) means replaced by something more up to date—not giving an explanation of something. (B)—in the context of ancient languages, a transcription would probably be a decoded version of

something. That would be the opposite of encoding something. (C)'s *obliterated* and *analysis* imply a contrast—wiping something out is different from figuring it out. In (E) *discovered* and *obfuscation* are more at odds than they are alike. Obfuscation means confusion, while a discovery usually sheds light on a situation.

6 **(E)** The clue here is the structure "quite normal and even (blank)"—the missing word has a more positive meaning than the word *normal.* Then we get, "I was therefore surprised," which tips us off to look for contrast. *Commendable* and *complimentary* in (A) are both positive. In (B), *odious* means hateful, so *odious* and *insulting* are both negative. *Conciliatory* in (C) means placating or reconciling, which fits in with *apologetic. Commonplace* and *typical* in (D) mean the same thing. Only correct choice (E) is left—*laudable* means praiseworthy while *derogatory* means belittling or detracting.

7 **(D)** Whatever we're doing to the burden of medical costs is causing the removal of the second blank, signaled by *thereby.* In (A), it doesn't make sense to say that to augment or add to the burden would remove a problem—it could make the problem worse. In (B), a *perquisite* is a reward over and above one's salary. But would eliminating a burden remove a perquisite? In (C), to ameliorate means to improve, but you can't talk about removing a major study of medical care. (D) is perfect. To assuage means to make less severe and an impediment is an obstacle. Assuaging the burden would remove an impediment to medical care, so (D)'s correct. As for (E), to clarify means to explain or make clear, and explaining the burden of medical costs wouldn't remove an explanation.

8 **(A)** A *novel* is a type of *book.* That's an easy bridge. In (A), is an epic a type of poem? Yes, an epic is a long narrative poem, so (A) is right. In (B), a house isn't a type of library. (C) is tempting—

tales and *fables* are related—but a fable is a kind of tale, not vice versa, so it's not parallel. In (D), a number is not a type of page and in (E) a play isn't a type of theater.

9 **(B)** *Ravenous* means extremely hungry—the second word is an extreme version of the first word. In (A), *desirous* means desiring or wanting something—it's not an extreme form of *thirsty.* (B) is perfect—*titanic* is an amplification of *large. Titanic* means gigantic, so (B) is the answer.

Eminent and *famous* in (C) mean the same thing. (D)'s *disoriented* and *dizzy* are close in meaning. To be disoriented means to have lost your bearings, and when you're dizzy, you feel as if you're going to fall down. (E)'s *obese* and *gluttonous* could be related, but don't have to be. *Gluttonous* comes from *gluttony*—it means excessive eating or drinking. Gluttony doesn't have to result in obesity and it's not an extreme form of it.

10 **(B)** A bouquet is an arrangement of flowers, so the first word will be an arrangement of the second word. The first word in (A) is archaic—a humidor is a container for tobacco—a container for tobacco is not the same as a formal arrangement of it. The next choice, (B), is more like it. A mosaic is made of tiles, just as a bouquet is made up of flowers. That's a good match. In (C), a tapestry is not made of color, it's made of threads woven to make a design. You can't argue that a tapestry is an arrangement of colors. (D) also has problems. A pile of blocks could be an arrangement. But a bouquet isn't just a group of flowers—it's a formal arrangement. In the same way, a mosaic is an orderly arrangement of tiles. A pile isn't a formal arrangement. What about *sacristy* and *vestment* in (E)? A sacristy is a room in a church where priests' clothes or vestments are kept, so vestments are stored in a sacristy. The correct answer is (B).

11 **(D)** *Quixotic* means impractical, after the hero of *Don Quixote*. A realist is a person who is especially realistic. *Realistic* is the opposite of *quixotic,* so a realist is never quixotic. In (A), pedantic people show off their learning. Many scholars are pedantic, so this won't work. In (B), a fool is foolish—a synonym for *idiotic.* The same relationship holds true for (C)—an idler is a lethargic person. (D) looks good—a tormentor is vicious or cruel. The opposite sort of person would be kinder and more sympathetic—a tormentor is never sympathetic. (E) *dyspeptic* means suffering from indigestion. A "diner" is someone who eats—some diners get dyspeptic, some don't, so (D)'s correct.

12 **(E)** A shard is a broken fragment of glass or crockery. Glass, when it shatters, creates shards, so a shard is a piece of broken glass. (E) shows the same analogy—a splinter is a piece of broken wood. As for the wrong choices, in (A), a grain is the basic unit that sand comes in, but you can't talk about breaking sand. (B)'s *morsel* means a bit of food, but a meal doesn't shatter into morsels. In (C), a rope is composed of strands and (D) a quilt is made from scraps. The correct answer is (E).

13 **(A)** The word *filter* is used as a verb. When you use a filter, an impurity is removed, so you filter to remove an impurity. The word *expurgate* in (A) means to censor to remove obscenities—you expurgate to remove an obscenity. To whitewash, (B), is to misrepresent a bad thing to make it look better. An infraction isn't removed by whitewashing it, it's only covered up, so (B) isn't parallel. In (C), perjury is the crime of lying under oath. To testify doesn't mean to remove a false statement. In (D) penance is something you do to atone for a sin, but you don't perform to remove penance. And in (E) you don't vacuum to remove a carpet. So (A) is correct.

14 **(A)** *Paraphrase* means restatement of a text using different words. *Verbatim* means word for word or exact. A paraphrase is not verbatim—the words are near opposites. The only choices opposite in meaning are *approximation* and *precise,* in (A). An approximation is an estimate, while something that's precise is exact, so an approximation is not precise. A description might or might not be vivid in (B). In (C), *apt* means appropriate, so a quotation could be apt. There's no relationship in (D), *interpretation* and *valid,* or in (E), *significance* and *uncertain.* (A) is correct.

15 **(E)** Even if you didn't know what *oncology* means, you might have guessed the study of something because of the *ology* ending, and judging from the other word it's probably the study of tumors. The choices look like sciences too. (A)'s pairing of *chronology* and *time* looks okay, but not dead on. There's a science called chronology, the science of arranging time into periods. Chronology is not exactly the study of time—it's a science involved with mapping events in time. Likewise, (B) is almost there. The *theo* in *theology* comes from the Greek word for god, and *theology* means the study of gods or religious beliefs. *Tenet,* on the other hand, means a particular belief or principle. It's too narrow to say theology is the study of tenets. We can eliminate (C) because *aural* is not the study of sound—that would be closer to *acoustics.* In (D), philology is a field that includes the study of literary history, language history, and systems of writing, not the study of religion. Taxonomy, (E), is the study of classification, the correct answer. *Taxonomy* is also used to refer specifically to the classification of organisms.

16 **(B)** *Intransigent* means unyielding—the opposite of *flexible.* Our bridge is "a person who is intransigent is lacking in flexibility." The only

pair that looks good is (B), *disinterested* and *partisanship*. One who's disinterested is unbiased—he doesn't have an interest in either side of a dispute. *Partisan* means partial to a particular party or cause. That's the opposite of disinterested. So partisanship, the quality of being biased, is lacking in a person who could be described as disinterested.

In (A), *transient* means transitory, so you wouldn't say that someone transient lacks mobility. In (C), *dissimilar* means not similar, along the same lines as *variation*. You can't say that something progressive lacks transition, so (D) is no good. The word *ineluctable* in (E) means inescapable, while *modality* is a longer way of saying mode.

Reading Passage: Questions 17–23

The longer of the two reading comp passages appears first. The author's main concern, the aim of science to derive a theory which describes particles and their forces as simply as possible, becomes apparent early in the first paragraph. Simplicity is so important that the author sets it up as a criterion for judging the specific theory of nature. Then the author outlines a recently developed theory which he considers to be a remarkable achievement for its frugality and level of detail. He then asserts that an even simpler theory is conceivable and goes on to mention one that promises at least a partial unification of elementary particles and forces. The last half of the second paragraph and the final paragraph describe this theory in greater detail.

17 **(D)** We need either a choice that describes the similarity between the theories, or one that falsifies information about them. (D) should raise your suspicions. The author acknowledged at the end of the first paragraph that the first theory could account for the entire observed hierarchy of material

structure. (D) is right, but let's look at the others.

(A) is a valid difference between the two theories—the second is presented as a simpler alternative to the first. (B) is also a real difference. The first theory encompass gravitation and the second unifies three of the four forces, which makes it a better theory, but it doesn't account for gravitation. The first theory includes leptons and quarks, while the second combines these two classes into just one, so (C) is valid. In a similar way, the second theory unifies three of the four forces outlined in the first theory, so (E) is valid. Again, it's (D).

18 **(B)** This question asks for the primary purpose, and we know that the author is concerned with theories that describe, simply and precisely, particles and their forces. The author's primary purpose is to describe attempts to develop a simplified theory of nature. Skimming through the choices, (B) looks good. (E) doesn't fit at all. You might say the author summarizes the theories describing matter, but he doesn't summarize all that is known about matter itself. As for (A), the author doesn't cite a misconception in either of the theories he describes. At most, he mentions ways in which the first could be simplified but this doesn't imply that there's a misconception. The author does refer to the second theory as a leading candidate for achieving unification, but predicting its success, (C), is far from his primary purpose. As for (D), although it's implied that scientists in general do prefer simpler theories, their reasons for this preference are never discussed. Again, it's (B) for Question 18.

19 **(E)** This question is a scattered detail question concerning quarks. In the first paragraph we're told that quarks are constituents of the proton and the neutron. It's reasonable, then, to say that quarks are the elementary building blocks of

protons and neutrons, option I. Since option I is correct, we can eliminate choices which exclude it, (B) and (D). The remaining choices are either I only, I and II only, or I, II, and III. You could skip II and go to III. If you're sure III is right, you can assume that II is also and pick (E). It turns out that III can be easily checked at the end of paragraph three, where the author states that a new theory incorporates the leptons and quarks into a single family or class, so option III is correct. For a complete list, let's look at option II. In the very first sentence the author tells us that elementary particles don't have an internal structure and since quarks are elementary particles, option II is indeed correct, and (E) is our answer.

20 (C) It should be clear that the author has some very definite criteria for judging the usefulness or worth of various theories of nature. As for option I, *simplicity* should leap off the page at you—it's what this passage is all about. We can eliminate (B) and (E). The author also takes the theory's completeness into consideration. He commends the first theory he describes because it accounts for the entire observed hierarchy of material structure and therefore option II is correct. We know that (C) must be correct because there is no I, II, and III choice. But let's look at III anyway. Does the author ever mention proving either of those two theories he describes? Proof is of no concern to him—there's no mention in the passage of any experiments, or of wanting to find experimental proof. So III is out and (C) it is.

21 (E) We've mentioned that the second theory doesn't include gravitation in its attempt to unify the four basic forces. We need the author's opinion about this omission. The author introduces the theory in the second paragraph, describing it as an ambitious theory that promises at least a partial unification of elementary particles and forces. The failure to include gravitation and achieve complete unifica-

tion doesn't dampen the author's enthusiasm and he seems to suggest that gravitation's omission can't be helped, at least at this stage. So, although the omission is a limitation—it prevents total unification—it is also unavoidable. It looks like (E) does the trick.

You could see the limitation as a defect, (A), but the author never gives the impression that the omission of gravitation disqualifies the theory. As for (B), *deviation* is a funny word—deviation from what? More important, we've already seen that the author doesn't consider the omission to be unjustified. For the same reason, (C) can be eliminated. If the omission of gravitation can't be avoided, then it certainly isn't a needless oversimplification. Finally (D) is out because there's no way that gravitation's omission could be an oversight. A scientist just forgot about one of the four basic forces when developing a theory of nature? No, the idea is that, for now at least, gravitation just can't be fit in, and (E) is correct.

22 (B) The passage begins with the author's discussion of the simplicity of elementary particles and the theories which describe them. In the third sentence, the author sets forth simplicity as a standard for judging theories of nature. In the rest of the passage, the author measures two specific theories against this standard. (B) summarizes this setup nicely and it's our answer. (A) is way off base. Although the author might be said to enumerate distinctions between how the two theories treat elementary particles, he doesn't enumerate distinctions among the particles. (C) is easy to eliminate—the author describes only two methods of grouping particles and forces—not three. As for (D), the author doesn't criticize the first theory he describes or call it inaccurate—he commends it. Finally, (E) goes overboard. As we mentioned in our discussion of option III in Question 20, the author is not interested in scientific verification. Nothing is ever mentioned about proving or verifying either of the theories he describes. Again, (B) is correct.

23 (C) This question shouldn't be difficult. It asks us to put ourselves in the author's shoes and figure out what sort of theory he would find superior to present theories. We already know—a simpler theory. The author's criteria for judging a theory are its simplicity and its ability to account for the largest possible number of known phenomena. Which choice represents a theory with one or both of these characteristics? (A) misrepresents the two theories described in the passage. The author says that the first theory could account for the entire observed hierarchy of material structure. The second does also, even though gravitation must be thrown in as a separate force. A theory that could account for a larger number of structures isn't what's needed.

As for (B), why would the author approve of a theory that reduces the four basic forces to two which are incompatible? (C) is on the right track. The author would prefer a theory that accounts for all matter with the fewest particles and forces and this is offered by (C), the correct answer. (D) is out because it wouldn't represent an improvement on currently existing theories. They account for gravitation, although they haven't yet unified it with the other three forces. Finally, (E) represents a step backwards. The current theories hypothesize that both protons and neutrons are formed by combinations of elementary particles. Again, it's (C).

Reading Passage: Questions 24–27

The second passage is short but dense, and the author doesn't arrive at her main point until the last sentence. We see that the author sets herself in opposition to presentist historians, people who believe that white abolitionists and suffragists comprise the abiding principles of equality and the equal right of all to life, liberty, and the pursuit of happiness.

Their evidence is presented in the first three sentences. First, a majority of both groups tried to assure people that the changes they advocated weren't revolutionary and served to support rather than to undermine the status quo. A certain group of abolitionists disclaimed miscegenationist intentions—they were careful to assert that their interest in obtaining freedom for blacks didn't mean they were advocating mixing of races. And finally, suffragists saw no conflict between racism or nativism and their movement's objectives. Presentist historians apparently think that, by denying any revolutionary intentions and miscegenationist intentions, and by justifying nativism and racism, both groups were undermining their own principles. And, because their objectives— the abolition of slavery and voting rights for all—go hand in hand with our present conception of equality, presentist historians think that both groups undermine the principle of equality at the same time. The author uses the same evidence to argue that the actions of both groups served not to show how far these groups deviated from a fixed principle of equality, but to show what the principle meant in their own generations. The author thinks that the principle of equality is not unchanging, but means different things for different generations and that presentist historians err when they judge these movements by our conception of equality.

24 (C) We need the author's main point, which we just formulated—the actions of abolitionists and suffragists demonstrate the meaning that equality had in their time. (C) expresses this, and it's the correct answer. (A) is wrong because it's the presentist historians who believe that the actions of the abolitionists and suffragists compromised their principles. (B) has nothing to do with the author's discussion. A comparison of beliefs never occurs. As for (D), the author charges presentist historians with misinterpreting abolitionist and suffragist ideology,

not with willfully misrepresenting it. Finally, (E) constitutes a criticism the author makes about presentist historians—that they impose their own value systems on the past, rather than interpreting actions in the appropriate historical context. Again, it's (C) for this question.

25 (A) We can infer something about the author's concept of the principle of equality—it's clear that the author thinks the principle of equality is not abiding. Rather, she thinks, it encompasses different things for people at different times. We can give the nod to option I, which eliminates (B), (C), and (E). Since the only choices left include option I only or options I and II only, option III can be eliminated. Option II—does the author suggest that the suffragists applied the principle of equality more consistently than abolitionists? No, if anything, she implies that they applied it equally consistently. We're left with (A) as our answer. We know option III can't be true—presentist historians say that abolitionists and suffragists compromised the principle of equality, not the author, who thinks their actions conform to their generation's conception of equality.

26 (C) This question deals with the logical structure of the author's argument, how she argues her case against the presentist historians. She uses the same evidence to support her views that they do, cites the actions of the suffragists and abolitionists, states that the presentist historians knew of these actions, then presents her own interpretation of these same actions. She's applying a different interpretation to the same set of facts, and (C) is our answer. The author doesn't cite any new evidence, so both (A) and (B) can be ruled out. As for (D), the author refutes not the accuracy of the historians' data but the accuracy of their interpretation. Finally, the author doesn't claim that the historians' argument is flawed by a

logical contradiction, (E). She claims instead that they erred by assuming that equality is an abiding value and by measuring the actions of past groups against this concept of equality. Again, it's (C) for this question.

27 (B) We need to know what the author suggests about the abolitionist movement. Well, in her references to this movement, the author mentions the non-Garrisonian abolitionists. If there were non-Garrisonian abolitionists, it seems reasonable to assume that Garrisonian abolitionists existed. Also, the author refers to a majority of white abolitionists who made certain denials. This implies that there was a minority of abolitionists who didn't make such denials and also that there were black abolitionists. In other words, the abolitionist movement was subdivided into different groups and these groups didn't always share identical ideologies. This corresponds closely to (B), the correct answer. As for (A), the passage does state that some abolitionists denied that they had revolutionary or miscegenationist intentions, but these denials don't seem to be an attempt to disguise their real intentions. (C) is wrong because the author thinks the abolitionists did live by their principles. As for (D), presentist historians might claim that abolitionists undermined their objectives by making certain disclaimers to the public. But even they wouldn't say that these disclaimers were the result of abolitionists misunderstanding their objectives. Finally, the passage makes no mention of radical factions within the abolitionist movement and the effects of abolitionists' actions on their movement's progress is never discussed, so (E) is out. Again, (B) is correct.

28 (C) Our first word is *undermine,* which means to weaken or cause to collapse, especially by secret means. The opposite would be something like *build up* or make *stronger.* The best choice here is (C), *bolster,* meaning to support. The only

other tempting choice was (A), *appreciate*, but a better opposite for appreciate would be *resent*.

29 **(B)** *Obsequious* means servile or submissive. The opposite of *obsequious* would be something like *snooty* or *arrogant*. *Haughty*, (B), fits perfectly. A haughty person is overly arrogant while an obsequious person is overly eager to please. None of the other choices comes close.

30 **(C)** The word *blanch* may be familiar to you if you cook. Foods like broccoli are blanched by plunging them in boiling water so they lose color. In the same way, a person might blanch from fear, shock, or dismay. Since *blanch* means to whiten or turn pale, the opposite would be to redden or blush. (C), *flush* is what we need. None of the other answer choices are particularly colorful.

31 **(A)** The word *dissipated* is a pejorative reference to someone devoted to the pursuit of pleasure—the opposite of dissipation is restraint or moderation. (A) is correct because *temperate* means moderate or self-restrained. None of the other answer choices have to do with moderation. *Inundated* means overwhelmed or deluged.

32 **(D)** *Fecundity* means fertility, the capacity for producing life, whether it be children or vegetation. Clearly the opposite would be (D), *sterility*, which refers to an inability to reproduce. None of the other choices comes close, and the only unusual word is (A)'s *levity*, which means silliness or frivolity.

33 **(D)** In Question 33, *encumber* means to block or weigh down. A good synonym would be *oppress*. The best opposite is (D), *disburden*, which means to free from oppression. *Animate* (A), means to make alive—its opposite would be something like deaden. To inaugurate, (B), is to begin or commence. To bleach is to pale or whiten and to obliterate means to erase or remove.

34 **(C)** The word *disseminate* isn't easy to figure out if you don't know it—it means to spread widely. Ideas, theories, and beliefs can all be disseminated. The opposite of spreading an idea is suppressing it, (C). None of the other choices works.

35 **(D)** *Restive* looks like the word rested, but the two don't mean the same thing at all. Restive can mean stubborn or restless. A mule that won't move is restive, as is a fidgety child. We need something like *obedient*, *quiet*, or *settled*, and it's *patient*, (D). *Morose* in (A) means gloomy. *Intangible* means untouchable or elusive. *Fatigued* means tired and the opposite of *curious* would be *indifferent*.

36 **(B)** If you didn't know what *syncopated* means, you might have guessed it had something to do with rhythm from the expression *out of sync*. That would lead you to (B), *normally accented*. *Syncopation* refers to a pattern or rhythm in which stress is shifted onto normally unaccented beats.

The opposite of (A)'s *carefully executed* would be *haphazard*, and (C)'s *brightly illuminated* is the opposite of *dim*. *Obscure* would be an antonym for (D)'s *easily understood*. *Justly represented*, in (E) isn't easy to match, but even if you couldn't eliminate all the choices, you could have at least narrowed the field.

37 **(C)** *Vituperative* means verbally abusive. The opposite of defaming someone with vituperative remarks would be praising them—(C)'s *laudatory* means expressing praise. As for the other choices, *lethal* means deadly and *incapacitated* means incapable or unfit. In (D), *insulated* means protected, as in *insulation*, and *prominent*, (E), means famous.

KAPLAN

38 **(B)** *Saturnine* is probably the hardest word in the section. It means heavy, gloomy, sluggish, so its opposite is cheerful or lively. The answer is (B), *ebullient* which means bubbling with enthusiasm or high spirited. (A)'s *magnanimous* means generous or high minded. *Finicky,* (C), means fussy or picky. The opposite of (D), *unnatural* is natural and (E), *impoverished* means poor.

39 **(E)** *Callous* means unfeeling, uncaring, but if this person has concern for the earthquake victims, her reputation must be an unfounded one, so the correct choice will mean *contradicted* or *proved false.* This is one of the meanings of *belied,* correct choice (E). (B), *rescinded,* is the second best answer. It means revoked or withdrawn, but you don't say that a reputation is rescinded. (A), (C), and (D) are the opposite of what we're looking for—they don't make sense in this context.

40 **(B)** *No longer* and *therefore* show strong contrast—something is done with the original plans because they are no longer something else. (B) expresses this contrast, *applicable . . . rejected,* and if we plug in these words, the plans could no longer be applied so they were tossed aside. In (A), there's no contrast between something being relevant, or pertinent, and its being adaptable, capable of being changed to fit a new situation. In (C), *expedient* means convenient—it makes no sense for something not expedient to be adopted or taken up. In (D), *appraised* means judged or rated, which doesn't follow from no longer being acceptable. In (E) it doesn't make sense to say that the plans were no longer capable or that the plans were allayed, or minimized—again, (B) is the best choice.

41 **(D)** The second half of the sentence is about each tiny layer of the surface of the cross-section of the sandstone. This must explain what the first part alludes to, so the first blank must mean *layered*—otherwise, what tiny layers is the author talking about? On this basis, (D) is the best answer since *stratified* means layered. In (A), a ridge isn't really a layer. In (B), a facet is a face or flat surface, so *multifaceted* can't be right. *Distinctive,* in (C) means distinguishing or individual. And *coarse* in (E) means rough. Looking at the second blank, *enlargement,* in (A), has nothing to do with the formation of the stone. In (B), if the phrase *angle of deposition* means anything at all, it's an obscure geological term and can't be what we want here. The remaining choices could refer to the time or place in which material is deposited. Since (D) has the best answer for the first blank and a possible answer for the second blank, it's correct.

42 **(D)** The phrase *and now* suggests that the second part of the sentence will say something consistent with the first part. Whatever the convict has always insisted upon, the new evidence must support his claim. (D) gets this connection right—*innocence . . vindicate.* To vindicate means to clear from an accusation, to prove innocent. The convict has always insisted upon his own innocence and now at last there is new evidence to vindicate him—this makes perfect sense and it's the answer. In (A), *defensiveness* means a tendency to defend oneself and *incarcerate* means to put in prison. In (B), *culpability* is guilt, as in the word *culprit,* and *exonerate* means to clear from guilt. In (C), to anathematize someone means to curse him or pronounce a strong sentence against him but that doesn't go with *blamelessness.* In (E) *contrition* is a sense of remorse, while to condemn someone means to pass judgment against him. This is probably second best, but it doesn't follow as logically as (D), so (D) is correct.

43 **(A)** The word *but* signals a contrast between the opinion of plate tectonics when the theory was first proposed, and the opinion of it now—either people disbelieved the theory at first and believe it now or vice versa. (A), *opposition . . grant* provides the contrast. If most geophysicists now grant its validity, they believe in it. That's the opposite of opposing it, so (A) is the answer. In (B), *consideration* is a neutral term—people are thinking about the theory, but it doesn't provide the necessary contrasts with *see,* which implies that physicists now recognize the validity of the theory. In (C), *acclamation* means loud praise and *boost* means to support enthusiastically—no contrast there. In (D), a *prognostication* is a prediction of the future, which doesn't make sense in this context and *learn its validity* doesn't make sense either, so (D) isn't a good choice. In (E), *contention* is argument and to bar means to exclude or forbid—there is no contrast with this pair. Again, (A) is the correct answer.

44 **(D)** *Despite* clues you in to a contrast between something professed, claimed or pretended, and reality, indicated by the glint in her eyes. A glint in someone's eye is a sign of strong interest, so *obsession* and *fascination,* in (A) and (D) are tempting. We want a contrast with strong interest, so the first word is something like disinterest. We find *indifference* in (D) and *obliviousness* in (C). Since both words in (D) fit, it must be correct. None of the others offers the kind of contrasts we need. There's no contrast between *intelligence* and *obsession,* in (A), between *interest* and *concern* in (B), or between *obliviousness* and *confusion* in (C). We get a contrast in (E) between *expertise* and *unfamiliarity,* but the words don't make sense—a glint in someone's eye isn't a sign of unfamiliarity.

45 **(C)** We're looking for something that goes with sacred scriptures and implies a formal system of belief, but something whose absence doesn't rule out a legacy of traditional religious practices and basic values. We can eliminate choices (A), (B), and (E) because if Shinto lacked followers, customs, or faith it wouldn't be a legacy of traditional religious practices and basic values. Relics, (D), are sacred objects but relics don't make something a formal system of beliefs. The best choice is (C)—a dogma is a formal religious belief.

46 **(E)** Something impeccable is perfect, it doesn't have a flaw. In (A) *impeachable* means subject to accusation, so something impeachable is not necessarily without crime. *Obstreperous,* in (B) means loud or unruly, not without permission. *Impetuous,* in (C) means rash or without care, rather than without warning. In (D), *moribund* means in the process of dying, so it's inappropriate to use *living.* In (E), *absurd* means without sense, so this is the correct answer.

47 **(D)** A seismograph is an instrument used to measure an earthquake, so we need another instrument used to measure something. In (A), a stethoscope is an instrument used to listen to a patient's chest. Only indirectly can this be used to measure a patient's health. In (B), a speedometer doesn't measure a truck—it measures the speed of any kind of vehicle. In (C), a telescope doesn't measure astronomy. A telescope is an instrument used to observe far away objects. In (D), a thermometer measures temperature, so this looks like a promising answer. In (E), an abacus is used in arithmetic as a calculator but it doesn't measure arithmetic. So (D) is the best answer.

48 **(D)** To guzzle is to drink very quickly, taking big gulps, so the relationship is one of speed or degree. In (A), *elucidate* and *clarify* mean to make clearer. One doesn't imply greater speed or volume

than the other. Similarly, with (B) to ingest is to eat or drink—it doesn't mean to eat in big bites. In (C), to boast and to describe are two unrelated ways of talking. In (D), to stride is to walk quickly, taking big steps, so this may be the answer. In (E), *condemn* is stronger than the first word *admonish,* meaning to rebuke—the opposite of how the stem pair is presented. So (D) is the best answer.

49 **(C)** An orator is a public speaker and *articulate* means able to express oneself well. You can form the bridge, "A successful orator is one who is articulate." With that in mind, (A) may seem tempting but the profession of soldier isn't defined as aspiring towards being merciless. In (B) a celebrity is a famous person, not by definition a talented one. (C) is good—a good judge has to be unbiased. It's safe to say that a biased judge is a bad judge in the same way that an inarticulate orator is a bad orator. In (D), a novice is a beginner—it wouldn't be unusual for a novice to be unfamiliar but that's not what makes a good novice. In (E), a dignitary is a person of high rank, and such a person doesn't need to be respectful. (C) is correct.

50 **(C)** A badge is the identification worn by a policeman. In (A), a placard is a sign carried by a demonstrator. There's a link here but a placard isn't an official ID and a demonstrator doesn't necessarily carry a placard. (B) is wrong because although there is a tradition for a sailor to have a tattoo, a tattoo isn't an official identification of a sailor. In (C) a convict wears a number on his uniform to identify him, so this is plausible. In (D) the pedigree of a dog is the dog's lineage or genealogy, not something worn by the dog as identification. In (E), even though a fingerprint may be used to identify a defendant, everybody has fingerprints. So the best answer is (C).

51 **(B)** To scrutinize means to observe intently, so the relationship is one of degree. In (A), to pique interest is to excite interest. The words mean the same thing. In (B) to beseech means to request with great fervor—this is more like it. In (C) to search is the process you go through to discover something. That's different from the stem pair. In (D) to grin is to smile broadly—this reverses the original pair. And in (E) to dive means to jump in a certain way or under certain conditions, not to jump intently. The best answer is (B).

52 **(D)** If you didn't know what *epicurean* means, you might have had trouble here, but you can still eliminate some choices. There must be some relationship between *epicurean* and *indulge.* Could (A) have the same relationship? No, because there really is no relationship between *frightened* and *ugly.* Something ugly doesn't necessarily frighten people. Same with (C)—there's no relationship between *hesitate* and *unproductive.* There are good relationships for the other choices but let's see if we can eliminate them. In (E) the relationship is that something comprehensible can be understood. Do you think that something epicurean can be indulged? That sounds odd—just about everyone and everything can be indulged.

In (B) *revocable* means something can be taken back, so the relationship is, "Something revocable can be retracted." That's the same relationship that we just saw in (E), another clue that they must be wrong. If (B) and (E) share the same relationship, they can't both be right, so they must both be wrong. That leaves us with (D) and there our relationship is something like someone vindictive is likely to revenge himself and that sounds better. In fact, an epicurean person is one who is likely to indulge himself, so (D) is correct here.

53 **(E)** *Diluvial* means having to do with a flood. You may have heard the word *antidiluvian,* meaning before the flood, Noah's flood, that is—in other words, a long time ago. So our bridge is "having to do with." In (A), *criminal* can mean having to do with crime but it doesn't mean having to do with punishment. In (B), *biological* means having to do with living things. Bacteria are living things but to define *biological* as having to do with bacteria would be too narrow. In (C), *judicial* means having to do with the administration of justice. A verdict is the decision about the guilt or innocence of a defendant, a small part of the judicial process. (D)'s *candescent* means giving off light rather than having to do with light. This leaves (E) and *cardiac* means having to do with the heart, so (E) is correct.

54 **(C)** It is in the nature of a sphinx to perplex. This comes from Greek mythology—the sphinx was a monster that asked a riddle that no one could answer. *Sphinx* can be used to mean anything that is difficult to understand, so our bridge is: "A sphinx is known for perplexing." In (A), an oracle is a soothsayer, someone who predicts the future—an oracle doesn't interpret. In (B), a prophet is someone who foretells the future. This may help someone to prepare but you don't say that a prophet is known for preparing. In (C), a siren can be a beautiful or a seductive woman who lures men. So (C) looks good—a siren lures in the same way that a sphinx perplexes. In (D), the role of a jester is to amuse, not necessarily to astound. In (E), a minotaur is a mythological monster—it didn't, by definition, anger someone. So (C) is correct.

Reading Passage: Questions 55–57

This reading comp passage is short and it's followed by three questions—the remaining passage will be long with eight questions. The style of this natural science passage is factual, descriptive and straightforward, although the discussion does get fairly detailed. The topic is clear from the first sentence: our knowledge of how fish schools are formed and how their structure is maintained. The next two sentences get more specific and express the author's main point—that, contrary to the previous theory, the structure of fish schools is not primarily dependent on vision.

The tone is objective, but it's worth noting that since the author is contrasting the new knowledge about lateral lines with older, outdated knowledge, he must be skeptical of the notion that vision is the primary means of forming and maintaining fish schools. The rest of the passage is a more technical report of how the schools are structured, how individual fish actually behave in forming schools—this is detail and the best way to deal with it is to read it attentively but more quickly than the earlier lines.

55 **(E)** This Roman numeral-format question focuses on detail. The stem is asking what the structure of fish schools depends on, and the options focus on the more technical elements in the last half of the passage. The author states that ideal positions of individual fish aren't maintained rigidly and this contradicts option I right away. The idea of random aggregation appears: the school formation results from a probabilistic arrangement that appears like a random aggregation, so the idea is that fish are positioned probabilistically, but not rigidly. Option II is true, repeating the idea in the next sentence that fish school structure is maintained by the preference of fish to have a certain distance from their neighbors. Option III is true, too. It's a paraphrase of the last two sentences, that each fish uses its vision and lateral line first to measure the speed of the other fish, then to adjust its own speed to conform, based primarily on the position and movements of other fish. So options II and III are true and (E) is the right choice.

56 **(C)** You know the primary purpose here is to present new ideas that challenge the emphasis of the old theory. So you're probably safe in assuming that the author's attitude toward the old idea will be at least somewhat negative. You can therefore cross off choices that sound neutral or positive, (B), (D), and (E). The negative choices are (A) and (C). (A) is out because it is much too extreme—the author is not offended or indignant, nor does he or she argue that vision is insignificant—quite the contrary. This leaves (C), the best choice. The author disagrees with the old theory since it overlooks the role of the lateral line, but the disagreement is tempered by an acknowledgment that the old theory did recognize the role of vision. So it's a qualified or measured disagreement—the adjective *considered* works well here. Again, the correct answer is (C).

57 **(C)** This question involves inference as the word *suggests* in the stem suggests. It refers to the latter, more detailed half of the passage, and that's where correct answer (C) is. It's logically suggested by the last couple of sentences where you're told that once it establishes its position, each fish uses its eyes and lateral line to measure the movements of nearby fish in order to maintain appropriate speed and position. Since the school is moving, each fish's adjustments must be ongoing and continuous, as (C) states. (A) is wrong because auditory organs aren't mentioned. Lateral lines correspond to a sense of touch, not hearing. (B) and (D) both have words that should strike you as improbable. Nothing suggests that each fish rigorously avoids any disruptive movements, (B), or that the fish would make sudden unexpected movements only in the presence of danger, (D). The idea in (E) also isn't mentioned. It's never suggested that a fish, once part of a school, completely loses its ability to act on its own. Again, (C) is our answer for Question 19.

Reading Passage: Questions 58–65

This passage is divided into three paragraphs. If you figure out what each paragraph covers, you've understood the passage's handful of ideas, plus you've sketched out a rough mental map. In this passage, the first 10 or 15 lines take you through the first paragraph and into the second and if you were careful you picked up the author's broad topic area (ancient Greek social anxiety), the style of the writing (dense and scholarly) and the tone or attitude (expository and neutral).

The second paragraph gives you the central point—what the Greeks apparently succeeded in doing was discovering a way of measuring and explaining chaotic experience so that chaos was no longer so threatening and anxiety producing. This recognition of order in the midst of chaos served as the basis of a spiritual ideal for the Greeks. So by the end of the second paragraph you have the author's central idea plus all the information about style, tone, and topics in the beginning. The first sentence of the last paragraph tells you the search for order and clarity in the midst of chaos is reflected especially in Greek philosophy. The rest of the paragraph is a description of how various philosophers and schools of philosophy offered solutions to the problem of finding order and measure in a disorderly world.

58 **(E)** This kind of primary purpose question is common, and here the right answer is (E). In this case, both the noun and the verb are right on the money. The verb is exactly right for this author's expository neutral tone, and a cultural phenomenon, the Greeks' perception of chaos and their solution to the problem, is what the author is describing. The verbs in (B) and (C), *challenge* and *question* eliminate them right away—no opinion is given but the author's own, and philosophy in (C) is discussed only

in the last paragraph. The noun phrase in (A), *conflicting viewpoints* is wrong. (D) is the most tempting—the author is looking at history and mentioning certain facts, but this misses the author's purpose, which is not to simply list facts but rather to describe and define something in the form of a thesis. Again, (E) is the correct answer.

59 **(A)** This is from the first sentence of the second paragraph and it's the central idea that's being focused on, that the discovery of this substratum helped bring a satisfying new sense of order into experience, thus reforming the Greeks' perception of worldly chaos. The choice that paraphrases this point is (A), the perception of constant change was altered by the idea of a permanent principle of order lying underneath it—this is the main point of the passage. (B) is out because severe social problems are never mentioned, at least not in any concrete way. As for (C), it misses the point made in the sentence the question refers to. The passage does refer to pain and bewilderment and to an earlier period of political turbulence, but this choice goes overboard with its notions of painful memories and national humiliation and so on. As for (D), a few lines into the second paragraph the author says directly that the discovery did much more than satisfy intellectual curiosity. And (E) also contradicts the author, distorting a detail at the end of the paragraph. It's not mysticism, but rationality and careful analysis that lead to order and clarity, so it's (A) for this question.

60 **(D)** The author is arguing in the second, third, and fourth sentences that the Greeks identified rational thought and spiritual ideals as inseparable. Rationality, order, measure, and so forth became equivalent to spiritual ideals for the Greeks. Toward the end of the second paragraph the author states that rationality and spirituality are not mutually exclusive. The choice that's most clearly consistent

with this is (D). As for (A), the passage never suggests that ordinary Greeks were unfamiliar with or uninterested in the concepts of rational thought and spiritual ideals. The passage suggests quite the contrary. (B) and (C) are both inconsistent with the passage as well. All the philosophers mentioned accepted the notion that rationality was the key, amounting to an ideal to understanding the world. (E) picks up on the mention of poetry at the beginning of the last paragraph, but the point there is that Greek poetry manifested the sense of cultural anxiety that philosophy tried to alleviate.

61 **(B)** This question is looking for the choice that isn't mentioned as reflecting the Greeks' anxieties about chaos. The one that's never mentioned is (B), that it was reflected in aspects of their religion. We don't actually learn anything about Greek religion in the passage—we just don't know and we certainly can't infer anything about specific aspects. Each of the other choices is mentioned specifically. (A) is implied in the long opening sentence of the first paragraph—the national psyche and historical experience both relate to national consciousness. (C), the sense of change in the physical world, is mentioned at the start of paragraph two. (D), the striving for order and philosophy, is discussed throughout the third paragraph. And finally (E), lyric poetry, is mentioned at the start of the paragraph as one place where the sense of anxiety was expressed directly.

62 **(C)** Your mental map should have taken you straight to the last paragraph—the Milesians are discussed in the first several sentences. (C) encapsulates what the passage says, that Milesians were interested primarily in understanding a fundamental order in nature, outside the disturbing world of human society. (A) gets it backwards—the Milesians apparently ignored questions that were inherently

human. (B) and (D) contradict the passage. None of the philosophies mentioned did what these choices suggest, either to sharply distinguish between rationality and spirituality, (B), or to integrate rationality and mysticism, (D). (E), finally, describes the approach of the Pythagoreans who were absorbed by the logic and order of mathematics, rather than by attempts to explain physical phenomena.

63 **(D)** You're being asked not the actual content, but the logical progression of the contents. Is he or she making a series of disconnected assertions? Making a point and backing it up with factual evidence, or what? What's the author up to logically in the lines referred to? In the preceding sentence the author is talking about the Greeks' discovery of order and measure, and that it helped them get a secure handle on chaotic experience. The discovery was a relief—its impact was almost religious in nature. In the next sentence, the author says that this recognition, discovery, of order and measure was much more than merely intellectually satisfying—it served as a basic part of their spiritual values. The author quotes Plato to support his point, to give an idea of the significance of measure.

In the last of the three sentences the author finishes up with a statement that pulls the strands of the thesis together and puts the basic point into clear cultural perspective. Rational definability or measure was never regarded by the Greeks as inconsistent with spirituality—(D) is the choice that describes things best. The problem with (A) is that the author isn't summarizing two viewpoints but discussing one thesis. As for (B), the author neither mentions evidence that weakens his thesis nor revises it. (C) is out because the author is not discussing two separate arguments that need to be reconciled by a third. It's just one argument that's the topic here. (E), finally, is wrong for the same reason—the author discusses one thesis only and never suggests any other.

64 **(A)** We know from our rough map of the structure that except for one reference to Plato in the middle of the second paragraph, philosophy is discussed only in the last paragraph, so that's where you'll find out about the Pythagoreans. The main thing about them was that they concentrated on mathematical abstraction. They shifted the focus in philosophy from the physical realm to the mathematical. The Milesians focused on physical phenomena, and that's the idea you see immediately in (A), the correct choice. (B) lists an idea that mentions the Pythagoreans—thinkers who came *after* the Pythagoreans focused on human behavior. (C) won't work because both of these schools and all other philosophies mentioned used rationality as the means to truth. (D) picks up what characterized the Milesians—we want the Pythagorean side of the contrast. (E) gets things backward—the Pythagoreans stressed mathematical theory over physical matter. Again, it's (A).

65 **(D)** The last sentence is saying that in all these various periods of Greek history and philosophy, the basic preoccupation of the Greeks was with the search for a kosmos. The term *kosmos* hasn't been used before, but because this sentence is at the end of the passage and because it's phrased as a summary, you should realize that the basic quest here must be the same one the author has been talking about all along. So this refers to the central problem for Greek society—how to find order and measure in a seemingly confusing and disorderly world. This search for a kosmos then is the passage's main idea, and correct choice (D) restates it.

(A) is out because the word *mystical* is incorrect, since the author states at the end of paragraph two that the Greeks stressed rationalism over mysticism. (B) and (E) are inconsistent with some major points. In (B), the idea that the Greeks would have regarded rationalism as sterile is completely wrong. And in (E), the ideals of

beauty and excellence, as mentioned in paragraph two, are preeminent and fundamental within the Greeks' world view. Finally, (C) talks about ending conflict among important schools of philosophy. This last sentence about the search for kosmos is talking about a quest you find in Greek thought as a whole, a much bigger topic than mere conflicts among philosophers.

66 **(A)** Enmity is the state of being an enemy—the opposite is friendship, (A). Reverence, (B), is great respect, the opposite of contempt. The opposite of boredom, (C), is interest. The opposite of (D), stylishness, is a lack of style, and the opposite of (E), awkwardness, is skillfulness.

67 **(B)** *Dilate* means expand and widen. The opposite is the word *contract*, so (B), shrink, is what we're looking for. (A), *enclose*, means to confine. The opposite of the word *hurry*, (C), is *delay*. *Inflate*, (D), means to expand or fill with air. The opposite of (E), *erase*, might be *preserve* or *set down*.

68 **(A)** A charlatan is a fraud or a quack. (A), genuine expert, is a possible answer. The opposite of a powerful leader, (B), is a follower or maybe a weak leader. The opposite of a false idol, (C), is a true god or a hero. The opposite of an unknown enemy, (D), is a known enemy, an unknown friend or a known friend. The opposite of (E), hardened villain, might be an innocent person or first offender. So it's (A).

69 **(C)** *Peripheral* means having to do with the periphery, the outer edge of something. The opposite of *peripheral* is *central*, (C). The opposite of (A), *civilized*, is *crude* or *savage*. (B), *partial*, means favoring or biased, or incomplete—it has lots of opposites but *peripheral* isn't one of them. *Harmed* is the opposite of *unharmed*, and the opposite of (E), *stable*, is *weak* or *inconstant*. (C) is correct.

70 **(E)** *Meritorious* means full of merit, deserving reward. Its opposite is *unpraiseworthy*, (E), the best choice. *Effulgent*, (A), means shiny—its opposite is *dull*. (B), *stationary*, means not moving. Neither (C), *uneven*, nor (D), *narrow-minded*, works, so (E) is correct.

71 **(C)** *Discharge* means to unburden, eject, or exude. However, it has a more specific meaning in military context: to release or remove someone from service. The opposite is to *enlist*, (C). The opposite of (A), *heal*, is *sicken*. The opposite of (B), *advance*, is *retreat*. (D) *penalize*, means to punish. The opposite of *delay*, (E), is *hasten*.

72 **(A)** A malediction is a curse. We want something like *benediction*, and we find *blessing* in (A). The opposite of *preparation*, (B), is *lack of preparation*. (C), *good omen*, has *bad omen* as its opposite. The opposite of (D), *liberation*, is *captivity*. The opposite of *pursuit*, (E), is tough, but it sure isn't *malediction*, so (A) is correct.

73 **(A)** *Mawkish* means sickeningly sentimental. *Unsentimental*, (A), is the answer here. The opposite of (B), *sophisticated*, is *naive* or *simple*. The opposite of *graceful*, (C), is *clumsy*. The opposite of *tense*, (D), is *relaxed*. There are various antonyms to *descriptive*, (E), but *mawkish* isn't one.

74 **(B)** Temerity is recklessness or foolish daring. Its opposite is *hesitancy* or *carefulness*. Blandness, (A), is a lack of character, not a lack of courage. (B), caution, fits—one with temerity lacks caution. The opposite of (C), *severity*, is *leniency*. The opposite of (D), *strength*, is *weakness*. Charm, (E), is personal appeal. The best answer is (B), caution.

75 **(C)** *Jejune* can mean immature or sopho-moric. The opposite would be *adult* or correct choice (C), *mature. Morose,* (A), means sad or moody. The opposite of *natural,* (B), is *artificial.* (D), *contrived,* means deliberately planned. Its opposite is *natural. Accurate,* (E), means precise or exact.

76 **(C)** *Vitiate* means to corrupt, put wrong, spoil, or make worse, and the opposite is *improve* or *correct.* The closest choice is *rectify,* (C). (A), *deaden,* is way off. The opposite of *trust,* (B), is *distrust* or *suspect.* The opposite of *drain,* (D), is *fill.* And the opposite of *amuse,* (E), is *bore* or *upset.*

Section Two—Quantitative

1 **(A)** To compare these two quantities, work column by column starting with the decimal point and working to the right. Both have a 0 in the tenths column, so no difference there. In the hundredths column, both have a 2, so we go to thousandths. Column A has a 6 and Column B has a 5—there are more thousandths in A than in B, so Column A is larger and (A) is the correct answer.

2 **(C)** Right triangles ABD and CDB share a hypotenuse, segment DB. The squared quantities should clue you to use the Pythagorean theorem. See that w and x are lengths of the legs of right triangle ABD. Side AD has length w, side AB has length x. Also, y and z are lengths of the legs of right triangle CDB. Side CD has length z, side CB has length y. Where a and b are lengths of the legs of a right triangle, and C is the length of the hypotenuse, $a^2 + b^2 = c^2$, so $w^2 + x^2$ = length BD^2. $y^2 + z^2$ also equals length DB^2, the quantities are equal and the answer is (C).

3 **(A)** We have $x + 4y = 6$ and $x = 2y$, and we want to compare x and y, so substitute $2y$ for x in the first equation. Using that information, solve for the other variable. Substitute $2y$ for x into $x + 4y = 6$ and get $2y + 4y = 6$ or $6y = 6$. Divide both sides by 6 and we get $y = 1$. If $y = 1$ and $x = 2y$ as the second equation tells us, x must equal 2. Since 2 is greater than 1, the quantity in Column A is greater.

4 **(B)** Question 4 looks hard—but you don't have to simplify to find the relationship. With positive numbers, you can square both without changing the relationship. That leaves you with $4^2 + 5^2$ in Column A and $3^2 + 6^2$ in Column B. 4^2 is 16, 5^2 is 25, 16 + 25 is 41. In Column B we have 3^2, that's 9 + 6^2, that's 36, 9 + 36 is 45. 45 is greater than 41, Column B is greater than Column A, and the answer is (B).

5 **(D)** Column A asks for the number of managerial employees—that's easy. There are 120 employees in the firm, and 25 percent of them are managerial. One-fourth of 120 is 30, the value of Column A.

Column B asks for two-thirds of the clerical employees. But we can't figure out how many workers are clerical workers, so we can't find two-thirds of that number. We can't determine a relationship, and the answer is choice (D).

6 **(B)** We have 12×1 over $12 + 1$ in Column A: 12×1 is 12 and $12 + 1$ is 13. So we have $\frac{12}{13}$ in Column A. In Column B we have $12 + 1 = 13$ in the numerator, and $12 \times 1 = 12$ in the denominator. So $\frac{12}{13}$ in Column A versus $\frac{13}{12}$ in Column B. Of course, $\frac{12}{13}$ is less than 1 while $\frac{13}{12}$ is greater than 1, and the answer is (B).

7 **(D)** You might suspect (D) because there are no variable restrictions. To make the columns look as much alike as you can, multiply out Column A. You'll get $a \times b$ or ab, plus $1 \times b$, plus $1 \times a$, plus 1×1 or 1. So you get $ab + a + b + 1$. Column B has $ab + 1$. We can subtract ab from both sides, and it won't change the relationship and we have $1 + a + b$ in Column A, and 1 in Column B. Subtract 1 from both sides and we have $a + b$ in Column A and 0 in Column B. But consider that a and b could be negative numbers. Since $a + b$ could be positive, negative, or zero, the answer is (D).

8 **(B)** In the two-digit number jk the value of digit j is twice the value of digit k. We have to compare the value of k in Column A with 6 in Column B. If you plug in 6 for k, then go back, you see that the value of digit j is twice digit k. We know that j isn't just a number—it's a digit, which means it's 0, 1, 2, 3, 4, 5, 6, 7, 8, or 9. So 12, twice the value of 6, can't be j. In other words, k has to be something less than 6, so the answer must be (B), the value in Column B is greater.

9 **(A)** We have a circle with right angle QPR as a central angle. The area of sector PQR is 4 and we're asked to compare the area of the circle with 4π. There's a shortcut—the right angle defines the sector, and you have the area of that sector. A 90° angle cuts off one-fourth of the circle. If you multiply by four, you have the area of the circle. So in Column A you have 4×4, and in Column B, you have 4π. π is about 3.14, and 4 is bigger than that, so Column A, 4×4, must be bigger than 4π, and the answer is (A).

10 **(C)** Henry purchased x apples and Jack purchased 10 apples less than one-third the number of apples Henry purchased. *One-third of* means the same as *one-third times* and the number of apples Henry purchased is x. So this boils down to $j = \frac{1}{3}x - 10$. You can plug this in for Column A. We have $\frac{1}{3}x - 10$ in Column A and in Column B we have $x - \frac{30}{3}$. Now you can clear the fraction in Column B. Let's split Column B to two fractions. $\frac{x}{3} - \frac{30}{3}$. We leave the $\frac{x}{3}$ alone and cancel the factor

of 3 from the numerator and denominator of $\frac{30}{3}$ and we're left with $\frac{x}{3} - 10$. What's $\frac{x}{3}$? It's one-third of x, so these two quantities are equal. Column A equals $\frac{1}{3}x - 10$, while Column B also equals $\frac{1}{3}x - 10$, so the answer is (C).

11 **(D)** You can suspect (D) because there are unrestricted variables. In Column A we have the volume of a rectangular solid with length 5 feet, width 4 feet, and height x feet. The formula is length times width times height, so we have 5 times 4 times x, or $20x$. In Column B we need the volume of rectangular solid with length 10 feet, width 8 feet, and height y feet. 10 times 8 times y gives you a volume of $80y$. Now you may think, I've got $20x$, and $80y$, so $80y$ must be bigger because there are more ys than xs. That would be true if x and y were close together, but the variables are unrestricted, and the answer is (D).

12 **(C)** We want to compare 50 with x, one of the angles formed by the intersection of ST and PT. Now angle QRS is labeled 80. We also know PQ and ST have the same length and QR and RS have the same length. If you add PQ and QR, you get PR. If you add ST and RS, you get RT. If you add equals to equals, you get equals, so $PQ + QR$ must be the same as $ST + RS$, which means that PR and RT are the same. You have isoceles triangle PRT and we're given one angle that has measure 80 and the second angle that has measure x. The angle measuring x is opposite equal side PR. That means the other angle must have the same measure, because it's opposite the other equal side. The sum of the interior angles in a triangle always equals 180°. $x + x + 80$ must equal 180, $2x$ must equal 100, $x = 50$. So x and 50 are equal, and the answer is (C).

13

(B) First we can cancel factors of 2 from $2 \times 16 \times 64$ on the left, and $2 \times 4n \times 256$ on the right. If we cancel a factor of 2 we have 16×64 on the left, $4n \times 256$ on the right. 64 goes into 256 four times, so let's cancel a factor of 64. That leaves us 16 on the left and $4n \times 4$ on the right. We can cancel a factor of 4 and we're left with 4 on the left and $4n$ on the right. If $4 = 4n$, n must equal 1, so we have $n = 1$ for Column A. Column B is 2, so the answer is (B).

14

(A) Try to solve the centered equation for y. First get all the ys on one side by adding $\frac{2}{y}$ to both sides. This gives $\frac{1}{3} = \frac{2}{y} - \frac{1}{2y}$. Multiplying both sides by y gives $\frac{y}{3} = 2 - \frac{1}{2} = 1\frac{1}{2}$. Multiplying both sides by 3 gives $y = 3 \times 1\frac{1}{2} = 4\frac{1}{2}$. So Column A is greater than Column B.

15

(D) The perimeter of ABC is 40 and the length of BC is 12, and we want to compare the length of AB with 14. In an isosceles triangle there are two sides with equal length, but we don't know whether side BC is one of those sides or not. If side BC is the unequal side, we have two unknown sides plus 12 and they have a sum of 40, the perimeter. The two remaining sides have a sum of 28, so each is 14. That would mean that AB and AC would have length 14. Then the answer would be (C). If BC is one of the equal sides, we have two sides length 12 and a third unknown side, and the sum is 40. $12 + 12$ is 24, so that the third side has length 16. AB could be one of the sides length 12, or the side length 12. There are three possible lengths for side AB—16, 14, and 12—so the answer is (D).

16

(B) Isolate $\frac{q}{p}$. Multiplying both sides of the equation by p gives $p - q = \frac{2p}{7}$. Subtracting p from both sides gives $-q = \frac{2p}{7} - p = \frac{2p}{7} - \frac{7p}{7} = -\frac{5p}{7}$. So $-q = -\frac{5p}{7}$, or $q = \frac{5p}{7}$. Dividing both sides by p gives $\frac{q}{p} = \frac{5}{7}$.

17

(D) We need the number that's a multiple of 8 and a factor of 72. Since 4 and 9 aren't multiples of 8, you can eliminate (A) and (B). 16 is, 24 is, but 36, (E), isn't. We're down to just 16 and 24; count by 16s and see if 16 is a factor of 72: 16, 32, 48, 64, 80. Well, that's not a factor of 72, so 24, (D) is correct.

18

(B) We need the value of $a + b + c$. We know that a, b, and c are exterior angles of our quadrilateral in the diagram and there's a fourth exterior angle which isn't labeled. But the measure of the interior angle next to it is given to us—it's 70°. The sum of the exterior angles of any figure is always 360°. So we can figure out the measure of the missing angle, then subtract it from 360 and get the sum of the other three. The unlabeled angle must be 110°. Now we know $110 + a + b + c = 360$, we subtract to get the sum of a, b, and c and we get 250, (B) as the correct answer.

19

(E) John has four ties, 12 shirts and three belts, and we need the number of days he can go without repeating. So we multiply the number of ties times the number of shirts times the number of belts. Four ties, 12 shirts—48 combinations. Multiply by three choices of belt, and you get 3×48 or 144 combinations, (E).

20 **(D)** Move the decimal points to the right until they disappear—but keep track of how many places you move the decimal. In (A) we have .00003 in the numerator. Move five places to the right to change it to 3. Then we change from .0007 to 70 in the denominator and we end up with $\frac{3}{70}$. In (B) we have .0008 on top, .0005 on the bottom—we get $\frac{8}{5}$. We have $\frac{70}{8}$ for (C). In (D), we end up with $\frac{60}{5}$ and $\frac{10}{8}$ in (E). Clearly (D), 12, is the largest value.

Graphs: Questions 21–25

21 **(E)** The bar graph doesn't give us the total number of general practice physicians, but if we add the number of males to the number of females, we get the total number of g.p. physicians. To find the percent who are male, we take the number of males and put it over the total number and that will give us our percent. We have about 2,000 women and about 23,000 men, making the total about 25,000. Well, if there are around 25,000 g.p. physicians altogether and 2,000 to 3,000 of them are female, that's what percent of 25,000? It's around 10 percent. About 22,500 are male, which gives us 90 percent, (E).

22 **(D)** We're looking for the lowest ratio of males to females so we have to get the smallest number of males and the largest number of females. Skimming the bar graphs, we can see that in pediatrics the female graph and the male graph are closer than any of the others. Pediatrics is (D), the correct answer.

23 **(C)** To refer to ages of physicians, we need to find the slice of the pie that goes from 45 to 54. It's 20 percent, but 20 percent of what? We're not looking for a percent, we're looking for a number of doctors. For general surgery the male bar goes up to about 35,000 and the female bar goes up to about 2,000—about 37,000 total. So 20 percent of 37,000 is the number of general surgery physicians between ages 45 and 54, inclusive. What's 20 percent of 37,000, or $\frac{1}{5}$ of 37,000? Well, let's see, $\frac{1}{5}$ of 35,000 is 7,000, $\frac{1}{5}$ of 2,000 is 400, making 7,400. (C) is 7,350, the correct answer.

24 **(E)** We'll have to find the total number of family practice physicians, which represents 7.5 percent of all the physicians in the United States, then we can find 100 percent of that number. The male bar of family practice physicians goes just over 36,000, so we'll say it's 36,000 plus. The number of females goes just over 6,000 so we'll call that 6,000 plus, so we have about 43,000 all together. This is 7.5 percent of all the physicians. 7.5 percent is awkward—it's three-quarters of 10 percent, which is $\frac{3}{4} \times \frac{1}{10}$, or $\frac{3}{40}$. So 43,000 is $\frac{3}{40}$ of the total number of doctors. To change 43,000 into the number of

total physicians, we multiply it by $\frac{40}{3}$. Think of it this way: we have an equation now, $\frac{3}{40}$ of the number we're looking for, we'll call it N, the number of physicians, equals 43,000. We want to get N by itself, so we have to get rid of that $\frac{3}{40}$. So we multiply by the reciprocal, $\frac{40}{3}$, and that leaves us N by itself on the left. But the hard part is multiplying $\frac{40}{3} \times 43,000$. What's $\frac{40}{3}$? It's $13\frac{1}{3}$ and that's easier to multiply. 13×43 is 559, so $13 \times 43,000$ is 559,000—you can look at your choices and estimate. Only one is close to 559,000—(E), 570,000, and we're going to add on to that, so (E) is the correct answer.

25 **(B)** How many male general surgeon physicians were under 35 years old? The pie chart breaks down general surgery physicians by age, so we'll be working with it. And, since we're looking for a number of general surgery physicians, we know that we're going to have to find the total number of general surgery physicians, then break it down according to the percentages on the pie chart.

We're told the number of female general surgery physicians in the under-35 category represented 3.5 percent of all the general surgery physicians. What this does is break that slice of the pie for under-35 into two smaller slices, one for men under 35 and one for women under 35. Now we know that the whole slice for under-35-year-olds is 30 percent of the total and we've just been told that the number of females under 35 is 3.5 percent of the total. So the difference between 30 percent and 3.5 percent must be the men in the under-35 category, which leaves 26.5 percent,

which we have to multiply by the total number of general surgery physicians.

We figured out in Question 23 that there were 37,000 total general surgery physicians, and 26.5 percent of those are men under 35. What's 26.5 percent of 37,000? One-quarter of 37,000 is 9,250 and that's very close to (A), but remember we've still got another 1.5 percent to go. One percent of 37,000 is 370 and half of that, or .5 percent will be 185, so if you add 370 and 185 to 9,250 you end up with a total of 9,805 which is very close to (B), the correct answer.

26 **(B)** We want to find the sum of the absolute value of three, the absolute value of –4 and the absolute value of 3 – 4. Well, the absolute value of 3 is 3, the absolute value of –4 is 4. What's the absolute value of 3 – 4? Do the subtraction inside the absolute value sign first, and we get –1. What's the absolute value of –1? It's 1, so we have 3 + 4 + 1 or 8 as our sum for Question 26, (B).

27 **(C)** This looks like a right triangle on a coordinate grid, but it's not a normal coordinate grid—the lines on the grid don't represent integer units, they represent units of less than an integer. Going up on the y-axis, we have .5, 1.0, 1.5, and 2.0, so the lines each represent half an integer and, going to the left, the lines are labeled –0.4, –0.8, –1.2, so these each represent .4, and yet the diagram's not drawn to scale. Going left to right, the vertical lines are actually farther apart than the horizontal lines, which represent more value on the number line.

Now, to find the area of the shaded region, a triangle, we need a base and a height. This is a right triangle because its base lies on the horizontal line on our grid and its height, the side to the right, lies right on a vertical line on the grid. What's the length of the bottom side? The far right end point is at –0.4 and the far left end point is at –1.6, and the difference is 1.2 units, so the base is 1.2. The lower right vertex has value .5 and the upper right vertex has value 2.0—the difference is 1.5, so that's the height. The base is 1.2 and the height is 1.5, and the formula for the area of a right triangle is one half base times height. We have 1.2 as our base, we can call that $\frac{6}{5}$, 1.5 is $\frac{3}{2}$ so the area is $\frac{1}{2} \times \frac{6}{5} \times \frac{3}{2}$— that's $\frac{9}{10}$ or .9, choice (C).

28 **(A)** You could find the number of tasks per hour from one computer, but that would add extra steps, because you want to find out how many computers you need to do a certain number of tasks in three hours. Well, if it can do 30 tasks in six hours, it can do 15 tasks in three hours. So, if you have two computers, that's 30 tasks, three is 45, four is 60, five is 75, six is 90. You can't get by with five because you have to get 80 tasks done, so you'll need six computers, (A).

29 **(C)** One side of triangle ABC is an edge of our cube, segment BC. But segments AB and AC aren't lengths of the edge of the cube or fractions of a length of an edge of the cube. Well, let's find the length of an edge of the cube. If the cube has volume

8, that's the length of an edge to the third power. Since 2 cubed is 8, the length of an edge of this cube is 2. We need AB and AC, and so we have to concentrate on smaller right triangles on the same face of the cube that includes triangle ABC.

On the upper left, directly above point A is an unlabeled vertex—let's call that point Y—and down below point A is an unlabeled vertex—we'll call that point X. Look at triangle AXC. It's a right triangle because angle AXC is one of the angles formed by two edges of a cube—and AC is its hypotenuse. AX is half an edge of the cube because point A's the midpoint of edge XY. That means that AX has length 1 and XC is an edge of the cube, so it has length 2. The legs of this right triangle are 1 and 2, so we can use the Pythagorean theorem to find the length of AC. AX^2 is 1^2, XC^2 is 2^2, 1^2 is 1, 2^2 is 4, the sum of 1 and 4 is 5, AC^2 is 5 and AC has length $\sqrt{5}$. AB is identical to AC because triangle AYB is identical to triangle AXC, so AB also has length $\sqrt{5}$ and the perimeter of ABC is $2 + 2\sqrt{5}$, choice (C).

30 **(D)** The catch here is that it's not which of the following is 850 percent *of* 8×10^3, it's which of the following is 850 percent *greater than* 8×10^3. Well, what's bigger, 850 percent of 1 or a number that's 850 percent greater than 1? 850 percent of 1 is 8.5×1 or $8\frac{1}{2}$. But a number that's 850 percent greater than 1 is 1 + 850 percent of 1, it's 1 + 8.5 or 9.5. So the number we want is $9.5 \times 8 \times 10^3$. $9.5 \times 8 = 76$, so the answer is 76×10^3, or 7.6×10^4, in scientific notation.

31 **(A)** We have to plug 1 in for x and solve the equation for y. Well, $x + 3$ is $1 + 3$—that's what's inside the parentheses and we do that first. We have $1 + 3 = 4$ inside the parentheses. $y = 4^2$, 4^2 is 16, and 16 is greater than 9, so the answer is (A).

32 **(B)** In both columns we'll use the basic formula: rate × time = distance. In Column A, 40 mph × 4 hours traveled gives you 160 miles. In Column B, 70 mph × $2\frac{1}{2}$ hours, $2 \times 70 = 140$, half of 70 is 35, and $140 + 35 = 175$ miles in Column B. 175 is greater than 160, so the answer is (B).

33 **(D)** This is intended to conjure up a picture of heavy cookies in one bag and light grapes in the other, but you can't assume that because cookies are usually bigger than grapes, these cookies weigh more than these grapes. Since you don't know how much each cookie and each grape weighs, you can't find the number of cookies or grapes, so it's (D).

34 **(C)** Here we have triangle ABC—base BC has been extended on one side so we have an exterior angle drawn in and labeled 120°. We want to compare side lengths AB and BC—in any triangle, the largest side will be opposite the largest angle, so we want to see which of these sides is opposite a larger angle. Since angle A is labeled 60°, is angle C less than, equal to, or greater than 60? Notice that the adjacent angle is 120°—the two together form a straight line, so their sum is 180°. $180 - 120 = 60$, so angle C is a 60° angle. Since the angles are equal, the sides are equal, and the answer is (C).

35 **(A)** Notice the way the diagram is set up—$a + b$ is the same as PQ. Our equation is $8a + 8b = 24$. Divide by 8. We end up with $a + b = 3$. PQ is 3 and since 3 is greater than 2, the answer is (A).

36 **(A)** All we know is that x is less than y but though we don't know their values, we may know enough to determine a relationship. In Column A we have $y - x$, the larger number minus the smaller number, so you must get a positive difference, even if both numbers are negative. In Column B you have the smaller number minus the larger number—the difference is the same except this time it is negative. So you can determine a relationship—you know the answer is (A), the quantity in Column A is always greater than the quantity in Column B.

37 **(D)** Remember, area equals $\frac{1}{2} \times$ base × height. Both triangles have the same height, because they have the same apex point A and each of them has as its base a part of line EB. So the one with the larger base has the larger area. Which is bigger, CB or DE? We have no way to figure it out. We are not given any relationships or lengths for any of those segments, so the answer is (D).

38

(B) There's one box that's in both rows—the one in the middle with value $\frac{2}{9}$. In fact, we have $\frac{1}{3} + \frac{2}{9} + y$ in the horizontal row, $x + \frac{2}{9} + \frac{4}{5}$ in the vertical row, and we are comparing x and y. Since $\frac{2}{9}$ is part of both rows, we can throw it out. So we have $\frac{1}{3} + y = \frac{4}{5} + x$. We have $\frac{1}{3} + y$ and that is the same as $\frac{4}{5} + x$. Since $\frac{4}{5}$ is greater than $\frac{1}{3}$, the number we add to $\frac{4}{5}$ has to be less than the number we add to $\frac{1}{3}$ for the sums to be the same. Since $\frac{4}{5}$ is greater than $\frac{1}{3}$, x must be less than y. The answer is (B).

39

(B) Looking at the fraction in Column A we have $\frac{1}{3} \times \frac{1}{4}$ in the numerator, $\frac{2}{3} \times \frac{1}{2}$ in the denominator. We can cancel the factor of $\frac{1}{3}$ from the numerator and denominator, right? Cancel a $\frac{1}{3}$ from each and you end up with $1 \times \frac{1}{4}$ in the numerator, $2 \times \frac{1}{2}$ in the denominator. Using the same approach, we can cancel a factor of $\frac{1}{4}$, so we're left with 1×1 in the numerator and 2×2 in the denominator, so the value of Column A is $\frac{1}{4}$. Now take a look at Column B. It's the reciprocal of the value in Column A. You have $\frac{2}{3} \times \frac{1}{2}$ in the numerator and $\frac{1}{3} \times \frac{1}{4}$ in the denominator. So you have $\frac{4}{1}$

as your value for Column B. With 4 in Column B and $\frac{1}{4}$ in Column A, the answer is (B).

40

(B) Make a map—if you have trouble with geometry, this will make it much easier. Eileen drives due north from town A to town B for 60 miles. Start at a point and draw a line straight up. Label the point you started at A and the point above it, B. Label 60 as the length of the distance from A to B. Next she drives due east from town B to town C for a distance of 80 miles. Start at point B, draw a line straight over to the right, call the right endpoint C, and label as 80 the distance BC. You have a right angle, angle ABC. Well, the distance from town A to town C is the hypotenuse of a right triangle if you draw line AC. The two legs are 60 and 80 and this is one of our Pythagorean ratios. It is a 6-8-10 triangle except this time it is 60-80-100. So the distance from A to C is 100 miles, the same as our value for Column A, so the answer is (B).

41

(C) Let's see if we can do something to make these look more alike by getting both sets of binomials so the $\sqrt{7}$s are in the front. We have $\sqrt{7} - 2$. Is that a positive or negative quantity? 2^2 is 4, 3^2 is 9 so $\sqrt{7}$ is between 2 and 3. We have $\sqrt{7} - 2$, that is positive, times $\sqrt{7} + 2$, that is positive again. Two positives in Column A and the product of two positives is always positive. What do we have in Column B? $2 - \sqrt{7}$, that is a negative number times negative $\sqrt{7} - 2$. $-\sqrt{7}$ is a negative, -2 is negative, that quantity is negative. You have the product of two negatives in Column B, but a product of two negatives is positive also, so you can't tell which is greater.

Let's see if we can make these quantities look more alike. With the last one on the right, $-\sqrt{7} - 2$, if we divide the whole thing by -1, we're left with a positive $\sqrt{7}$ and a positive 2, $\sqrt{7} + 2$. On the right in Column B we have $(2 - \sqrt{7}) \times (-1) \times (\sqrt{7} + 2)$, and $(\sqrt{7} + 2)$ is also in Column A, so we can cancel. Those two factors are the same, right? We have $-1 \times (2 - \sqrt{7})$. Let's distribute again. What is -1×2? It's -2. What is $-1 \times -\sqrt{7}$? It's $+\sqrt{7}$ so we end up with $+\sqrt{7} + -2$ or $\sqrt{7} - 2$. It's exactly the same as the factor in Column A. So the quantities are equal and the answer is (C).

42 **(B)** We can see from our diagram that r and s are the coordinates of a point on our line. We have a line on the graph with one point with coordinates $(\frac{5}{2}, \frac{7}{2})$. The line also goes through the origin $(0, 0)$, so what can we figure out about this line? Well, draw in the line $x = y$, a line which makes a 45° angle with the x-axis that goes from the lower left to the upper right—you notice that it goes through the point $(\frac{5}{2}, \frac{5}{2})$, because any point on line $x = y$ has the same x coordinate and y coordinate. Point $(\frac{5}{2}, \frac{7}{2})$ falls above point $(\frac{5}{2}, \frac{5}{2})$, because the y coordinate is greater, it's above the $x = y$ line. Similarly, point (r,s) lies above the $x = y$ line so the y coordinate is greater than the x coordinate and the y coordinate of that point is s. Where we have coordinates (r,s), r is the x coordinate, s is the y coordinate; s is greater than r in this case. The answer is (B), the quantity in Column B is greater.

43 **(D)** We have $1 - (\frac{1}{4})$ to the x power in Column A and we have 0.95 in Column B. That's a bizarre comparison, isn't it? Converting Column B, 0.95 into fraction form, $0.95 = \frac{95}{100} = \frac{19}{20}$. What do we have in Column A if $x = 1$? We have $1 - \frac{1}{4}$ or $\frac{3}{4}$. $\frac{19}{20}$ is greater than $\frac{3}{4}$. What happens if we have $x = 2$? Column A becomes $1 - (\frac{1}{4})^2$. $\frac{1}{4}^2$ is $\frac{1}{4} \times \frac{1}{4}$ or $1 - \frac{1}{16}$ is $\frac{15}{16}$. So what's bigger, $\frac{15}{16}$ or $\frac{19}{20}$? Still $\frac{19}{20}$, but as x gets larger and we multiply $\frac{1}{4}$ times itself more times, the amount that we're taking away from 1 is going to get smaller and we'll be taking less than $\frac{1}{20}$ away from 1 as soon as we get to $x = 3$. $\frac{1}{4}$ to the third power is $\frac{1}{64}$ and at that point Column A becomes $\frac{63}{64}$. What is bigger, $\frac{19}{20}$ or $\frac{63}{64}$? Well, $\frac{63}{64}$ is bigger, it is closer to 1, and there are two possible relationships here. If x is 1 or 2, Column B is greater. If x is 3 or larger, Column A is greater. The answer is choice (D).

44 **(A)** If you draw in some diameters in the circles, you will see that PS is equal to one diameter, and PQ is equal to two diameters. Let one diameter be d. The perimeter of $PQRS$ is then $PS + PQ + SR + QR = 60$. The circumference of a circle is

πd, where d is a diameter. Since we have two circles, the combined circumferences is $2 \times \pi d = 2\pi d$. Since π is greater than 3 (it's about 3.14), the value in Column A is greater than $6d$ in Column B.

45 **(B)** I hope you didn't try to figure out the exact values of each of these. Instead, if you look at Column B and Column A, they look sort of alike because they both have 3 in terms of a power. What is 3^{20}? It's 3×3^{19} right? So we can have $3^{19} + 3^{19} + 3^{19}$ in Column B. In Column A we have $3^{17} + 3^{18} + 3^{19}$. We can subtract 3^{19} from both sides and we're left with $3^{17} + 3^{18}$ in Column A and $3^{19} + 3^{19}$ in Column B. We know that 3^{19} is bigger than 3^{17} or 3^{18} so we know that $3^{19} + 3^{19}$ is bigger than $3^{17} + 3^{18}$. The answer is (B), the quantity in Column B is greater.

46 **(E)** We have $4 + y = 14 - 4y$ and we want to solve for y. We can isolate the ys on one side of the equal sign by adding $4y$ to both sides, giving us $4 + 5y = 14$. Subtracting the 4 from both sides we get $5y = 10$. Divide both sides by 5 and get $y = 2$, (E).

47 **(D)** Let's go the quickest, most obvious route and use the common denominator method. With $\frac{4}{5}$ and $\frac{5}{4}$, the denominator that we will use is easy to find; just use 5×4 or 20. $\frac{4}{5}$ is $\frac{16}{20}$ and $\frac{5}{4}$ is $\frac{25}{20}$. $\frac{16}{20} + \frac{25}{20}$ is $\frac{41}{20}$, which is (D).

48 **(E)** First we need to find m. We are told that $3m$ is 81. Well, 81 is 9×9. 9 is 3^2. So we have $3^2 \times 3^2 = 81$ or $3 \times 3 \times 3 \times 3 = 81$. How many factors of 3 are there in 81? There are 4, so m has the value 4. Now what's 4^3? 4×4 is 16. 16×4 is 64. So (E) is correct, 64 is m^3.

49 **(B)** We are looking for the area in square meters of square C. Now notice we have one side of square B butted up against one side of square A—they're not the same length, but the difference in their lengths is made up by the length of a side of square C. One side of square B + one side of square C = one side of square A. We can figure out the length of the side of A and length of the side of B, which will let us figure out the length of side of C. That is what we need to figure out the area of square C. The area of a square is its side squared. The area of square A is 81, so it has a side of $\sqrt{81} = 9$. The area of square B is 49, so it has sides of length $\sqrt{49} = 7$. So $9 = 7 + C$, so C must have length 2. So we have 2 as the length of the side of square C, 2^2 is 4, there are 4 square meters in gardening area C, and the answer is (B).

50 **(E)** We can figure out how many students scored exactly 85. Twenty-three scored 85 or over, and 18 scored over 85. So $23 - 18$ or 5 students scored exactly 85 on the exam, but that's no help. How many students scored less than 85? We don't know—we can't answer this question. It's (E), it can't be determined.

51 **(C)** We're asked in which year the energy use in country Y was closest to 650 million kilowatt hours, so we just have to follow the jagged line which represents energy use from left to right until we encounter a vertical line representing a year in which we're close to 650 million. The one year in which this is true is 1970. In no other year are we as close, so (C), 1970, is our answer.

52 **(C)** In order to find how many categories had energy use greater than 150 million kilowatts, you have to find out how many total kilowatts were used in that year using the line graph. You see that there were 600 million kilowatts used in 1965. What is the relationship of 150 million kilowatts to 600 million kilowatts? It's 25 percent of 600 million kilowatts, so we're looking for categories with more than 25 percent of the energy use for 1965. How many categories exceeded 25 percent? Just two, government and industrial. So our answer is (C).

53 **(D)** We can estimate quite a bit from our graph. If we look at our line chart, we can see that as time goes on, energy use goes up pretty steadily. It went up sharply between 1960 and 1965, then more gradually from 1965 to 1980. Because in more recent years the overall use was much greater, if the percent of industrial use was about the same over all the years, then as the overall use increases, the amount used for industrial purposes will increase also. Let's take a quick look at the bar graph and see if that is the case. Was the percent being used for industrial use about the same? Well, it didn't fluctuate much from 1960 to 1970, but in 1975 industrial use jumped significantly as a percent of the total, then shrank significantly going to 1980. The most likely answer is 1975, and if you find 40 percent of 690 million, your amount for 1975, you get 276 million kilowatt hours. Then if you find 20 percent of 710 million, your amount for 1980, you only get 142 million kilowatt hours, so (D), 1975, is the correct answer.

54 **(D)** What we are going to do for 1960 and 1965 is find the per capita personal use, then find the percent decrease from 1960 to 1965. To do that, we have to plug in a value for the population of country *Y* for 1960. Let's use 100 million for the '60s population. The per capita use in 1960 is the total personal use, which is 30 percent of 500 million, that's 150 million. We know that 150 million, the total personal use, equals 100 million, the population × the per capita use. The per capita use is $\frac{3}{2}$ or 1.5. Going on to 1965, we are told the population increased by 20 percent, so in 1965 the population was 120 million people. What was the total personal use of energy? It was a little bit less than 20 percent of our total 600 million so we'll call it 20 percent of 600 million, or 120 million. If total personal use is 120 million and we have 120 million people, that's one kilowatt hour per person. What's the percent decrease? It's a decrease of $\frac{1}{3}$, $33\frac{1}{3}$ percent. But remember, in 1965, they were using a little more energy for personal use than we figured. The correct answer must be a little greater than $33\frac{1}{3}$ percent, so 40 percent, (D), is the correct answer.

55 **(A)** Statement I says farm use of energy increased between 1960 and 1980. In 1960, 500 million. In 1980, 710 million kilowatt hours were used. What was the percent of farm use in 1960? It was 30 percent of the total in 1960 and a little bit less than 30 percent, around 28 percent, in 1980. The percent is very close together while the whole has become much larger from 1960 to 1980, so 30 percent of 500 million is less than 28 percent of 710 million. Farm use of energy did go up in that 20-year period and Statement I is going to be part of our answer. That eliminates two answer choices, (B) and (D).

How about Statement II? This one is harder. It says that in 1980, industrial use of energy was greater than industrial use of energy in 1965. But what was it in 1965? Industrial use of energy in 1965 was 30 percent

of 600 million. We got the percent from the bar graph, the total from the line chart. Okay, 30 percent of 600 million is 180 million. But what about 1980? In 1980 industrial use of energy was 20 percent of a larger whole, 710 million kilowatt hours. Well, 20 percent of 710 is 142 million. That's less than 180 million, isn't it? In fact, industrial use of energy went down from 1965 to 1980, so this can't be inferred from the graph and it's not part of our answer. That cuts out (C) and (E), leaving choice (A), I only. Statement III is another easy one to eliminate because it says more people were employed by the government of country *Y* in 1980 than in 1960. These graphs deal only with energy use, not with employment, so it's irrelevant and we can eliminate it. Only Statement I can be inferred, and (A) is correct.

56 (E) The average is $\dfrac{\text{The sum of terms}}{\text{The number of terms}}$. Here we have $y - z$ and the other number, which we will call x. The average of x and $y - z$ is $3y$, so $3y = \dfrac{x + y - z}{z}$. Multiplying both sides by 2 gives $6y = x + y - z$. Subtracting $y - z$ from both sides gives $5y + z = x$. So the other number, x, is $5y + z$, answer choice (E).

57 (B) We're told that the area of triangle *ABC* is 35 and in our diagram we're given a height for triangle *ABC*. If we use *AC* as the base of the triangle, the perpendicular distance from segment *AC* up to point *B* is 7, so we can find the length of *AC*. When we find the length *AC*, the base of triangle *ABC*, what do we have? We have the hypotenuse of

right triangle *ABC*. Given the hypotenuse and the length of leg *AD*, which is given in the diagram as 6, we'll be able to find the third leg of the triangle, side *DC*, which is what we're looking for. Okay, going back to triangle *ABC* where we started, the area is 35 and the height is 7. The area of a triangle is $\frac{1}{2}$ base × height, so $\frac{1}{2}$ base × height is 35, $\frac{1}{2}$ × 7 × length *AC* is 35. That means 7 × length *AC* is 70, so *AC* must have length 10. Now we can look at right triangle *ADC*. Here is a right triangle with one leg of length 6, the hypotenuse of length 10 and the third side unknown; what we have is a 6-blank-10 right triangle. That's one of our famous Pythagorean ratios—it's a 6-8-10 triangle. So *DC* must have length 2×4, or 8, (B).

58 (A) We're trying to find the shortest distance in meters a person would have to walk to go from point *A* to a point on side *BC* of the triangular field represented in our diagram. In order to get the shortest distance from side *BC* up to point *A*, we want to draw a perpendicular line from point *A* down to side *BC*. That will divide up the triangular field into right triangles. Let's draw in the path from point *A* down to segment *BC* and call the new vertex we make point *D*. We just created two smaller right triangles, *ADC* and *ADB*. Now our diagram tells us that length *BC* is 160 meters and *AB* is 100 meters—*AC* is also 100 meters. Now each of these two right triangles has 100 meters as the length of its hypotenuse. What does that tell you about triangle *ABC* ? *AB* and *AC* have the same lengths, so this is an isosceles triangle. That means that when you drew in the perpendicular

distance from A down to D, you split that isosceles triangle into two identical right triangles. Length BD is the same as length BC. So each of them is half of 160 meters, or 80 meters each. We have right triangles with hypotenuses of length 100 meters each and one leg of each these right triangles is 80 meters. This is a 3-4-5 right triangle, with each member of the ratio multiplied by 80. So AD must have length 60, and the minimum distance is 60 meters, (A).

59

(D) We're told the ratio of $2a$ to b is eight times the ratio of b to a. That's awkward to keep track of in English—it's a little easier to write fractions. The ratio of $2a{:}b$ equals $2\frac{a}{b}$. So $2\frac{a}{b} = 8(\frac{b}{a})$. We're asked to find what $\frac{b}{a}$ could be; that may tell you there's more than one possible value for $\frac{b}{a}$, but let's start with the equation we just put together using translation and isolate $\frac{b}{a}$. To do that, we'll divide both sides of the equation by 8, which is the same as multiplying by $\frac{1}{8}$. So now we have $\frac{1}{8} \times 2\frac{a}{b} = \frac{b}{a}$. Well, what is $\frac{1}{8} \times 2\frac{a}{b}$? It's $\frac{a}{4b}$. So $\frac{a}{4b} = \frac{b}{a}$. We need to multiply both sides of the equation right now on both sides by $\frac{b}{a}$. It'll be more complicated on the right side but simpler on the left because the as and bs on the left side will cancel out, and you'll be left with $\frac{1}{4}$. On the right you have

$\frac{b}{a} \times \frac{b}{a} = \frac{b^2}{a^2}$. So we have $\frac{1}{4} = \frac{b^2}{a^2}$. So $\frac{b}{a}$ could represent positive or negative $\frac{1}{2}$.

60

(D) A dentist earns n dollars for each filling plus x dollars for every 15 minutes. So the money is figured in two different ways; dollars for each filling and dollars per hour, represented in terms of 15 minutes. Our result will be a two-part answer choice. If you can figure out one part, it will let you eliminate some choices. She put in 21 fillings. She makes n dollars for each, so she gets $21n$ dollars for fillings. You can eliminate (B) and (E) because (B) has only $14n$ in it, and (E) has $\frac{21}{4}n$ dollars in it. That narrows our choices to (A), (C), and (D).

How about the hourly rate? The dentist works 14 hours in a week. Does that mean she makes $14x$ dollars? No, because the rate is dollars for every 15 minutes. Now if she makes x dollars for every 15 minutes and 15 minutes is $\frac{1}{4}$ of an hour, then we have to multiply that rate by 4 to get the rate per hour, it's $4x$ dollars per hour. Well, $4x$ times 14 hours is $56x$, so (D), $56x + 21n$, is correct.

Section Three—Analytical

Game 1: Questions 1–3

The Action
Five spices to arrange: A, C, N, S, and T.

The Rules
1) *N* must be harvested before *T*.

2) Cloves must be harvested immediately after allspice—no other spice can come between them.

3) *S* can't be first.

There are no overlaps here—no spices are mentioned by more than one rule. We know from Rule 1 that T isn't first, from Rule 2 we know C isn't first and from Rule 3 we know that *S* isn't first, so the first spice is either N or A.

1 **(A)** Let's take Rule 3 first and try to throw out answer choices. (B) has sage in the number one spot, so it's out. Rule 2 gives us our *AC* unit—*A* has to be immediately before *C*—choice (C) has *A* first and *C* fourth, and (D) has *C* before *A*, so they're both out. Rule 1 gives us our answer—*N* is before *T*, and choice (E) has *N* coming after *T*. (E)'s out, so it's (A) for this question.

2 **(D)** If we put *N* fourth, what are the consequences? Rule 1 springs to mind—nutmeg is before thyme. If *N* is fourth, *T* must be fifth. Now we have our *AC* unit, and we have *S*. We know *S* can't be first, and it can't be second either, since we can't separate *A* and *C*. *S* is third, so we've completed our sequence—*A*, *C*, *S*, *N*, *T*. Choice (D) must be false, since cloves can't be immediately followed by nutmeg. We need cloves, sage, and nutmeg second, third, and fourth. We don't have to go through the others as

long as you know that only sequence *A*, *C*, *S*, *N*, and *T* will work, and, again, it's (D) for Question 2.

3 **(C)** First, put *S* in the second slot and look for a spot for *A*—it's joined with *C*, which limits our options. *A* can't be first since *C* would then be second and the second slot is taken, so eliminate choices (A) and (B). *A* can't be last since *C* comes after *A*, so scratch (D) and (E), leaving (C). So (C), third and fourth, is correct, and that's it for that game, which was short and quite easy.

4 **(B)** We need a statement that weakens or has no effect on the logic. The conclusion here is that the way a judge came into his job often determines the result of a case. Judges decide differently depending on their term length—short termers think in light of political influences, while lifelong judges rely on a tradition of judicial wisdom. The author provides no supporting evidence, so the wrong choices will likely be evidence that strengthens the argument.

(A) gives support, asserting that if judges want to keep their jobs, they're likely to be swayed by voters, to improve their election chances. (C) shows that short termers rule in ways that the voters approve of. In (E), we find that only short term judges use pollsters—people who track public opinion. So (A), (C) and (E) lend credence to the allegations about the political sensitivity of short-term judges. (D) supports the viewpoint on lifelong judges, saying that appointed judges show consistency, implying that those judges turn a blind eye to politics. We're left with (B). If long termers act on their political knowledge, they're as fickle as the short termers. If they don't act on their political knowledge, it's simply irrelevant. (B) doesn't support the argument, and it's correct for this question.

5 **(E)** The author believes that a climate of peace has been created by spending on weapons systems and supports this claim by indicating that a decrease in conflicts can be attributed to the robust deployment of weapons systems. The author sees a causal connection between defense readiness maintained by greater spending and the lower number of attacks. (E) is the assumption underlying this connection. The author must assume that had defense spending not gone up, the number of attacks would have increased.

(A), in its reference to the causes of military action, is outside the scope. In (B), the author doesn't tell us that more defense spending has prevented military actions, and it is certainly not a necessary assumption. There's no claim about the future of peace for (C). Finally, (D)'s equation of weapons and personnel is silly—if the author has an opinion on this issue, he's keeping it to himself, so we can't ascribe this view to him. (E)'s the answer for Question 5.

6 **(C)** Within five years it will be cheaper to buy tuners and amplifiers separately instead of buying an integrated receiver. Previously, a receiver combining both tuner and amp was cheaper than the two purchased separately. What has changed? In other words, what's the basis of the author's claim? It's a recent trend showing that the average retail prices of tuners and amps have declined 20 percent and 35 percent respectively, while the average retail price of an integrated receiver has declined only 12 percent. But percentages often can't be compared unless you know the actual numbers. Try plugging in numbers. If tuners and amps each used to cost $1,000 apiece, while receivers used to cost only $100 apiece, then the 20 percent decline in the tuner and the 35 percent decline in the amp over the past two years wouldn't have brought them near the cost of a receiver. Tuners would cost $800 and amps $650 while a 12 percent

decline in the price of a receiver would bring its price to under $90. In five years a receiver will probably still be the better bargain. While the author implies that the price gap has been closing in the past two years, we don't know how much it has closed or the rate at which it will close in the next five years, so (C) is correct.

(A), (B), (D), and (E) don't help you decide the significance of the percent decline in the average prices. The life expectancy of stereo equipment, (A), is outside the scope. As for (B), this doesn't tell you anything about which component costs less—outside the scope. So is (D)—sales projections tell you nothing about the actual cost of the equipment. (E) is even farther out in left field in talking about the consumers rather than the equipment, so it's (C) for this question.

Game 2: Questions 7–10

The Action
Eight articles—five must be selected. Let's put theater articles in capitals and dance articles in lower case: *F, G, H, J, k, l, m, o.*

The Rules
1) We need at least three lowercase letters, so we either have three lowercase and two capitals, or four lowercase and one capital. If you're told that one lowercase isn't chosen, you can circle the other three. This kind of unpacking of rules can help you make realizations at the top of the game.

2) If *J* is chosen, *m* can't be. Put this together with Rule 1, and you see that if you cross out *m*, you know you can circle *k, l,* and *o,* and if you choose *m,* you have to cross out *J.*

3) If you choose *F,* you must choose *J.* Note: This doesn't mean that if you have *J* you have *F.*

Rule Overview

Both Rule 2 and Rule 3 mention *J*, and you can combine them. If you have *J*, you can't have *m* and in order to have *F*, you must have *J*, so this tells us that you can't have *F* and *m* in the same group.

7 **(E)** Question 7 tells us that *m* isn't chosen, and since *m* is a lowercase letter you have to choose the three remaining ones. So cross out *m*, circle *k*, *l*, and *o* and you need go no further. We're asked which must be chosen—(E) suggests *k*, and it's our answer. (A) through (D) are all capital letters, so they're out and it's (E) for this question.

8 **(C)** Let's make out a roster, circle *J*, and see what happens. Rule 2 tells us that if *J* is chosen, *m* is not, and if we cross out *m* we circle *k*, *l* and *o*. We have *J*, *k*, *l*, and *o* circled. Any of the remaining letters can be fifth—*F*, *G*, or *H*. There are three possible groupings—*J*, *k*, *l*, *o*, *F*; *J*, *k*, *l*, *o*, *G* and *J*, *k*, *l*, *o*, *H*. So (C), three, is correct.

9 **(A)** We just know that the choice of one article will make only one group of published articles acceptable, so let's start with the entities that we know the most about. As we saw in Question 8, choosing *J* yields three possible groups, not just one, so we can eliminate (C). *m*, (E), was mentioned in Rule 2 and in our deduced rule about *F* and *m* being incompatible. If you circle *m*, you have to cross out *J* and *F*, but the rest is wide open, so we can eliminate (E). What about *F*? If you circle *F*, Rule 3 says you have to circle *J*, and when you circle *J*, Rule 2 says you have to cross out *m*, and once you cross out *m*, you have to circle *k*, *l*, and *o*, and those are your five articles, *F*, *J*, *k*, *l*, *o*, so (A) is correct for this question.

10 **(E)** If you circle *G*, what follows? Nothing obvious, so our best bet is to try out the choices by attempting to disprove them. (A)—must it follow if *G* is chosen that *J* is not chosen? Well, what happens if *J* is chosen? If *G* and *J* are circled, we need three lowercase letters to satisfy Rule 1, and *k*, *l*, and *o* will fit the bill quite nicely, since we know we can't have *m* according to Rule 2, so as a group, *G*, *J*, *k*, *l*, *o* is perfectly acceptable. We can have *G* and *J*, so we can eliminate (A). (B) looks a little wordy—let's jump to (C), which says *H* is not chosen. If you circle *G* and *H*, you could fill out an acceptable group—you could have *G*, *H*, *k*, *l*, *m* or *G*, *H*, *l*, *m*, *o* or *G*, *H*, *k*, *l*, *o*, etcetera. Let's jump to (E) which says *F* isn't chosen. Well, if we circle *F* along with *G*, we know that we have to circle *J*, since Rule 3 says if *F* is chosen, *J* is chosen. But we have to choose five, and we've got three capitals circled here—there'd be no way to get three lowercase letters. If *G* is chosen, *F* can't be chosen—since we can disprove the statement in (E), choice (E) is correct.

For the record let's look at choices (B) and (D). The grouping *G*, *J*, *k*, *l*, *o* shows that (D) needn't be true and the grouping *G*, *k*, *l*, *m*, *o* shows that (B) needn't be true.

Game 3: Questions 11–14

The Action

This oddball game involves matching up dogs and trainers, then moving them around according to different commands. We have three trainers, *L*, *M*, and *O* for short and three dogs with the same initials—*l*, *m*, and *o*. We have three rooms, rooms 1, 2, and 3. We start with an initial room assignment for each trainer and each dog: *Ll* in Room 1, *Mm* in Room 2, and *Oo* in Room 3.

The Rules

1) Command W involves trainers only—the trainer in Room 1 moves to Room 2, the trainer in Room 2 moves to Room 3, and the trainer in Room 3 moves to Room 1.

2) With command *X*, the dogs in Rooms 1 and 2 switch places, and the dog in Room 3 stays put.

3) When command *Y* is called, the dogs in Rooms 2 and 3 switch places.

4) When command *Z* is called, the dogs in Rooms 1 and 3 switch places.

5) Command *A* is trickier—the dogs return to their original trainers. So *l* returns to *L*, *m* to *M* and *o* to *O*.

11

(B) Question 11 is basic—*Ll* in Room 1, *Mm* in Room 2, and *Oo* in Room 3. Command *W* is called, and Luis in Room 1 moves to Room 2, Molly in Room 2 moves to Room 3, and Oprah in Room 3 moves to Room 1. So we have in Room 1, *Ol*—the lowercase letters, the dogs, have stayed put—in Room 2 *Lm* and in Room 3 *Mo*. Which choice is true of that arrangement? (B), Molly is in Room 3. Yes, she's there with Onyx.

Looking at the wrong choices, (A) is out because Oprah is with Lassie, not with Mugs. (C) is out because Molly is with Onyx. (D) is out since Luis is in Room 2, and (E) is out because Luis is with Mugs, not with Onyx.

12

(C) We need to get Onyx to Room 2 from Room 3. One *Y* would do it—the dogs in Rooms 2 and 3 would switch, but that's not a choice. The only single command here is the one call of *W* in choice (A), but that moves trainers, not dogs, so it's out. The two calls of *X* suggested by (B) would leave Onyx where she started in Room 3. What about (C), two *W*s, then one *A*? Well, the *W* command moves only trainers so we'd have Oprah in Room 1, Luis in Room 2, and Molly in Room 3. A second call of *W* would move Molly to Room 1, Oprah to Room 2,

and Luis to Room 3. Then if we got a call of *A*, all of the dogs would seek out their original trainers—Mugs with Molly in Room 1, Onyx with Oprah in Room 2 and Lassie with Luis in Room 3—and Onyx would end up in Room 2. That's what we're looking for.

Let's see why choices (D) and (E) are wrong. (D) suggests two *W*s and one *Z*. The two *W*s give us *Ml* in Room 1, *Om* in Room 2, and *Lo* in Room 3. The *Z* would switch the dogs in Rooms 1 and 3, putting Onyx in Room 1, not in Room 2. The same for (E)—it puts Onyx in Room 1 because the two *X* calls cancel each other out and the *Z* would switch Lassie and Onyx—we're trying to get Onyx into Room 2, not Room 1. So (C) is correct for Question 12.

13

(B) There's a long way and a short way to do Question 13—the long way is to try out each sequence and see which one yields *O* and *l* in Room 2. The short, and smart way, is to think it through—see where *O* and *l* are in the original arrangement and see what commands would move them to Room 2. Thus, to get Oprah from her original position in Room 3 to the desired position in Room 2 you'd have to use *W*, the only one that moves trainers, twice. The first *W* moves Oprah to Room 1, and the second *W* moves her to Room 2. We also want to get Lassie from Room 1 to Room 2, and to do that we need one *X*. Do any of the choices have two *W*s and one *X*? (B) does, and it's correct for Question 13.

Let's look through the wrong choices. (A)'s sequence of *X*, *Y*, and *W* puts Oprah in Room 1 and Lassie in Room 3. (C)'s suggestion of *Z*, *W*, and *A* puts Oprah again in Room 1, but Lassie in Room 2. (D) puts *O* and *l* together but in Room 1 rather than Room 2. (E) puts Oprah in Room 2, but Lassie's in Room 3. Again, it's (B).

14

(C) We can answer this question the same way we answered Question 13. We want to get Molly and Onyx in Room 1, Oprah and Mugs in Room 2, and Luis and Lassie in Room 3. Let's deal with the trainers first. In order to get from *L, M, O* to *M, O, L*, you call *W* twice. Call *W* once, you'll get *O, L, M* in rooms 1, 2, and 3 respectively—call it again and you'll get *M, O, L*, which is what we want. Now for the dogs—how to go from *l, m, o* to *o, m, l*? One call of *Z* will do it. Command *Z* will make the dogs in Rooms 1 and 3 change places, giving us *o, m, l* as desired.

But two *W*s and one *Z* isn't a choice, so there must be some other way to get this arrangement. What if we tried calling *A* between the two calls of *W*? The first *W* would give us Oprah in 1, Luis in 2, and Molly in 3 with the dogs in their original positions. But if you called *A*, the dogs would seek out their original trainers, so you'd have Oprah and Onyx in Room 1, Luis and Lassie in Room 2, and Molly and Mugs in Room 3. Keeping in mind that we'll have a second *W* for the trainers, what will line up the dogs? *Y* will, putting Mugs in Room 2 and Lassie in Room 3 and we'd have our dogs all set. Our second *W* will get the trainers in the proper place, so we've got our answer, *W, A, Y, W*, choice (C).

Game 4: Questions 15–18

The Action

Seven people to distribute in eight apartments—one will remain empty. You can think of it as seven people and one nonperson to be distributed in the eight apartments, which are arranged in two levels of four adjacent apartments. The action of the game is taking your people, *P, Q, R, S, T, V, W,* and *E* for the empty apartment, and filling them in the eight apartments. We'll make A level the top and B level the bottom.

The Rules

1) *W* lives directly above *S*, which tells us that there has to be another apartment. *W* is on the *A* level and *S* is on the *B* level.

2) *S* and *Q* are on different levels, so *Q* must be on *A* level.

3) *P* and *T* are adjacent on the same level. Adjacent means it could be either *TP* or *PT*.

4) This jibes nicely with Rule 3—it tells us that *T* is not in one of the end apartments—*T* is not in Apartment 1 or in apartment 4. We know from Rule 3 that *P* is in an Apartment on one side of *T*, and Rule 4 tells us someone else is on the other side of *T*. So *T* is in either Apartment 2 or Apartment 3.

5) *W* is next to the empty apartment on the same level. We know that *W* is on level A, so the empty apartment is on level A.

Recap: So far we have *W*, the empty apartment, and *Q* on level A, *S* on level B—who's still up in the air? *P, T, V,* and *R*. But we know that *P* and *T* have to be adjacent on the same level, and they can't be on level A, since there are three apartments filled there, so by deduction we know that *P* and *T* are on level B. *V* and *R* are left—one will be on A, the other on B.

15

(A) Who must be on level B? Well, the only definite person on level B among the choices is *P*, choice (A) the correct answer. As for the others, choice *Q*, (B), and the empty apartment, (E), are both on Level A. As for choices (C) and (D), *R* and *V*, we know that one's on Level A, the other on B, but we don't know which is which. Again (A) is the correct answer.

16

(E) Here we learn about *W*—*W* lives in Apartment A-2 (Level A, Apartment 2). With *W* there, we put *S* directly underneath in Apartment B-2. We also know that the empty apart-

ment is either A-1 or A-3 so that it's adjacent to *W* on the same level. *T* has to be in either B-2 or B-3, because it can't be on the end, so since B-2 is occupied by *S, T* has to go in B-3 and we can put P in B-4, so (E) is correct.

As for (D), *P* is in Apartment 4 on B level, not on A level. And with choices (A) and (C), we have no more light on the *R* and *V* issue here. We know one's on Level A, the other's on B, but that's all we know. As for (B), we know the empty apartment could be either 1 or 3 on Level A. Again, (E) is correct.

17 **(E)** We're told that *R* is in A-3 and that *R* is directly above *P*. So *P* is in B-3, and we need a place for *V*. Once we have *P* in B-3, we know that since *T* can't be on the end so it can't be in B-1 or B-4, it has to be next to *P* in B-2. Now turn to your other large unit, the *W, S,* empty unit. Either *W* and *S* could be in the two Apartment 1s or in the two Apartment 4s. But they can't be in the two Apartment 4s since the empty apartment has to be next to *W* on the same level. The apartment next to A-4 is A-3 and that's occupied by *R*. So *W* and *S* have to be in A-1 and B-1 respectively, and the empty apartment goes next to *W* in A-2. *Q* would go into A-4 and we've got the place for *V*—only B-4 is left. That's (E), which is the correct answer.

18 **(C)** Put Q in A-2, directly above *T*, which is in B-2. The question asks about A-1, so we'll keep an eye on it as we make our deductions.

Now we turn to our *W, S,* empty apartment unit. Since *W* and the empty apartment have to be adjacent on Level A, there's only one place for them—in A-3 and 4, in either order. So only two people could go in A-1. *Q, W* and the empty apartment are out of the running, and *T, P,* and *S* are out because they're on Level B. It's between *V* and *R*—but since we're only looking for possibility here, either one will do. *R* is

not an answer choice but *V* is—it's (C), which is correct for Question 18.

Game 5: Questions 19–22

The Action
Seven people to distribute into three groups. Let's put senior members in capitals, *F, G, H,* and the junior members and applicants in lowercase letters, *k, m* and *p, r* —the only distinction that's significant is senior members from the rest since there has to be one senior member in each game. There are three different groups—backgammon with two people, chess with two, and dominoes with three.

The Rules
1) Seven people—seven slots.
2) We have three senior members, so we have one capital letter in each group.
3) *G* and *r* don't play the same game.
4) *H* and *p* must play the same game. They could play backgammon or chess or they could be two-thirds of the domino group, so if *H* and someone other than *p* are playing the same game (or *p* and someone else), they must play dominoes.
5) *m* doesn't play dominoes.

19 **(D)** Try Rule 5—both (A) and (C) violate it by having *m* play dominoes. Rule 4 looks helpful—*H* and *p* play the same game, but all the remaining choices, (B), (D), and (E) comply. Rule 3 tells us that *G* and *r* can't play the same game—(E) has them together. Rule 2 says we need a senior in each game. (B) has *F* and *H,* two seniors both playing dominoes and no senior playing chess, so (B) is out.

20 **(D)** Neither *k* nor *r* is a senior and we need a senior for each game—and only dominoes has three people playing it. Who can be the senior? It can't be *H*, since *H* is with *p*. It can't be *G* either since

G and *r* can't play the same game, so it's *F*. *F*, *k,* and *r* are playing dominoes, making (D) correct.

21 **(A)** Since *r* is lowercase we know that the other backgammon player must be a senior. *G* and *r* can't play together, so *G*'s out. Since *H* and *p* are together, it can't be *H*. It must be *F*, so *F* and *r* are playing backgammon.

Now go to your largest unit—the *Hp* unit. Either *H* and *p* are the entire chess group or two-thirds of the domino group. If they play chess, *G*, *m*, and *k* play dominoes but *m* can't play dominoes, so this won't work and *H* and *p* must play dominoes. With *F* in backgammon and *H* in dominoes the other senior, *G*, must play chess with *m*, and *k* must play dominoes with *H* and *p*. When *r* plays backgammon, there's only one grouping possible, so (A), one, is the correct answer for this question.

22 **(B)** Which group can't play the same game? What about *H* and *r* ? Remember when *H* is with someone else, *p* comes along so we'd have *H*, *p*, and *r* together, and they'd have to play dominoes. That would leave, say, *F* and *m* playing backgammon and *G* and *k* playing chess—no problem, so (A) can't be our answer. Nix (E), *p* and *r*, since we know that *H*, *p*, and *r* playing dominoes is okay. What about (B)? Well, *k* and *m* are both lowercase, and we need a capital for each, so *k*, *m*, plus a capital would play dominoes (since there are three of them.) But *m* can't play dominoes, so *k* and *m* can't play together, making (B) correct for Question 22.

For the record, *F* and *m*, (C), would be fine. We could have *F* and *m* playing backgammon, *G* and *k*

playing chess and *H*, *p,* and *r* playing dominoes. We've seen (D)'s suggestion of *G* and *m* before. *G* and *m* could play backgammon, *F* and *k* chess, and *H*, *p*, and *r* dominoes. The impossible pair is (B), the correct answer.

23 **(D)** According to the passage, when foreign aid money is tied, nation A gives money to nation B with the understanding that B will use the money only to buy A's products. That way, nation A makes most of its money back. The author says that European nations are phasing out this practice in order to avoid criticism leveled at other donors, "notably Japan." The inference to be drawn here is that Japan has been criticized for tying its foreign aid, so (D) is the inference we're looking for.

(A) isn't inferable because the passage discusses only *one* non-European nation, Japan, and its foreign aid policy. (E) says the same thing, that non-European nations are out for their own profit—one comment about Japan doesn't let you make sweeping inferences about non-European nations.

Choices (B) and (C) make statements of opinion—(B) about the role of ethical considerations and (C) about how to help underdeveloped countries—the author doesn't make any policy recommendations so (B) and (C) are wrong.

24 **(D)** The author argues that we must accept inconvenience if we want to secure the well-being of our world. Most pollution is caused by vehicle fuel and, according to the author, it "must be cut regardless of the costs." That's best summarized by (D). We must do what's necessary, no matter how drastic and costly, to save the environment.

The closest choice is probably (B), but the if/then statement in (B) argues that a lower rate of car use would be sufficient to drastically reduce pollution.

The author doesn't say that driving less is *sufficient* to cut pollution, but rather that it is *necessary* to cut pollution. (C) brings in the use of fuel by industry, which is outside the scope here. (A)'s saying that we've got to go back to the nineteenth century is too extreme to describe this argument. Finally, (E) states a causal relationship not implied in the stimulus—that people overuse their cars because they don't care about the environment. Again, (D) is correct.

25 (A) When the commissary serves fish on workdays, all the actors eat there. When the commissary doesn't serve fish on workdays, none of the actors eat there. What happens on a nonworkday? We don't know. So you're asked what it means when all the actors are eating in the commissary. One possibility is that it's a workday and they're serving fish, but there's still the question of nonworkdays. If all we know is that the actors are eating in the commissary, it's a workday and fish is served or it's not a workday and fish may or may not be served on nonworkdays. No choice says exactly this, but (A) comes the closest. (A) gives you three possibilities—that it's not a workday, that the commissary is serving fish, which would have to be true on a workday, or both, that it's a nonworkday and fish is being served. So (A) is the correct answer for this question.

(E) was tricky—it lists only one of the possible types of days when the actors could be found in the commissary, the workday with fish. (B) also doesn't say anything about nonworkdays. (C) and (D) have the opposite problem, claiming that any day the actors are in the commissary is a nonworkday, but we know that's not true because the actors always eat in the commissary on workdays when they're serving fish. (A) is correct.

Game 6: Questions 26–29

The Action
A lock has to be opened by pressing a combination of symbols. Each combination has five symbols: four letters and one number.

The Rules
1) This rule sets limits on the game—you'll be working with two basic options. Option 1 is the first situation described, where the number is second in the sequence. Option 2 is the second situation, where the number is the third symbol in the sequence.
2) Put a mark underneath the fourth and fifth space in each option.
3) We're dealing with Option 2, the one with the number in the third position. In Option 2, the fifth symbol must be either *B* or *D*, so we can write that under the fifth space in Option 2, *B* or *D* only.
4) The third symbol is a letter so you're dealing with, Option 1. In that combination there can't be any *F*s or *G*'s.
5) There are different ways to take note of this—some of you may have chosen to circle the rule, others of you might have put something into your scratch work. Just don't forget it!

Recap: In Option 1, the number appears in the second place, there can't be an *F* or a *G*, and the fourth and fifth letters can't be the same. In Option 2, the number is in the third place, the fifth must be *B* or *D*, and the fourth place and the fifth place can't be the same. And the overall rule for the game is that the first letter has to be the letter that's closest to the beginning of the alphabet.

26 **(D)** The quickest way to get through this is to take the rules and check the choices against them. Rule 1 eliminates (E) since it has the number in the fourth position. (C) can be thrown out because it has G in the fourth and the fifth positions—a Rule 2 no-no. Move on to Rule 3: The only one left with a number in the third position is (A) and (A) obeys Rule 3. Rule 4 applies to (B) and (D) because it talks about having a number in the second position—you can't have any *F*s or *G*s and (B) violates that rule with an *F* in the first position. Rule 5 eliminates (A). The letter *B* at the end of (A) violates Rule 5. (A) begins with the letter *E*, meaning that all the rest of the letters must appear later in the alphabet than *E*, and *B* doesn't. That leaves correct choice (D).

27 **(E)** You can eliminate (B) and D) right off—numbers only appear in the second or the third position. Then (C) goes, since you can't start a sequence with *Y*—think about the alphabet— *X*, *Y*, *Z*. Only one letter comes after *Y*, so you'd be left with only one letter to fill the rest of the sequence, and Rule 2 says that you can't have the fourth and fifth symbols the same. That leaves with (A), *F*, and (E), *E*. Seeing the F should make you suspicious—in Option 1 you can't use *F* at all, and in Option 2, where you've got the number in third place. In Option 2 you must end the sequence with *B* or *D*. If you start with *F*, you begin with a letter later in the alphabet than *B* or *D*, violating Rule 5. So *F* can't begin the sequence in Option 2 and *E*, (E), is the correct answer.

28 **(A)** You're told about a combination with *B* first and *G* fourth. When you see *G*, you know you'll be dealing with Option 2 only because Option 1 can't contain *G*s. In Option 2, *B* is first, and you don't know what's second. The number is third, *G* is fourth, and fifth is either *B* or *D*, since in Option 2 fifth is either *B* or *D*. You've used *B* first, so you're left with only *D* for the fifth position. Your sequence has *B* first, you don't know what's second, a number third, *G* fourth, and *D* fifth. *D* has to be fifth and only (A) gives you that option, so it's correct.

29 **(D)** The first thing is to figure out why the combination isn't acceptable. You have a number third so it's Option 2. You end with *B* or *D* and this one ends with *F*, so we need to switch that *F* for a *B* or a *D* and make the sequence correct. (A) offers to replace *F* with *B* but remember Rule 5—the first symbol must be closest to the beginning of the alphabet. If you replace *F* with *B* you break that rule because the sequence begins with *C*. The only way to put *B* at the end of this and still have an acceptable sequence would be to change the *C* to an *A* but you can't do that, so (A) won't work. Skim down to (D), replace the *F* with a *D*, which works with the rules for Option 2, the sequence ends with a *D*, and *D* comes later in the alphabet than *C*, so you're obeying Rule 5. So (D) is correct.

As for the others, if you do (B) and reverse the *C* and *P* you have Option 2 ending with an *F*—not acceptable. With (C) you reverse the *Q* and the 8 and you have Option 1 with the number second—but you can't have an *F* in Option 1, so (C) won't work. Finally, (E) says replace the *C* with an *A*. Well, we talked about that when we were talking about (A). If you did (A) and (E) together you'd have an acceptable sequence but replacing the *C* with an *A* does not solve the problem of having the *F* at the end which isn't acceptable when the number is third. So (D) is correct.

30 **(E)** The first sentence is evidence: some scientists argue that our planet may support more sources of protein than we think if presently unfished areas have as many fish in them as do the areas presently fished. The author then concludes that we can provide protein to the whole world even if

population grows at its present rate. But his conclusion isn't expressed conditionally, and his evidence is. He's just *assuming* that there are as many fish in the unfished part of the ocean as there are in the fished part. Furthermore, he's assuming that the scientists are right, that those fish would mean the planet supports more protein than we believe. To conclude that we can feed the hungry masses, the author must assume that we can get the fish in the unfished areas, but maybe those areas are unfished because we can't fish them. Since we need to weaken the argument, we need an answer choice that contradicts one of these assumptions. Correct choice (E) denies the last one. Choice (E) says it will take 30 years before we can fish those areas, so we can't ensure the availability of protein to everyone over the next two decades.

As for the others, (A) supplies the view of some scientists that fish are less plentiful in the unfished areas. This isn't as damaging as (E). (B) is close, but the cost of the technology isn't as damning an obstacle as its availability. (C) focuses on cost. The author doesn't argue that the world can be fed *cheaply*. As for (D), cutting the population growth certainly doesn't weaken the argument. So (E) is correct.

31 (A) It's possible for travelers to enter and remain in the Republic for anywhere from one to 59 days, but there's at least one condition for stays of longer than seven days. Although it's possible to stay for more than seven days, if you do so, you need a special visa. Correct choice (A) is simply a statement in which the antecedent, or if-clause, of the original is affirmed and the consequent, or then-clause, flows from it, just as it's supposed to. If a traveler wants to stay 14 days, then a special visa is required. This jibes perfectly with the if/then statement in the stimulus, so (A) it is.

(B) and (D) are wrong because each implies that some travelers don't need visas. We don't know that travelers staying fewer than seven days don't need

anything. Maybe all travelers to the Republic need visas of some kind, so neither (B) nor (D) must be true. (C) fails by bringing up the topic of whether or not people have the visas they require. All the stimulus tells us is when a special visa is required. (E) comes out of thin air. Nothing precludes the possibility that every person in the Republic needs a visa, even those just passing through, so (E) isn't inferable from the stimulus, and (A) is correct.

32 (B) The conclusion here is that the U.S. economy continues to grow and prosper. As evidence, the author cites the expansion over the last 15 years of the service sector, where last year alone 500,000 Americans found employment. She assumes that this growth correlates to growth in the economy. But what if declines in other sectors offset the growth in service? If, as correct choice (B) says, growth in the service sector can be at least partly attributed to a decline in the manufacturing and heavy industry sectors, then growth in the service sector can't be a reliable indicator of growth in the overall economy.

(A) tends to support the conclusion—job offers imply health, contributing to a sense that the economy isn't in bad shape. (C) doesn't do much to affect the author's conclusion. Just because the American economy isn't sluggish doesn't mean it's growing and prospering. (D) can be eliminated because the author is claiming that the American economy is prospering—she isn't claiming that it's prospering more than ever. Finally, (E) weakens the argument a bit, suggesting that some of the evidence for the claim of economic growth and health isn't as central as the author believes. But using the service sector as a barometer of economic growth may be valid, regardless of the doubt (E) casts on how much of that growth is caused by the service sector. Since (E)'s ability to weaken the argument is dubious while (B)'s is certain, it's (B) for this question.

Game 7: Questions 33–37

The Action

Three bells, a high, a medium, and a low, and eight rings. You ring the low bell three times, the medium bell three times, and the high bell two times.

The Rules

1) The sixth ring is the medium bell—put it in.
2) You'll have to split the low rings up—they will always be separated by medium and high rings.
3) The two high bells will stick together throughout the game.

33 **(B)** Starting with Rule 1, (C) puts a low bell sixth, which can't be true. Rule 2 won't let us ring the low bell twice in succession, so dump (D) and (E). And Rule 3 eliminates (A) by splitting up the high bells, leaving us with (B), the correct answer.

34 **(E)** We have to ring the high bell fifth, so we've got to ring the other high bell fourth—the two high bells have to stay together and we've got the medium bell ringing sixth. Now we have to split up the low bells. We'll have to put two before this high-high-medium set, and one after. So we'll put one low first, one low third, then fill the space between with a medium. The beginning looks like this—low-medium-low-high-high-medium. You've got one low and one medium left. Does it matter which goes seventh and which eighth? No, either way would be acceptable. (A), the low bell is first, yes. (B) the medium bell is second, yes. (C), the low bell is third, yes. (D), the high bell is fourth, yes. (E), the low bell is rung seventh—it could be true, but it could be eighth. So (E) could be true, but it doesn't have to be true, so it's correct.

35 **(D)** We're told that we have a medium bell fourth and we know that there's a medium bell sixth. Sketching it out, we have eight spots for bells to be rung, and medium bells fourth and sixth. When can't the high bell be rung? Remember when we're talking about the high bell, we're talking about both high bells because they stay together. So you can't ring a high bell in five because there's no room for the other high bell next to it.

(A) puts the high bell first, and you'd ring the other high bell second and split up the low bells third, fifth, and seventh. (B) puts the high bell second and you can use the same sequence and ring a high bell second. (C) puts a high bell third, the other high bell second, and splits up the low bells by placing one first, one fifth, and one seventh. And (E) puts the high bell eighth. You can ring the other high bell seventh and still split up the lows by placing one first, one third, and one fifth. So (D) is the correct answer.

36 **(D)** If you sketch out a sequence, one of the things you can see is that the third, fourth, and fifth group buts up against the medium bell in sixth. So all high and medium bells would be unacceptable in third, fourth, and fifth because you'd have a solid group of high and medium in the center and no way to split up those low bells on the ends. (D) has a high in third, a medium in fourth, and a medium in fifth, and it's the correct answer.

As for the wrong choices, (A) has high, medium, and low in third, fourth, and fifth. That works if we put a low first, the other high second, next to the high bell in third, and a medium and a low in seventh and eighth, if the high bells are together and the lows are split up. (B) has low, medium, and low, in third, fourth, and fifth. That makes it easy to split up the lows—we can put the third one in first and we still have two spaces at the end to keep the two highs together. (C) has the two highs in third and fourth

and a low in fifth. We put a low first and a medium second next to the high in space three. That leaves seventh and eighth to put the other low and the other medium. Finally, (E) has high, low, and medium in third, fourth, and fifth. You put the high bell in second so it's next to first and third, and you put one low first and the other in seventh or eighth, with the other medium to keep them split up. (D) is correct.

37

(C) (A) mentions ringing the high bell first. This may have "rung a bell" because, having done Questions 9 and 10, we've discussed whether or not it's possible to ring the high bell first—yes, it's acceptable. If you remembered that, you don't need to work out a sequence again. Let's skip to (D)—it says the high bell is fourth—you know that this is all right from Question 9. How about (E), ringing the low bell fifth? When we worked on Question 10 in trying out the possibilities we put the low bell fifth, so you know that this is acceptable. (B) says the low bell is rung second—what I did was put the low bell second and the medium first, put the two highs third and fourth, a low fifth, medium sixth, low seventh, medium eighth. That's acceptable, so (C) is correct—you can't ring the medium bell third. If you have a medium bell ringing third and another sixth, you have three groups of two spaces, first and second, fourth and fifth, seventh and eighth. One of those pairs has to contain the high bells but then you have three low bells to split up, and no way to do that. So the correct answer is (C).

Game 8: Questions 38–42

The Action

Try a simple tack—break it into two flow charts, one following the Priority 1 mail and one following the Priority 2 mail—it's actually much simpler.

Information overload? Here's a breakdown: six cities, two types of memos, and the basic idea that they're sent from the head office to the branches. The second introductory paragraph: any branch that receives a memo from the head office has to pass it on to at least one other branch. The other branch can pass it on but it doesn't have to.

The real key to your work on this game is the second set of rules about which branches can send memos to which other branches. Look at the last rule—it says that Fresno can't send memos to any other branches. What that means in terms of the game is something very simple: Fresno is a dead end. Let's look at the questions.

38

(B) One thing to notice is that you have to consider both Priority 1 and Priority 2, because both are sent from home to Atlanta. Check out both flow charts, and you notice that when memos go to Atlanta, in both cases, the next place they go is to Caracas. So the memos must leave Atlanta and go to Caracas, and that makes (B) correct. All of the wrong choices are only "could be trues"—you could send the memo on to Beijing, you could send it on to Dakar, to Edinburgh, to Fresno —but you don't have to.

39

(E) Four of the choices describe routes of travel that the memo *could* have followed and one describes a route that the memo *could not* have followed and that's (E), a Priority 2 memo initially sent to Beijing. Take a look at the Priority 2 flowchart—start at the home office, send the memo

to Beijing, and then send it where? The only place that you can send it is Fresno and Fresno is a dead end. So it can't be a memo sent from Edinburgh to Fresno, so (E)'s correct. All the others work. (A), you can send a Priority 1 memo from Atlanta to Caracas to Beijing to Edinburgh to Fresno. (B), you can send a Priority 1 memo to Dakar to Caracas to Beijing to Edinburgh and then to Fresno. (C) is fine, a Priority 1 memo can be sent to Dakar, and then to Caracas, Beijing, Edinburgh, and Fresno. (D) is also fine, a Priority 2 memo can be sent to Atlanta to Caracas to Dakar to Edinburgh and then to Fresno.

40
(A) If it wasn't originally sent to Atlanta, where was it sent? With Priority 2, the only places a memo can go from home are Atlanta and Beijing. If it didn't go to Atlanta, it went to Beijing. This will ring a bell because we're following the same path that we followed in the last question. The only place a Priority 2 memo can go after Beijing is Fresno, a dead end. Only two branches, Beijing and Fresno, could have seen the memo, and (A) is correct.

41
(C) If the memo didn't go through Atlanta, where did it go? A Priority 1 memo would go to Dakar and a Priority 2 memo would go to Beijing. We want it to end up in Edinburgh—does that ring a bell? In Priority 2 we're dealing with the same path—a Priority 2 memo starting at home and going to Beijing then goes to Fresno, a dead end, so it can't do what this question asks. So all you have to do is concentrate on your Priority 1 system and see how a memo would go from Dakar to Edinburgh. After it goes to Dakar, the only place it can go is to Caracas, and from Caracas you could send it back to Dakar—but you want it to move toward Edinburgh. Send it to Beijing and the only place it can go is Edinburgh. So you take a Priority 1 memo and send it from home to Dakar to Caracas to Beijing then to Edinburgh. The question asks you

how many branches saw this memo besides Edinburgh. Dakar, Caracas, and Beijing, that's three, and the answer is (C).

42
(D) Here you'll have to try out both Priority 1 and Priority 2 memos. (A) asks if you can go from Atlanta to Caracas to Beijing. Yes, in both Priority 1 and Priority 2. (B) talks about going from Atlanta to Caracas to Beijing to Edinburgh. You know that both Priority 1 and Priority 2 you can go from Atlanta to Caracas to Beijing. Can you keep going to Edinburgh? Yes, in Priority 1—that's where you go from Beijing with Priority 1 memos. So (B) won't do it. As for (C), in Priority 2 you can go from Atlanta to Caracas to Dakar to Edinburgh. (D) suggests sending a memo from Beijing to Edinburgh to Fresno. A Priority 1 memo can't go to Beijing from the head office. The only way to get a Priority 1 memo to Beijing is through Atlanta or Dakar to Caracas and then on to Beijing. So (D) can't be the complete path of a Priority 1 memo from the home office because it can't start in Beijing.

As far as Priority 2 goes, when a Priority 2 memo leaves the head office and goes to Beijing, the only other place it can go is Fresno, the dead end. (D) describes a path that's impossible for both Priority 1 and Priority 2, so it's correct. (E) suggests sending from Dakar to Caracas to Beijing. That works in Priority 1; you can start in the home office, go to Dakar to Caracas to Beijing.

Game 9: Questions 43–47

The Action
You have to sequence trophies on Shelves 1, 2, and 3, from top to bottom.

The Rules

2) Rule 2 seems to be the most helpful so let's look at it first. *F* must be on the shelf immediately above the shelf that *L* is on. You have two basic options. In Option 1 you place *F* on Shelf 1 and L on Shelf 2. With Option 2 you put *F* on Shelf 2 and *L* on Shelf 3.

1) In Option 1, we can write next to Shelf 2 "no *J*," and in Option 2, we can write next to Shelf 3 "no *J*."

3) No shelf can hold all three bowling trophies.

4) *K* can't be on Shelf 2—that's for either option.

43

(B) *G* and *H* are on Shelf 2, so if you remember that three bowling trophies can't be on the same shelf, this tells us that we must work with Option 1. If you put *G* and *H* on Shelf 2 in Option 2, you'd be breaking Rule 3—you'd have all three bowling trophies on the same shelf. So you'll have *F* on Shelf 1, and *L*, *G*, and *H* on Shelf 2. What must be true? Take a look at (B), *L* is on Shelf 2. Yes, we just went through the deduction whereby you realize you must use Option 1 in which *F* is on Shelf 1 and *L* is on Shelf 2. So (B) is the correct answer.

44

(D) Right away we realize that you can't use Option 2 here because Option 2 already has a tennis trophy on Shelf 3, *L*, so you will work with Option 1, *F* on the first shelf and *L* on the second shelf. You know that neither *J* nor *K* can appear on Shelf 2 in Option 1. *J* and *K* are tennis trophies, so if the question specifies that you can't have a tennis trophy on Shelf 3 and you can't have these two trophies on Shelf 2, then the only place for them is on Shelf 1. In other words, *K* and *J* must be on the same shelf, so (D) is correct.

45

(C) This question is directing you to Option 2, because you already know that *J* isn't allowed on Shelf 2 in Option 1. With Option 2 you know that *F* must appear on Shelf 2, so (C) is correct.

46

(D) In only one option can Shelf 1 remain empty, Option 2. The rest of the question says, "Which of the following must be false?" which means "Which of the following arrangements won't work?" First, let's look at the basic situation. We have Option 2 and we have *F* on 2 and *L* on 3, and Shelf 1 remains empty. That tells us that we can do something with *J* and *K*. We know in Option 2 that *J* can't go on Shelf 3 and Shelf 1 is empty, so the only other place for it is Shelf 2. We know that *K* can't be on Shelf 2 and Shelf 1 is empty, so the only home for K is Shelf 3. So we have Shelf 1 empty, Shelf 2 with *F* and *J*, and Shelf 3 with *L* and *K*. What to do with *G* and *H*? The only thing we can't do is put them on Shelf 2 because that would violate Rule 3. So if we keep them together we have to put them on Shelf 3. If we split them up, we can put *G* on 2 and *H* on Shelf 3 or vice versa.

(A), can we put *H* and *F* on the same shelf? Sure, we've already said we can put one of *G* and *H* on Shelf 2 and one on Shelf 3. (B), can we put exactly three trophies on Shelf 2? Sure, we just did with (A). We put *F*, *J*, and *H* together on Shelf 2 and that left us with *L*, *K*, and *G* together on Shelf 3. (C)—can we put *G* and *H* on the same shelf? Yes, as long as they're on Shelf 3 and not on Shelf 2. (D), can we put exactly two trophies on Shelf 3? We have on Shelf 3, *L* and *K*. To have exactly two trophies on Shelf 3, we would put both *G* and *H* somewhere else and we can't put *G* and *H* together on Shelf 2 because that would violate Rule 3. So (D) is our answer here—it's the thing we can't do. (E), can we put *G* and *K* on the same shelf? Yes, whether *G* is alone or together with *H*, it's possible to do this.

47 **(A)** This is hard because the if-clause doesn't narrow it down to one of the two options. *L* and *G* can be on the same shelf in both options, which makes your work more complicated. In both options there's just one empty shelf—in Option 1 it's Shelf 3, and in Option 2 it's Shelf 1. Let's see if we can make any more deductions about both options. In Option 1, if we have to leave Shelf 3 empty, we can figure out what to do with *K* and *J* because they can't be on Shelf 2 and Shelf 3 is empty, so Shelf 1 has *F*, *K*, and *J* and Shelf 2 has *L* and *G* and the only thing left is *H*, on either Shelf 1 or Shelf 2. In Option 2 we know that *K* can't be on Shelf 2, and Shelf 1 has to be empty, so the only place for *K* is Shelf 3. *J* can't be on Shelf 3 in Option 2, Shelf 1 is empty, so *J* is on Shelf 2. So we end up with *F* and *J* on Shelf 2, *L*, *K*, and *G* on Shelf 3 and Shelf 1 empty, and *H* is a floater.

For the answer to be correct, it must be true in both options—you hit pay dirt right away, because (A) is correct. It says if *H* is on Shelf 3, then *J* is on Shelf 2. The only way to put *H* on Shelf 3 is Option 2, where Shelf 3 is open. You can put *H* on Shelf 3, and in Option 2, *J* is on Shelf 2, so (A)'s correct. (B) describes *K* and *L* as being on the same shelf, but that's true only in Option 2. (C) says if *H* is on Shelf 2, *J* is on Shelf 3, but *J* is never on Shelf 3. (D) has *F* and *K* on the same shelf; that's true in Option 1 only and not in Option 2. And (E) has *J* on Shelf 2. That's Option 2, but it goes on to say that *H* is on Shelf 1, and in Option 2 Shelf 1 is empty. Again, (A) is correct.

48 **(E)** What if someone prefers the look of finished furniture over the look of painted furniture? Would that factor outweigh the person's desire to reduce work time and costs? We don't know—the author assumes that only the three factors he discusses—work time, cost, and longevity—determine a person's decision to paint rather than to finish.

(E) says more or less the same thing and is our answer. As for (A), the author concludes that some people might prefer painting because it costs less and it saves work time, not because it is necessarily better than finishing. (B) is a distortion of the author's conclusions. The author needn't assume that most people will consider saving time and cost more important than longevity. (C) is wrong because it falls outside the scope. The discussion is limited only to people who will paint or finish—it doesn't include people who will do neither. As for (D), the author doesn't assume that work time, cost, and longevity are equally important factors in deciding whether to paint or to finish. Choice (E) is correct.

49 **(C)** We need to find evidence that will strengthen the zoologists' conclusion, so we want to establish some connection between cubs living in captivity and an inability to hunt successfully in the wild. (C) does the trick—if cubs raised in captivity could hunt successfully in the wild, it would suggest that aggressive play is not a factor in learning to hunt. But (C) demonstrates that Cowonga lion cubs raised in captivity can't hunt successfully in the wild. Unless there are other differences, the aggressive play could very well be the cause of this. (A) doesn't strengthen a connection between hunting and the lack of aggressive play—maybe the wild cubs would be equally successful at hunting if they didn't play aggressively. As for (B), that other predatory animals also engage in aggressive play when young doesn't mean that this play is necessary for successful hunting in later life. (D) is irrelevant—just because the skills used in play are similar to the skills necessary for hunting doesn't mean that cubs learn the hunting skills through play. And (E) doesn't strengthen the zoologists' conclusion—it simply repeats the part of the evidence they cite in support of their argument. So (C) is correct.

50

(A) Two assumptions hold this argument together. First, the author decides that the survey means that the student body has become more religious. Then she decides that this is what has reduced cheating. So we'll look for a choice that suggests that either increased attendance at religious services or reduced cheating could be attributed to factors other than these. We get the former in (A). If most students attend services for social reasons, then this majority isn't attending because of increased religiosity, and this would destroy the author's primary assumption. (B) would *strengthen* the author's argument since it sums up her second assumption. If the students had really become religious, the author would be justified in asserting that the religiosity was a factor in the decrease of cheating. (C) lists the change in exam procedures made 15 years ago, but the survey compares attendance today with attendance 10 years ago, and the author's implicitly speaking of the last decade. (D) tries to attack the author's evidence, positing that not all students responded to the survey. But a survey just needs a sufficiently representative sample. (E) takes us way out of the ballpark—who said cheating was a major problem? All we know is that it's been massively reduced. The answer is (A).

APPENDICES

APPENDICES

ROOT LIST

The Kaplan Root List can boost your knowledge of GRE-level words, and that can help you get more questions right. No one can predict exactly which words will show up on your test, but there are certain words that the test makers favor. The Root List gives you the component parts of many typical GRE words. Knowing these words can help you because you may run across them on your GRE. Also, becoming comfortable with the types of words that pop up will reduce your anxiety about the test.

Knowing roots can help you in two more ways. First, instead of learning one word at a time, you can learn a whole group of words that contain a certain root. They'll be related in meaning, so if you remember one, it will be easier for you to remember others. Second, roots can often help you decode an unknown GRE word. If you recognize a familiar root, you could get a good enough grasp of the word to answer the question.

GET BACK TO YOUR ROOTS

Most of the words we use every day have their origins in simple roots. Once you know the root, it's much easier to figure out what a strange word means.

Take a look through this appendix. Many of the roots are easy to learn, and they'll help you on the test.

The Kaplan Root List

❑ A, AN—not, without
amoral, atrophy, asymmetrical, anarchy, anesthetic, anonymity, anomaly

❑ AB, A—from, away, apart
abnormal, abdicate, aberration, abhor, abject, abjure, ablution, abnegate, abortive, abrogate, abscond, absolve, abstemious, abstruse, annul, avert, aversion

❑ AC, ACR—sharp, sour
acid, acerbic, exacerbate, acute, acuity, acumen, acrid, acrimony

❑ AD, A—to, toward
adhere, adjacent, adjunct, admonish, adroit, adumbrate, advent, abeyance, abet, accede, accretion, acquiesce, affluent, aggrandize, aggregate, alleviate, alliteration, allude, allure, ascribe, aspersion, aspire, assail, assonance, attest

❑ ALI, ALTR—another
alias, alienate, inalienable, altruism

❑ AM, AMI—love
amorous, amicable, amiable, amity

❑ AMBI, AMPHI—both
ambiguous, ambivalent, ambidextrous, amphibious

❑ AMBL, AMBUL—walk
amble, ambulatory, perambulator, somnambulist

❑ ANIM—mind, spirit, breath
animal, animosity, unanimous, magnanimous

❑ ANN, ENN—year
annual, annuity, superannuated, biennial, perennial

❑ ANTE, ANT—before
antecedent, antediluvian, antebellum, antepenultimate, anterior, antiquity, antiquated, anticipate

❑ ANTHROP—human
anthropology, anthropomorphic, misanthrope, philanthropy

❑ ANTI, ANT—against, opposite
antidote, antipathy, antithesis, antacid, antagonist, antonym

❑ AUD—hear
audio, audience, audition, auditory, audible

❑ AUTO—self
autobiography, autocrat, autonomous

❑ BELLI, BELL—war
belligerent, bellicose, antebellum, rebellion

❑ BENE, BEN—good
benevolent, benefactor, beneficent, benign

❑ BI—two
bicycle, bisect, bilateral, bilingual, biped

❑ BIBLIO—book
Bible, bibliography, bibliophile

❑ BIO—life
biography, biology, amphibious, symbiotic, macrobiotics

❑ BURS—money, purse
reimburse, disburse, bursar

❑ CAD, CAS, CID—happen, fall
accident, cadence, cascade, deciduous

❑ CAP, CIP—head
captain, decapitate, capitulate, precipitous, precipitate

❑ CAP, CAPT, CEPT, CIP—take, hold, seize
capable, capacious, recapitulate, captivate, deception, intercept, precept, inception, anticipate, emancipation, incipient, percipient, cede, precede, accede, recede, antecedent, intercede, secede, cession

❑ CARN—flesh
carnal, carnage, carnival, carnivorous, incarnate

❑ CED, CESS—yield, go
cease, cessation, incessant, cede

❑ CHROM—color
chrome, chromatic, monochrome

❑ CHRON—time
chronology, chronic, anachronism

❑ CIDE—murder
suicide, homicide, regicide, patricide

❑ CIRCUM—around
circumference, circumlocution, circumnavigate, circumscribe, circumspect, circumvent

❑ CLIN, CLIV—slope
incline, declivity, proclivity

❑ CLUD, CLUS, CLAUS, CLOIS—shut, close
conclude, reclusive, claustrophobia, cloister, preclude, occlude

❑ CO, COM, CON—with, together
coeducation, coagulate, coalesce, coerce, cogent, cognate, collateral, colloquial, colloquy, commensurate, commodious, compassion, compatriot, complacent, compliant, complicity, compunction, concerto, conciliatory, concord, concur, condone, conflagration, congeal, congenial, congenital, conglomerate, conjure, conjugal, conscientious, consecrate, consensus, consonant, constrained, contentious, contrite, contusion, convalescence, convene, convivial, convoke, convoluted, congress

❑ COGN, GNO—know
recognize, cognition, cognizance, incognito, diagnosis, agnostic, prognosis, gnostic, ignorant

❑ CONTRA—against
controversy, incontrovertible, contravene

❑ CORP—body
corpse, corporeal, corpulence

❑ COSMO, COSM—world
cosmopolitan, cosmos, microcosm, macrocosm

❑ CRAC, CRAT—rule, power
democracy, bureaucracy, theocracy, autocrat, aristocrat, technocrat

❑ CRED—trust, believe
incredible, credulous, credence

❑ CRESC, CRET—grow
crescent, crescendo, accretion

❑ CULP—blame, fault
culprit, culpable, inculpate, exculpate

❑ CURR, CURS—run
current, concur, cursory, precursor, incursion

❑ DE—down, out, apart
depart, debase, debilitate, declivity, decry, deface,

defamatory, defunct, delegate, demarcation, demean, demur, deplete, deplore, depravity, deprecate, deride, derivative, desist, detest, devoid

❑ DEC—ten, tenth
decade, decimal, decathlon, decimate

❑ DEMO, DEM—people
democrat, demographics, demagogue, epidemic, pandemic, endemic

❑ DI, DIURN—day
diary, diurnal, quotidian

❑ DIA—across
diagonal, diatribe, diaphanous

❑ DIC, DICT—speak
diction, interdict, predict, abdicate, indict, verdict

❑ DIS, DIF, DI—not, apart, away
disaffected, disband, disbar, disburse, discern, discordant, discredit, discursive, disheveled, disparage, disparate, dispassionate, dispirit, dissemble, disseminate, dissension, dissipate, dissonant, dissuade, distend, differentiate, diffidence, diffuse, digress, divert

❑ DOC, DOCT—teach
doctrine, docile, doctrinaire

❑ DOL—pain
condolence, doleful, dolorous, indolent

❑ DUC, DUCT—lead
seduce, induce, conduct, viaduct, induct

❑ EGO—self
ego, egoist, egocentric

❑ EN, EM—in, into
enter, entice, encumber, endemic, ensconce, enthrall, entreat, embellish, embezzle, embroil, empathy

❑ ERR—wander
erratic, aberration, errant

❑ EU—well, good
eulogy, euphemism, euphony, euphoria, eurythmics, euthanasia

❑ EX, E—out, out of
exit, exacerbate, excerpt, excommunicate, exculpate, execrable, exhume, exonerate, exorbitant, exorcise, expatriate, expedient, expiate, expunge, expurgate, extenuate, extort, extremity, extricate, extrinsic, exult, evoke, evict, evince, elicit, egress, egregious

❑ FAC, FIC, FECT, FY, FEA—make, do
factory, facility, benefactor, malefactor, fiction, fictive, beneficent, affect, confection, refectory, magnify, unify, rectify, vilify, feasible

❑ FAL, FALS—deceive
false, infallible, fallacious

❑ FERV—boil
fervent, fervid, effervescent

❑ FID—faith, trust
confident, diffidence, perfidious, fidelity

❑ FLU, FLUX—flow
fluent, flux, affluent, confluence, effluvia, superfluous

❑ FORE—before
forecast, foreboding, forestall

❏ FRAG, FRAC—break
fragment, fracture, diffract, fractious, refract

❏ FUS—pour
profuse, infusion, effusive, diffuse

❏ GEN—birth, class, kin
generation, congenital, homogeneous, heterogeneous, ingenious, engender, progenitor, progeny

❏ GRAD, GRESS—step
graduate, gradual, retrograde, centigrade, degrade, gradation, gradient, progress, congress, digress, transgress, ingress, egress

❏ GRAPH, GRAM—writing
biography, bibliography, epigraph, grammar, epigram

❏ GRAT—pleasing
grateful, gratitude, gratis, ingrate, congratulate, gratuitous, gratuity

❏ GRAV, GRIEV—heavy
grave, gravity, aggravate, grieve, aggrieve, grievous

❏ GREG—crowd, flock
segregate, gregarious, egregious, congregate, aggregate

❏ HABIT, HIBIT—have, hold
habit, inhibit, cohabit, habitat

❏ HAP—by chance
happen, haphazard, hapless, mishap

❏ HELIO, HELI—sun
heliocentric, helium, heliotrope, aphelion, perihelion

❏ HETERO—other
heterosexual, heterogeneous, heterodox

❏ HOL—whole
holocaust, catholic, holistic

❏ HOMO—same
homosexual, homogenize, homogeneous, homonym

❏ HOMO—man
homo sapiens, homicide, bonhomie

❏ HYDR—water
hydrant, hydrate, dehydration

❏ HYPER—too much, excess
hyperactive, hyperbole, hyperventilate

❏ HYPO—too little, under
hypodermic, hypothermia, hypochondria, hypothesis, hypothetical

❏ IN, IG, IL, IM, IR—not
incorrigible, indefatigable, indelible, indubitable, inept, inert, inexorable, insatiable, insentient, insolvent, insomnia, interminable, intractable, incessant, inextricable, infallible, infamy, innumerable, inoperable, insipid, intemperate, intrepid, inviolable, ignorant, ignominious, ignoble, illicit, illimitable, immaculate, immutable, impasse, impeccable, impecunious, impertinent, implacable, impotent, impregnable, improvident, impassioned, impervious, irregular, invade, inaugurate, incandescent, incarcerate, incense, indenture, induct, ingratiate, introvert, incarnate, inception, incisive, infer

❏ IN, IL, IM, IR—in, on, into
infusion, ingress, innate, inquest, inscribe, insinuate, inter, illustrate, imbue, immerse, implicate, irrigate, irritate

❑ INTER—between, among
intercede, intercept, interdiction, interject, interlocutor, interloper, intermediary, intermittent, interpolate, interpose, interregnum, interrogate, intersect, intervene

❑ INTRA, INTR—within
intrastate, intravenous, intramural, intrinsic

❑ IT, ITER—between, among
transit, itinerant, reiterate, transitory

❑ JECT, JET—throw
eject, interject, abject, trajectory, jettison

❑ JOUR—day
journal, adjourn, sojourn

❑ JUD—judge
judge, judicious, prejudice, adjudicate

❑ JUNCT, JUG—join
junction, adjunct, injunction, conjugal, subjugate

❑ JUR—swear, law
jury, abjure, adjure, conjure, perjure, jurisprudence

❑ LAT—side
lateral, collateral, unilateral, bilateral, quadrilateral

❑ LAV, LAU, LU—wash
lavatory, laundry, ablution, antediluvian

❑ LEG, LEC, LEX—read, speak
legible, lecture, lexicon

❑ LEV—light
elevate, levitate, levity, alleviate

❑ LIBER—free
liberty, liberal, libertarian, libertine

❑ LIG, LECT—choose, gather
eligible, elect, select

❑ LIG, LI, LY—bind
ligament, oblige, religion, liable, liaison, lien, ally

❑ LING, LANG—tongue
lingo, language, linguistics, bilingual

❑ LITER—letter
literate, alliteration, literal

❑ LITH—stone
monolith, lithograph, megalith

❑ LOQU, LOC, LOG—speech, thought
eloquent, loquacious, colloquial, colloquy, soliloquy, circumlocution, interlocutor, monologue, dialogue, eulogy, philology, neologism

❑ LUC, LUM—light
lucid, illuminate, elucidate, pellucid, translucent

❑ LUD, LUS—play
ludicrous, allude, delusion, allusion, illusory

❑ MACRO—great
macrocosm, macrobiotics

❑ MAG, MAJ, MAS, MAX—great
magnify, majesty, master, maximum, magnanimous, magnate, magnitude

❑ MAL—bad
malady, maladroit, malevolent, malodorous

❏ MAN—hand
manual, manuscript, emancipate, manifest

❏ MAR—sea
submarine, marine, maritime

❏ MATER, MATR—mother
maternal, matron, matrilineal

❏ MEDI—middle
intermediary, medieval, mediate

❏ MEGA—great
megaphone, megalomania, megaton, megalith

❏ MEM, MEN—remember
memory, memento, memorabilia, reminisce

❏ METER, METR, MENS—measure
meter, thermometer, perimeter, metronome,
commensurate

❏ MICRO—small
microscope, microorganism, microcosm, microbe

❏ MIS—wrong, bad, hate
misunderstand, misanthrope, misapprehension,
misconstrue, misnomer, mishap

❏ MIT, MISS—send
transmit, emit, missive

❏ MOLL—soft
mollify, emollient, mollusk

❏ MON, MONIT—warn
admonish, monitor, premonition

❏ MONO—one
monologue, monotonous, monogamy, monolith,
monochrome

❏ MOR—custom, manner
moral, mores

❏ MOR, MORT—dead
morbid, moribund, mortal, amortize

❏ MORPH—shape
amorphous, anthropomorphic, metamorphosis,
morphology

❏ MOV, MOT, MOB, MOM—move
remove, motion, mobile, momentum, momentous

❏ MUT—change
mutate, mutability, immutable, commute

❏ NAT, NASC—born
native, nativity, natal, neonate, innate, cognate,
nascent, renascent, renaissance

❏ NAU, NAV—ship, sailor
nautical, nauseous, navy, circumnavigate

❏ NEG—not, deny
negative, abnegate, renege

❏ NEO—new
neoclassical, neophyte, neologism, neonate

❏ NIHIL—none, nothing
annihilation, nihilism

❏ NOM, NYM—name
nominate, nomenclature, nominal, cognomen,
misnomer, ignominious, antonym, homonym,
pseudonym, synonym, anonymity

❏ NOX, NIC, NEC, NOC—harm
obnoxious, noxious, pernicious, internecine,
innocuous

❏ NOV—new
novelty, innovation, novitiate

❏ NUMER—number
numeral, numerous, innumerable, enumerate

❏ OB—against
obstruct, obdurate, obfuscate, obnoxious, obsequious, obstinate, obstreperous, obtrusive

❏ OMNI—all
omnipresent, omnipotent, omniscient, omnivorous

❏ ONER—burden
onerous, onus, exonerate

❏ OPER—work
operate, cooperate, inoperable

❏ PAC—peace
pacify, pacifist, pacific

❏ PALP—feel
palpable, palpitation

❏ PAN—all
panorama, panacea, panegyric, pandemic, panoply

❏ PATER, PATR—father
paternal, paternity, patriot, compatriot, expatriate, patrimony, patricide, patrician

❏ PATH, PASS—feel, suffer
sympathy, antipathy, empathy, apathy, pathos, impassioned

❏ PEC—money
pecuniary, impecunious, peculation

❏ PED, POD—foot
pedestrian, pediment, expedient, biped, quadruped, tripod

❏ PEL, PULS—drive
compel, compelling, expel, propel, compulsion

❏ PEN—almost
peninsula, penultimate, penumbra

❏ PEND, PENS—hang
pendant, pendulous, compendium, suspense, propensity

❏ PER—through, by, for, throughout
perambulator, percipient, perfunctory, permeable, perspicacious, pertinacious, perturbation, perusal, perennial, peregrinate

❏ PER—against, destruction
perfidious, pernicious, perjure

❏ PERI—around
perimeter, periphery, perihelion, peripatetic

❏ PET—seek, go toward
petition, impetus, impetuous, petulant, centripetal

❏ PHIL—love
philosopher, philanderer, philanthropy, bibliophile, philology

❏ PHOB—fear
phobia, claustrophobia, xenophobia

❏ PHON—sound
phonograph, megaphone, euphony, phonetics, phonics

❏ PLAC—calm, please
placate, implacable, placid, complacent

☐ PON, POS—put, place
postpone, proponent, exponent, preposition, posit, interpose, juxtaposition, depose

☐ PORT—carry
portable, deportment, rapport

☐ POT—drink
potion, potable

☐ POT—power
potential, potent, impotent, potentate, omnipotence

☐ PRE—before
precede, precipitate, preclude, precocious, precursor, predilection, predisposition, preponderance, prepossessing, presage, prescient, prejudice, predict, premonition, preposition

☐ PRIM, PRI—first
prime, primary, primal, primeval, primordial, pristine

☐ PRO—ahead, forth
proceed, proclivity, procrastinator, profane, profuse, progenitor, progeny, prognosis, prologue, promontory, propel, proponent, propose, proscribe, protestation, provoke

☐ PROTO—first
prototype, protagonist, protocol

☐ PROX, PROP—near
approximate, propinquity, proximity

☐ PSEUDO—false
pseudoscientific, pseudonym

☐ PYR—fire
pyre, pyrotechnics, pyromania

☐ QUAD, QUAR, QUAT—four
quadrilateral, quadrant, quadruped, quarter, quarantine, quaternary

☐ QUES, QUER, QUIS, QUIR—question
quest, inquest, query, querulous, inquisitive, inquiry

☐ QUIE—quiet
disquiet, acquiesce, quiescent, requiem

☐ QUINT, QUIN—five
quintuplets, quintessence

☐ RADI, RAMI—branch
radius, radiate, radiant, eradicate, ramification

☐ RECT, REG—straight, rule
rectangle, rectitude, rectify, regular

☐ REG —king, rule
regal, regent, interregnum

☐ RETRO—backward
retrospective, retroactive, retrograde

☐ RID, RIS—laugh
ridiculous, deride, derision

☐ ROG—ask
interrogate, derogatory, abrogate, arrogate, arrogant

☐ RUD—rough, crude
rude, erudite, rudimentary

☐ RUPT—break
disrupt, interrupt, rupture

☐ SACR, SANCT—holy
sacred, sacrilege, consecrate, sanctify, sanction, sacrosanct

❏ SCRIB, SCRIPT, SCRIV—write
scribe, ascribe, circumscribe, inscribe, proscribe, script, manuscript, scrivener

❏ SE—apart, away
separate, segregate, secede, sedition

❏ SEC, SECT, SEG—cut
sector, dissect, bisect, intersect, segment, secant

❏ SED, SID—sit
sedate, sedentary, supersede, reside, residence, assiduous, insidious

❏ SEM—seed, sow
seminar, seminal, disseminate

❏ SEN—old
senior, senile, senescent

❏ SENT, SENS—feel, think
sentiment, nonsense, assent, sentient, consensus, sensual

❏ SEQU, SECU—follow
sequence, sequel, subsequent, obsequious, obsequy, non sequitur, consecutive

❏ SIGN—mark, sign
signal, designation, assignation

❏ SIM, SEM—similar, same
similar, semblance, dissemble, verisimilitude

❏ SIN—curve
sine curve, sinuous, insinuate

❏ SOL—sun
solar, parasol, solarium, solstice

❏ SOL—alone
solo, solitude, soliloquy, solipsism

❏ SOMN—sleep
insomnia, somnolent, somnambulist

❏ SON—sound
sonic, consonance, dissonance, assonance, sonorous, resonate

❏ SOPH—wisdom
philosopher, sophistry, sophisticated, sophomoric

❏ SPEC, SPIC—see, look
spectator, circumspect, retrospective, perspective, perspicacious

❏ SPER—hope
prosper, prosperous, despair, desperate

❏ SPERS, SPAR—scatter
disperse, sparse, aspersion, disparate

❏ SPIR—breathe
respire, inspire, spiritual, aspire, transpire

❏ STRICT, STRING—bind
strict, stricture, constrict, stringent, astringent

❏ STRUCT, STRU—build
structure, construe, obstruct

❏ SUB—under
subconscious, subjugate, subliminal, subpoena, subsequent, subterranean, subvert

❑ SUMM—highest
summit, summary, consummate

❑ SUPER, SUR—above
supervise, supercilious, supersede, superannuated, superfluous, insurmountable, surfeit

❑ SURGE, SURRECT—rise
surge, resurgent, insurgent, insurrection

❑ SYN, SYM—together
synthesis, sympathy, synonym, syncopation, synopsis, symposium, symbiosis

❑ TACIT, TIC—silent
tacit, taciturn, reticent

❑ TACT, TAG, TANG—touch
tact, tactile, contagious, tangent, tangential, tangible

❑ TEN, TIN, TAIN—hold, twist
detention, tenable, tenacious, pertinacious, retinue, retain

❑ TEND, TENS, TENT—stretch
intend, distend, tension, tensile, ostensible, contentious

❑ TERM—end
terminal, terminus, terminate, interminable

❑ TERR—earth, land
terrain, terrestrial, extraterrestrial, subterranean

❑ TEST—witness
testify, attest, testimonial, testament, detest, protestation

❑ THE—god
atheist, theology, apotheosis, theocracy

❑ THERM—heat
thermometer, thermal, thermonuclear, hypothermia

❑ TIM—fear, frightened
timid, intimidate, timorous

❑ TOP—place
topic, topography, utopia

❑ TORP—stiff, numb
torpedo, torpid, torpor

❑ TORT—twist
distort, extort, tortuous

❑ TOX—poison
toxic, toxin, intoxication

❑ TRACT—draw
tractor, intractable, protract

❑ TRANS—across, over, through, beyond
transport, transgress, transient, transitory, translucent, transmutation

❑ TREM, TREP—shake
tremble, tremor, tremulous, trepidation, intrepid

❑ TURB—shake
disturb, turbulent, perturbation

❑ UMBR—shadow
umbrella, umbrage, adumbrate, penumbra

❑ UNI, UN—one
unify, unilateral, unanimous

❑ URB—city
urban, suburban, urbane

❑ VAC—empty
vacant, evacuate, vacuous

❑ VAL, VAIL—value, strength
valid, valor, ambivalent, convalescence, avail, prevail, countervail

❑ VEN, VENT—come
convene, contravene, intervene, venue, convention, circumvent, advent, adventitious

❑ VER—true
verify, verity, verisimilitude, veracious, aver, verdict

❑ VERB—word
verbal, verbose, verbiage, verbatim

❑ VERT, VERS—turn
avert, convert, pervert, revert, incontrovertible, divert, subvert, versatile, aversion

❑ VICT, VINC—conquer
victory, conviction, evict, evince, invincible

❑ VID, VIS—see
evident, vision, visage, supervise

❑ VIL—base, mean
vile, vilify, revile

❑ VIV, VIT—life
vivid, vital, convivial, vivacious

❑ VOC, VOK, VOW—call, word
vocal, equivocate, vociferous, convoke, evoke, invoke, avow

❑ VOL—wish
voluntary, malevolent, benevolent, volition

❑ VOLV, VOLUT—turn, roll
revolve, evolve, convoluted

❑ VOR—eat
devour, carnivore, omnivorous, voracious

MATH REFERENCE

The math on the GRE covers a lot of ground—from basic algebra to symbol problems to geometry.

Don't let yourself be intimidated. We've highlighted the 75 most important concepts that you need and listed them in this appendix. Although you probably learned most of this stuff in high school, this list is a great way to refresh your memory.

The GRE math tests your understanding of a relatively limited number of mathematical concepts. It is possible to learn all the math you need to know for the GRE in a short time. In fact, you've seen it all before. Listed on the following pages are 75 things you need to know for the GRE, divided into three levels.

Level 1 is the most basic. You couldn't answer any GRE math questions if you didn't know Level 1 math. Most people preparing to take the GRE are already pretty good at Level 1 math. Look over the Level 1 list below just to make sure you're comfortable with the basics.

Level 2 is the place for most people to start their review of math. These skills and formulas come into play quite frequently on the GRE, especially in the medium and hard questions. If you're like a lot of students, your Level 2 math is probably rusty.

Level 3 is the hardest math you'll find on the GRE. These are skills and formulas that you might find difficult. Don't spend a lot of time on Level 3 if you still have gaps in Level 2. But once you've about mastered Level 2, then tackling Level 3 can put you over the top.

Level 1
(Math You Probably Already Know)

1. How to add, subtract, multiply, and divide WHOLE NUMBERS

2. How to add, subtract, multiply, and divide FRACTIONS

3. How to add, subtract, multiply, and divide DECIMALS

4. How to convert FRACTIONS TO DECIMALS and DECIMALS TO FRACTIONS

5. How to add, subtract, multiply, and divide POSITIVE AND NEGATIVE NUMBERS

6. How to plot points on the NUMBER LINE

7. How to plug a number into an ALGEBRAIC EXPRESSION

8. How to SOLVE a simple EQUATION

9. How to add and subtract LINE SEGMENTS

10. How to find the THIRD ANGLE of a TRIANGLE, given the other two angles

Level 2 (Math You Might Need to Review)

11. How to use the PERCENT FORMULA

Identify the part, the percent, and the whole.

$$Part = percent \times whole$$

> **HINT: You'll usually find the part near the word *is* and the whole near the word *of*.**

Example: (Find the part)

What is 12 percent of 25?

Setup:

$Part = \frac{12}{100} \times 25 = 3$

Example: (Find the percent)

45 is what percent of 9?

Setup:

$45 = Percent \times 9 = 5 \times 9$

$Percent = 5 \times 100\% = 500\%$

Example: (Find the whole)

15 is $\frac{3}{5}$ percent of what number?

Setup:

$\frac{3}{5}$ percent $= \frac{3}{500}$

$15 = \frac{3}{500} \times$ whole

Whole = 2,500

12. How to use the PERCENT INCREASE/DECREASE FORMULAS

Identify the original whole and the amount of increase/decrease.

$$Percent\ increase = \frac{amount\ of\ increase}{original\ whole} \times 100\%$$

$$Percent\ decrease = \frac{amount\ of\ decrease}{original\ whole} \times 100\%$$

Example:

The price goes up from $80 to $100. What is the percent increase?

Setup:

$Percent\ increase = \frac{20}{80} \times 100\% = 25\%$

> **HINT: Be sure to use the original whole—not the new whole—for the base.**

13. How to predict whether a sum, difference, or product will be ODD or EVEN

Don't bother memorizing the rules. Just take simple numbers like 1 and 2 and see what happens.

Example:

If m is even and n is odd, is the product mn odd or even?

Setup:

Say $m = 2$ and $n = 1$.
2×1 is even, so mn is even.

14. How to recognize MULTIPLES OF 2, 3, 4, 5, 6, 9, and 10

2: Last digit is even.
3: Sum of digits is multiple of 3.
4: Last two digits are a multiple of 4.
5: Last digit is 5 or 0.
6: Sum of digits is multiple of 3 and last digit is even.
9: Sum of digits is multiple of 9.
10: Last digit is 0.

15. How to find a COMMON FACTOR

Break both numbers down to their prime factors to see what they have in common. Then multiply the shared prime factors to find all common factors.

Example:

What factors greater than 1 do 135 and 225 have in common?

Setup:

First find the prime factors of 135 and 225. $135 = 3 \times 3 \times 3 \times 5$, and $225 = 3 \times 3 \times 5 \times 5$. The shared prime factors are two 3s and a 5. Multiply 3, 3, and 5 in every possible combination to find all common multiples: $3 \times 3 = 9$, $3 \times 5 = 15$, and $3 \times 3 \times 5 = 45$.

16. How to find a COMMON MULTIPLE

The product is the easiest common multiple to find. If the two numbers have any factors in common, you can divide them out of the product to get a lower common multiple.

Example

What is the least common multiple of 28 and 42?

Setup:

The product $28 \times 42 = 1,176$ is a common multiple, but not the least. $28 = 2 \times 2 \times 7$, and $42 = 2 \times 3 \times 7$. They share a 2 and a 7, so divide the product by 2 and then by 7. $1,176 \div 2 = 588$. $588 \div 7 = 84$. The least common multiple is 84.

17. How to find the AVERAGE

$$Average = \frac{sum\ of\ terms}{number\ of\ terms}$$

18. How to use the AVERAGE to find the SUM

$$Sum = (average) \times (number\ of\ terms)$$

Example:

17.5 is the average (arithmetic mean) of 24 numbers. What is the sum?

Setup:

Sum = $17.5 \times 24 = 420$

19. How to find the AVERAGE of CONSECUTIVE NUMBERS

The average of evenly spaced numbers is simply the average of the smallest number and the largest number. The average of all the integers from 13 to 77, for example, is the same as the average of 13 and 77:

$$\frac{13 + 77}{2} = \frac{90}{2} = 45$$

20. How to COUNT CONSECUTIVE NUMBERS

The number of integers from A to B inclusive is $B - A + 1$.

Example:

How many integers are there from 73 through 419, inclusive?

Setup:

$419 - 73 + 1 = 347$

HINT: Don't forget to add 1.

---

21. How to find the SUM OF CONSECUTIVE NUMBERS

Sum = (average) × (number of terms)

Example:

What is the sum of the integers from 10 through 50, inclusive?

Setup:

Average = $(10 + 50) \div 2 = 30$

Number of terms = $50 - 10 + 1 = 41$

Sum = $30 \times 41 = 1,230$

22. How to find the MEDIAN

Put the numbers in numerical order and take the middle number. (If there's an even number of numbers, the average of the two numbers in the middle is the median.)

Example:

What is the median of 88, 86, 57, 94, and 73?

Setup:

Put the numbers in numerical order and take the middle number:

$$57, 73, 86, 88, 94$$

The median is 86. (If there's an even number of numbers, take the average of the two in the middle.)

23. How to find the MODE

Take the number that appears most often. For example, if your test scores were 88, 57, 68, 85, 98, 93, 93, 84, and 81, the mode of the scores is 93 because it appears more often than any other score. (If there's a tie for most often, then there's more than one mode.)

24. How to use actual numbers to determine a RATIO

To find a ratio, put the number associated with *of* on the top and the word associated with *to* on the bottom.

$$Ratio = \frac{of}{to}$$

The ratio of 20 oranges to 12 apples is $\frac{20}{12}$, or $\frac{5}{3}$.

25. How to use a ratio to determine an ACTUAL NUMBER

Set up a proportion.

Example:

The ratio of boys to girls is 3 to 4. If there are 135 boys, how many girls are there?

Setup:

$$\frac{3}{4} = \frac{135}{x}$$
$$3 \times x = 4 \times 135$$
$$x = 180$$

26. How to use actual numbers to determine a RATE

Identify the quantities and the units to be compared. Keep the units straight.

Example:

Anders typed 9,450 words in $3\frac{1}{2}$ hours. What was his rate in words per minute?

Setup:

First convert $3\frac{1}{2}$ hours to 210 minutes. Then set up the rate with words on top and minutes on bottom:

$$\frac{9,450 \text{ words}}{210 \text{ minutes}} = 45 \text{ words per minute}$$

HINT: The unit before *per* goes on top, and the unit after *per* goes on the bottom.

27. How to deal with TABLES, GRAPHS, AND CHARTS

Read the question and all labels extra carefully. Ignore extraneous information and zero in on what the question asks for.

28. How to count the NUMBER OF POSSIBILITIES

Forget about combinations and permutations formulas. You won't need them on the GRE. The number of possibilities is generally so small that the best approach is just to write them out systematically and count them.

Example:

How many three-digit numbers can be formed with the digits 1, 3, and 5?

Setup:

Write them out. Be systematic so you don't miss any: 135, 153, 315, 351, 513, 531. Count them: six possibilities.

29. How to calculate a simple PROBABILITY

$$Probability = \frac{number\ of\ favorable\ outcomes}{total\ number\ of\ possible\ outcomes}$$

30. How to work with new SYMBOLS

If you see a symbol you've never seen before, don't freak out: it's a made-up symbol. Everything you need to know is in the question stem. Just follow the instructions.

31. How to SIMPLIFY POLYNOMIALS

First multiply to eliminate all parentheses. Then combine like terms.

32. How to FACTOR certain POLYNOMIALS

Learn to spot these classic factorables:

$$ab + ac = a(b + c)$$
$$a^2 + 2ab + b^2 = (a + b)^2$$
$$a^2 - 2ab + b^2 = (a - b)^2$$
$$a^2 - b^2 = (a - b)(a + b)$$

33. How to solve for one variable IN TERMS OF ANOTHER

To find x "in terms of" y : isolate x on one side, leaving y as the only variable on the other.

34. How to solve an INEQUALITY

Treat it much like an equation—adding, subtracting, multiplying, and dividing both sides by the same thing. Just remember to reverse the inequality sign if you multiply or divide by a negative number.

35. How to TRANSLATE ENGLISH INTO ALGEBRA

Look for the key words and systematically turn phrases into algebraic expressions and sentences into equations.

HINT: Be extra careful of the order you place numbers in when subtraction is called for.

---

36. How to find an ANGLE formed by INTERSECTING LINES

Vertical angles are equal. Adjacent angles add up to 180°.

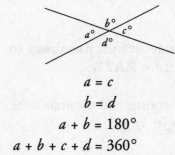

$$a = c$$
$$b = d$$
$$a + b = 180°$$
$$a + b + c + d = 360°$$

KAPLAN

37. How to find an angle formed by a TRANSVERSAL across PARALLEL LINES

All the acute angles are equal. All the obtuse angles are equal. An acute plus an obtuse equals 180°.

Example:

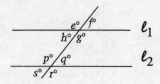

ℓ_1 is parallel to ℓ_2

$e = g = p = r$

$f = h = q = s$

$e + q = g + s = 180°$

HINT: Forget about the terms *alternate interior*, *alternate exterior*, and *corresponding* angles. The GRE never uses them.

---

38. How to find the AREA of a TRIANGLE

$$Area = \frac{1}{2}(base)(height)$$

Example:

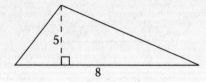

Setup:

$Area = \frac{1}{2}(5)(8) = 20$

HINT: You might have to construct an altitude, as we did in the triangle above.

---

39. How to work with ISOSCELES TRIANGLES

Isosceles triangles have two equal sides and two equal angles. If a GRE question tells you that a triangle is isoceles, you can bet that you'll need to use that information to find the length of a side or a measure of an angle.

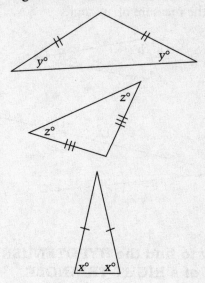

40. How to work with EQUILATERAL TRIANGLES

Equilateral triangles have three equal sides and three 60° angles. If a GRE question tells you that a triangle is equilateral, you can bet that you'll need to use that information to find the length of a side or a measure of an angle.

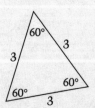

41. How to work with SIMILAR TRIANGLES

In similar triangles, corresponding angles are equal and corresponding sides are proportional. If a GRE question tells you that triangles are similar, you'll probably need that information to find the length of a side or the measure of an angle.

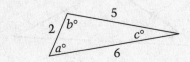

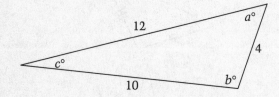

42. How to find the HYPOTENUSE or a LEG of a RIGHT TRIANGLE

Pythagorean theorem: $a^2 + b^2 = c^2$

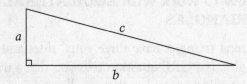

HINT: Most right triangles on the GRE are "special" right triangles (see below), so you can often bypass the Pythagorean theorem.

---

43. How to spot "SPECIAL" RIGHT TRIANGLES

3-4-5
5-12-13
30-60-90
45-45-90

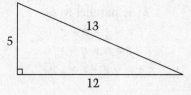

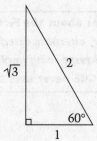

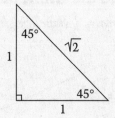

HINT: Learn to spot "special" right triangles—the less you have to calculate the Pythagorean theorem, the more time you save.

KAPLAN

44. How to find the PERIMETER of a RECTANGLE

Perimeter = 2(length + width)

Example:

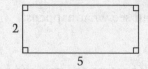

Setup:

Perimeter = 2(2 + 5) = 14

45. How to find the AREA of a RECTANGLE

Area = (length)(width)

Example:

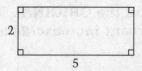

Setup:

Area = 2 × 5 = 10

46. How to find the AREA of a SQUARE

Area = (side)²

Example:

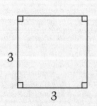

Setup: Area = 3^2 = 9

47. How to find the CIRCUMFERENCE of a CIRCLE

Circumference = 2πr

Example:

Setup:

Circumference = 2π(5) = 10π

48. How to find the AREA of a CIRCLE

Area = πr²

Example:

Setup:

Area = π × 5² = 25π

49. How to find the DISTANCE BETWEEN POINTS on the coordinate plane

If two points have the same *x*s or the same *y*s—that is, they make a line segment that is parallel to an axis—all you have to do is subtract the numbers that are different.

Example:

What is the distance from (2, 3) to (–7, 3)?

Setup:

The *y*s are the same, so just subtract the *x*s.

$$2 - (-7) = 9$$

If the points have different *x*s and different *y*s, make a right triangle and use the Pythagorean theorem.

Example:

What is the distance from (2, 3) to (–1, –1)?

Setup:

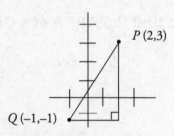

It's a 3-4-5 triangle!

$$PQ = 5$$

HINT: Look for "special" right triangles.

---

50. How to find the SLOPE of a LINE

$$Slope = \frac{rise}{run} = \frac{change\ in\ y}{change\ in\ x}$$

Example:

What is the slope of the line that contains the points (1, 2) and (4, –5)?

Setup:

$$Slope = \frac{2 - (-5)}{1 - 4} = -\frac{7}{3}$$

Level 3 (Math You Might Find Difficult)

51. How to determine COMBINED PERCENT INCREASE/DECREASE

Start with 100 and see what happens.

Example:

A price rises by 10 percent one year and by 20 percent the next. What's the combined percent increase?

Setup:

Say the original price is $100.

Year one: $100 + (10% of 100) = 100 + 10 = 110.

Year two: 110 + (20% of 110) = 110 + 22 = 132.

From 100 to 132—That's a 32 percent increase.

52. How to find the ORIGINAL WHOLE before percent increase/decrease

Example:

After decreasing by 5 percent, the population is now 57,000. What was the original population?

Setup:

.95 × (Original Population) = 57,000

Original Population = 57,000 ÷ .95 = 60,000

53. How to solve a REMAINDERS problem

Pick a number that fits the given conditions and see what happens.

Example:

When n is divided by 7, the remainder is 5. What is the remainder when $2n$ is divided by 7?

Setup:

Find a number that leaves a remainder of 5 when divided by 7. A good choice would be 12. If $n = 12$, then $2n = 24$, which, when divided by 7, leaves a remainder of 3.

54. How to solve a DIGITS problem

Use a little logic—and some trial and error.

Example:

If A, B, C, and D represent distinct digits in the addition problem below, what is the value of D?

$$AB$$
$$+ BA$$
$$\overline{CDC}$$

Setup:

Two 2-digit numbers will add up to at most something in the 100s, so $C = 1$. B plus A in the units' column gives a 1, and since it can't simply be that $B + A = 1$, it must be that $B + A = 11$, and a 1 gets carried. In fact, A and B can be just about any pair of digits that add up to 11 (3 and 8, 4 and 7, etcetera), but it

doesn't matter what they are, they always give you the same thing for D :

$$
\begin{array}{cc}
47 & 83 \\
+74 & +38 \\
\hline
121 & 121
\end{array}
$$

55. How to find a WEIGHTED AVERAGE

Give each term the appropriate "weight."

Example:

The girls' average score is 30. The boys' average score is 24. If there are twice as many boys as girls, what is the overall average?

Setup:

$$\text{Weighted Avg.} = \frac{1 \times 30 + 2 \times 24}{3} = \frac{78}{3} = 26$$

HINT: Don't just average the averages.

---

56. How to find the NEW AVERAGE when a number is added or deleted

Use the sum of the terms of the old average to help you find the new average.

Example:

Michael's average score after four tests is 80. If he scores 100 on the fifth test, what's his new average?

Setup:

Find the original sum from the original average:
Original sum = $4 \times 80 = 320$
Add the fifth score to make the new sum:
New sum = $320 + 100 = 420$
Find the new average from the new sum:
New average = $\frac{420}{5} = 84$

57. How to use the ORIGINAL AVERAGE and NEW AVERAGE to figure out WHAT WAS ADDED OR DELETED

Use the sums.

Number added = (new sum) − (original sum)

Number deleted = (original sum) − (new sum)

Example:

The average of five numbers is 2. After one number is deleted, the new average is −3. What number was deleted?

Setup:

Find the original sum from the original average:

Original sum = 5 × 2 = 10

Find the new sum from the new average:

New sum = 4 × (−3) = −12

The difference between the original sum and the new sum is the answer.

Number deleted = 10 − (−12) = 22

58. How to find an AVERAGE RATE

Convert to totals.

$$\text{Average } A \text{ per } B = \frac{\text{Total } A}{\text{Total } B}$$

Example:

If the first 500 pages have an average of 150 words per page, and the remaining 100 pages have an average of 450 words per page, what is the average number of words per page for the entire 600 pages?

Setup:

Total pages = 500 + 100 = 600

Total words = 500 × 150 + 100 × 450 = 120,000

$$\text{Average words per page} = \frac{120,000}{600} = 200$$

To find an average speed, you also convert to totals.

$$\text{Average speed} = \frac{\text{total distance}}{\text{total time}}$$

Example: Rosa drove 120 miles one way at an average speed of 40 miles per hour and returned by the same 120-mile route at an average speed of 60 miles per hour. What was Rosa's average speed for the entire 240-mile round trip?

Setup: To drive 120 miles at 40 mph takes 3 hours. To return at 60 mph takes 2 hours. The total time, then, is 5 hours.

$$\text{Average speed} = \frac{240 \text{ miles}}{5 \text{ hours}} = 48 \text{ mph}$$

HINT: Don't just average the rates.

---

59. How to determine a COMBINED RATIO

Multiply one or both ratios by whatever you need to in order to get the terms they have in common to match.

Example:

The ratio of *a* to *b* is 7:3. The ratio of *b* to *c* is 2:5. What is the ratio of *a* to *c*?

Setup:

Multiply each member of *a:b* by 2 and multiply each member of *b:c* by 3 and you get *a:b* = 14:6 and *b:c* = 6:15. Now that the *b*'s match, you can just take *a* and *c* and say *a:c* = 14:15.

60. How to MULTIPLY/DIVIDE POWERS

Add/subtract the exponents.

Example:

$$x^b \times x^b = x^{(a+b)}$$
$$2^3 \times 2^4 = 2^7$$

Example:

$$\frac{x^c}{x^d} = x^{(c-d)}$$

$$\frac{5^6}{5^2} = 5^4$$

61. How to RAISE A POWER TO A POWER

Multiply the exponents.

Example:

$$(x^a)^b = x^{ab}$$
$$(3^4)^5 = 3^{20}$$

62. How to ADD, SUBTRACT, MULTIPLY, and DIVIDE ROOTS

You can add/subtract roots only when the parts inside the $\sqrt{}$ are identical.

Example:

$$\sqrt{2} + 3\sqrt{2} = 4\sqrt{2}$$
$$\sqrt{2} - 3\sqrt{2} = -2\sqrt{2}$$

$\sqrt{2} + \sqrt{3}$—cannot be combined.

To multiply/divide roots, deal with what's inside the $\sqrt{}$ and outside the $\sqrt{}$ separately.

Example:

$$(2\sqrt{3})(7\sqrt{5}) = (2 \times 7)\,(\sqrt{3 \times 5}) = 14\sqrt{15}$$

$$\frac{10\sqrt{21}}{5\sqrt{3}} = \frac{10}{5}\sqrt{\frac{21}{3}} = 2\sqrt{7}$$

63. How to SIMPLIFY A SQUARE ROOT

Look for perfect squares (4, 9, 16, 25, 36...) inside the $\sqrt{}$. Factor them out and "unsquare" them.

Example:

$$\sqrt{48} = \sqrt{16} \times \sqrt{3} = 4\sqrt{3}$$
$$\sqrt{180} = \sqrt{36} \times \sqrt{5} = 6\sqrt{5}$$

64. How to solve certain QUADRATIC EQUATIONS

Forget the quadratic formula. Manipulate the equation (if necessary) into the "_____ = 0" form, factor the left side, and break the quadratic into two simple equations.

Example:

$$x^2 + 6 = 5x$$
$$x^2 - 5x + 6 = 0$$
$$(x-2)(x-3) = 0$$
$$x - 2 = 0 \ \text{ or } \ x - 3 = 0$$
$$x = 2 \text{ or } 3$$

Example:

$$x^2 = 9$$

$$x = 3 \text{ or } -3$$

**HINT: Watch out for x^2.
There can be two solutions.**

---

65. How to solve MULTIPLE EQUATIONS

When you see two equations with two variables on the GRE, they're probably easy to combine in such a way that you get something closer to what you're looking for.

Example:

If $5x - 2y = -9$ and $3y - 4x = 6$, what is the value of $x + y$?

Setup:

The question doesn't ask for x and y separately, so don't solve for them separately if you don't have to. Look what happens if you just rearrange a little and "add" the equations:

$$5x - 2y = -9$$
$$-4x + 3y = 6$$
$$\overline{}$$
$$x + y = -3$$

**HINT: Don't do more work than you
have to. Look for the shortcut.**

---

66. How to find the MAXIMUM and MINIMUM lengths for a SIDE of a TRIANGLE

If you know two sides of a triangle, you know that the third side is between the difference and the sum.

Example:

The length of one side of a triangle is 7. The length of another side is 3. What is the range of possible lengths for the third side?

Setup:

The third side is greater than the difference ($7 - 3 = 4$) and less than the sum ($7 + 3 = 10$).

67. How to find one angle or the sum of all the ANGLES of a REGULAR POLYGON

Sum of the interior angles in a polygon with n *sides =*
$(n - 2) \times 180$

Degree measure of one angle in a Regular Polygon with
n *sides =* $\dfrac{(n-2) \times 180}{n}$

Example:

What is the measure of one angle of a regular pentagon?

Setup:

Plug $n = 5$ into the formula:

$$\text{Degree measure of one angle} =$$
$$\frac{(5-2) \times 180}{5} = \frac{540}{5} = 108$$

68. How to find the LENGTH of an ARC

Think of an arc as a fraction of the circle's circumference.

$$Length\ of\ arc = \frac{n}{360} \times 2\pi r$$

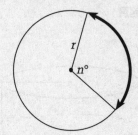

69. How to find the AREA of a SECTOR

Think of a sector as a fraction of the circle's area.

$$Area\ of\ sector = \frac{n}{360} \times \pi r^2$$

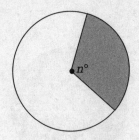

70. How to find the dimensions or area of an INSCRIBED or CIRCUMSCRIBED FIGURE

Look for the connection. Is the diameter the same as a side or a diagonal?

Example:

If the area of the square is 36, what is the circumference of the circle?

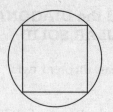

Setup:

To get the circumference, you need the diameter or radius. The circle's diameter is also the square's diagonal, which (it's a 45-45-90 triangle!) is $6\sqrt{2}$.

$$Circumference = \pi(diameter) = 6\pi\sqrt{2}$$

71. How to find the VOLUME of a RECTANGULAR SOLID

$$Volume = length \times width \times height$$

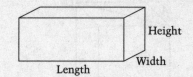

72. How to find the SURFACE AREA of a RECTANGULAR SOLID

To find the surface area of a rectangular solid, you have to find the area of each face and add them together. Here's the formula.

Surface area =
2(length × width + length × height + width × height)

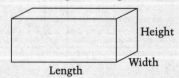

73. How to find the DIAGONAL of a RECTANGULAR SOLID

Use the Pythagorean theorem twice, unless you spot "special" triangles.

Example: What is the length of AG?

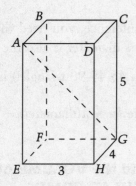

Setup: Draw diagonal AC.

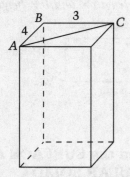

ABC is a 3-4-5 triangle, so $AC = 5$. Now look at triangle ACG:

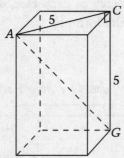

ACG is another special triangle, so you don't need to use the Pythagorean theorem. ACG is a 45-45-90, so $AG = 5\sqrt{2}$.

74. How to find the VOLUME of a CYLINDER

$Volume = \pi r^2 h$

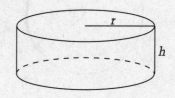

75. How to find the VOLUME of a SPHERE

$Volume = \frac{4}{3}\pi r^3$

Notes

Notes

Notes

Notes

Notes

Notes

Want more information about our services, products, or the nearest Kaplan center?

Call our nationwide toll-free numbers:

1-800-KAP-TEST for information on our live courses, private tutoring and admissions consulting
1-800-KAP-ITEM for information on our products
1-888-KAP-LOAN* for information on student loans

(outside the U.S.A., call **1-212-262-4980**)

Connect with us in cyberspace:

On AOL, keyword:"Kaplan"
On the World Wide Web, go to: **http://www.kaplan.com**
Via e-mail: info@kaplan.com

Write to:

Kaplan Educational Centers
888 Seventh Avenue
New York, NY 10106

Educational Centers

Kaplan Educational Centers is one of the nation's premier education companies, providing individuals with a full range of resources to achieve their educational and career goals. Kaplan, celebrating its 60th anniversary, is a wholly owned subsidiary of the Washington Post Company.

TEST PREPARATION AND ADMISSIONS CONSULTING

Kaplan's nationally recognized test prep courses cover more than 20 standardized tests, including secondary school, college, and graduate school entrance exams and foreign language and professional licensing exams. In addition, Kaplan offers private tutoring and comprehensive one-to-one admissions and application advice for students applying to law and business school.

SCORE! EDUCATIONAL CENTERS

SCORE! after-school learning centers help K-8 students build confidence, academic and goal-setting skills in a motivating, sports-oriented environment. Our cutting-edge interactive curriculum continually assesses and adapts to each child's academic needs and learning style. Enthusiastic Academic Coaches serve as positive role models creating a high-energy atmosphere where learning is exciting and fun.

KAPLAN LEARNING SERVICES

Kaplan Learning Services provides customized assessment, education, and training programs to elementary and high schools, universities and businesses to help students and employees reach their academic and career goals.

KAPLAN PROGRAMS FOR INTERNATIONAL STUDENTS AND PROFESSIONALS

Kaplan services international students and professionals in the U.S. through *Access America*, a series of intensive English language programs. These programs are offered at Kaplan City Centers and four new campus-based centers in California, Washington and New York via Kaplan/LCP International Institute. Kaplan and Kaplan/LCP offer specialized services to sponsors including placement at top American universities, fellowship management, academic monitoring and reporting, and financial administration.

KAPLAN PUBLISHING

Kaplan Books, a joint imprint with Simon & Schuster, publishes titles in test preparation, admissions, education, career development and life skills; Kaplan and *Newsweek* jointly publish the popular guides, **How to Get Into College** and **How to Choose a Career & Graduate School**. *SCORE!* and *Newsweek* have teamed up to publish **How to Help Your Child Succeed in School**.

KAPLOAN

Students may obtain information and advice about educational loans for college and graduate school through **KapLoan** (Kaplan Student Loan Information Program). Through an affiliation with one of the nation's largest student loan providers, **KapLoan** helps direct students and their families through the often bewildering financial aid process.

KAPLAN INTERACTIVE

Kaplan InterActive delivers award-winning educational products and services including Kaplan's best-selling **Higher Score** test-prep software and sites on the internet (http://www.kaplan.com) and America Online. Kaplan and Cendant Software jointly offer educational software for the K-12 retail and school markets.

KAPLAN CAREER SERVICES

Kaplan helps students and graduates find jobs through Kaplan Career Services, the leading provider of career fairs in North America. The division includes **Crimson & Brown Associates**, the nation's leading diversity recruitment and publishing firm, and **The Lendman Group and Career Expo,** both of which help clients identify highly sought-after technical personnel, and sales and marketing professionals.

COMMUNITY OUTREACH

Kaplan provides educational resources to thousands of financially disadvantaged students annually working closely with educational institutions, not-for-profit groups, government agencies and other grass roots organizations on a variety of national and local support programs. Kaplan enriches local communities by employing high school, college, and graduate students, creating valuable work experiences for vast numbers of young people each year.

CPSIA information can be obtained
at www.ICGtesting.com
Printed in the USA
FFHW011723201118
49520680-53885FF

9 781305 08